Craftsman 2011 NATIONAL PLUMBING & HVAC ESTIMATOR

By James A. Thomson

Includes inside the back cover:

- An estimating CD with all the costs in this book, plus,
- An estimating program that makes it easy to use these costs,
- An interactive video guide to the National Estimator program,
- A program that converts your estimates into invoices,
- A program that exports your estimates to QuickBooks Pro.

Quarterly price updates on the Web are free and automatic all during 2011. You'll be prompted when it's time to collect the next update. A connection to the Web is required.

Download all of Craftsman's most popular costbooks for one low price with the Craftsman Site License. http://CraftsmanSiteLicense.com

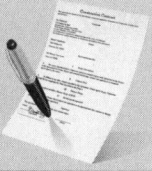

- ● Turn your estimate into a bid.
- ● Turn your bid into a contract.
- ● ConstructionContractWriter.com

Craftsman Book Company
6058 Corte del Cedro / P.O. Box 6500 / Carlsbad, CA 92018

Acknowledgments

The sample "Standard Form Subcontract" and "Subcontract Change Order" forms used in the final section of this book are reprinted with the permission of the publisher, the Associated General Contractors of America (National Office), 1957 E Street NW, Washington, District of Columbia 20006.

Cover design by: *Bill Grote*
Cover photography by *Jim Karnik Films*

© 2010 Craftsman Book Company
ISBN 978-1-57218-248-6
Published October 2010 for the year 2011.

Contents

How to Use This Book

This 2011 National Plumbing & HVAC Estimator is a guide to estimating labor and material costs for plumbing, heating, ventilating and air conditioning systems in residential, commercial and industrial buildings.

 Inside the back cover of this book you'll find an envelope with a compact disk. The disk has National Estimator, an easy-to-use estimating program with all the cost estimates in this book. Insert the CD in your computer and wait a few seconds. Installation should begin automatically. (If not, click Start, Settings, Control Panel, double-click Add/Remove Programs and Install.) Select Show Me from the installation menu and Dan will show you how to use National Estimator. When Show Me is complete, select Install Program. When the National Estimator program has been installed, click Help on the menu bar, click Contents, click Print all Topics, click File and click Print Topic to print a 40-page instruction manual for National Estimator.

Costs in This Manual will apply within a few percent on a wide variety of projects. Using the information given on the pages that follow will explain how to use these costs and suggest procedures to follow when compiling estimates. Reading the remainder of this section will help you produce more reliable estimates for plumbing and HVAC work.

 **Manhour Estimates in This Book** will be accurate for some jobs and inaccurate for others. No manhour estimate fits all jobs because every construction project is unique. Expect installation times to vary widely from job to job, from crew to crew, and even for the same crew from day to day.

There's no way to eliminate all errors when making manhour estimates. But you can minimize the risk of a major error by:

1. Understanding what's included in the manhour estimates in this book, and

2. Adjusting the manhour estimates in this book for unusual job conditions.

The Craft@Hrs Column. Manhour estimates in this book are listed in the column headed *Craft@Hrs.* For example, on page 19 you'll see an estimate for installing a 6 gallon hot water heater. In the *Craft@Hrs* column opposite 6 gallon you'll see:

P1@.500

To the left of the @ symbol you see an abbreviation for the recommended work crew.

Page 7 shows the wage rates and craft codes used in this book.

To the right of the @ symbol you see a number. The number is the estimated manhours (not crew hours) required to install each unit of material listed. In the case of a 6 gallon hot water heater, P1@.500 means that .500 manhours are required to install 1 hot water heater.

 **Costs in the Labor $ Column** are based on manhour estimates in the Craft@Hrs column. Multiply the manhour estimate by the assumed hourly labor cost to find the installation cost in the *Labor $* column. For example, .500 manhours times $35.76 (the average wage for crew P1) is $17.88.

Quarterly price updates on the Web are free and automatic all during 2011. You'll be prompted when it's time to collect the next update. A connection to the Web is required.

Manhour Estimates include all productive labor normally associated with installing the materials described. These estimates assume normal conditions: experienced craftsmen working on reasonably well planned and managed new construction with fair to good productivity. Labor estimates also assume that materials are standard grade, appropriate tools are on hand, work done by other crafts is adequate, layout and installation are relatively uncomplicated, and working conditions don't slow progress.

All manhour estimates include tasks such as:

- Unloading and storing construction materials, tools and equipment on site.

- Working no more than two floors above or below ground level.

- Working no more than 10 feet above an uncluttered floor.

- Normal time lost due to work breaks.

- Moving tools and equipment from a storage area or truck not more than 200 feet from the work area.

- Returning tools and equipment to the storage area or truck at the end of the day.

- Planning and discussing the work to be performed.

- Normal handling, measuring, cutting and fitting.

- Regular cleanup of construction debris.

- Infrequent correction or repairs required because of faulty installation.

If the work you're estimating won't be done under these conditions, you need to apply a correction factor to adjust the manhour estimates in this book to fit your job.

Applying Correction Factors. Analyze your job carefully to determine whether a labor correction factor is needed. Failure to consider job conditions is probably the most common reason for inaccurate estimates.

Use one or more of the recommended correction factors in Table 1 to adjust for unusual job conditions. To make the adjustment, multiply the manhour estimate by the appropriate conversion factor. On some jobs, several correction factors may be needed. A correction factor less than 1.00 means that favorable working conditions will reduce the manhours required.

Supervision Expense to the installing contractor is not included in the labor cost. The cost of supervision and non-productive labor varies widely from job to job. Calculate the cost of supervision and non-productive labor and add this to the estimate.

Hourly Labor Costs also vary from job to job. This book assumes an average manhour labor cost of $41.85 for plumbers and $41.04 for sheet metal workers. If these hourly labor costs are not accurate for your jobs, adjust the labor costs up or down by an appropriate percentage. Instructions on the next page explains how to make these adjustments. If you're using the National Estimator disk, it's easy to set your own wage rates.

Hourly labor costs in this book include the basic wage, fringe benefits, the employer's contribution to welfare, pension, vacation and apprentice funds, and all tax and insurance charges based on wages. Table 2 at the top of the next page shows how hourly labor costs in this book were calculated. It's important that you understand what's included in the figures in each of the six columns in Table 2. Here's an explanation:

Column 1, the base wage per hour, is the craftsman's hourly wage. These figures are representative of what many contractors are paying plumbers, sheet metal workers and helpers in 2011.

Column 2, taxable fringe benefits, includes vacation pay, sick leave and other taxable benefits. These fringe benefits average about 5.15% of the base wage for many plumbing and HVAC contractors. This benefit is in addition to the base wage.

Condition	Correction Factor
Work in large open areas, no partitions	.85
Prefabrication under ideal conditions, bench work	.90
Large quantities of repetitive work	.90
Very capable tradesmen	.95
Work 300' from storage area	1.03
Work 400' from storage area	1.05
Work 500' from storage area	1.07
Work on 3rd through 5th floors	1.05
Work on 6th through 9th floors	1.10
Work on 10th through 13th floors	1.15
Work on 14th through 17th floors	1.20
Work on 18th through 21st floors	1.25
Work over 21 floors	1.35
Work in cramped shafts	1.30
Work in commercial kitchens	1.10
Work above a sloped floor	1.25
Work in attic space	1.50
Work in crawl space	1.20
Work in a congested equipment room	1.20
Work 15' above floor level	1.10
Work 20' above floor level	1.20
Work 25' above floor level	1.30
Work 30' above floor level	1.40
Work 35' to 40' above floor level	1.50

Table 1   Recommended Correction Factors

Column 3, insurance and employer-paid taxes in percent, shows the insurance and tax rate for the craft workers. The cost of insurance in this column includes workers' compensation and contractor's casualty and liability coverage. Insurance rates vary widely from state to state and depend on a contractor's loss experience. Note that taxes and insurance increase the hourly labor cost by approximately 30%. There is no legal way to avoid these costs.

Column 4, insurance and employer taxes in dollars, shows the hourly cost of taxes and insurance. Insurance and taxes are paid on the costs in both columns 1 and 2.

Column 5, non-taxable fringe benefits, includes employer paid non-taxable benefits such as medical coverage and tax-deferred pension and profit sharing plans. These fringe benefits average 4.55% of the base wage for many plumbing and HVAC contractors.

Column Number	1	2	3	4	5	6
Craft	Base wage per hour	Taxable fringe benefits (at 5.15% of base wage)	Insurance and employer taxes (%)	Insurance and employer taxes ($)	Non-taxable fringe benefits (at 4.55% of base wage)	Total hourly cost used in this book
Laborer	20.55	1.06	32.93%	7.12	0.94	29.67
Plumber	30.90	1.59	24.47%	7.95	1.41	41.85
Sheet Metal Worker	29.90	1.54	26.21%	8.24	1.36	41.04
Operating Engineer	29.35	1.51	25.42%	7.84	1.34	40.04
Sprinkler Fitter	30.35	1.56	25.28%	8.07	1.38	41.36
Electrician	30.27	1.56	20.04%	6.38	1.38	39.59
Cement Mason	26.09	1.34	23.35%	6.40	1.19	35.02

Craft Code	Crew Composition	Average Hourly Cost per Manhour
ER	4 building plumbers, 2 building laborers, 1 operating engineer	38.11
SN	4 building sheet metal workers, 2 building laborers, 1 operating engineer	37.65
P1	1 building plumber and 1 building laborer	35.76
ST	1 sprinkler fitter	41.36
SK	4 sprinkler fitters, 2 building laborers, 1 operating engineer	37.83
SL	1 sprinkler fitter and 1 laborer	35.52
S2	1 building sheet metal worker, 1 building laborer	35.36
BE	1 electrician	39.59
CF	1 cement mason	35.02
SW	1 sheet metal worker	41.04

Table 2 Labor Costs Used in This Book

The employer pays no taxes or insurance on these benefits.

Column 6, the total hourly cost in dollars, is the sum of columns 1, 2, 4, and 5. The labor costs in Column 6 were used to compute costs in the Labor $ column of this book.

Adjusting Costs in the Labor $ Column. The hourly labor costs used in this book may apply within a few percent on many of your jobs. But wage rates may be much higher or lower in some areas. If the hourly costs shown in Column 6 of Table 2 are not accurate for your work, adjust labor costs to fit your jobs.

For example, suppose your hourly labor costs are as follows:

Plumber	$19.00
Laborer	$16.00
Total hourly crew cost	$35.00

Your average cost per manhour would be $17.50 ($35.00 per crew hour divided by 2 because this is a crew of two).

A labor cost of $17.50 is about 50% of the $35.76 labor cost used for crew P1. Multiply costs in the Labor $ column by .50 to find your estimated cost.

For example, notice on page 19 that the labor cost for installing a 6 gallon hot water heater is $17.90 each. If installed by your plumbing crew working at an average cost of $17.50 per manhour, your estimated cost would be 50% of $17.90 or $8.95 per heater.

Adjusting the labor costs in this book will make your estimates much more accurate. Making adjustments to labor costs is both quick and easy if you use the National Estimator disk.

Equipment Cost will vary according to need and application. It typically is $110 per hour for a 10-ton hydraulic truck-mounted crane.

Material Costs in this manual are intended to reflect what medium- to low-volume contractors will be paying in 2011 after applying normal discounts. These costs include charges for delivery to within 25 to 30 miles of the supplier.

Overhead and Profit for the installing contractor are not included in the costs in this manual unless specifically identified in the text. Markup can vary widely with local economic conditions, competition and the installing contractor's operating expenses. Add the markup that's appropriate for your company, the job and the competitive environment.

How Accurate Are These Figures? As accurate as possible considering that the editors don't know your material suppliers, haven't seen the plans or specifications, don't know what building code applies or where the job is, had to project material costs at least six months into the future, and had no record of how much work the crew that will be assigned to the job can handle.

You wouldn't bid a job under those conditions. And I don't claim that all plumbing and HVAC work is done at these prices.

Estimating Is an Art, not a science. There is no one price that applies on all jobs. On many jobs the range between high and low bid will be 10% or more. There's room for legitimate disagreement on what the correct costs are, even when complete plans and specifications are available, the date and site are established, and labor and material costs are identical for all bidders.

No estimate fits all jobs. Good estimates are custom made for a particular project and a single contractor through judgment, analysis and experience. This book is not intended as a substitute for judgment, analysis and sound estimating practice. It's an aid in developing an informed opinion of cost, not an answer book.

Additional Costs to Consider

Here's a checklist of additional costs to consider before submitting any bid.

1. Sales taxes

2. Mobilization costs

3. Payment and performance bond costs

4. Permits and fees

5. Storage container rental costs

6. Utility costs

7. Tool costs

8. Callback costs during warranty period

9. Demobilization costs

Exclusions and Clarifications

Neither the job specifications nor the contract may identify exactly what work should be included in the plumbing and HVAC bid. Obviously, you have to identify what work is included in the job.

The most efficient way to define the scope of the work is to prepare a list of tasks not normally performed by your company and attach that list to each bid submitted. Here's a good list of work that should be excluded from your bid.

Your Bid Should Exclude

Final cleaning of plumbing fixtures

Backings for plumbing fixtures

Toilet room accessories

Electrical work, including motor starters

Electrical wiring and conduit over 100 volts

Temporary utilities

Painting, priming and surface preparation

Structural cutting, patching or repairing

Fire protection and landscape sprinklers

Equipment supports

Surveying and layout of control lines

Removal or stockpiling of excess soil

Concrete work, including forming and rebar

Setting of equipment furnished by others

Equipment, unless shown, and personnel hoisting

Wall and floor blockouts

Pitch pockets

The costs of performance or payment bonds

Site utilities

Asbestos removal or disposal

Contaminated soil removal or disposal

Major increases in copper material prices

Fire dampers not shown on the plans

Your Bid Should Include

Trash sweep-up only. Others haul it away

Site utilities from building to property line only

Piping to 5 feet outside the building only

Plumbing & HVAC permits for your work only

Beware of Price Changes

There's no way to be sure what prices will be in three to six months. All labor, equipment, material and subcontract prices in a bid should be based on costs anticipated when the project is expected to be built, not when the estimate is compiled. That presents a problem. Except for the installation of underground utilities, most plumbing and HVAC work is done six months to a year after the bid is submitted. When possible, get price protection in writing from your suppliers and subcontractors. If your suppliers and subs won't guarantee prices, include an escalation allowance in your bid to cover anticipated price increases.

Material Pricing Conditions

All equipment and material prices quoted by your vendors will be conditional. They usually don't include sales tax and are subject to specific payment and shipping terms. Every estimator should understand the meaning of common shipping terms. They define who pays the freight and who has responsibility for processing freight-damage claims. Here's a summary of important conditions you should understand.

F.O.B. Factory (Free On Board at the Factory): Title passes to the buyer when the goods are delivered by the seller to the freight carrier. The buyer pays the freight and is responsible for freight-damage claims.

F.O.B. Factory F.F.A. (Free On Board at the Factory, Full Freight Allowed): The title passes to the buyer when the goods are delivered by the seller to the freight carrier. The seller pays the freight charges, but the buyer is responsible for freight-damage claims.

F.O.B. (city of destination) (Free On Board to your city): The title passes to the buyer when the goods are delivered by the seller to the freight terminal in the city, or nearest city, of destination. The seller pays the freight and is responsible for freight-damage claims to the terminal. The buyer pays the freight charge and is responsible for freight-damage claims from the terminal to the final destination.

F.O.B. Job Site (Free On Board at job site, or contractor's shop): The title passes to the buyer when the goods are delivered to the job site (or shop). The seller pays the freight and is responsible for freight-damage claims.

F.A.S. Port [of a specific city] (Free Alongside Ship at the nearest port): The title passes to the buyer when goods are delivered to the ship dock or port terminal. The seller pays the freight and is responsible for freight-damage claims to the ship dock or port terminal only. The buyer pays the freight and is responsible for freight-damage claims from the ship dock or port terminal to the designated delivery point.

Obviously, it's to your advantage to instruct all vendors to quote costs F.O.B. the job site or your shop.

Reducing Costs

Most construction specifications allow the use of alternative equipment and materials. It's the estimator's responsibility to select the most cost-effective products. Research and compare your costs before making any decisions. Avoid selecting any material or equipment simply because that's what you've always done.

Don't recommend plastic products such as ABS, PVC, or polypropylene pipe or corrugated flexible ducts until you've checked local code requirements. Most building codes prohibit use of these materials inside public buildings such as schools, care centers and hospitals.

It's wise to select 100% factory-packaged equipment. Beware of equipment labeled "Some assembly required." Field labor costs for mounting loose coils, motors and similar equipment are very high.

Value Engineering

Let's suppose you've submitted a combined plumbing and HVAC bid for $233,000. Your cutthroat competitor put in a bid at $4,000 less, $229,000. Obviously there's no way you're going to get the job. Right?

Not so fast! Maybe value engineering can help you win that contract — while fattening your profit margin.

Suppose the proposal you submitted had two parts. Part I is the bid for $233,000, based entirely on job plans and specs, just the way they were written. But appended to your proposal is Part II, a list of suggestions for saving money without sacrificing any of the capacity or quality designed into the system. Here's an example of what might be in Part II:

1. Deduct for providing pipe hanger spacings per UPC in lieu of specified spacings:
$1,750.00

2. Deduct for reducing heating hot water pipe sizes by using 40 degrees F Delta T in lieu of specified 20 degrees F Delta T: $4,600.00

3. Deduct for providing pressure/temperature taps at air handling units, pumps and chillers in lieu of specified thermometers and pressure gauges:
$875.00

4. Deduct for eliminating water treatment in closed piping systems: $1,800.00

5. Deduct for piping chilled and heating hot water pumps in parallel in lieu of providing 100% stand-by pumps: $2,900.00

Total deductions: **$11,925.00**

Adopting these suggestions would make you low bidder by nearly $8,000. A saving like that will be tempting to most owners, especially if the owner understands that your suggestions result in a system that is every bit as good and maybe better than the system as originally designed.

You're not offering to undercut the competition. Far from it. You're using knowledge and experience to create better value for the owner. That's called value engineering and it's likely to win the respect of nearly all cost-conscious owners.

Notice that reducing costs is only part of what value engineering is all about. You don't cut costs at the expense of system quality, integrity, capacity or performance.

Don't waste your time, and your client's, by offering to substitute cheaper or lower-quality fixtures or equipment. Any cutthroat contractor with a price list can do that. Recommend the use of inferior materials and you'll be associated with the inferior goods you promote. Some owners consider even the suggestion to be insulting.

The recommendations you make (like most of those in the example) will require design changes. You can expect to be examined (or even challenged) on these points. Be ready to explain and defend each of your suggestions. Convince the client (or the design engineer) that your ideas are based on sound engineering principles and you're well on the way to winning the owner's confidence and the contract.

Now, let's go back to the list and see how we might justify the five value engineering recommendations.

1. **Pipe Hanger Spacing.** The pipe hanger spacings recommended in the Uniform Plumbing Code (UPC) are calculated by experienced, professional structural engineers. The safety factors used in these calculations are very conservative. They've been widely used for many years and have proved to be more than adequate. There's no need for more hangers than the UPC requires.

2. **Changing HHW Delta T.** In hydronic heating systems, heat measured in Btus is pumped to terminal units. The proposed change of the Delta T, from 20 degrees F to 40 degrees F, has no effect whatsoever on how many Btus the system delivers. You're not changing anything but the volume of water being pumped. At lower volume levels, the size of the pump, the pipe and the pipe insulation can all be reduced. Not one of these changes will affect the system's ability to transmit heat. Furthermore, operating costs will also drop, since less pump horsepower will be needed to run the smaller pump.

3. **Thermometers/Pressure Gauges.** Thermometers and pressure gauges installed on or near vibrating machinery have a very short life expectancy. Gauges quickly lose accuracy under harsh conditions. Readings will become less and less reliable. That's potentially dangerous. You can avoid this problem by using insertion-type pressure/temperature taps instead. Store these sensitive gauges in a desk drawer or a tool crib when not in use. Safely stored, they're protected from damage. They'll give accurate readings longer and won't need to be replaced as often. And they're simple to use. Just insert a gauge in one of the conveniently located taps. Make the reading, then remove the gauge and put it away.

4. **Water Treatment.** ITT Bell & Gossett has done studies on corrosion in closed hydronic systems that have a make-up water rate of no more than 5% per year. These studies show that corrosion virtually stops when entrained air is either removed or depleted. No water treatment is needed in this closed system.

5. **100% Standby Pumps.** Two pumps piped and operated in parallel are more economical. Even if one pump fails, the other pump can maintain delivery at 75 to 80% of the designed flow rate. That's usually adequate for emergency operation.

These cost-saving ideas are small, but could tip the balance in your favor. I hope they demonstrate the potential that value engineering has when bidding jobs. Any time you're compiling an estimate, keep an eye out for ways to save money or reduce the owner's cost. Jot a note to yourself about each potential saving you identify. Before submitting the bid, make a list of your alternate suggestions. Maybe best of all, markup on your value engineering suggestions can be higher than your normal markup. If value engineering can cut costs by $10,000, maybe as much as $4,000 of that should end up in your pocket!

Value Engineering: Surplus Materials

Value engineering doesn't begin and end with job plans and specs. Value engineering means getting the most value at the least cost, no matter whether it's value to the owner or value to the contractor. Smart mechanical contractors learn to build extra value into their jobs by controlling shrinkage of materials. Nearly every significant plumbing and HVAC job ends with at least some surplus material on hand. Material left over when the job is done tends to be discarded as waste or hauled off the job in the back of a truck that doesn't have your company name on the door. And why not? It's surplus – not needed. The owner didn't need it. So now it's up for grabs.

Not quite. Let's consider who actually owns that surplus material. When your company has been paid, every piece of material your crew installed belongs to the building owner. But what about those fittings, hangers and valves delivered to the job site but never actually used? Almost certainly, those materials were included in your bid. So aren't they the property of the owner? Not in my opinion. The owner contracted for a mechanical system and (presumably) has one. Unless it's a cost-plus job or a labor-only job, the owner didn't buy materials delivered to the job site. The owner bought a mechanical system and has one – completely separate and apart from any surplus materials. In my mind, the property owner has no more claim to left-over materials than the same owner would have claim to labor hours not expended or equipment not used on the same job.

Unless there's some provision in your contract to the contrary, surplus material belongs to the installing contractor. But your right to that material and the chance of actually getting it back to your shop are two very different propositions. I see recovery of surplus material as a training issue. As a matter of company policy, make it clear to your crews that surplus material belongs to your company. The supervisor on every job should be accountable for recovery of excess material. Every significant job will have at least some surplus. Accounting for that surplus should be part of your routine close-out procedure. Fortunately, it's not difficult. I'll explain.

Control of surplus materials begins with a good checklist, or form. I recommend the Materials, Equipment and Tool form, "MET" for short. A blank MET form appears following this section. Your MET should show both what's delivered to the job site (material, equipment and tools) and surplus "drops" returned to your shop at project close-out. A MET ensures that the estimator, the shop inventory manager and your field supervisor are on the same page. Your MET establishes accountability. Nothing falls through the cracks. Job input equals job output plus returns. Everything delivered to the job and not expended should be returned to your shop.

Here's how it works:

1. Based on the estimate that won you the job, the items needed are purchased for the job and staged for delivery to the job site.

2. As materials, equipment and tools are delivered to the job site, your supervisor completes the first three columns of the MET form: Description, Quantity and Date.

3. As work is completed, the same supervisor completes the four columns under Returned to Inventory: Quantity Returned, Date, Status Code and Value. The status code will be either "RS" (Returned and Salvaged) or "RN" (Returned New).

4. Back at your shop, both RS and RN materials should be restored to inventory.

5. If your company has an inventory manager, have that manager assign the return value to each item returned. If you're using QuickBooks Pro, the "Adjust Inventory" feature can handle this task quite easily. Add two new categories under "Inventory Stock on Hand by Vendor." The first new category is Returned Salvage. The second is Returned New. Be sure the value of RS materials includes the cost of any reconditioning done to restore salvaged materials (such as pumps and boilers) to serviceable condition.

6. Comparing MET deployed to the job site with MET returned to inventory yields MET actually used on the job. That's a very important number to every plumbing and HVAC estimator. Be sure actual usage gets entered on the Project Summary form.

7. When the take-off on your next estimate is complete, compare that materials list with a summary of RS and RN materials on hand from prior jobs.

8. Evaluate which returned materials can be redeployed on the new job.

9. It's a management decision to either (1) charge the new job for the cost of RS and RN materials already on hand, or (2) consider materials on hand as "free" and a competitive advantage in winning the new bid. Either way, RN and RS materials are an asset to your company.

Plumbing and HVAC materials are expensive. Every mechanical contractor has an interest in MET tracking. Everyone in your company should be aware of the need for good materials management. Used correctly, the MET form in this book can help engineer more value into your jobs.

Maximizing the Value of Old Estimates

There should be two profits in every job. The first is money in the bank — a return on time and expenses. The second is what you learn from the job — primarily by comparing the estimate you made with what turns out to be your actual cost. On some jobs, the value of lessons learned may outweigh net revenue.

Every plumbing and HVAC contractor has marginal jobs. That's normal. What *shouldn't* be normal is repeating mistakes. The best way to avoid trouble in your future is to keep track of your past. Keeping old estimates available for reference can help prevent errors on new estimates.

As your file of completed estimates grows, organization becomes more important. You need an easy way to find similar projects with the same components and comparable scope of work. If your estimating file is in QuickBooks Pro, searching by keyword may be enough. Otherwise, I recommend creating a short summary for each completed job, and an index that references all summaries available for comparison. You'll find a blank Project Summary form at the end of this section. To make reference easier, create an index by type of job and equipment used. You may choose to use an alphabetical index based on client name or project ID.

How to complete the Project Summary form is obvious. The many ways to use this form may not be so obvious, so here are a few pointers.

1. Use your index of Project Summary forms to find completed jobs most similar to the job you're bidding. Believe it or not, Project Summary forms with the widest margin of error will be most useful. Ask yourself: Who worked on those projects? Who was the field superintendent? Who were the vendors? Did the errors result from poor estimating or the poor performance of vendors, supervisors or crews? The most common estimating errors occur when (a) inspecting the job site, (b) examining the plans or (c) reading the specifications. What did you miss and why? Look for pitfalls to avoid in the job now being estimated. Identify the biggest two or three mistakes made when bidding that job. Make a notation about each on the Project Summary form.

2. Now look at your bid for the current job. Which mistakes made on a prior job might you expect on this job? Concentrate on the big three oversights to avoid: Inspecting the job site; examining the plans; and reading the specifications.

3. Unless there's a major error in take-off, your estimate of material costs should be within about 5 percent of the actual costs of materials. However, it's common for labor cost estimates to vary 20 percent or more from actual labor costs. This is precisely where data from old jobs comes in handy. If your Project Summary files show that some project types are consistent money-losers, either shift your company's focus to another class of work, factor more contingency into your bids, or find some way to wring inefficiencies out of the labor component. Poor staging, delivery and retrieval procedures drag down labor productivity on any job.

4. Use your file of Project Summary forms to spot any common thread that runs through either money-making jobs or money-losing jobs. For example, if the names of certain subcontractors or vendors are prominent on low-margin jobs, maybe there's a relationship between your profit margin and choice of subs and suppliers. Even the best and most reliable vendors can become complacent if not challenged occasionally.

5. Project Summary forms should note changes and extras identified after the contract was signed — both for which your company was paid and changes done without additional compensation. Projects with changes and extras that exceed about 4 percent of the contract price deserve special scrutiny. Jobs with changes beyond about 4 percent aren't good for business, at least in my opinion. Nearly all changes have a negative impact on your job schedule and require a disproportionate investment of management resources. Too many changes can antagonize the owner and design staff, even if they were responsible for the altered plans. You may know of a mechanical contractor with a reputation for capitalizing on change orders. But I've rarely seen a job plagued with changes that turned into a money-maker for anyone — except the attorneys. Your file of Project Summary forms will show job types that carry change order risk. Before finalizing and submitting any bid, consider whether the job will get mired in disputes over changes and extras. If similar jobs have ended on the courthouse steps, factor that risk into your estimate.

Utility of a Project Summary forms file is limited only by your ingenuity. The important point is to keep and organize the source of your second profit available on every job. What you learn can be more valuable than what you earn.

The Estimating Procedure

Every plumbing and HVAC estimator works under deadline pressure. You'll seldom have the luxury of spending as much time as you would like on an estimate. Estimators who aren't organized waste valuable time and tend to make careless errors. Try to be well-organized and consistent in your approach to estimating. For most projects, I recommend that you follow the procedures listed below and in the order listed:

1. Get a second set of project drawings and specifications for use by your suppliers and subcontractors. Remember that your subs and suppliers need access to the plans and specs and time to prepare their quotes.

2. Study the plans and specs carefully. Highlight important items. Make a list of specific tasks that require labor unit correction factors. The estimate is never complete until you're totally familiar with the project and the applicable construction codes.

3. Get the general contractor or owner to identify the proposed construction schedule and subcontractor lay-down (storage) area. Work schedule and site conditions always affect your costs.

4. Contact all potential suppliers and subcontractors as early as possible. Set a time when each can come to your office to make their take-offs from the spare set of contract documents.

When this important preliminary work is done, or in progress, it's time to begin your detailed take-off.

Guidelines for Good Estimating

You can compile estimates on a legal pad, a printed estimating form or on a computer. Regardless of the method, these guidelines will apply:

List Each Cost Separately on your take-off sheet. Don't combine system estimates, even if the materials are the same type. A combined system estimate may have to be completely redone if materials for one system are changed at a later date. Use the Estimate Detail Sheet on page 16 if you don't already have a good material take-off form.

Use Engineer's Identification Numbers when listing equipment. The word pump without any other description is ambiguous when there are several pumps included in the project.

Don't Forget Labor Adjustment factors if your labor costs are significantly higher or lower than the costs used in this book. See instructions on page 7 for adjusting labor costs.

Use Colored Pencils or highlighters to mark the items you've taken off and listed. Use a different color for each piping or ducting system.

Log Telephone Quotes and other important phone conversations on a telephone quote form. See the sample on page 18.

Project Estimated Costs for labor, material and equipment to the time when the work is expected to be done, not when the job is being estimated.

The only good estimate is a complete estimate. You've probably heard this saying, "He who makes the most mistakes is likely to be low bidder, and live to regret it."

Preparing the Proposal

It's both common courtesy and good business practice to deliver an unpriced copy of your bid or proposal letter to the general contractor three or four days before the bid deadline date. This gives the contractor time to study your proposal and obtain alternate pricing for items you may have excluded. To avoid misunderstandings, make sure your proposals include, as a minimum, the following elements:

1. The complete name and address of the proposed project.

2. Specification title and issue date.

3. A complete listing of drawings and their issue or revision date.

4. A complete list of addenda and their dates of issue.

5. A list of specification section numbers covered by your proposal.

6. A list of exclusions, clarifications and assumptions.

Your final bid can be phoned in or sent by fax, but it should reach the general contractor or owner no more than five or ten minutes before the bid deadline. Prices submitted too early may have to be revised because of last-minute price changes by subcontractors or suppliers.

MET Worksheet
Material, Equipment and Tool Delivery and Surplus Return Record

Project ID _____   Job Location _____

Supervisor _____   Start Date _____

Description of Material, Equipment or Tool Delivered or Returned	Delivered to Job Site		Returned to Inventory			
	Quantity Delivered	Date Delivered	Quantity Returned	Date Returned	Status Code RN or RS	Value at Return

PROJECT SUMMARY

Project ID _____ Job Location _____

Short description _____

Supervisor _____

Index ID _____ Start Date _____

Estimator _____ Client _____

Major vendors _____ Subcontractors _____

_____

Sources of cost deviation _____

_____

_____

Related Projects by ID Number _____

Thumbnail Summary	Labor	Material	Equipment	Subcontract	Deployed RN/RS	Total
Actual cost						
Estimate						
Over/(Under)						
Full Summary						
Bid amount						
Estimated cost						
Projected profit						
Cost overrun						
Bid profit						
Change orders						
Cost of changes						
Total profit						
Total profit with RN/RS						
Redeployment						

15

Estimate Detail Sheet

Data carried forward from Take-Off Quantity Survey Sheet(s)

Company/Department _____ Estimator _____ Date _____

Project _____ Checked by _____ Date _____

Address _____ Notes: _____

Job description _____ Estimate # _____

CSI Division/Account _____ Estimate due _____

Item Description	Quantity	Unit	Crew @ MH/Unit	Manhours Ext.	Materials		Labor		Equipment		Subcontract		Total $
					Unit $	Ext. $	Unit $	Ext. $	Unit $	Ext. $	Unit $	Ext. $	
Totals This Sheet				Manhours	Material $		Labor $		Equipment $		Subcontract $		Total $

Carry totals forward to Estimate Summary Sheet _____ Estimate # _____ Estimate Detail Sheet _____ of _____

Quotation Sheet

Job: _____

Supplier: _____

Salesperson: _____ **Phone No:** _____

Per Plans/Specs: _____ **Freight:** _____ **Terms:** _____

Description	Delivery Time	Price

By: _____

Record of Telephone Conversation

Date: _____ Time: _____ Project: _____

Telecon with: _____

Company: _____ Phone No: _____

Subject: _____

Details of Conversation: _____

_____

_____

_____

_____

_____

_____

_____

_____

_____

_____

_____

_____

_____

_____

_____

_____

_____

_____

_____

_____

_____

_____

By: _____

Description	Craft@Hrs	Unit	Material $	Labor $	Equipment $	Total $

Electric domestic hot water heater, (residential).
Set in place only (floor models). Make additional allowances for pipe and electrical connections. (See below)

Description	Craft@Hrs	Unit	Material $	Labor $	Equipment $	Total $
6 gallon						
1.5 KW/110V	P1@.500	Ea	287.00	17.90	—	304.90
10 gallon						
1.5 KW/110V	P1@.500	Ea	320.00	17.90	—	337.90
15 gallon						
1.5 KW/110V	P1@.750	Ea	337.00	26.80	—	363.80
20 gallon						
1.5 KW/110V	P1@.750	Ea	349.00	26.80	—	375.80
30 gallon						
1.5 KW/110V	P1@1.00	Ea	372.00	35.80	—	407.80
40 gallon						
1.5 KW/110V	P1@1.20	Ea	390.00	42.90	—	432.90
50 gallon						
3 KW/110V	P1@1.30	Ea	420.00	46.50	—	466.50
12 gallon						
3 KW/220V	P1@.500	Ea	338.00	17.90	—	355.90
20 gallon						
3 KW/220V	P1@.750	Ea	370.00	26.80	—	396.80
30 gallon						
3 KW/220V	P1@1.00	Ea	405.00	35.80	—	440.80
40 gallon						
3 KW/220V	P1@1.20	Ea	440.00	42.90	—	482.90
50 gallon						
3 KW/220V	P1@1.30	Ea	470.00	46.50	—	516.50

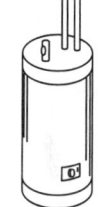

Electric domestic hot water heater, (commercial), 208/240 volt.
Set in place only. Make additional allowances for pipe and electrical connections. (See below)

Description	Craft@Hrs	Unit	Material $	Labor $	Equipment $	Total $
96 gallon, 12 kw	P1@1.50	Ea	1,880.00	53.60	—	1,933.60
96 gallon, 18 kw	P1@1.50	Ea	2,540.00	53.60	—	2,593.60
96 gallon, 36 kw	P1@1.50	Ea	2,640.00	53.60	—	2,693.60
120 gallon, 18 kw	P1@2.00	Ea	2,700.00	71.50	—	2,771.50
120 gallon, 36 kw	P1@2.00	Ea	2,780.00	71.50	—	2,851.50
120 gallon, 54 kw	P1@2.00	Ea	3,300.00	71.50	—	3,371.50
120 gallon, 63 kw	P1@2.00	Ea	3,560.00	71.50	—	3,631.50

Gas-fired domestic hot water heater, (residential).
Set in place only, Make additional allowances for pipe and combustion venting connections. (See below)

Description	Craft@Hrs	Unit	Material $	Labor $	Equipment $	Total $
30 gallon	P1@1.00	Ea	384.00	35.80	—	419.80
40 gallon	P1@1.00	Ea	622.00	35.80	—	657.80
50 gallon	P1@1.50	Ea	706.00	53.60	—	759.60

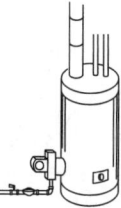

Description	Craft@Hrs	Unit	Material $	Labor $	Equipment $	Total $

Gas-fired domestic hot water heater, (commercial), standard efficiency. Set in place only, Make additional allowances for pipe and combustion venting connections. (See below)

Description	Craft@Hrs	Unit	Material $	Labor $	Equipment $	Total $
50 gal./95 gph	P1@2.00	Ea	1,720.00	71.50	—	1,791.50
67 gal./106 gph	P1@2.00	Ea	2,040.00	71.50	—	2,111.50
76 gal./175 gph	P1@2.00	Ea	2,730.00	71.50	—	2,801.50
91 gal./291 gph	P1@2.00	Ea	3,290.00	71.50	—	3,361.50

Gas-fired domestic hot water heater, (commercial), energy miser. Set in place only, Make additional allowances for pipe and combustion venting connections. (See below)

Description	Craft@Hrs	Unit	Material $	Labor $	Equipment $	Total $
50 gal./95 gph	P1@2.00	Ea	4,060.00	71.50	—	4,131.50
67 gal./106 gph	P1@2.00	Ea	4,240.00	71.50	—	4,311.50
76 gal./175 gph	P1@2.00	Ea	5,260.00	71.50	—	5,331.50
91 gal./291 gph	P1@2.00	Ea	6,240.00	71.50	—	6,311.50

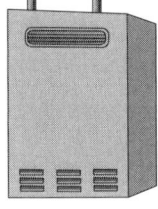

Tankless natural gas water heaters. Ambient pressure. DOE and UL rated. For residential, multi-dwelling and light commercial potable water applications. Add the cost of piping, tempering valve, circulating pump, controls, and electrical connection, post-installation inspection by both the fire marshal and the mechanical inspector to validate federal, state and local energy tax credits or energy tax credit offsets. For larger arrays (laundries, institutional facilities, food processing plants), develop an estimate based on the required capacity and multiply these costs by the number of heaters required. Rated in Btu's and gallons per minute capacity. (1 Mbh = 1,000 Btu's)

Description	Craft@Hrs	Unit	Material $	Labor $	Equipment $	Total $
19.5–140 Mbh, .75–5.8 Gpm	P1@16.0	Ea	1,450.00	572.00	—	2,022.00
11-199 Mbh, .5-7 Gpm	P1@20.0	Ea	1,700.00	715.00	—	2,415.00
25-235 Mbh .75-9.6 Gpm	P1@20.0	Ea	1,700.00	715.00	—	2,415.00

Tankless electric point-of-use water heaters. Ambient pressure, DOE and UL rated. For residential, multi-dwelling and light commercial potable water applications. Cost does not include piping, tempering valve, circulating pump, controls, storage tank, electrical connection, stack or stack liner. Add the cost of post-installation inspection by the mechanical inspector to validate federal, state and local energy tax credits or energy tax credit offsets. In rated gallons per minute capacity.

Description	Craft@Hrs	Unit	Material $	Labor $	Equipment $	Total $
5.5 Kw/40 Amp, .75-2 Gpm	P1@4.00	Ea	336.00	143.00	—	479.00
9.5 Kw/50 Amp .75-2.5 Gpm	P1@4.25	Ea	399.00	152.00	—	551.00
19 Kw/100 Amp 1-3.5 Gpm	P1@4.50	Ea	662.00	161.00	—	823.00
28 Kw/120 Amp 1.5-5 Gpm	P1@4.75	Ea	1,210.00	170.00	—	1,380.00

Description	Craft@Hrs	Unit	Material $	Labor $	Equipment $	Total $

Domestic hot water heater connection assembly. Includes supply, return, recirculation and relief piping and fittings (copper), relief and isolation valves. Make additional allowances for gas and venting connections where applicable.

Description	Craft@Hrs	Unit	Material $	Labor $	Equipment $	Total $
3/4" residential	P1@1.75	Ea	293.00	62.60	—	355.60
3/4" commercial	P1@2.25	Ea	394.00	80.50	—	474.50
1" commercial	P1@2.75	Ea	691.00	98.30	—	789.30
1¼" commercial	P1@3.50	Ea	848.00	125.00	—	973.00
1½" commercial	P1@3.75	Ea	881.00	134.00	—	1,015.00
2" commercial	P1@4.50	Ea	942.00	161.00	—	1,103.00
2½" commercial	P1@5.75	Ea	1,960.00	206.00	—	2,166.00
3" commercial	P1@6.50	Ea	3,000.00	232.00	—	3,232.00

Domestic water heater combustion vent connection. Make additional allowances for piping distances greater than 25'.

Description	Craft@Hrs	Unit	Material $	Labor $	Equipment $	Total $
2" b-vent	P1@.090	LF	4.77	3.22	—	7.99
3" b-vent	P1@.100	LF	5.91	3.58	—	9.49
4" b-vent	P1@.110	LF	7.86	3.93	—	11.79
6" b-vent	P1@.130	LF	8.95	4.65	—	13.60
Tankless Heater						
Vent kit	P1@2.50	Ea	335.00	89.40	—	424.40
Power vent kit	P1@2.00	Ea	857.00	71.50	—	928.50

Water Softener Controllers

Description	Craft@Hrs	Unit	Material $	Labor $	Equipment $	Total $

Water softener, time clock controller. Including brine tank, brine well and pick up tube. Labor includes setting in place, connecting the unit to an existing domestic water distribution system, start up and testing.

Description	Craft@Hrs	Unit	Material $	Labor $	Equipment $	Total $
20,000 grain water softener, TCC	P1@4.50	Ea	505.00	161.00	—	666.00
30,000 grain water softener, TCC	P1@4.50	Ea	538.00	161.00	—	699.00
45,000 grain water softener, TCC	P1@4.50	Ea	598.00	161.00	—	759.00
50,000 grain water softener, TCC	P1@4.75	Ea	675.00	170.00	—	845.00
60,000 grain water softener, TCC	P1@4.75	Ea	797.00	170.00	—	967.00
75,000 grain water softener, TCC	P1@5.00	Ea	855.00	179.00	—	1,034.00
90,000 grain water softener, TCC	P1@5.50	Ea	1,155.00	197.00	—	1,352.00
120,000 grain water softener, TCC	P1@5.75	Ea	1,245.00	206.00	—	1,451.00

Water softener, mechanically metered controller. Including brine tank, brine well and pick up tube. Labor includes setting in place, connecting the unit to an existing domestic water distribution system, start up and testing.

Description	Craft@Hrs	Unit	Material $	Labor $	Equipment $	Total $
20,000 grain water softener, MMC	P1@4.50	Ea	655.00	161.00	—	816.00
30,000 grain water softener, MMC	P1@4.50	Ea	683.00	161.00	—	844.00
45,000 grain water softener, MMC	P1@4.50	Ea	744.00	161.00	—	905.00
50,000 grain water softener, MMC	P1@4.75	Ea	820.00	170.00	—	990.00
60,000 grain water softener, MMC	P1@4.75	Ea	958.00	170.00	—	1,128.00
75,000 grain water softener, MMC	P1@5.00	Ea	1,015.00	179.00	—	1,194.00
90,000 grain water softener, MMC	P1@5.50	Ea	1,315.00	197.00	—	1,512.00
120,000 grain water softener, MMC	P1@5.75	Ea	1,406.00	206.00	—	1,612.00

Description	Craft@Hrs	Unit	Material $	Labor $	Equipment $	Total $

Water softener, electronically metered controller. Including brine tank, brine well and pick up tube. Labor includes setting in place, connecting the unit to an existing domestic water distribution system, start up and testing.

Description	Craft@Hrs	Unit	Material $	Labor $	Equipment $	Total $
20,000 grain water softener, EMC	P1@4.50	Ea	695.00	161.00	—	856.00
30,000 grain water softener, EMC	P1@4.50	Ea	715.00	161.00	—	876.00
45,000 grain water softener, EMC	P1@4.50	Ea	785.00	161.00	—	946.00
50,000 grain water softener, EMC	P1@4.75	Ea	860.00	170.00	—	1,030.00
60,000 grain water softener, EMC	P1@4.75	Ea	998.00	170.00	—	1,168.00
75,000 grain water softener, EMC	P1@5.00	Ea	1,055.00	179.00	—	1,234.00
90,000 grain water softener, EMC	P1@5.50	Ea	1,354.00	197.00	—	1,551.00
120,000 grain water softener, EMC	P1@5.75	Ea	1,446.00	206.00	—	1,652.00

Water softener accessories

Description	Craft@Hrs	Unit	Material $	Labor $	Equipment $	Total $
By-pass valve	P1@.400	Ea	67.00	14.30	—	81.30
Manifold Adapter kit	P1@.200	Ea	18.00	7.15	—	25.15
Turbulator	P1@.400	Ea	33.00	14.30	—	47.30

Iron filter, electronically meter controller. Manganese green sand filter. Labor includes setting in place, connecting the unit to an existing domestic water distribution system, start up and testing.

Description	Craft@Hrs	Unit	Material $	Labor $	Equipment $	Total $
42,000 Iron filter, (1.5 cf media), 5 gpm,	P1@4.00	Ea	657.00	143.00	—	800.00
65,000 Iron filter, (2.0 cf media), 6 gpm	P1@4.50	Ea	777.00	161.00	—	938.00
84,000 Iron filter, (2.5 cf media), 8 gpm	P1@4.75	Ea	830.00	170.00	—	1,000.00
Replacement green sand media	P1@1.20	CF	38.00	42.90	—	80.90

Iron filter accessories

Description	Craft@Hrs	Unit	Material $	Labor $	Equipment $	Total $
By-pass valve	P1@.400	Ea	67.00	14.30	—	81.30
Air vent	P1@.200	Ea	53.00	7.15	—	60.15
Air controller	P1@.400	Ea	60.00	14.30	—	74.30

Water Softener Accessories

Description	Craft@Hrs	Unit	Material $	Labor $	Equipment $	Total $

Combination iron filter/water softener. Zeolite resins soften water and remove iron and manganese. Controller automatically controls PH level. Labor includes setting in place, connecting the unit to an existing domestic water distribution system, start up and testing.

Description	Craft@Hrs	Unit	Material $	Labor $	Equipment $	Total $
40,000 Iron filter, 1.3 cf media,	P1@4.00	Ea	1,270.00	143.00	—	1,413.00
60,000 Iron filter, 1.7 cf media,	P1@4.50	Ea	1,375.00	161.00	—	1,536.00
80,000 Iron filter, 2.5 cf media,	P1@4.75	Ea	1,995.00	170.00	—	2,165.00

Hot water softener, time clock controller. Brass valve construction. Designed for 150 F. maximum operating temperature. Includes brine tank, brine well and pick up tube. Labor includes setting in place, connecting the unit to an existing domestic water distribution system, start up and testing.

Description	Craft@Hrs	Unit	Material $	Labor $	Equipment $	Total $
20,000 grain hot water softener,	P1@4.50	Ea	1,595.00	161.00	—	1,756.00
30,000 grain hot water softener,	P1@4.50	Ea	1,695.00	161.00	—	1,856.00
40,000 grain hot water softener,	P1@4.50	Ea	1,760.00	161.00	—	1,921.00
60,000 grain hot water softener,	P1@4.75	Ea	2,088.00	170.00	—	2,258.00

Pressure tank, fiberglass wound. Labor includes setting in place, connecting the tank to a domestic water distribution system and testing.

Description	Craft@Hrs	Unit	Material $	Labor $	Equipment $	Total $
Fiberglass pressure tank, 20 gallon	P1@2.00	Ea	249.00	71.50	—	320.50
Fiberglass pressure tank, 30 gallon	P1@2.00	Ea	280.00	71.50	—	351.50
Fiberglass pressure tank, 80 gallon	P1@2.75	Ea	455.00	98.30	—	553.30
Fiberglass pressure tank, 120 gallon	P1@3.50	Ea	600.00	125.00	—	725.00
Brass tank tee assembly, 3/4"	P1@3.50	Ea	30.00	125.00	—	155.00
Brass tank tee assembly, 1"	P1@3.50	Ea	56.00	125.00	—	181.00
Brass tank tee assembly, 1-1/4"	P1@3.50	Ea	96.00	125.00	—	221.00

Description	Craft@Hrs	Unit	Material $	Labor $	Equipment $	Total $

Ultra-violet water disinfection unit. Stainless steel reactor, audible and visible alarm, lamp end-of-life indicator and 7 day override. Gpm rating at 30,000 mj/m2 output. Labor includes setting in place, connecting to the water system and testing.

Description	Craft@Hrs	Unit	Material $	Labor $	Equipment $	Total $
UV system, 1 gpm, 1/4" in/out,	P1@3.00	Ea	187.00	107.00	—	294.00
UV system, 6 gpm, 1/2" in/out,	P1@3.00	Ea	363.00	107.00	—	470.00
UV system, 8 gpm, 3/4" in/out,	P1@4.00	Ea	421.00	143.00	—	564.00
UV system, 12 gpm, 3/4" in/out,	P1@4.00	Ea	537.00	143.00	—	680.00
UV replacement lamp, 20W, 1 gpm	P1@.750	Ea	41.80	26.80	—	68.60
UV replacement lamp, 32W, 6 gpm	P1@.750	Ea	47.30	26.80	—	74.10
UV replacement lamp, 39W, 8-12 gpm	P1@.750	Ea	60.50	26.80	—	87.30
UV replacement ballast, 420 Mv/110V	P1@1.00	Ea	183.00	35.80	—	218.80

Kitchen equipment booster heater

Description	Craft@Hrs	Unit	Material $	Labor $	Equipment $	Total $
1,000 watt	P1@4.00	Ea	516.00	143.00	—	659.00

Dishwasher

Description	Craft@Hrs	Unit	Material $	Labor $	Equipment $	Total $
Built-in	P1@5.00	Ea	750.00	179.00	—	929.00

Garbage disposal

Description	Craft@Hrs	Unit	Material $	Labor $	Equipment $	Total $
1/2 HP	P1@2.00	Ea	157.00	71.50	—	228.50
3/4 HP	P1@2.00	Ea	266.00	71.50	—	337.50

Grease and oil interceptor

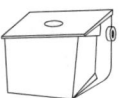

Description	Craft@Hrs	Unit	Material $	Labor $	Equipment $	Total $
4 GPM	P1@4.00	Ea	312.00	143.00	—	455.00
10 GPM	P1@5.00	Ea	509.00	179.00	—	688.00
15 GPM	P1@7.00	Ea	759.00	250.00	—	1,009.00
20 GPM	P1@8.00	Ea	919.00	286.00	—	1,205.00

Hair and lint interceptor

Description	Craft@Hrs	Unit	Material $	Labor $	Equipment $	Total $
1½"	P1@.650	Ea	196.00	23.20	—	219.20
2"	P1@.750	Ea	278.00	26.80	—	304.80

All bronze 3/4" to 1-1/2" in-line NPT pump

Description	Craft@Hrs	Unit	Material $	Labor $	Equipment $	Total $
1/12 HP	P1@1.50	Ea	466.00	53.60	—	519.60
1/6 HP	P1@1.50	Ea	695.00	53.60	—	748.60
1/4 HP	P1@1.50	Ea	814.00	53.60	—	867.60

Kitchen Equipment Connections

Description	Craft@Hrs	Unit	Material $	Labor $	Equipment $	Total $
Kitchen appliance gas trim						
1/2"	P1@1.15	Ea	33.80	41.10	—	74.90
3/4"	P1@1.30	Ea	61.80	46.50	—	108.30
1"	P1@1.60	Ea	71.60	57.20	—	128.80
1¼"	P1@2.10	Ea	118.00	75.10	—	193.10
1½"	P1@2.50	Ea	149.00	89.40	—	238.40
2"	P1@3.00	Ea	200.00	107.00	—	307.00
Hot and cold water supply						
1/2"	P1@1.10	Ea	36.00	39.30	—	75.30
3/4"	P1@1.55	Ea	51.00	55.40	—	106.40
1"	P1@1.90	Ea	69.30	67.90	—	137.20
1¼"	P1@2.50	Ea	97.70	89.40	—	187.10
1½"	P1@3.00	Ea	123.00	107.00	—	230.00
Continuous waste						
2-part	P1@.250	Ea	51.20	8.94	—	60.14
3-part	P1@.350	Ea	86.10	12.50	—	98.60
4-part	P1@.450	Ea	112.00	16.10	—	128.10
Indirect waste						
1/2"	P1@1.05	Ea	9.94	37.50	—	47.44
3/4"	P1@1.50	Ea	16.70	53.60	—	70.30
1"	P1@1.90	Ea	26.90	67.90	—	94.80
1¼"	P1@2.15	Ea	39.80	76.90	—	116.70
1½"	P1@2.60	Ea	52.40	93.00	—	145.40
2"	P1@3.00	Ea	80.20	107.00	—	187.20
Kitchen fixture waste tailpiece						
1½"	P1@.100	Ea	11.50	3.58	—	15.08
Kitchen fixture trap with solder bushing						
1½"	P1@.250	Ea	38.40	8.94	—	47.34
2"	P1@.300	Ea	53.50	10.70	—	64.20

Description	Craft@Hrs	Unit	Material $	Labor $	Equipment $	Total $

Water closet, floor mounted, flush tank, white vitreous china, lined tank. Complete with trim. Make additional allowances for rough-in. Based on American Standard Cadet series. ADA means American Disabilities Act compliant. (Wheelchair accessible)

Description	Craft@Hrs	Unit	Material $	Labor $	Equipment $	Total $
Round bowl	P1@2.10	Ea	229.00	75.10	—	304.10
Elongated bowl	P1@2.10	Ea	277.00	75.10	—	352.10
ADA, 18" high	P1@2.10	Ea	376.00	75.10	—	451.10

Water closet, floor mounted, flush valve, white vitreous china. Complete with trim. Make additional allowances for rough-in. Based on American Standard. ADA means American Disabilities Act compliant. (Wheelchair accessible)

Description	Craft@Hrs	Unit	Material $	Labor $	Equipment $	Total $
Elongated bowl	P1@2.60	Ea	358.00	93.00	—	451.00
Elongated bowl, ADA 18" high	P1@2.60	Ea	426.00	93.00	—	519.00
Elongated bowl with a bedpan cleanser	P1@4.10	Ea	622.00	147.00	—	769.00
Elongated bowl, ADA 18" high with a bedpan cleanser	P1@4.10	Ea	672.00	147.00	—	819.00

Water closet, wall hung, flush valve, white vitreous china. Complete with fixture carrier and all trim. Make additional allowances for rough-in. Based on American Standard Afwall series.

Description	Craft@Hrs	Unit	Material $	Labor $	Equipment $	Total $
Elongated bowl	P1@3.55	Ea	572.00	127.00	—	699.00
Elongated bowl with electronic flush valve	P1@3.80	Ea	1,020.00	136.00	—	1,156.00
Elongated bowl with bedpan cleanser	P1@5.05	Ea	825.00	181.00	—	1,006.00
Electronic flush valve, add	P1@.600	Ea	449.00	21.50	—	470.50

Urinal, wall hung, flush valve, white vitreous china. Complete with trim. Make additional allowances for rough-in.

Description	Craft@Hrs	Unit	Material $	Labor $	Equipment $	Total $
Siphon jet type	P1@3.15	Ea	572.00	113.00	—	685.00
Wash out type	P1@3.10	Ea	466.00	111.00	—	577.00
Wash down type	P1@3.00	Ea	328.00	107.00	—	435.00
Urinal carrier, add	P1@.600	Ea	106.00	21.50	—	127.50
Electronic flush valve, add	P1@.600	Ea	371.00	21.50	—	392.50

Urinal, stall type, flush valve, white vitreous china. Complete with trim. Make additional allowances for rough-in.

Description	Craft@Hrs	Unit	Material $	Labor $	Equipment $	Total $
Stall urinal	P1@5.00	Ea	1,120.00	179.00	—	1,299.00

Plumbing Fixtures

Description	Craft@Hrs	Unit	Material $	Labor $	Equipment $	Total $

Lavatory, wall hung, white vitreous china. Complete with trim and fixture carrier. Make additional allowances for rough-in. ADA means American Disabilities Act compliant. (Wheelchair accessible)

Description	Craft@Hrs	Unit	Material $	Labor $	Equipment $	Total $
Wall hung lav	P1@2.70	Ea	475.00	96.60	—	571.60
Wall hung, ADA	P1@2.70	Ea	694.00	96.60	—	790.60
Fixture carrier	P1@.500	Ea	98.00	17.90	—	115.90

Countertop lavatory, white. Complete with trim. Make additional allowances for rough-in.

Description	Craft@Hrs	Unit	Material $	Labor $	Equipment $	Total $
Vitreous china	P1@2.00	Ea	352.00	71.50	—	423.50
Enameled steel	P1@2.00	Ea	297.00	71.50	—	368.50
Acrylic	P1@2.00	Ea	217.00	71.50	—	288.50

Bathtub, white, 60" x 32". Complete with trim, including shower head. Make additional allowances for rough-in.

Description	Craft@Hrs	Unit	Material $	Labor $	Equipment $	Total $
Enameled steel	P1@2.50	Ea	496.00	89.40	—	585.40
Cast iron	P1@3.50	Ea	718.00	125.00	—	843.00
Fiberglass	P1@2.50	Ea	459.00	89.40	—	548.40
Acrylic	P1@2.50	Ea	490.00	89.40	—	579.40

Tub and shower combination, fiberglass, white. Complete with trim. Make additional allowances for rough-in.

Description	Craft@Hrs	Unit	Material $	Labor $	Equipment $	Total $
One piece	P1@4.50	Ea	1,020.00	161.00	—	1,181.00
Two piece (reno)	P1@5.50	Ea	1,300.00	197.00	—	1,497.00
Four piece (reno)	P1@6.25	Ea	1,390.00	224.00	—	1,614.00

Shower stall, white, 36" x 36". Complete with trim. Make additional allowances for rough-in.

Description	Craft@Hrs	Unit	Material $	Labor $	Equipment $	Total $
Fiberglass One piece	P1@3.50	Ea	636.00	125.00	—	761.00
Fiberglass Three piece	P1@4.25	Ea	818.00	152.00	—	970.00
Acrylic One piece	P1@3.50	Ea	948.00	125.00	—	1,073.00
Acrylic Three piece	P1@4.25	Ea	1,240.00	152.00	—	1,392.00

Shower basin, 36" x 36". Complete with trim (faucet, shower head and strainer). Make additional allowances for rough-in.

Description	Craft@Hrs	Unit	Material $	Labor $	Equipment $	Total $
Fiberglass	P1@2.50	Ea	432.00	89.40	—	521.40
Acrylic	P1@2.50	Ea	466.00	89.40	—	555.40
Molded stone	P1@2.65	Ea	449.00	94.80	—	543.80

Description	Craft@Hrs	Unit	Material $	Labor $	Equipment $	Total $

Kitchen sink, double compartment. Complete with trim. Make additional allowances for rough-in.

Description	Craft@Hrs	Unit	Material $	Labor $	Equipment $	Total $
Stainless steel	P1@2.15	Ea	350.00	76.90	—	426.90
Cast iron	P1@2.50	Ea	459.00	89.40	—	548.40
Acrylic	P1@2.15	Ea	416.00	76.90	—	492.90

Kitchen sink, single compartment. Complete with trim. Make additional allowances for rough-in.

Description	Craft@Hrs	Unit	Material $	Labor $	Equipment $	Total $
Stainless steel	P1@2.00	Ea	296.00	71.50	—	367.50
Cast iron	P1@2.10	Ea	342.00	75.10	—	417.10
Acrylic	P1@2.00	Ea	309.00	71.50	—	380.50

Bar sink. Complete with trim. Make additional allowances for rough-in.

Description	Craft@Hrs	Unit	Material $	Labor $	Equipment $	Total $
Stainless steel	P1@2.00	Ea	266.00	71.50	—	337.50
Acrylic	P1@2.00	Ea	179.00	71.50	—	250.50

Exam room sink. Complete with trim. Make additional allowances for rough-in.

Description	Craft@Hrs	Unit	Material $	Labor $	Equipment $	Total $
Stainless steel	P1@2.10	Ea	384.00	75.10	—	459.10
Acrylic	P1@2.10	Ea	328.00	75.10	—	403.10

Laboratory sink. Complete with trim. Make additional allowances for rough-in.

Description	Craft@Hrs	Unit	Material $	Labor $	Equipment $	Total $
Stainless steel	P1@2.25	Ea	441.00	80.50	—	521.50
Acrylic	P1@2.25	Ea	384.00	80.50	—	464.50

Laundry sink, double compartment. Complete with trim. Make additional allowances for rough-in.

Description	Craft@Hrs	Unit	Material $	Labor $	Equipment $	Total $
Cast iron	P1@3.50	Ea	515.00	125.00	—	640.00
Acrylic	P1@2.25	Ea	226.00	80.50	—	306.50

Laundry sink, single compartment. Complete with trim. Make additional allowances for rough-in.

Description	Craft@Hrs	Unit	Material $	Labor $	Equipment $	Total $
Cast iron	P1@2.75	Ea	446.00	98.30	—	544.30
Acrylic	P1@2.00	Ea	157.00	71.50	—	228.50

Mop sink, floor mounted, 36" x 24". Complete with trim. Make additional allowances for rough-in.

Description	Craft@Hrs	Unit	Material $	Labor $	Equipment $	Total $
Molded stone	P1@2.65	Ea	629.00	94.80	—	723.80
Terrazzo	P1@2.65	Ea	693.00	94.80	—	787.80
Acrylic	P1@2.35	Ea	483.00	84.00	—	567.00

Plumbing Fixtures

Description	Craft@Hrs	Unit	Material $	Labor $	Equipment $	Total $

Slop sink, enameled cast iron with P trap, standard. Complete with trim. Make additional allowances for rough-in.

Description	Craft@Hrs	Unit	Material $	Labor $	Equipment $	Total $
Slop sink with P trap, Std.	P1@3.50	Ea	1,030.00	125.00	—	1,155.00

Floor sink, recessed, enameled steel, white. Add 40% to material prices for acid-resisting finish. Complete with strainer. Make additional allowances for rough-in.

Description	Craft@Hrs	Unit	Material $	Labor $	Equipment $	Total $
9" x 9"	P1@1.00	Ea	88.40	35.80	—	124.20
12" x 12"	P1@1.00	Ea	103.00	35.80	—	138.80
15" x 15"	P1@1.15	Ea	70.10	41.10	—	111.20
18" x 18"	P1@1.25	Ea	91.40	44.70	—	136.10
24" x 24"	P1@1.50	Ea	124.00	53.60	—	177.60

Drinking fountain, refrigerated, stainless steel. Complete with trim. Make additional allowances for rough-in. ADA means American Disabilities Act compliant. (Wheelchair accessible)

Description	Craft@Hrs	Unit	Material $	Labor $	Equipment $	Total $
Free standing	P1@2.00	Ea	1,120.00	71.50	—	1,191.50
Semi recessed	P1@2.50	Ea	1,470.00	89.40	—	1,559.40
Fully recessed	P1@2.50	Ea	2,540.00	89.40	—	2,629.40
Wall hung	P1@2.00	Ea	1,040.00	71.50	—	1,111.50
Wall hung, ADA	P1@2.50	Ea	2,540.00	89.40	—	2,629.40

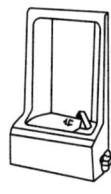

Drinking fountain, non-refrigerated. Complete with trim. Make additional allowances for rough-in. ADA means American Disabilities Act compliant. (Wheelchair accessible) S.S. means stainless steel.

Description	Craft@Hrs	Unit	Material $	Labor $	Equipment $	Total $
Recessed, china	P1@2.50	Ea	854.00	89.40	—	943.40
Wall hung, china	P1@2.00	Ea	486.00	71.50	—	557.50
Recessed, S.S.	P1@2.50	Ea	970.00	89.40	—	1,059.40
Wall hung, S.S.	P1@2.00	Ea	518.00	71.50	—	589.50
ADA, S.S.	P1@2.50	Ea	892.00	89.40	—	981.40

Description	Craft@Hrs	Unit	Material $	Labor $	Equipment $	Total $

Commercial plumbing fixture rough-in. Includes type L copper supply pipe and DWV copper (to 2½") or cast iron (MJ) DWV (over 2½") drain and vent piping. Make additional allowances for plumbing fixtures and trim. Use these costs for preliminary estimates.

Description	Craft@Hrs	Unit	Material $	Labor $	Equipment $	Total $
Water closet, wall hung, flush valve, with carrier	P1@2.25	Ea	857.00	80.50	—	937.50
Water closet, wall hung, flush valve, no carrier	P1@1.95	Ea	312.00	69.70	—	381.70
Water closet, floor mounted, flush valve	P1@2.75	Ea	693.00	98.30	—	791.30
Water closet, floor mounted, tank type	P1@2.25	Ea	531.00	80.50	—	611.50
Bidet	P1@2.00	Ea	370.00	71.50	—	441.50
Urinal, wall hung, flush valve, with carrier	P1@3.10	Ea	932.00	111.00	—	1,043.00
Urinal, wall hung, flush valve, without carrier	P1@2.35	Ea	531.00	84.00	—	615.00
Lavatory, wall hung, with carrier	P1@2.40	Ea	769.00	85.80	—	854.80
Lavatory	P1@1.90	Ea	370.00	67.90	—	437.90
Sink	P1@1.90	Ea	399.00	67.90	—	466.90
Bath tub	P1@2.35	Ea	570.00	84.00	—	654.00
Shower	P1@2.60	Ea	668.00	93.00	—	761.00
Mop sink	P1@2.40	Ea	473.00	85.80	—	558.80
Slop sink	P1@2.60	Ea	339.00	93.00	—	432.00
Laundry tub	P1@1.95	Ea	402.00	69.70	—	471.70
Wash fountain	P1@2.10	Ea	434.00	75.10	—	509.10
Lab sink, glass drainage	P1@3.80	Ea	1,710.00	136.00	—	1,846.00
Lab sink, acid resistant plastic drainage	P1@2.65	Ea	271.00	94.80	—	365.80
Drinking fountain	P1@2.20	Ea	294.00	78.70	—	372.70
Emergency eyewash and shower	P1@1.75	Ea	112.00	62.60	—	174.60
Washing machine	P1@2.25	Ea	431.00	80.50	—	511.50

Description	Craft@Hrs	Unit	Material $	Labor $	Equipment $	Total $

Commercial plumbing fixture group rough-in. Includes type L copper supply pipe and DWV copper (to 2½") or cast iron (MJ) DWV (over 2½") drain and vent piping. Make additional allowances for plumbing fixtures and trim. Use these costs for preliminary estimates.

Description	Craft@Hrs	Unit	Material $	Labor $	Equipment $	Total $
3 piece washroom group	P1@5.50	Ea	1,060.00	197.00	—	1,257.00
3 piece washroom group back to back	P1@9.75	Ea	1,930.00	349.00	—	2,279.00
Kitchen sink, back to back	P1@2.15	Ea	566.00	76.90	—	642.90
Battery of water closets, floor mounted, tank type, per water closet	P1@1.75	Ea	441.00	62.60	—	503.60
Battery of water closets, floor mounted, flush valve, per water closet	P1@2.20	Ea	576.00	78.70	—	654.70
Battery of water closets, wall hung, flush valve, with carrier, per water closet	P1@1.80	Ea	758.00	64.40	—	822.40
Battery of water closets, wall hung, flush valve, without carrier, per water closet	P1@1.50	Ea	222.00	53.60	—	275.60
Battery of urinals, wall hung, flush valve with carrier, per urinal	P1@2.45	Ea	911.00	87.60	—	998.60
Battery of urinals, wall hung, flush valve without carrier, per urinal	P1@1.90	Ea	470.00	67.90	—	537.90
Battery of lavatory basins, wall hung, with carrier, per lavatory	P1@2.00	Ea	708.00	71.50	—	779.50
Battery of lavatory basins, without carrier, per lavatory	P1@1.50	Ea	320.00	53.60	—	373.60

Residential plumbing fixture rough-in. Includes polyethylene (PE) supply pipe and ABS DWV drain and vent piping. Make additional allowances for plumbing fixtures and trim. Use these costs for preliminary estimates.

Description	Craft@Hrs	Unit	Material $	Labor $	Equipment $	Total $
Water closet, floor mounted, tank type	P1@2.00	Ea	122.00	71.50	—	193.50
Bidet	P1@1.85	Ea	91.90	66.20	—	158.10
Lavatory	P1@1.75	Ea	91.90	62.60	—	154.50
Counter sink	P1@1.75	Ea	101.00	62.60	—	163.60
Bathtub	P1@2.10	Ea	91.90	75.10	—	167.00
Shower	P1@2.45	Ea	134.00	87.60	—	221.60
Laundry tub	P1@1.75	Ea	83.70	62.60	—	146.30
Washing machine	P1@2.00	Ea	104.00	71.50	—	175.50

Copper, Type K with Brazed Joints

Type K hard-drawn copper pipe with wrought copper fittings and brazed joints is used in a wide variety of plumbing and HVAC systems such as potable water, heating hot water, chilled water, compressed air, refrigerant and A.C. condensate.

Brazed joints are those made with silver or other alloy filler metals having melting points at, or above, 1,000 degrees F. Maximum working pressure/temperature relationships for brazed joints are approximately as follows:

Maximum Working Pressure (PSIG)*			
Water Temperature	Nominal Pipe Size (inches)		
(degrees F.)	Up to 1	1¼ to 2	2½ to 4
Up to 350	270	190	150

For copper pipe and solder-type fittings using brazing alloys melting at, or above 1,000 degrees F.

This section has been arranged to save the estimator's time by including all normally-used system components such as pipe, fittings, valves, hanger assemblies, riser clamps and miscellaneous items under one heading. Additional items can be found under "Plumbing and Piping Specialties." The cost estimates in this section are based on the conditions, limitations and wage rates described in the section "How to Use This Book" beginning on page 5.

Description	Craft@Hrs	Unit	Material $	Labor $	Equipment $	Total $

Type K copper pipe with brazed joints, installed horizontally. Complete installation including six to twelve wrought copper tees and six to twelve wrought copper elbows every 100', and hangers spaced to meet plumbing code (a tee & elbow every 8.3' for ½" dia., a tee and elbow every 16.6' for 4" dia.).

Description	Craft@Hrs	Unit	Material $	Labor $	Equipment $	Total $
1/2"	P1@.117	LF	3.41	4.18	—	7.59
3/4"	P1@.127	LF	5.97	4.54	—	10.51
1"	P1@.147	LF	9.53	5.26	—	14.79
1¼"	P1@.167	LF	13.00	5.97	—	18.97
1½"	P1@.187	LF	16.00	6.69	—	22.69
2"	P1@.217	LF	27.20	7.76	—	34.96
2½"	P1@.247	LF	44.80	8.83	—	53.63
3"	P1@.277	LF	51.60	9.91	—	61.51
4"	P1@.317	LF	94.50	11.30	—	105.80

Copper, Type K with Brazed Joints

Description	Craft@Hrs	Unit	Material $	Labor $	Equipment $	Total $

Type K copper pipe with brazed joints, installed risers. Complete installation including a wrought copper reducing tee every floor and a riser clamp every other floor.

Description	Craft@Hrs	Unit	Material $	Labor $	Equipment $	Total $
1/2"	P1@.064	LF	3.08	2.29	—	5.37
3/4"	P1@.074	LF	5.55	2.65	—	8.20
1"	P1@.079	LF	8.48	2.83	—	11.31
1¼"	P1@.084	LF	11.70	3.00	—	14.70
1½"	P1@.104	LF	13.80	3.72	—	17.52
2"	P1@.134	LF	23.60	4.79	—	28.39
2½"	P1@.154	LF	35.30	5.51	—	40.81
3"	P1@.174	LF	49.50	6.22	—	55.72
4"	P1@.194	LF	77.30	6.94	—	84.24

Type K copper pipe with brazed joints, pipe only

Description	Craft@Hrs	Unit	Material $	Labor $	Equipment $	Total $
1/2"	P1@.032	LF	2.45	1.14	—	3.59
3/4"	P1@.035	LF	4.53	1.25	—	5.78
1"	P1@.038	LF	5.72	1.36	—	7.08
1¼"	P1@.042	LF	7.65	1.50	—	9.15
1½"	P1@.046	LF	8.33	1.64	—	9.97
2"	P1@.053	LF	14.90	1.90	—	16.80
2½"	P1@.060	LF	19.40	2.15	—	21.55
3"	P1@.066	LF	25.30	2.36	—	27.66
4"	P1@.080	LF	45.30	2.86	—	48.16

Type K copper 45 degree ell C x C with brazed joints

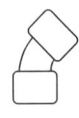

Description	Craft@Hrs	Unit	Material $	Labor $	Equipment $	Total $
1/2"	P1@.129	Ea	.99	4.61	—	5.60
3/4"	P1@.181	Ea	1.88	6.47	—	8.35
1"	P1@.232	Ea	5.54	8.30	—	13.84
1¼"	P1@.284	Ea	7.50	10.20	—	17.70
1½"	P1@.335	Ea	11.10	12.00	—	23.10
2"	P1@.447	Ea	18.30	16.00	—	34.30
2½"	P1@.550	Ea	46.70	19.70	—	66.40
3"	P1@.654	Ea	61.60	23.40	—	85.00
4"	P1@.860	Ea	119.00	30.80	—	149.80

Type K copper 90 degree ell C x C with brazed joints

Description	Craft@Hrs	Unit	Material $	Labor $	Equipment $	Total $
1/2"	P1@.129	Ea	.57	4.61	—	5.18
3/4"	P1@.181	Ea	1.21	6.47	—	7.68
1"	P1@.232	Ea	4.21	8.30	—	12.51
1¼"	P1@.284	Ea	6.83	10.20	—	17.03
1½"	P1@.335	Ea	10.40	12.00	—	22.40
2"	P1@.447	Ea	18.60	16.00	—	34.60
2½"	P1@.550	Ea	38.60	19.70	—	58.30
3"	P1@.654	Ea	48.80	23.40	—	72.20
4"	P1@.860	Ea	104.00	30.80	—	134.80

Description	Craft@Hrs	Unit	Material $	Labor $	Equipment $	Total $

Type K copper 90 degree ell Ftg x C with brazed joints

Description	Craft@Hrs	Unit	Material $	Labor $	Equipment $	Total $
1/2"	P1@.129	Ea	1.21	4.61	—	5.82
3/4"	P1@.181	Ea	3.99	6.47	—	10.46
1"	P1@.232	Ea	5.98	8.30	—	14.28
1¼"	P1@.284	Ea	8.33	10.20	—	18.53
1½"	P1@.335	Ea	12.50	12.00	—	24.50
2"	P1@.447	Ea	25.00	16.00	—	41.00
2½"	P1@.550	Ea	52.50	19.70	—	72.20
3"	P1@.654	Ea	59.00	23.40	—	82.40
4"	P1@.860	Ea	159.00	30.80	—	189.80

Type K copper tee C x C x C with brazed joints

Description	Craft@Hrs	Unit	Material $	Labor $	Equipment $	Total $
1/2"	P1@.156	Ea	.98	5.58	—	6.56
3/4"	P1@.204	Ea	2.64	7.30	—	9.94
1"	P1@.263	Ea	9.98	9.40	—	19.38
1¼"	P1@.322	Ea	15.80	11.50	—	27.30
1½"	P1@.380	Ea	23.00	13.60	—	36.60
2"	P1@.507	Ea	36.80	18.10	—	54.90
2½"	P1@.624	Ea	74.60	22.30	—	96.90
3"	P1@.741	Ea	115.00	26.50	—	141.50
4"	P1@.975	Ea	192.00	34.90	—	226.90

Copper, Type K with Brazed Joints

Description	Craft@Hrs	Unit	Material $	Labor $	Equipment $	Total $

Type K copper branch reducing tee C x C x C with brazed joints

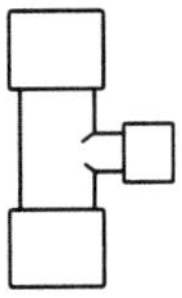

Description	Craft@Hrs	Unit	Material $	Labor $	Equipment $	Total $
1/2" x 3/8"	P1@.129	Ea	1.98	4.61	—	6.59
3/4" x 1/2"	P1@.182	Ea	4.62	6.51	—	11.13
1" x 1/2"	P1@.209	Ea	11.70	7.47	—	19.17
1" x 3/4"	P1@.234	Ea	11.70	8.37	—	20.07
1¼" x 1/2"	P1@.241	Ea	17.90	8.62	—	26.52
1¼" x 3/4"	P1@.264	Ea	17.90	9.44	—	27.34
1¼" x 1"	P1@.287	Ea	17.90	10.30	—	28.20
1½" x 1/2"	P1@.291	Ea	21.80	10.40	—	32.20
1½" x 3/4"	P1@.307	Ea	21.80	11.00	—	32.80
1½" x 1"	P1@.323	Ea	21.80	11.60	—	33.40
1½" x 1¼"	P1@.339	Ea	21.80	12.10	—	33.90
2" x 1/2"	P1@.348	Ea	29.30	12.40	—	41.70
2" x 3/4"	P1@.373	Ea	29.30	13.30	—	42.60
2" x 1"	P1@.399	Ea	29.30	14.30	—	43.60
2" x 1¼"	P1@.426	Ea	29.30	15.20	—	44.50
2" x 1½"	P1@.452	Ea	29.30	16.20	—	45.50
2½" x 1/2"	P1@.468	Ea	87.80	16.70	—	104.50
2½" x 3/4"	P1@.485	Ea	87.80	17.30	—	105.10
2½" x 1"	P1@.502	Ea	87.80	18.00	—	105.80
2½" x 1½"	P1@.537	Ea	87.80	19.20	—	107.00
2½" x 2"	P1@.555	Ea	87.80	19.80	—	107.60
3" x 1¼"	P1@.562	Ea	99.80	20.10	—	119.90
3" x 2"	P1@.626	Ea	104.00	22.40	—	126.40
3" x 2½"	P1@.660	Ea	104.00	23.60	—	127.60
4" x 1¼"	P1@.669	Ea	180.00	23.90	—	203.90
4" x 1½"	P1@.718	Ea	180.00	25.70	—	205.70
4" x 2"	P1@.768	Ea	180.00	27.50	—	207.50
4" x 2½"	P1@.819	Ea	180.00	29.30	—	209.30
4" x 3"	P1@.868	Ea	180.00	31.00	—	211.00

Description	Craft@Hrs	Unit	Material $	Labor $	Equipment $	Total $

Type K copper reducer with brazed joints

Description	Craft@Hrs	Unit	Material $	Labor $	Equipment $	Total $
1/2" x 3/8"	P1@.148	Ea	.71	5.29	—	6.00
3/4" x 1/2"	P1@.155	Ea	1.47	5.54	—	7.01
1" x 3/4"	P1@.206	Ea	2.41	7.37	—	9.78
1¼" x 1/2"	P1@.206	Ea	4.05	7.37	—	11.42
1¼" x 3/4"	P1@.232	Ea	4.05	8.30	—	12.35
1¼" x 1"	P1@.258	Ea	4.05	9.23	—	13.28
1½" x 1/2"	P1@.232	Ea	6.19	8.30	—	14.49
1½" x 3/4"	P1@.257	Ea	6.19	9.19	—	15.38
1½" x 1"	P1@.283	Ea	6.19	10.10	—	16.29
1½" x 1¼"	P1@.308	Ea	6.19	11.00	—	17.19
2" x 1/2"	P1@.287	Ea	9.60	10.30	—	19.90
2" x 3/4"	P1@.277	Ea	9.60	9.91	—	19.51
2" x 1"	P1@.338	Ea	9.60	12.10	—	21.70
2" x 1¼"	P1@.365	Ea	9.60	13.10	—	22.70
2" x 1½"	P1@.390	Ea	9.60	13.90	—	23.50
2½" x 1"	P1@.390	Ea	26.20	13.90	—	40.10
2½" x 1¼"	P1@.416	Ea	26.20	14.90	—	41.10
2½" x 1½"	P1@.442	Ea	26.20	15.80	—	42.00
2½" x 2"	P1@.497	Ea	26.20	17.80	—	44.00
3" x 1¼"	P1@.468	Ea	30.60	16.70	—	47.30
3" x 1½"	P1@.493	Ea	30.60	17.60	—	48.20
3" x 2"	P1@.548	Ea	30.60	19.60	—	50.20
3" x 2½"	P1@.600	Ea	30.60	21.50	—	52.10
4" x 2"	P1@.652	Ea	59.00	23.30	—	82.30
4" x 2½"	P1@.704	Ea	59.00	25.20	—	84.20
4" x 3"	P1@.755	Ea	59.00	27.00	—	86.00

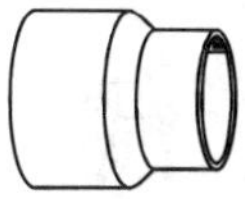

Type K copper adapter C x MPT

Description	Craft@Hrs	Unit	Material $	Labor $	Equipment $	Total $
1/2"	P1@.090	Ea	1.92	3.22	—	5.14
3/4"	P1@.126	Ea	2.75	4.51	—	7.26
1"	P1@.162	Ea	6.83	5.79	—	12.62
1¼"	P1@.198	Ea	12.80	7.08	—	19.88
1½"	P1@.234	Ea	14.30	8.37	—	22.67
2"	P1@.312	Ea	18.50	11.20	—	29.70
2½"	P1@.384	Ea	60.10	13.70	—	73.80
3"	P1@.456	Ea	99.80	16.30	—	116.10

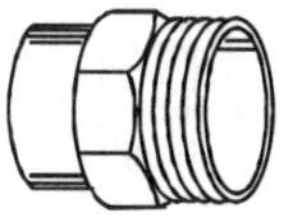

Type K copper adapter C x FPT

Description	Craft@Hrs	Unit	Material $	Labor $	Equipment $	Total $
1/2"	P1@.090	Ea	2.93	3.22	—	6.15
3/4"	P1@.126	Ea	5.30	4.51	—	9.81
1"	P1@.162	Ea	8.03	5.79	—	13.82
1¼"	P1@.198	Ea	16.80	7.08	—	23.88
1½"	P1@.234	Ea	20.40	8.37	—	28.77
2"	P1@.312	Ea	31.30	11.20	—	42.50
2½"	P1@.384	Ea	78.80	13.70	—	92.50
3"	P1@.456	Ea	109.00	16.30	—	125.30

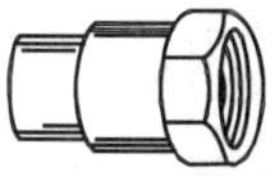

Copper, Type K with Brazed Joints

Description	Craft@Hrs	Unit	Material $	Labor $	Equipment $	Total $

Type K copper flush bushing with brazed joints

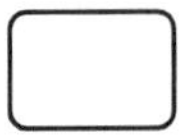

Description	Craft@Hrs	Unit	Material $	Labor $	Equipment $	Total $
1/2" x 3/8"	P1@.148	Ea	1.79	5.29	—	7.08
3/4" x 1/2"	P1@.155	Ea	3.04	5.54	—	8.58
1" x 3/4"	P1@.206	Ea	7.88	7.37	—	15.25
1" x 1/2"	P1@.206	Ea	9.00	7.37	—	16.37
1¼" x 1"	P1@.258	Ea	9.38	9.23	—	18.61
1½" x 1¼"	P1@.308	Ea	11.90	11.00	—	22.90
2" x 1½"	P1@.390	Ea	22.40	13.90	—	36.30

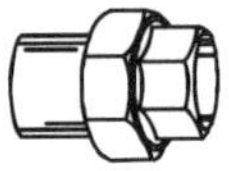

Type K copper union with brazed joints

Description	Craft@Hrs	Unit	Material $	Labor $	Equipment $	Total $
1/2"	P1@.146	Ea	7.14	5.22	—	12.36
3/4"	P1@.205	Ea	8.55	7.33	—	15.88
1"	P1@.263	Ea	19.10	9.40	—	28.50
1¼"	P1@.322	Ea	34.80	11.50	—	46.30
1½"	P1@.380	Ea	42.80	13.60	—	56.40
2"	P1@.507	Ea	63.80	18.10	—	81.90

Type K copper dielectric union with brazed joints

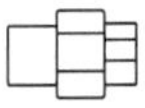

Description	Craft@Hrs	Unit	Material $	Labor $	Equipment $	Total $
1/2"	P1@.146	Ea	6.70	5.22	—	11.92
3/4"	P1@.205	Ea	6.70	7.33	—	14.03
1"	P1@.263	Ea	13.80	9.40	—	23.20
1¼"	P1@.322	Ea	23.60	11.50	—	35.10
1½"	P1@.380	Ea	32.50	13.60	—	46.10
2"	P1@.507	Ea	44.90	18.10	—	63.00

Type K copper pressure cap with brazed joints

Description	Craft@Hrs	Unit	Material $	Labor $	Equipment $	Total $
1/2"	P1@.083	Ea	.58	2.97	—	3.55
3/4"	P1@.116	Ea	1.19	4.15	—	5.34
1"	P1@.149	Ea	2.69	5.33	—	8.02
1¼"	P1@.182	Ea	3.86	6.51	—	10.37
1½"	P1@.215	Ea	5.63	7.69	—	13.32
2"	P1@.286	Ea	10.20	10.20	—	20.40
2½"	P1@.352	Ea	25.70	12.60	—	38.30
3"	P1@.418	Ea	40.20	14.90	—	55.10
4"	P1@.550	Ea	61.70	19.70	—	81.40

Type K copper coupling with brazed joints

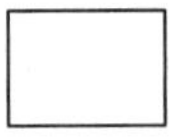

Description	Craft@Hrs	Unit	Material $	Labor $	Equipment $	Total $
1/2"	P1@.129	Ea	.28	4.61	—	4.89
3/4"	P1@.181	Ea	.70	6.47	—	7.17
1"	P1@.232	Ea	2.54	8.30	—	10.84
1¼"	P1@.284	Ea	4.45	10.20	—	14.65
1½"	P1@.335	Ea	6.44	12.00	—	18.44
2"	P1@.447	Ea	9.83	16.00	—	25.83
2½"	P1@.550	Ea	23.00	19.70	—	42.70
3"	P1@.654	Ea	35.10	23.40	—	58.50
4"	P1@.860	Ea	68.90	30.80	—	99.70

Description	Craft@Hrs	Unit	Material $	Labor $	Equipment $	Total $

Class 125 bronze body gate valve, brazed joints

Description	Craft@Hrs	Unit	Material $	Labor $	Equipment $	Total $
1/2"	P1@.240	Ea	15.40	8.58	—	23.98
3/4"	P1@.300	Ea	19.30	10.70	—	30.00
1"	P1@.360	Ea	27.10	12.90	—	40.00
1¼"	P1@.480	Ea	34.30	17.20	—	51.50
1½"	P1@.540	Ea	46.10	19.30	—	65.40
2"	P1@.600	Ea	77.00	21.50	—	98.50
2½"	P1@1.00	Ea	128.00	35.80	—	163.80
3"	P1@1.50	Ea	184.00	53.60	—	237.60

Class 125 iron body gate valve, flanged joints

Description	Craft@Hrs	Unit	Material $	Labor $	Equipment $	Total $
2"	P1@.500	Ea	295.00	17.90	—	312.90
2½"	P1@.600	Ea	398.00	21.50	—	419.50
3"	P1@.750	Ea	432.00	26.80	—	458.80
4"	P1@1.35	Ea	634.00	48.30	—	682.30
5"	P1@2.00	Ea	1,210.00	71.50	—	1,281.50
6"	P1@2.50	Ea	1,210.00	89.40	—	1,299.40

Class 125 bronze body globe valve, with brazed joints

Description	Craft@Hrs	Unit	Material $	Labor $	Equipment $	Total $
1/2"	P1@.240	Ea	28.60	8.58	—	37.18
3/4"	P1@.300	Ea	38.10	10.70	—	48.80
1"	P1@.360	Ea	65.60	12.90	—	78.50
1¼"	P1@.480	Ea	92.30	17.20	—	109.50
1½"	P1@.540	Ea	125.00	19.30	—	144.30
2"	P1@.600	Ea	202.00	21.50	—	223.50

Class 125 iron body globe valve, flanged joints

Description	Craft@Hrs	Unit	Material $	Labor $	Equipment $	Total $
2½"	P1@.650	Ea	334.00	23.20	—	357.20
3"	P1@.750	Ea	401.00	26.80	—	427.80
4"	P1@1.35	Ea	541.00	48.30	—	589.30

200 PSIG iron body butterfly valve, lug type, lever operated

Description	Craft@Hrs	Unit	Material $	Labor $	Equipment $	Total $
2"	P1@.450	Ea	128.00	16.10	—	144.10
2½"	P1@.450	Ea	132.00	16.10	—	148.10
3"	P1@.550	Ea	139.00	19.70	—	158.70
4"	P1@.550	Ea	173.00	19.70	—	192.70

200 PSIG iron body butterfly valve, wafer type, lever operated

Description	Craft@Hrs	Unit	Material $	Labor $	Equipment $	Total $
2"	P1@.450	Ea	116.00	16.10	—	132.10
2½"	P1@.450	Ea	119.00	16.10	—	135.10
3"	P1@.550	Ea	128.00	19.70	—	147.70
4"	P1@.550	Ea	155.00	19.70	—	174.70

Copper, Type K with Brazed Joints

Description	Craft@Hrs	Unit	Material $	Labor $	Equipment $	Total $

Class 125 bronze body 2-piece ball valve, brazed joints

Description	Craft@Hrs	Unit	Material $	Labor $	Equipment $	Total $
1/2"	P1@.240	Ea	7.53	8.58	—	16.11
3/4"	P1@.300	Ea	10.00	10.70	—	20.70
1"	P1@.360	Ea	17.30	12.90	—	30.20
1¼"	P1@.480	Ea	28.90	17.20	—	46.10
1½"	P1@.540	Ea	39.00	19.30	—	58.30
2"	P1@.600	Ea	49.40	21.50	—	70.90
3"	P1@1.50	Ea	338.00	53.60	—	391.60
4"	P1@1.80	Ea	443.00	64.40	—	507.40

Class 125 bronze body swing check valve, brazed joints

Description	Craft@Hrs	Unit	Material $	Labor $	Equipment $	Total $
1/2"	P1@.240	Ea	16.90	8.58	—	25.48
3/4"	P1@.300	Ea	24.20	10.70	—	34.90
1"	P1@.360	Ea	31.70	12.90	—	44.60
1¼"	P1@.480	Ea	45.00	17.20	—	62.20
1½"	P1@.540	Ea	63.50	19.30	—	82.80
2"	P1@.600	Ea	106.00	21.50	—	127.50
2½"	P1@1.00	Ea	212.00	35.80	—	247.80
3"	P1@1.50	Ea	305.00	53.60	—	358.60

Class 125 iron body swing check valve, flanged joints

Description	Craft@Hrs	Unit	Material $	Labor $	Equipment $	Total $
2"	P1@.500	Ea	148.00	17.90	—	165.90
2½"	P1@.600	Ea	188.00	21.50	—	209.50
3"	P1@.750	Ea	233.00	26.80	—	259.80
4"	P1@1.35	Ea	343.00	48.30	—	391.30

Class 125 iron body silent check valve, wafer type

Description	Craft@Hrs	Unit	Material $	Labor $	Equipment $	Total $
2"	P1@.500	Ea	108.00	17.90	—	125.90
2½"	P1@.600	Ea	120.00	21.50	—	141.50
3"	P1@.750	Ea	138.00	26.80	—	164.80
4"	P1@1.35	Ea	185.00	48.30	—	233.30

Class 125 bronze body strainer, threaded joints

Description	Craft@Hrs	Unit	Material $	Labor $	Equipment $	Total $
1/2"	P1@.230	Ea	26.40	8.22	—	34.62
3/4"	P1@.260	Ea	34.70	9.30	—	44.00
1"	P1@.330	Ea	42.50	11.80	—	54.30
1¼"	P1@.440	Ea	59.30	15.70	—	75.00
1½"	P1@.495	Ea	89.30	17.70	—	107.00
2"	P1@.550	Ea	154.00	19.70	—	173.70

Class 125 iron body strainer, flanged joints

Description	Craft@Hrs	Unit	Material $	Labor $	Equipment $	Total $
2"	P1@.500	Ea	118.00	17.90	—	135.90
2½"	P1@.600	Ea	132.00	21.50	—	153.50
3"	P1@.750	Ea	152.00	26.80	—	178.80
4"	P1@1.35	Ea	260.00	48.30	—	308.30

Description	Craft@Hrs	Unit	Material $	Labor $	Equipment $	Total $

Installation of 2-way control valve, threaded joints

Description	Craft@Hrs	Unit	Material $	Labor $	Equipment $	Total $
1/2"	P1@.210	Ea	—	7.51	—	7.51
3/4"	P1@.275	Ea	—	9.83	—	9.83
1"	P1@.350	Ea	—	12.50	—	12.50
1¼"	P1@.430	Ea	—	15.40	—	15.40
1½"	P1@.505	Ea	—	18.10	—	18.10
2"	P1@.675	Ea	—	24.10	—	24.10
2½"	P1@.830	Ea	—	29.70	—	29.70
3"	P1@.990	Ea	—	35.40	—	35.40

Installation of 3-way control valve, threaded joints

Description	Craft@Hrs	Unit	Material $	Labor $	Equipment $	Total $
1/2"	P1@.260	Ea	—	9.30	—	9.30
3/4"	P1@.365	Ea	—	13.10	—	13.10
1"	P1@.475	Ea	—	17.00	—	17.00
1¼"	P1@.575	Ea	—	20.60	—	20.60
1½"	P1@.680	Ea	—	24.30	—	24.30
2"	P1@.910	Ea	—	32.50	—	32.50
2½"	P1@1.12	Ea	—	40.10	—	40.10
3"	P1@1.33	Ea	—	47.60	—	47.60

Companion flange, 150 pound cast brass

Description	Craft@Hrs	Unit	Material $	Labor $	Equipment $	Total $
2"	P1@.290	Ea	352.00	10.40	—	362.40
2½"	P1@.380	Ea	436.00	13.60	—	449.60
3"	P1@.460	Ea	436.00	16.40	—	452.40
4"	P1@.600	Ea	695.00	21.50	—	716.50

Bolt and gasket sets

Description	Craft@Hrs	Unit	Material $	Labor $	Equipment $	Total $
2"	P1@.500	Ea	3.86	17.90	—	21.76
2½"	P1@.650	Ea	4.51	23.20	—	27.71
3"	P1@.750	Ea	7.67	26.80	—	34.47
4"	P1@1.00	Ea	12.80	35.80	—	48.60

Thermometer with well

Description	Craft@Hrs	Unit	Material $	Labor $	Equipment $	Total $
7"	P1@.250	Ea	159.00	8.94	—	167.94
9"	P1@.250	Ea	164.00	8.94	—	172.94

Dial type pressure gauge

Description	Craft@Hrs	Unit	Material $	Labor $	Equipment $	Total $
2½"	P1@.200	Ea	32.70	7.15	—	39.85
3½"	P1@.200	Ea	43.60	7.15	—	50.75

Copper, Type K with Brazed Joints

Description	Craft@Hrs	Unit	Material $	Labor $	Equipment $	Total $
Pressure/temperature tap						
tap	P1@.150	Ea	15.50	5.36	—	20.86
Hanger with swivel assembly						
1/2"	P1@.250	Ea	3.74	8.94	—	12.68
3/4"	P1@.250	Ea	4.13	8.94	—	13.07
1"	P1@.250	Ea	4.33	8.94	—	13.27
1¼"	P1@.300	Ea	4.44	10.70	—	15.14
1½"	P1@.300	Ea	4.77	10.70	—	15.47
2"	P1@.300	Ea	4.98	10.70	—	15.68
2½"	P1@.350	Ea	6.78	12.50	—	19.28
3"	P1@.350	Ea	8.37	12.50	—	20.87
4"	P1@.350	Ea	9.19	12.50	—	21.69
Riser clamp						
1/2"	P1@.100	Ea	2.41	3.58	—	5.99
3/4"	P1@.100	Ea	3.55	3.58	—	7.13
1"	P1@.100	Ea	3.59	3.58	—	7.17
1¼"	P1@.105	Ea	4.31	3.75	—	8.06
1½"	P1@.110	Ea	4.56	3.93	—	8.49
2"	P1@.115	Ea	4.84	4.11	—	8.95
2½"	P1@.120	Ea	5.06	4.29	—	9.35
3"	P1@.120	Ea	5.51	4.29	—	9.80
4"	P1@.125	Ea	7.02	4.47	—	11.49

Type K hard-drawn copper pipe with wrought copper fittings and soft-soldered joints is used in a wide variety of plumbing and HVAC systems such as potable water, heating hot water, chilled water and A.C. condensate.

Soft-soldered joints are those made with solders having melting points in the 350 degree F. to 500 degree F. range. Maximum working pressure/temperature relationships for soft-soldered joints are approximately as follows:

Maximum Working Pressures (PSIG)*				
Soft-solder Type	Water Temperature (degrees F.)	Nominal Pipe Size (inches)		
		Up to 1	1¼ to 2	2½ to 4
50-50 tin-lead**	100	200	175	150
	150	150	125	100
	200	100	90	75
	250	85	75	50
95-5 tin-antimony	100	500	400	300
	150	400	350	275
	200	300	250	200
	250	200	175	150

*For copper pipe and solder-type fittings using soft-solders melting at approximately 350 degrees F. to 500 degrees F.
**The use of any solder containing lead is not allowed in potable water systems.

This section has been arranged to save the estimator's time by including all normally-used system components such as pipe, fittings, valves, hanger assemblies, riser clamps and miscellaneous items under one heading. Additional items can be found under "Plumbing and Piping Specialties." The cost estimates in this section are based on the conditions, limitations and wage rates described in the section "How to Use This Book" beginning on page 5.

Description	Craft@Hrs	Unit	Material $	Labor $	Equipment $	Total $

Type K copper pipe with soft-soldered joints installed horizontally.
Complete installation including six to twelve wrought copper tees and six to twelve wrought copper elbows every 100', and hangers spaced to meet plumbing code (*a tee & elbow every 8.3' for ½" dia., a tee and elbow every 16.6' for 4" dia.*)

Description	Craft@Hrs	Unit	Material $	Labor $	Equipment $	Total $
1/2"	P1@.112	LF	3.41	4.01	—	7.42
3/4"	P1@.122	LF	5.97	4.36	—	10.33
1"	P1@.142	LF	9.53	5.08	—	14.61
1¼"	P1@.162	LF	13.00	5.79	—	18.79
1½"	P1@.182	LF	16.00	6.51	—	22.51
2"	P1@.212	LF	27.20	7.58	—	34.78
2½"	P1@.242	LF	44.80	8.65	—	53.45
3"	P1@.272	LF	51.60	9.73	—	61.33
4"	P1@.312	LF	94.50	11.20	—	105.70

Copper, Type K with Soft-Soldered Joints

Description	Craft@Hrs	Unit	Material $	Labor $	Equipment $	Total $

Type K copper pipe with soft-soldered joints, installed risers.

Complete installation including a wrought copper reducing tee every floor and a riser clamp every other floor.

Description	Craft@Hrs	Unit	Material $	Labor $	Equipment $	Total $
1/2"	P1@.062	LF	3.08	2.22	—	5.30
3/4"	P1@.072	LF	5.55	2.57	—	8.12
1"	P1@.077	LF	8.48	2.75	—	11.23
1¼"	P1@.082	LF	11.70	2.93	—	14.63
1½"	P1@.102	LF	13.80	3.65	—	17.45
2"	P1@.132	LF	23.60	4.72	—	28.32
2½"	P1@.152	LF	35.30	5.44	—	40.74
3"	P1@.172	LF	49.50	6.15	—	55.65
4"	P1@.192	LF	77.30	6.87	—	84.17

Type K copper pipe with soft-soldered joints, pipe only

Description	Craft@Hrs	Unit	Material $	Labor $	Equipment $	Total $
1/2"	P1@.032	LF	2.45	1.14	—	3.59
3/4"	P1@.035	LF	4.53	1.25	—	5.78
1"	P1@.038	LF	5.72	1.36	—	7.08
1¼"	P1@.042	LF	7.65	1.50	—	9.15
1½"	P1@.046	LF	8.33	1.64	—	9.97
2"	P1@.053	LF	14.90	1.90	—	16.80
2½"	P1@.060	LF	19.40	2.15	—	21.55
3"	P1@.066	LF	25.30	2.36	—	27.66
4"	P1@.080	LF	45.30	2.86	—	48.16

Type K copper 45 degree ell C x C with soft-soldered joints

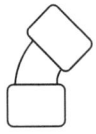

Description	Craft@Hrs	Unit	Material $	Labor $	Equipment $	Total $
1/2"	P1@.107	Ea	.99	3.83	—	4.82
3/4"	P1@.150	Ea	1.86	5.36	—	7.22
1"	P1@.193	Ea	5.49	6.90	—	12.39
1¼"	P1@.236	Ea	7.48	8.44	—	15.92
1½"	P1@.278	Ea	11.00	9.94	—	20.94
2"	P1@.371	Ea	18.10	13.30	—	31.40
2½"	P1@.457	Ea	46.60	16.30	—	62.90
3"	P1@.543	Ea	61.40	19.40	—	80.80
4"	P1@.714	Ea	119.00	25.50	—	144.50

Description	Craft@Hrs	Unit	Material $	Labor $	Equipment $	Total $

Type K copper 90 degree ell C x C with soft-soldered joints

Description	Craft@Hrs	Unit	Material $	Labor $	Equipment $	Total $
1/2"	P1@.107	Ea	.59	3.83	—	4.42
3/4"	P1@.150	Ea	1.21	5.36	—	6.57
1"	P1@.193	Ea	4.18	6.90	—	11.08
1¼"	P1@.236	Ea	6.83	8.44	—	15.27
1½"	P1@.278	Ea	10.30	9.94	—	20.24
2"	P1@.371	Ea	18.50	13.30	—	31.80
2½"	P1@.457	Ea	38.30	16.30	—	54.60
3"	P1@.543	Ea	48.50	19.40	—	67.90
4"	P1@.714	Ea	104.00	25.50	—	129.50

Type K copper 90 degree ell Ftg. x C with soft-soldered joint

Description	Craft@Hrs	Unit	Material $	Labor $	Equipment $	Total $
1/2"	P1@.107	Ea	.91	3.83	—	4.74
3/4"	P1@.150	Ea	1.99	5.36	—	7.35
1"	P1@.193	Ea	5.93	6.90	—	12.83
1¼"	P1@.236	Ea	8.25	8.44	—	16.69
1½"	P1@.278	Ea	12.40	9.94	—	22.34
2"	P1@.371	Ea	25.00	13.30	—	38.30
2½"	P1@.457	Ea	52.40	16.30	—	68.70
3"	P1@.543	Ea	58.70	19.40	—	78.10
4"	P1@.714	Ea	119.00	25.50	—	144.50

Type K copper tee C x C x C with soft-soldered joints

Description	Craft@Hrs	Unit	Material $	Labor $	Equipment $	Total $
1/2"	P1@.129	Ea	.98	4.61	—	5.59
3/4"	P1@.181	Ea	2.64	6.47	—	9.11
1"	P1@.233	Ea	9.90	8.33	—	18.23
1¼"	P1@.285	Ea	15.80	10.20	—	26.00
1½"	P1@.337	Ea	22.70	12.10	—	34.80
2"	P1@.449	Ea	36.60	16.10	—	52.70
2½"	P1@.552	Ea	74.60	19.70	—	94.30
3"	P1@.656	Ea	114.00	23.50	—	137.50
4"	P1@.863	Ea	190.00	30.90	—	220.90

Copper, Type K with Soft-Soldered Joints

Description	Craft@Hrs	Unit	Material $	Labor $	Equipment $	Total $

Type K copper branch reducing tee C x C x C with soft-soldered joints

Description	Craft@Hrs	Unit	Material $	Labor $	Equipment $	Total $
1/2" x 3/8"	P1@.129	Ea	1.98	4.61	—	6.59
3/4" x 1/2"	P1@.182	Ea	4.20	6.51	—	10.71
1" x 1/2"	P1@.195	Ea	8.25	6.97	—	15.22
1" x 3/4"	P1@.219	Ea	8.25	7.83	—	16.08
1¼" x 1/2"	P1@.225	Ea	12.60	8.05	—	20.65
1¼" x 3/4"	P1@.247	Ea	12.60	8.83	—	21.43
1¼" x 1"	P1@.268	Ea	12.60	9.58	—	22.18
1½" x 1/2"	P1@.272	Ea	15.40	9.73	—	25.13
1½" x 3/4"	P1@.287	Ea	15.40	10.30	—	25.70
1½" x 1"	P1@.302	Ea	15.40	10.80	—	26.20
1½" x 1¼"	P1@.317	Ea	15.40	11.30	—	26.70
2" x 1/2"	P1@.325	Ea	20.60	11.60	—	32.20
2" x 3/4"	P1@.348	Ea	20.60	12.40	—	33.00
2" x 1"	P1@.373	Ea	20.60	13.30	—	33.90
2" x 1¼"	P1@.398	Ea	20.60	14.20	—	34.80
2" x 1½"	P1@.422	Ea	20.60	15.10	—	35.70
2½" x 1/2"	P1@.437	Ea	62.10	15.60	—	77.70
2½" x 3/4"	P1@.453	Ea	62.10	16.20	—	78.30
2½" x 1"	P1@.469	Ea	62.10	16.80	—	78.90
2½" x 1½"	P1@.502	Ea	62.10	18.00	—	80.10
2½" x 2"	P1@.519	Ea	62.10	18.60	—	80.70
3" x 1¼"	P1@.525	Ea	70.20	18.80	—	89.00
3" x 2"	P1@.585	Ea	72.90	20.90	—	93.80
3" x 2½"	P1@.617	Ea	72.90	22.10	—	95.00
4" x 1¼"	P1@.625	Ea	95.30	22.40	—	117.70
4" x 1½"	P1@.671	Ea	95.30	24.00	—	119.30
4" x 2"	P1@.718	Ea	95.30	25.70	—	121.00
4" x 2½"	P1@.765	Ea	95.30	27.40	—	122.70
4" x 3"	P1@.811	Ea	95.30	29.00	—	124.30

Description	Craft@Hrs	Unit	Material $	Labor $	Equipment $	Total $

Type K copper reducer with soft-soldered joints

Description	Craft@Hrs	Unit	Material $	Labor $	Equipment $	Total $
1/2" x 3/8"	P1@.124	Ea	.74	4.43	—	5.17
3/4" x 1/2"	P1@.129	Ea	1.37	4.61	—	5.98
1" x 3/4"	P1@.172	Ea	2.41	6.15	—	8.56
1¼" x 1/2"	P1@.172	Ea	4.05	6.15	—	10.20
1¼" x 3/4"	P1@.193	Ea	4.05	6.90	—	10.95
1¼" x 1"	P1@.215	Ea	4.05	7.69	—	11.74
1½" x 1/2"	P1@.193	Ea	6.19	6.90	—	13.09
1½" x 3/4"	P1@.214	Ea	6.19	7.65	—	13.84
1½" x 1"	P1@.236	Ea	6.19	8.44	—	14.63
1½" x 1¼"	P1@.257	Ea	6.19	9.19	—	15.38
2" x 1/2"	P1@.239	Ea	9.60	8.55	—	18.15
2" x 3/4"	P1@.261	Ea	9.60	9.33	—	18.93
2" x 1"	P1@.282	Ea	9.60	10.10	—	19.70
2" x 1¼"	P1@.304	Ea	9.60	10.90	—	20.50
2" x 1½"	P1@.325	Ea	9.60	11.60	—	21.20
2½" x 1"	P1@.325	Ea	26.20	11.60	—	37.80
2½" x 1¼"	P1@.347	Ea	26.20	12.40	—	38.60
2½" x 1½"	P1@.368	Ea	26.20	13.20	—	39.40
2½" x 2"	P1@.414	Ea	26.20	14.80	—	41.00
3" x 2"	P1@.457	Ea	30.60	16.30	—	46.90
3" x 2½"	P1@.500	Ea	30.60	17.90	—	48.50
4" x 2"	P1@.543	Ea	59.00	19.40	—	78.40
4" x 2½"	P1@.586	Ea	59.00	21.00	—	80.00
4" x 3"	P1@.629	Ea	59.00	22.50	—	81.50

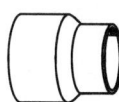

Type K copper adapter C x MPT with soft-soldered joint

Description	Craft@Hrs	Unit	Material $	Labor $	Equipment $	Total $
1/2"	P1@.075	Ea	1.76	2.68	—	4.44
3/4"	P1@.105	Ea	2.75	3.75	—	6.50
1"	P1@.134	Ea	6.83	4.79	—	11.62
1¼"	P1@.164	Ea	12.80	5.86	—	18.66
1½"	P1@.194	Ea	14.30	6.94	—	21.24
2"	P1@.259	Ea	18.50	9.26	—	27.76
2½"	P1@.319	Ea	60.10	11.40	—	71.50
3"	P1@.378	Ea	99.80	13.50	—	113.30

Copper, Type K with Soft-Soldered Joints

Description	Craft@Hrs	Unit	Material $	Labor $	Equipment $	Total $

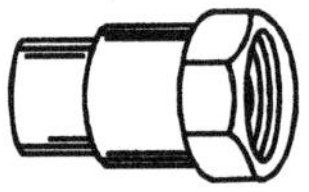

Type K copper adapter C x FPT with soft-soldered joint

Description	Craft@Hrs	Unit	Material $	Labor $	Equipment $	Total $
1/2"	P1@.075	Ea	2.93	2.68	—	5.61
3/4"	P1@.105	Ea	5.30	3.75	—	9.05
1"	P1@.134	Ea	8.03	4.79	—	12.82
1¼"	P1@.164	Ea	16.80	5.86	—	22.66
1½"	P1@.194	Ea	20.40	6.94	—	27.34
2"	P1@.259	Ea	31.30	9.26	—	40.56
2½"	P1@.319	Ea	78.80	11.40	—	90.20
3"	P1@.378	Ea	109.00	13.50	—	122.50

Type K copper flush bushing with soft-soldered joints

Description	Craft@Hrs	Unit	Material $	Labor $	Equipment $	Total $
1/2" x 3/8"	P1@.124	Ea	1.79	4.43	—	6.22
3/4" x 1/2"	P1@.129	Ea	3.04	4.61	—	7.65
1" x 3/4"	P1@.172	Ea	7.88	6.15	—	14.03
1" x 1/2"	P1@.172	Ea	9.00	6.15	—	15.15
1¼" x 1"	P1@.215	Ea	9.38	7.69	—	17.07
1½" x 1¼"	P1@.257	Ea	11.90	9.19	—	21.09
2" x 1½"	P1@.325	Ea	22.40	11.60	—	34.00

Type K copper union with soft-soldered joint

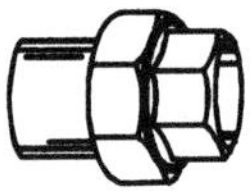

Description	Craft@Hrs	Unit	Material $	Labor $	Equipment $	Total $
1/2"	P1@.121	Ea	7.14	4.33	—	11.47
3/4"	P1@.170	Ea	8.55	6.08	—	14.63
1"	P1@.218	Ea	19.10	7.80	—	26.90
1¼"	P1@.267	Ea	34.80	9.55	—	44.35
1½"	P1@.315	Ea	42.80	11.30	—	54.10
2"	P1@.421	Ea	63.80	15.10	—	78.90

Type K copper dielectric union with soft-soldered joint

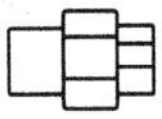

Description	Craft@Hrs	Unit	Material $	Labor $	Equipment $	Total $
1/2"	P1@.121	Ea	6.70	4.33	—	11.03
3/4"	P1@.170	Ea	6.70	6.08	—	12.78
1"	P1@.218	Ea	13.80	7.80	—	21.60
1¼"	P1@.267	Ea	23.60	9.55	—	33.15
1½"	P1@.315	Ea	32.50	11.30	—	43.80
2"	P1@.421	Ea	44.90	15.10	—	60.00

Type K copper cap with soft-soldered joint

Description	Craft@Hrs	Unit	Material $	Labor $	Equipment $	Total $
1/2"	P1@.069	Ea	.58	2.47	—	3.05
3/4"	P1@.096	Ea	1.19	3.43	—	4.62
1"	P1@.124	Ea	2.69	4.43	—	7.12
1¼"	P1@.151	Ea	3.86	5.40	—	9.26
1½"	P1@.178	Ea	5.63	6.37	—	12.00
2"	P1@.237	Ea	10.20	8.48	—	18.68
2½"	P1@.292	Ea	25.70	10.40	—	36.10
3"	P1@.347	Ea	40.20	12.40	—	52.60
4"	P1@.457	Ea	61.70	16.30	—	78.00

Description	Craft@Hrs	Unit	Material $	Labor $	Equipment $	Total $

Type K copper coupling with soft-soldered joints

Description	Craft@Hrs	Unit	Material $	Labor $	Equipment $	Total $
1/2"	P1@.107	Ea	.37	3.83	—	4.20
3/4"	P1@.150	Ea	.93	5.36	—	6.29
1"	P1@.193	Ea	2.54	6.90	—	9.44
1¼"	P1@.236	Ea	4.45	8.44	—	12.89
1½"	P1@.278	Ea	6.44	9.94	—	16.38
2"	P1@.371	Ea	9.83	13.30	—	23.13
2½"	P1@.457	Ea	23.00	16.30	—	39.30
3"	P1@.543	Ea	35.10	19.40	—	54.50
4"	P1@.714	Ea	68.90	25.50	—	94.40

Class 125 bronze body gate valve, soft-soldered joints

Description	Craft@Hrs	Unit	Material $	Labor $	Equipment $	Total $
1/2"	P1@.200	Ea	15.30	7.15	—	22.45
3/4"	P1@.249	Ea	19.30	8.90	—	28.20
1"	P1@.299	Ea	27.10	10.70	—	37.80
1¼"	P1@.398	Ea	34.30	14.20	—	48.50
1½"	P1@.448	Ea	45.90	16.00	—	61.90
2"	P1@.498	Ea	76.70	17.80	—	94.50
2½"	P1@.830	Ea	127.00	29.70	—	156.70
3"	P1@1.24	Ea	183.00	44.30	—	227.30

Class 125 iron body gate valve, flanged joints

Description	Craft@Hrs	Unit	Material $	Labor $	Equipment $	Total $
2"	P1@.500	Ea	295.00	17.90	—	312.90
2½"	P1@.600	Ea	398.00	21.50	—	419.50
3"	P1@.750	Ea	432.00	26.80	—	458.80
4"	P1@1.35	Ea	634.00	48.30	—	682.30
5"	P1@2.00	Ea	1,210.00	71.50	—	1,281.50
6"	P1@2.50	Ea	1,210.00	89.40	—	1,299.40

Class 125 bronze body globe valve, with soft-soldered joints

Description	Craft@Hrs	Unit	Material $	Labor $	Equipment $	Total $
1/2"	P1@.200	Ea	28.60	7.15	—	35.75
3/4"	P1@.249	Ea	38.10	8.90	—	47.00
1"	P1@.299	Ea	65.60	10.70	—	76.30
1¼"	P1@.398	Ea	92.30	14.20	—	106.50
1½"	P1@.448	Ea	125.00	16.00	—	141.00
2"	P1@.498	Ea	202.00	17.80	—	219.80

Class 125 iron body globe valve, flanged ends

Description	Craft@Hrs	Unit	Material $	Labor $	Equipment $	Total $
2½"	P1@.600	Ea	334.00	21.50	—	355.50
3"	P1@.750	Ea	401.00	26.80	—	427.80
4"	P1@1.35	Ea	538.00	48.30	—	586.30

Description	Craft@Hrs	Unit	Material $	Labor $	Equipment $	Total $

200 PSIG iron body butterfly valve, lug type, lever operated

Description	Craft@Hrs	Unit	Material $	Labor $	Equipment $	Total $
2"	P1@.450	Ea	128.00	16.10	—	144.10
2½"	P1@.450	Ea	132.00	16.10	—	148.10
3"	P1@.550	Ea	139.00	19.70	—	158.70
4"	P1@.550	Ea	173.00	19.70	—	192.70

200 PSIG iron body butterfly valve, wafer type, lever operated

Description	Craft@Hrs	Unit	Material $	Labor $	Equipment $	Total $
2"	P1@.450	Ea	116.00	16.10	—	132.10
2½"	P1@.450	Ea	119.00	16.10	—	135.10
3"	P1@.550	Ea	128.00	19.70	—	147.70
4"	P1@.550	Ea	155.00	19.70	—	174.70

Class 125 bronze body 2-piece ball valve, soft-soldered joints

Description	Craft@Hrs	Unit	Material $	Labor $	Equipment $	Total $
1/2"	P1@.200	Ea	7.53	7.15	—	14.68
3/4"	P1@.249	Ea	8.98	8.90	—	17.88
1"	P1@.299	Ea	17.30	10.70	—	28.00
1¼"	P1@.398	Ea	28.90	14.20	—	43.10
1½"	P1@.448	Ea	39.00	16.00	—	55.00
2"	P1@.498	Ea	49.40	17.80	—	67.20
3"	P1@1.24	Ea	338.00	44.30	—	382.30
4"	P1@1.45	Ea	443.00	51.90	—	494.90

Class 125 bronze body swing check valve, soft-soldered joints

Description	Craft@Hrs	Unit	Material $	Labor $	Equipment $	Total $
1/2"	P1@.200	Ea	16.90	7.15	—	24.05
3/4"	P1@.249	Ea	24.20	8.90	—	33.10
1"	P1@.299	Ea	31.70	10.70	—	42.40
1¼"	P1@.398	Ea	45.00	14.20	—	59.20
1½"	P1@.448	Ea	63.50	16.00	—	79.50
2"	P1@.498	Ea	106.00	17.80	—	123.80
2½"	P1@.830	Ea	212.00	29.70	—	241.70
3"	P1@1.24	Ea	305.00	44.30	—	349.30

Class 125 iron body swing check valve, flanged ends

Description	Craft@Hrs	Unit	Material $	Labor $	Equipment $	Total $
2"	P1@.500	Ea	148.00	17.90	—	165.90
2½"	P1@.600	Ea	188.00	21.50	—	209.50
3"	P1@.750	Ea	233.00	26.80	—	259.80
4"	P1@1.35	Ea	343.00	48.30	—	391.30

Class 125 iron body silent check valve, wafer type

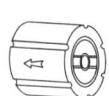

Description	Craft@Hrs	Unit	Material $	Labor $	Equipment $	Total $
2"	P1@.500	Ea	108.00	17.90	—	125.90
2½"	P1@.600	Ea	120.00	21.50	—	141.50
3"	P1@.750	Ea	138.00	26.80	—	164.80
4"	P1@1.35	Ea	185.00	48.30	—	233.30

Description	Craft@Hrs	Unit	Material $	Labor $	Equipment $	Total $

Class 125 bronze body strainer, threaded ends

Description	Craft@Hrs	Unit	Material $	Labor $	Equipment $	Total $
1/2"	P1@.230	Ea	26.40	8.22	—	34.62
3/4"	P1@.260	Ea	34.70	9.30	—	44.00
1"	P1@.330	Ea	42.50	11.80	—	54.30
1¼"	P1@.440	Ea	59.30	15.70	—	75.00
1½"	P1@.495	Ea	89.30	17.70	—	107.00
2"	P1@.550	Ea	154.00	19.70	—	173.70

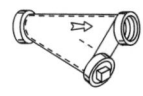

Class 125 iron body strainer, flanged ends

Description	Craft@Hrs	Unit	Material $	Labor $	Equipment $	Total $
2"	P1@.500	Ea	118.00	17.90	—	135.90
2½"	P1@.600	Ea	132.00	21.50	—	153.50
3"	P1@.750	Ea	152.00	26.80	—	178.80
4"	P1@1.35	Ea	260.00	48.30	—	308.30

Installation of copper 2-way control valve, threaded joints

Description	Craft@Hrs	Unit	Material $	Labor $	Equipment $	Total $
1/2"	P1@.210	Ea	—	7.51	—	7.51
3/4"	P1@.275	Ea	—	9.83	—	9.83
1"	P1@.350	Ea	—	12.50	—	12.50
1¼"	P1@.430	Ea	—	15.40	—	15.40
1½"	P1@.505	Ea	—	18.10	—	18.10
2"	P1@.675	Ea	—	24.10	—	24.10
2½"	P1@.830	Ea	—	29.70	—	29.70
3"	P1@.990	Ea	—	35.40	—	35.40

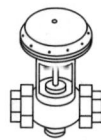

Installation of copper 3-way control valve, threaded joints

Description	Craft@Hrs	Unit	Material $	Labor $	Equipment $	Total $
1/2"	P1@.260	Ea	—	9.30	—	9.30
3/4"	P1@.365	Ea	—	13.10	—	13.10
1"	P1@.475	Ea	—	17.00	—	17.00
1¼"	P1@.575	Ea	—	20.60	—	20.60
1½"	P1@.680	Ea	—	24.30	—	24.30
2"	P1@.910	Ea	—	32.50	—	32.50
2½"	P1@1.12	Ea	—	40.10	—	40.10
3"	P1@1.33	Ea	—	47.60	—	47.60

Companion flange, 150 pound cast brass

Description	Craft@Hrs	Unit	Material $	Labor $	Equipment $	Total $
2"	P1@.290	Ea	352.00	10.40	—	362.40
2½"	P1@.380	Ea	436.00	13.60	—	449.60
3"	P1@.460	Ea	436.00	16.40	—	452.40
4"	P1@.600	Ea	695.00	21.50	—	716.50

Copper, Type K with Soft-Soldered Joints

Description	Craft@Hrs	Unit	Material $	Labor $	Equipment $	Total $
Bolt and gasket sets						
2"	P1@.500	Ea	3.92	17.90	—	21.82
2½"	P1@.650	Ea	4.60	23.20	—	27.80
3"	P1@.750	Ea	7.84	26.80	—	34.64
4"	P1@1.00	Ea	13.10	35.80	—	48.90
Thermometer with well						
7"	P1@.250	Ea	159.00	8.94	—	167.94
9"	P1@.250	Ea	164.00	8.94	—	172.94
Dial type pressure gauge						
2½"	P1@.200	Ea	32.70	7.15	—	39.85
3½"	P1@.200	Ea	43.60	7.15	—	50.75
Pressure/temperature tap						
tap	P1@.150	Ea	15.50	5.36	—	20.86
Hanger with swivel assembly						
1/2"	P1@.250	Ea	3.74	8.94	—	12.68
3/4"	P1@.250	Ea	4.13	8.94	—	13.07
1"	P1@.250	Ea	4.33	8.94	—	13.27
1¼"	P1@.300	Ea	4.44	10.70	—	15.14
1½"	P1@.300	Ea	4.77	10.70	—	15.47
2"	P1@.300	Ea	4.98	10.70	—	15.68
2½"	P1@.350	Ea	6.78	12.50	—	19.28
3"	P1@.350	Ea	8.37	12.50	—	20.87
4"	P1@.350	Ea	9.19	12.50	—	21.69
Riser clamp						
1/2"	P1@.100	Ea	2.41	3.58	—	5.99
3/4"	P1@.100	Ea	3.55	3.58	—	7.13
1"	P1@.100	Ea	3.59	3.58	—	7.17
1¼"	P1@.105	Ea	4.33	3.75	—	8.08
1½"	P1@.110	Ea	4.58	3.93	—	8.51
2"	P1@.115	Ea	4.84	4.11	—	8.95
2½"	P1@.120	Ea	5.09	4.29	—	9.38
3"	P1@.120	Ea	5.52	4.29	—	9.81
4"	P1@.125	Ea	7.03	4.47	—	11.50

Type L hard-drawn copper pipe with wrought copper fittings and brazed joints is used in a wide variety of plumbing and HVAC systems such as potable water, heating hot water, chilled water, compressed air, refrigerant and A.C. condensate.

Brazed joints are those made with silver or other alloy filler metals having melting points at, or above, 1,000 degrees F. Maximum working pressure/temperature relationships for brazed joints are approximately as follows:

Maximum Working Pressure (PSIG)*			
Water Temperature (degrees F.)	Nominal Pipe Size (inches)		
	Up to 1	1¼ to 2	2½ to 4
Up to 350	270	190	150

*For copper pipe and solder-type fittings using brazing alloys
melting at, or above, 1,000 degrees F.

This section has been arranged to save the estimator's time by including all normally-used system components such as pipe, fittings, valves, hanger assemblies, riser clamps and miscellaneous items under one heading. Additional items can be found under "Plumbing and Piping Specialties." The cost estimates in this section are based on the conditions, limitations and wage rates described in the section "How to Use This Book" beginning on page 5.

Description	Craft@Hrs	Unit	Material $	Labor $	Equipment $	Total $

Type L copper pipe with brazed joints installed horizontally. Complete installation including six to twelve wrought copper tees and six to twelve wrought copper elbows every 100', and hangers spaced to meet plumbing code (*a tee & elbow every 8.3' for ½" dia., a tee and elbow every 16.6' for 4" dia.*)

Description	Craft@Hrs	Unit	Material $	Labor $	Equipment $	Total $
1/2"	P1@.116	LF	2.34	4.15	—	6.49
3/4"	P1@.126	LF	3.62	4.51	—	8.13
1"	P1@.146	LF	6.50	5.22	—	11.72
1¼"	P1@.166	LF	9.23	5.94	—	15.17
1½"	P1@.186	LF	12.40	6.65	—	19.05
2"	P1@.216	LF	22.40	7.72	—	30.12
2½"	P1@.246	LF	39.40	8.80	—	48.20
3"	P1@.276	LF	54.60	9.87	—	64.47
4"	P1@.316	LF	78.80	11.30	—	90.10

Copper, Type L with Brazed Joints

Description	Craft@Hrs	Unit	Material $	Labor $	Equipment $	Total $

Type L copper pipe with brazed joints, installed risers. Complete installation including a wrought copper reducing tee every floor and a riser clamp every other floor.

Description	Craft@Hrs	Unit	Material $	Labor $	Equipment $	Total $
1/2"	P1@.063	LF	2.30	2.25	—	4.55
3/4"	P1@.073	LF	3.64	2.61	—	6.25
1"	P1@.073	LF	6.16	2.61	—	8.77
1¼"	P1@.083	LF	8.70	2.97	—	11.67
1½"	P1@.103	LF	10.90	3.68	—	14.58
2"	P1@.133	LF	28.50	4.76	—	33.26
2½"	P1@.153	LF	30.80	5.47	—	36.27
3"	P1@.173	LF	39.90	6.19	—	46.09
4"	P1@.193	LF	59.30	6.90	—	66.20

Type L copper pipe with brazed joints, pipe only

Description	Craft@Hrs	Unit	Material $	Labor $	Equipment $	Total $
1/2"	P1@.032	LF	1.76	1.14	—	2.90
3/4"	P1@.035	LF	2.79	1.25	—	4.04
1"	P1@.038	LF	3.87	1.36	—	5.23
1¼"	P1@.042	LF	5.54	1.50	—	7.04
1½"	P1@.046	LF	7.08	1.64	—	8.72
2"	P1@.053	LF	12.20	1.90	—	14.10
2½"	P1@.060	LF	17.70	2.15	—	19.85
3"	P1@.066	LF	24.00	2.36	—	26.36
4"	P1@.080	LF	37.70	2.86	—	40.56

Type L copper 45 degree ell C x C with brazed joints

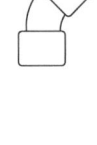

Description	Craft@Hrs	Unit	Material $	Labor $	Equipment $	Total $
1/2"	P1@.129	Ea	.99	4.61	—	5.60
3/4"	P1@.181	Ea	1.88	6.47	—	8.35
1"	P1@.232	Ea	5.54	8.30	—	13.84
1¼"	P1@.284	Ea	7.50	10.20	—	17.70
1½"	P1@.335	Ea	11.10	12.00	—	23.10
2"	P1@.447	Ea	18.30	16.00	—	34.30
2½"	P1@.550	Ea	46.70	19.70	—	66.40
3"	P1@.654	Ea	61.60	23.40	—	85.00
4"	P1@.860	Ea	119.00	30.80	—	149.80

Type L copper 90 degree ell C x C with brazed joints

Description	Craft@Hrs	Unit	Material $	Labor $	Equipment $	Total $
1/2"	P1@.129	Ea	.57	4.61	—	5.18
3/4"	P1@.181	Ea	1.21	6.47	—	7.68
1"	P1@.232	Ea	4.21	8.30	—	12.51
1¼"	P1@.284	Ea	6.83	10.20	—	17.03
1½"	P1@.335	Ea	10.40	12.00	—	22.40
2"	P1@.447	Ea	18.60	16.00	—	34.60
2½"	P1@.550	Ea	38.60	19.70	—	58.30
3"	P1@.654	Ea	48.80	23.40	—	72.20
4"	P1@.860	Ea	104.00	30.80	—	134.80

Description	Craft@Hrs	Unit	Material $	Labor $	Equipment $	Total $

Type L copper 90 degree ell Ftg. x C with brazed joint

Description	Craft@Hrs	Unit	Material $	Labor $	Equipment $	Total $
1/2"	P1@.129	Ea	.91	4.61	—	5.52
3/4"	P1@.181	Ea	2.00	6.47	—	8.47
1"	P1@.232	Ea	5.98	8.30	—	14.28
1¼"	P1@.284	Ea	8.33	10.20	—	18.53
1½"	P1@.335	Ea	12.50	12.00	—	24.50
2"	P1@.447	Ea	25.00	16.00	—	41.00
2½"	P1@.550	Ea	52.50	19.70	—	72.20
3"	P1@.654	Ea	59.00	23.40	—	82.40
4"	P1@.860	Ea	119.00	30.80	—	149.80

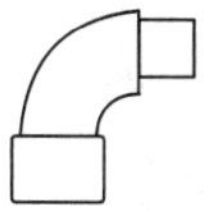

Type L copper tee C x C x C with brazed joints

Description	Craft@Hrs	Unit	Material $	Labor $	Equipment $	Total $
1/2"	P1@.156	Ea	.98	5.58	—	6.56
3/4"	P1@.204	Ea	2.64	7.30	—	9.94
1"	P1@.263	Ea	9.98	9.40	—	19.38
1¼"	P1@.322	Ea	15.80	11.50	—	27.30
1½"	P1@.380	Ea	23.00	13.60	—	36.60
2"	P1@.507	Ea	36.80	18.10	—	54.90
2½"	P1@.624	Ea	74.60	22.30	—	96.90
3"	P1@.741	Ea	115.00	26.50	—	141.50
4"	P1@.975	Ea	192.00	34.90	—	226.90

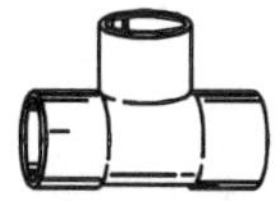

Type L copper branch reducing tee C x C x C with brazed joints

Description	Craft@Hrs	Unit	Material $	Labor $	Equipment $	Total $
1/2" x 3/8"	P1@.129	Ea	1.98	4.61	—	6.59
3/4" x 1/2"	P1@.182	Ea	4.20	6.51	—	10.71
1" x 1/2"	P1@.209	Ea	8.25	7.47	—	15.72
1" x 3/4"	P1@.234	Ea	8.25	8.37	—	16.62
1¼" x 1/2"	P1@.241	Ea	12.60	8.62	—	21.22
1¼" x 3/4"	P1@.264	Ea	12.60	9.44	—	22.04
1¼" x 1"	P1@.287	Ea	12.60	10.30	—	22.90
1½" x 1/2"	P1@.291	Ea	15.40	10.40	—	25.80
1½" x 3/4"	P1@.307	Ea	15.40	11.00	—	26.40
1½" x 1"	P1@.323	Ea	15.40	11.60	—	27.00
1½" x 1¼"	P1@.339	Ea	15.40	12.10	—	27.50
2" x 1/2"	P1@.348	Ea	20.60	12.40	—	33.00
2" x 3/4"	P1@.373	Ea	20.60	13.30	—	33.90
2" x 1"	P1@.399	Ea	20.60	14.30	—	34.90
2" x 1¼"	P1@.426	Ea	20.60	15.20	—	35.80
2" x 1½"	P1@.452	Ea	20.60	16.20	—	36.80
2½" x 1/2"	P1@.468	Ea	62.10	16.70	—	78.80
2½" x 3/4"	P1@.485	Ea	62.10	17.30	—	79.40
2½" x 1"	P1@.502	Ea	62.10	18.00	—	80.10
2½" x 1½"	P1@.537	Ea	62.10	19.20	—	81.30
2½" x 2"	P1@.555	Ea	62.10	19.80	—	81.90
3" x 1¼"	P1@.562	Ea	70.20	20.10	—	90.30
3" x 2"	P1@.626	Ea	72.90	22.40	—	95.30
3" x 2½"	P1@.660	Ea	72.90	23.60	—	96.50
4" x 1¼"	P1@.669	Ea	127.00	23.90	—	150.90
4" x 1½"	P1@.718	Ea	127.00	25.70	—	152.70
4" x 2"	P1@.768	Ea	127.00	27.50	—	154.50
4" x 2½"	P1@.819	Ea	127.00	29.30	—	156.30
4" x 3"	P1@.868	Ea	127.00	31.00	—	158.00

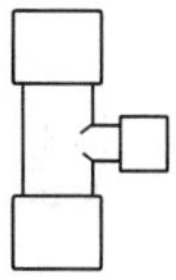

Description	Craft@Hrs	Unit	Material $	Labor $	Equipment $	Total $

Type L copper reducer with brazed joints

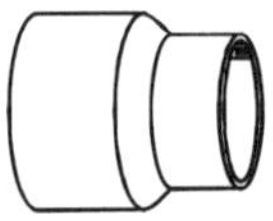

Description	Craft@Hrs	Unit	Material $	Labor $	Equipment $	Total $
1/2" x 3/8"	P1@.148	Ea	.71	5.29	—	6.00
3/4" x 1/2"	P1@.155	Ea	1.34	5.54	—	6.88
1" x 3/4"	P1@.206	Ea	2.41	7.37	—	9.78
1¼" x 1/2"	P1@.206	Ea	4.05	7.37	—	11.42
1¼" x 3/4"	P1@.232	Ea	4.05	8.30	—	12.35
1¼" x 1"	P1@.258	Ea	4.05	9.23	—	13.28
1½" x 1/2"	P1@.232	Ea	6.19	8.30	—	14.49
1½" x 3/4"	P1@.257	Ea	6.19	9.19	—	15.38
1½" x 1"	P1@.283	Ea	6.19	10.10	—	16.29
1½" x 1¼"	P1@.308	Ea	6.19	11.00	—	17.19
2" x 1/2"	P1@.287	Ea	9.60	10.30	—	19.90
2" x 3/4"	P1@.277	Ea	9.60	9.91	—	19.51
2" x 1"	P1@.338	Ea	9.60	12.10	—	21.70
2" x 1¼"	P1@.365	Ea	9.60	13.10	—	22.70
2" x 1½"	P1@.390	Ea	9.60	13.90	—	23.50
2½" x 1"	P1@.390	Ea	26.20	13.90	—	40.10
2½" x 1¼"	P1@.416	Ea	26.20	14.90	—	41.10
2½" x 1½"	P1@.442	Ea	26.20	15.80	—	42.00
2½" x 2"	P1@.497	Ea	26.20	17.80	—	44.00
3" x 1¼"	P1@.468	Ea	30.60	16.70	—	47.30
3" x 1½"	P1@.493	Ea	30.60	17.60	—	48.20
3" x 2"	P1@.548	Ea	30.60	19.60	—	50.20
3" x 2½"	P1@.600	Ea	30.60	21.50	—	52.10
4" x 2"	P1@.652	Ea	59.00	23.30	—	82.30
4" x 2½"	P1@.704	Ea	59.00	25.20	—	84.20
4" x 3"	P1@.755	Ea	59.00	27.00	—	86.00

Type L copper adapter C x MPT with brazed joint

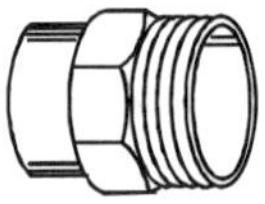

Description	Craft@Hrs	Unit	Material $	Labor $	Equipment $	Total $
1/2"	P1@.090	Ea	1.76	3.22	—	4.98
3/4"	P1@.126	Ea	2.75	4.51	—	7.26
1"	P1@.162	Ea	6.83	5.79	—	12.62
1¼"	P1@.198	Ea	12.80	7.08	—	19.88
1½"	P1@.234	Ea	14.30	8.37	—	22.67
2"	P1@.312	Ea	18.50	11.20	—	29.70
2½"	P1@.384	Ea	60.10	13.70	—	73.80
3"	P1@.456	Ea	99.80	16.30	—	116.10

Type L copper adapter C x FPT with brazed joint

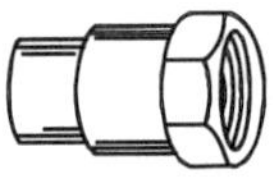

Description	Craft@Hrs	Unit	Material $	Labor $	Equipment $	Total $
1/2"	P1@.090	Ea	2.93	3.22	—	6.15
3/4"	P1@.126	Ea	5.30	4.51	—	9.81
1"	P1@.162	Ea	8.03	5.79	—	13.82
1¼"	P1@.198	Ea	16.80	7.08	—	23.88
1½"	P1@.234	Ea	20.40	8.37	—	28.77
2"	P1@.312	Ea	31.30	11.20	—	42.50
2½"	P1@.384	Ea	78.80	13.70	—	92.50
3"	P1@.456	Ea	109.00	16.30	—	125.30

Description	Craft@Hrs	Unit	Material $	Labor $	Equipment $	Total $

Type L copper flush bushing with brazed joints

Description	Craft@Hrs	Unit	Material $	Labor $	Equipment $	Total $
1/2" x 3/8"	P1@.148	Ea	1.79	5.29	—	7.08
3/4" x 1/2"	P1@.155	Ea	3.04	5.54	—	8.58
1" x 3/4"	P1@.206	Ea	7.88	7.37	—	15.25
1" x 1/2"	P1@.206	Ea	9.00	7.37	—	16.37
1¼" x 1"	P1@.258	Ea	9.38	9.23	—	18.61
1½" x 1¼"	P1@.308	Ea	11.90	11.00	—	22.90
2" x 1½"	P1@.390	Ea	22.40	13.90	—	36.30

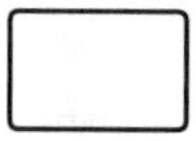

Type L copper union with brazed joints

Description	Craft@Hrs	Unit	Material $	Labor $	Equipment $	Total $
1/2"	P1@.146	Ea	7.14	5.22	—	12.36
3/4"	P1@.205	Ea	8.55	7.33	—	15.88
1"	P1@.263	Ea	19.10	9.40	—	28.50
1¼"	P1@.322	Ea	34.80	11.50	—	46.30
1½"	P1@.380	Ea	42.80	13.60	—	56.40
2"	P1@.507	Ea	63.80	18.10	—	81.90

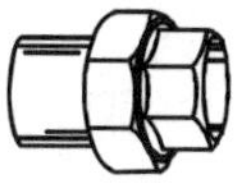

Type L copper dielectric union with brazed joints

Description	Craft@Hrs	Unit	Material $	Labor $	Equipment $	Total $
1/2"	P1@.146	Ea	6.70	5.22	—	11.92
3/4"	P1@.205	Ea	6.70	7.33	—	14.03
1"	P1@.263	Ea	13.80	9.40	—	23.20
1¼"	P1@.322	Ea	23.60	11.50	—	35.10
1½"	P1@.380	Ea	32.50	13.60	—	46.10
2"	P1@.507	Ea	44.90	18.10	—	63.00

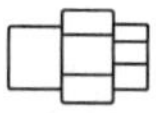

Type L copper cap with brazed joints

Description	Craft@Hrs	Unit	Material $	Labor $	Equipment $	Total $
1/2"	P1@.083	Ea	.58	2.97	—	3.55
3/4"	P1@.116	Ea	1.19	4.15	—	5.34
1"	P1@.149	Ea	2.69	5.33	—	8.02
1¼"	P1@.182	Ea	3.86	6.51	—	10.37
1½"	P1@.215	Ea	5.63	7.69	—	13.32
2"	P1@.286	Ea	10.20	10.20	—	20.40
2½"	P1@.352	Ea	25.70	12.60	—	38.30
3"	P1@.418	Ea	40.20	14.90	—	55.10
4"	P1@.550	Ea	61.70	19.70	—	81.40

Type L copper coupling with brazed joints

Description	Craft@Hrs	Unit	Material $	Labor $	Equipment $	Total $
1/2"	P1@.129	Ea	.40	4.61	—	5.01
3/4"	P1@.181	Ea	.93	6.47	—	7.40
1"	P1@.232	Ea	2.54	8.30	—	10.84
1¼"	P1@.284	Ea	4.45	10.20	—	14.65
1½"	P1@.335	Ea	6.44	12.00	—	18.44
2"	P1@.447	Ea	9.83	16.00	—	25.83
2½"	P1@.550	Ea	23.00	19.70	—	42.70
3"	P1@.654	Ea	35.10	23.40	—	58.50
4"	P1@.860	Ea	68.90	30.80	—	99.70

Description	Craft@Hrs	Unit	Material $	Labor $	Equipment $	Total $

Class 125 bronze body gate valve, brazed joints

Description	Craft@Hrs	Unit	Material $	Labor $	Equipment $	Total $
1/2"	P1@.240	Ea	15.40	8.58	—	23.98
3/4"	P1@.300	Ea	19.30	10.70	—	30.00
1"	P1@.360	Ea	27.10	12.90	—	40.00
1¼"	P1@.480	Ea	34.30	17.20	—	51.50
1½"	P1@.540	Ea	46.10	19.30	—	65.40
2"	P1@.600	Ea	77.00	21.50	—	98.50
2½"	P1@1.00	Ea	128.00	35.80	—	163.80
3"	P1@1.50	Ea	184.00	53.60	—	237.60

Class 125 iron body gate valve, flanged joints

Description	Craft@Hrs	Unit	Material $	Labor $	Equipment $	Total $
2"	P1@.500	Ea	295.00	17.90	—	312.90
2½"	P1@.600	Ea	398.00	21.50	—	419.50
3"	P1@.750	Ea	432.00	26.80	—	458.80
4"	P1@1.35	Ea	634.00	48.30	—	682.30
5"	P1@2.00	Ea	1,210.00	71.50	—	1,281.50
6"	P1@2.50	Ea	1,210.00	89.40	—	1,299.40

Class 125 bronze body globe valve, with brazed joints

Description	Craft@Hrs	Unit	Material $	Labor $	Equipment $	Total $
1/2"	P1@.240	Ea	28.60	8.58	—	37.18
3/4"	P1@.300	Ea	38.10	10.70	—	48.80
1"	P1@.360	Ea	65.60	12.90	—	78.50
1¼"	P1@.480	Ea	92.30	17.20	—	109.50
1½"	P1@.540	Ea	125.00	19.30	—	144.30
2"	P1@.600	Ea	202.00	21.50	—	223.50

Class 125 iron body globe valve, flanged joints

Description	Craft@Hrs	Unit	Material $	Labor $	Equipment $	Total $
2½"	P1@.650	Ea	334.00	23.20	—	357.20
3"	P1@.750	Ea	401.00	26.80	—	427.80
4"	P1@1.35	Ea	541.00	48.30	—	589.30

200 PSIG iron body butterfly valve, lug type, lever operated

Description	Craft@Hrs	Unit	Material $	Labor $	Equipment $	Total $
2"	P1@.450	Ea	128.00	16.10	—	144.10
2½"	P1@.450	Ea	132.00	16.10	—	148.10
3"	P1@.550	Ea	139.00	19.70	—	158.70
4"	P1@.550	Ea	173.00	19.70	—	192.70

200 PSIG iron body butterfly valve, wafer type, lever operated

Description	Craft@Hrs	Unit	Material $	Labor $	Equipment $	Total $
2"	P1@.450	Ea	116.00	16.10	—	132.10
2½"	P1@.450	Ea	119.00	16.10	—	135.10
3"	P1@.550	Ea	128.00	19.70	—	147.70
4"	P1@.550	Ea	155.00	19.70	—	174.70

Description	Craft@Hrs	Unit	Material $	Labor $	Equipment $	Total $

Class 125 bronze body 2-piece ball valve, brazed joints

Description	Craft@Hrs	Unit	Material $	Labor $	Equipment $	Total $
1/2"	P1@.240	Ea	7.53	8.58	—	16.11
3/4"	P1@.300	Ea	10.00	10.70	—	20.70
1"	P1@.360	Ea	17.30	12.90	—	30.20
1¼"	P1@.480	Ea	28.90	17.20	—	46.10
1½"	P1@.540	Ea	39.00	19.30	—	58.30
2"	P1@.600	Ea	49.40	21.50	—	70.90
3"	P1@1.50	Ea	338.00	53.60	—	391.60
4"	P1@1.80	Ea	443.00	64.40	—	507.40

Class 125 bronze body swing check valve, brazed joints

Description	Craft@Hrs	Unit	Material $	Labor $	Equipment $	Total $
1/2"	P1@.240	Ea	16.90	8.58	—	25.48
3/4"	P1@.300	Ea	24.20	10.70	—	34.90
1"	P1@.360	Ea	31.70	12.90	—	44.60
1¼"	P1@.480	Ea	45.00	17.20	—	62.20
1½"	P1@.540	Ea	63.50	19.30	—	82.80
2"	P1@.600	Ea	106.00	21.50	—	127.50
2½"	P1@1.00	Ea	212.00	35.80	—	247.80
3"	P1@1.50	Ea	305.00	53.60	—	358.60

Class 125 iron body swing check valve, flanged joints

Description	Craft@Hrs	Unit	Material $	Labor $	Equipment $	Total $
2"	P1@.500	Ea	148.00	17.90	—	165.90
2½"	P1@.600	Ea	188.00	21.50	—	209.50
3"	P1@.750	Ea	233.00	26.80	—	259.80
4"	P1@1.35	Ea	343.00	48.30	—	391.30

Class 125 iron body silent check valve, wafer type

Description	Craft@Hrs	Unit	Material $	Labor $	Equipment $	Total $
2"	P1@.500	Ea	108.00	17.90	—	125.90
2½"	P1@.600	Ea	120.00	21.50	—	141.50
3"	P1@.750	Ea	138.00	26.80	—	164.80
4"	P1@1.35	Ea	185.00	48.30	—	233.30

Class 125 bronze body strainer, threaded joints

Description	Craft@Hrs	Unit	Material $	Labor $	Equipment $	Total $
1/2"	P1@.230	Ea	26.40	8.22	—	34.62
3/4"	P1@.260	Ea	34.70	9.30	—	44.00
1"	P1@.330	Ea	42.50	11.80	—	54.30
1¼"	P1@.440	Ea	59.30	15.70	—	75.00
1½"	P1@.495	Ea	89.30	17.70	—	107.00
2"	P1@.550	Ea	154.00	19.70	—	173.70

Class 125 iron body strainer, flanged joints

Description	Craft@Hrs	Unit	Material $	Labor $	Equipment $	Total $
2"	P1@.500	Ea	118.00	17.90	—	135.90
2½"	P1@.600	Ea	132.00	21.50	—	153.50
3"	P1@.750	Ea	152.00	26.80	—	178.80
4"	P1@1.35	Ea	260.00	48.30	—	308.30

Description	Craft@Hrs	Unit	Material $	Labor $	Equipment $	Total $

Installation of 2-way control valve, threaded joints

Description	Craft@Hrs	Unit	Material $	Labor $	Equipment $	Total $
1/2"	P1@.210	Ea	—	7.51	—	7.51
3/4"	P1@.275	Ea	—	9.83	—	9.83
1"	P1@.350	Ea	—	12.50	—	12.50
1¼"	P1@.430	Ea	—	15.40	—	15.40
1½"	P1@.505	Ea	—	18.10	—	18.10
2"	P1@.675	Ea	—	24.10	—	24.10
2½"	P1@.830	Ea	—	29.70	—	29.70
3"	P1@.990	Ea	—	35.40	—	35.40

Installation of 3-way control valve, threaded joints

Description	Craft@Hrs	Unit	Material $	Labor $	Equipment $	Total $
1/2"	P1@.260	Ea	—	9.30	—	9.30
3/4"	P1@.365	Ea	—	13.10	—	13.10
1"	P1@.475	Ea	—	17.00	—	17.00
1¼"	P1@.575	Ea	—	20.60	—	20.60
1½"	P1@.680	Ea	—	24.30	—	24.30
2"	P1@.910	Ea	—	32.50	—	32.50
2½"	P1@1.12	Ea	—	40.10	—	40.10
3"	P1@1.33	Ea	—	47.60	—	47.60

Companion flange, 150 pound cast brass

Description	Craft@Hrs	Unit	Material $	Labor $	Equipment $	Total $
2"	P1@.290	Ea	348.00	10.40	—	358.40
2½"	P1@.380	Ea	429.00	13.60	—	442.60
3"	P1@.460	Ea	429.00	16.40	—	445.40
4"	P1@.600	Ea	683.00	21.50	—	704.50

Bolt and gasket sets

Description	Craft@Hrs	Unit	Material $	Labor $	Equipment $	Total $
2"	P1@.500	Ea	3.86	17.90	—	21.76
2½"	P1@.650	Ea	4.51	23.20	—	27.71
3"	P1@.750	Ea	7.67	26.80	—	34.47
4"	P1@1.00	Ea	12.80	35.80	—	48.60

Thermometer with well

Description	Craft@Hrs	Unit	Material $	Labor $	Equipment $	Total $
7"	P1@.250	Ea	159.00	8.94	—	167.94
9"	P1@.250	Ea	164.00	8.94	—	172.94

Dial type pressure gauge

Description	Craft@Hrs	Unit	Material $	Labor $	Equipment $	Total $
2½"	P1@.200	Ea	33.10	7.15	—	40.25
3½"	P1@.200	Ea	43.90	7.15	—	51.05

Description	Craft@Hrs	Unit	Material $	Labor $	Equipment $	Total $

Pressure/temperature tap

Description	Craft@Hrs	Unit	Material $	Labor $	Equipment $	Total $
tap	P1@.150	Ea	15.70	5.36	—	21.06

Hanger with swivel assembly

Description	Craft@Hrs	Unit	Material $	Labor $	Equipment $	Total $
1/2"	P1@.250	Ea	3.74	8.94	—	12.68
3/4"	P1@.250	Ea	4.13	8.94	—	13.07
1"	P1@.250	Ea	4.33	8.94	—	13.27
1¼"	P1@.300	Ea	4.44	10.70	—	15.14
1½"	P1@.300	Ea	4.77	10.70	—	15.47
2"	P1@.300	Ea	4.98	10.70	—	15.68
2½"	P1@.350	Ea	6.78	12.50	—	19.28
3"	P1@.350	Ea	8.37	12.50	—	20.87
4"	P1@.350	Ea	9.19	12.50	—	21.69

Riser clamp

Description	Craft@Hrs	Unit	Material $	Labor $	Equipment $	Total $
1/2"	P1@.100	Ea	2.41	3.58	—	5.99
3/4"	P1@.100	Ea	3.55	3.58	—	7.13
1"	P1@.100	Ea	3.59	3.58	—	7.17
1¼"	P1@.105	Ea	4.33	3.75	—	8.08
1½"	P1@.110	Ea	4.58	3.93	—	8.51
2"	P1@.115	Ea	4.84	4.11	—	8.95
2½"	P1@.120	Ea	5.09	4.29	—	9.38
3"	P1@.120	Ea	5.52	4.29	—	9.81
4"	P1@.125	Ea	7.03	4.47	—	11.50

Copper, Type L with Soft-Soldered Joints

Description	Craft@Hrs	Unit	Material $	Labor $	Equipment $	Total $

Type L hard-drawn copper pipe with wrought copper fittings and soft-soldered joints is used in a wide variety of plumbing and HVAC systems such as potable water, heating hot water, chilled water, and A.C. condensate.

Soft-soldered joints are those made with solders having melting points in the 350 degree F. to 500 degree F. range. Maximum working temperature/pressure relationships for soft-soldered joints are approximately as follows:

Maximum Working Pressures (PSIG)*				
Soft-solder Type	Water Temperature (degrees F.)	Nominal Pipe Size (inches)		
		Up to 1	1¼ to 2	2½ x 4
50-50 tin-lead**	100	200	175	150
	150	150	125	100
	200	100	90	75
	250	85	75	50
95-5 tin-antimony	100	500	400	300
	150	400	350	275
	200	300	250	200
	250	200	175	150

*For copper pipe and solder-type fittings using soft-solders melting at approximately 350 degrees F. to 500 degrees F.

**The use of any solder containing lead is not allowed in potable water systems.

This section has been arranged to save the estimator's time by including all normally-used system components such as pipe, fittings, valves, hanger assemblies, riser clamps and miscellaneous items under one heading. Additional items can be found under "Plumbing and Piping Specialties." The cost estimates in this section are based on the conditions, limitations and wage rates described in the section "How to Use This Book" beginning on page 5.

Description	Craft@Hrs	Unit	Material $	Labor $	Equipment $	Total $

Type L copper pipe with soft-soldered joints installed horizontally.

Complete installation including eight wrought copper tees and eight wrought copper elbows every 100', and hangers spaced to meet plumbing code. Use these figures for preliminary estimates.

1/2"	P1@.111	LF	2.16	3.97	—	6.13
3/4"	P1@.121	LF	3.35	4.33	—	7.68
1"	P1@.141	LF	6.06	5.04	—	11.10
1¼"	P1@.161	LF	8.63	5.76	—	14.39
1½"	P1@.181	LF	11.60	6.47	—	18.07
2"	P1@.211	LF	21.00	7.55	—	28.55
2½"	P1@.241	LF	36.40	8.62	—	45.02
3"	P1@.271	LF	50.90	9.69	—	60.59
4"	P1@.311	LF	73.40	11.10	—	84.50

Type L copper pipe with soft-soldered joints, installed risers.

Complete installation including a wrought copper reducing tee every floor and a riser clamp every other floor.

1/2"	P1@.061	LF	2.34	2.18	—	4.52
3/4"	P1@.071	LF	3.64	2.54	—	6.18
1"	P1@.076	LF	6.16	2.72	—	8.88
1¼"	P1@.081	LF	8.70	2.90	—	11.60
1½"	P1@.101	LF	10.90	3.61	—	14.51
2"	P1@.131	LF	28.50	4.68	—	33.18
2½"	P1@.151	LF	30.80	5.40	—	36.20
3"	P1@.171	LF	39.90	6.11	—	46.01
4"	P1@.191	LF	59.30	6.83	—	66.13

Type L copper pipe with soft-soldered joints, pipe only

1/2"	P1@.032	LF	1.76	1.14	—	2.90
3/4"	P1@.035	LF	2.79	1.25	—	4.04
1"	P1@.038	LF	3.87	1.36	—	5.23
1¼"	P1@.042	LF	5.54	1.50	—	7.04
1½"	P1@.046	LF	7.08	1.64	—	8.72
2"	P1@.053	LF	12.20	1.90	—	14.10
2½"	P1@.060	LF	17.70	2.15	—	19.85
3"	P1@.066	LF	24.00	2.36	—	26.36
4"	P1@.080	LF	37.70	2.86	—	40.56

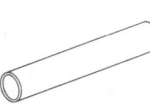

Type L copper 45 degree ell C x C with soft-soldered joints

1/2"	P1@.107	Ea	.99	3.83	—	4.82
3/4"	P1@.150	Ea	1.88	5.36	—	7.24
1"	P1@.193	Ea	6.77	6.90	—	13.67
1¼"	P1@.236	Ea	9.16	8.44	—	17.60
1½"	P1@.278	Ea	11.10	9.94	—	21.04
2"	P1@.371	Ea	18.30	13.30	—	31.60
2½"	P1@.457	Ea	46.70	16.30	—	63.00
3"	P1@.543	Ea	61.60	19.40	—	81.00
4"	P1@.714	Ea	119.00	25.50	—	144.50

Type L copper 90 degree ell C x C with soft-soldered joints

1/2"	P1@.107	Ea	.57	3.83	—	4.40
3/4"	P1@.150	Ea	1.21	5.36	—	6.57
1"	P1@.193	Ea	4.21	6.90	—	11.11
1¼"	P1@.236	Ea	6.83	8.44	—	15.27
1½"	P1@.278	Ea	10.40	9.94	—	20.34
2"	P1@.371	Ea	18.60	13.30	—	31.90
2½"	P1@.457	Ea	38.60	16.30	—	54.90
3"	P1@.543	Ea	48.80	19.40	—	68.20
4"	P1@.714	Ea	104.00	25.50	—	129.50

Copper, Type L with Soft-Soldered Joints

Description	Craft@Hrs	Unit	Material $	Labor $	Equipment $	Total $

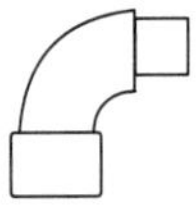

Type L copper 90 degree ell Ftg. x C with soft-soldered joint

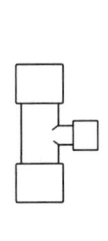

Description	Craft@Hrs	Unit	Material $	Labor $	Equipment $	Total $
1/2"	P1@.107	Ea	1.21	3.83	—	5.04
3/4"	P1@.150	Ea	3.99	5.36	—	9.35
1"	P1@.193	Ea	5.98	6.90	—	12.88
1¼"	P1@.236	Ea	8.33	8.44	—	16.77
1½"	P1@.278	Ea	12.50	9.94	—	22.44
2"	P1@.371	Ea	25.00	13.30	—	38.30
2½"	P1@.457	Ea	52.50	16.30	—	68.80
3"	P1@.543	Ea	59.00	19.40	—	78.40
4"	P1@.714	Ea	159.00	25.50	—	184.50

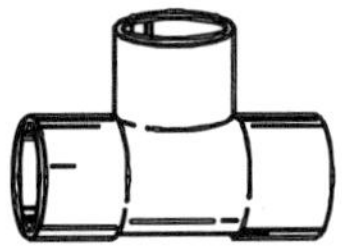

Type L copper tee C x C x C with soft-soldered joints

Description	Craft@Hrs	Unit	Material $	Labor $	Equipment $	Total $
1/2"	P1@.129	Ea	.98	4.61	—	5.59
3/4"	P1@.181	Ea	2.64	6.47	—	9.11
1"	P1@.233	Ea	9.98	8.33	—	18.31
1¼"	P1@.285	Ea	15.80	10.20	—	26.00
1½"	P1@.337	Ea	23.00	12.10	—	35.10
2"	P1@.449	Ea	36.80	16.10	—	52.90
2½"	P1@.552	Ea	74.60	19.70	—	94.30
3"	P1@.656	Ea	115.00	23.50	—	138.50
4"	P1@.863	Ea	192.00	30.90	—	222.90

Type L copper branch reducing tee C x C x C with soft-soldered joints

Description	Craft@Hrs	Unit	Material $	Labor $	Equipment $	Total $
1/2" x 3/8"	P1@.129	Ea	1.98	4.61	—	6.59
3/4" x 1/2"	P1@.182	Ea	4.20	6.51	—	10.71
1" x 1/2"	P1@.195	Ea	8.25	6.97	—	15.22
1" x 3/4"	P1@.219	Ea	8.25	7.83	—	16.08
1¼" x 1/2"	P1@.225	Ea	12.60	8.05	—	20.65
1¼" x 3/4"	P1@.247	Ea	12.60	8.83	—	21.43
1¼" x 1"	P1@.268	Ea	12.60	9.58	—	22.18
1½" x 1/2"	P1@.272	Ea	15.40	9.73	—	25.13
1½" x 3/4"	P1@.287	Ea	15.40	10.30	—	25.70
1½" x 1"	P1@.302	Ea	15.40	10.80	—	26.20
1½" x 1¼"	P1@.317	Ea	15.40	11.30	—	26.70
2" x 1/2"	P1@.325	Ea	20.60	11.60	—	32.20
2" x 3/4"	P1@.348	Ea	20.60	12.40	—	33.00
2" x 1"	P1@.373	Ea	20.60	13.30	—	33.90
2" x 1¼"	P1@.398	Ea	20.60	14.20	—	34.80
2" x 1½"	P1@.422	Ea	20.60	15.10	—	35.70
2½" x 1/2"	P1@.437	Ea	62.10	15.60	—	77.70
2½" x 3/4"	P1@.453	Ea	62.10	16.20	—	78.30
2½" x 1"	P1@.469	Ea	62.10	16.80	—	78.90
2½" x 1½"	P1@.502	Ea	62.10	18.00	—	80.10
2½" x 2"	P1@.519	Ea	62.10	18.60	—	80.70
3" x 1¼"	P1@.525	Ea	70.20	18.80	—	89.00
3" x 2"	P1@.585	Ea	72.90	20.90	—	93.80
3" x 2½"	P1@.617	Ea	72.90	22.10	—	95.00
4" x 1¼"	P1@.625	Ea	127.00	22.40	—	149.40
4" x 1½"	P1@.671	Ea	127.00	24.00	—	151.00
4" x 2"	P1@.718	Ea	127.00	25.70	—	152.70
4" x 2½"	P1@.765	Ea	127.00	27.40	—	154.40
4" x 3"	P1@.811	Ea	127.00	29.00	—	156.00

Description	Craft@Hrs	Unit	Material $	Labor $	Equipment $	Total $

Type L copper reducer with soft-soldered joints

Description	Craft@Hrs	Unit	Material $	Labor $	Equipment $	Total $
1/2" x 3/8"	P1@.124	Ea	.71	4.43	—	5.14
3/4" x 1/2"	P1@.129	Ea	1.34	4.61	—	5.95
1" x 3/4"	P1@.172	Ea	2.41	6.15	—	8.56
1¼" x 1/2"	P1@.172	Ea	4.05	6.15	—	10.20
1¼" x 3/4"	P1@.193	Ea	4.05	6.90	—	10.95
1¼" x 1"	P1@.215	Ea	4.05	7.69	—	11.74
1½" x 1/2"	P1@.193	Ea	6.19	6.90	—	13.09
1½" x 3/4"	P1@.214	Ea	6.19	7.65	—	13.84
1½" x 1"	P1@.236	Ea	6.19	8.44	—	14.63
1½" x 1¼"	P1@.257	Ea	6.19	9.19	—	15.38
2" x 1/2"	P1@.239	Ea	9.60	8.55	—	18.15
2" x 3/4"	P1@.261	Ea	9.60	9.33	—	18.93
2" x 1"	P1@.282	Ea	9.60	10.10	—	19.70
2" x 1¼"	P1@.304	Ea	9.60	10.90	—	20.50
2" x 1½"	P1@.325	Ea	9.60	11.60	—	21.20
2½" x 1"	P1@.325	Ea	26.20	11.60	—	37.80
2½" x 1¼"	P1@.347	Ea	26.20	12.40	—	38.60
2½" x 1½"	P1@.368	Ea	26.20	13.20	—	39.40
2½" x 2"	P1@.414	Ea	26.20	14.80	—	41.00
3" x 2"	P1@.457	Ea	30.60	16.30	—	46.00
3" x 2½"	P1@.500	Ea	30.60	17.90	—	48.50
4" x 2"	P1@.543	Ea	59.00	19.40	—	78.40
4" x 2½"	P1@.586	Ea	59.00	21.00	—	80.00
4" x 3"	P1@.629	Ea	59.00	22.50	—	81.50

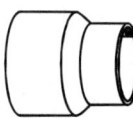

Type L copper adapter C x MPT with soft-soldered joint

Description	Craft@Hrs	Unit	Material $	Labor $	Equipment $	Total $
1/2"	P1@.075	Ea	1.76	2.68	—	4.44
3/4"	P1@.105	Ea	2.75	3.75	—	6.50
1"	P1@.134	Ea	6.83	4.79	—	11.62
1¼"	P1@.164	Ea	12.80	5.86	—	18.66
1½"	P1@.194	Ea	14.30	6.94	—	21.24
2"	P1@.259	Ea	18.50	9.26	—	27.76
2½"	P1@.319	Ea	60.10	11.40	—	71.50
3"	P1@.378	Ea	99.80	13.50	—	113.30

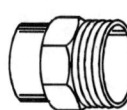

Type L copper adapter C x FPT with soft-soldered joint

Description	Craft@Hrs	Unit	Material $	Labor $	Equipment $	Total $
1/2"	P1@.075	Ea	2.93	2.68	—	5.61
3/4"	P1@.105	Ea	5.30	3.75	—	9.05
1"	P1@.134	Ea	8.03	4.79	—	12.82
1¼"	P1@.164	Ea	16.80	5.86	—	22.66
1½"	P1@.194	Ea	20.40	6.94	—	27.34
2"	P1@.259	Ea	31.30	9.26	—	40.56
2½"	P1@.319	Ea	78.80	11.40	—	90.20
3"	P1@.378	Ea	109.00	13.50	—	122.50

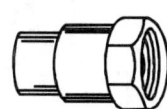

Copper, Type L with Soft-Soldered Joints

Description	Craft@Hrs	Unit	Material $	Labor $	Equipment $	Total $

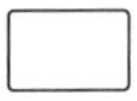

Type L copper flush bushing with soft-soldered joints

Description	Craft@Hrs	Unit	Material $	Labor $	Equipment $	Total $
1/2" x 3/8"	P1@.124	Ea	1.79	4.43	—	6.22
3/4" x 1/2"	P1@.129	Ea	3.04	4.61	—	7.65
1" x 3/4"	P1@.172	Ea	7.88	6.15	—	14.03
1" x 1/2"	P1@.172	Ea	9.00	6.15	—	15.15
1¼" x 1"	P1@.215	Ea	9.38	7.69	—	17.07
1½" x 1¼"	P1@.257	Ea	11.90	9.19	—	21.09
2" x 1½"	P1@.325	Ea	22.40	11.60	—	34.00

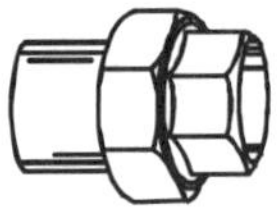

Type L copper union with soft-soldered joint

Description	Craft@Hrs	Unit	Material $	Labor $	Equipment $	Total $
1/2"	P1@.121	Ea	7.14	4.33	—	11.47
3/4"	P1@.170	Ea	8.55	6.08	—	14.63
1"	P1@.218	Ea	19.10	7.80	—	26.90
1¼"	P1@.267	Ea	34.80	9.55	—	44.35
1½"	P1@.315	Ea	42.80	11.30	—	54.10
2"	P1@.421	Ea	63.80	15.10	—	78.90

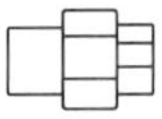

Type L copper dielectric union with soft-soldered joint

Description	Craft@Hrs	Unit	Material $	Labor $	Equipment $	Total $
1/2"	P1@.121	Ea	6.70	4.33	—	11.03
3/4"	P1@.170	Ea	6.70	6.08	—	12.78
1"	P1@.218	Ea	13.80	7.80	—	21.60
1¼"	P1@.267	Ea	23.60	9.55	—	33.15
1½"	P1@.315	Ea	32.50	11.30	—	43.80
2"	P1@.421	Ea	44.90	15.10	—	60.00

Type L copper cap with soft-soldered joint

Description	Craft@Hrs	Unit	Material $	Labor $	Equipment $	Total $
1/2"	P1@.069	Ea	.58	2.47	—	3.05
3/4"	P1@.096	Ea	1.19	3.43	—	4.62
1"	P1@.124	Ea	2.69	4.43	—	7.12
1¼"	P1@.151	Ea	3.86	5.40	—	9.26
1½"	P1@.178	Ea	5.63	6.37	—	12.00
2"	P1@.237	Ea	10.20	8.48	—	18.68
2½"	P1@.292	Ea	25.70	10.40	—	36.10
3"	P1@.347	Ea	40.20	12.40	—	52.60
4"	P1@.457	Ea	61.70	16.30	—	78.00

Type L copper coupling with soft-soldered joints

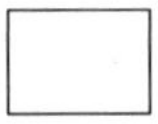

Description	Craft@Hrs	Unit	Material $	Labor $	Equipment $	Total $
1/2"	P1@.107	Ea	.40	3.83	—	4.23
3/4"	P1@.150	Ea	.93	5.36	—	6.29
1"	P1@.193	Ea	2.54	6.90	—	9.44
1¼"	P1@.236	Ea	4.45	8.44	—	12.89
1½"	P1@.278	Ea	6.44	9.94	—	16.38
2"	P1@.371	Ea	9.83	13.30	—	23.13
2½"	P1@.457	Ea	23.00	16.30	—	39.30
3"	P1@.543	Ea	35.10	19.40	—	54.50
4"	P1@.714	Ea	68.90	25.50	—	94.40

Description	Craft@Hrs	Unit	Material $	Labor $	Equipment $	Total $
Class 125 bronze body gate valve, soft-soldered joints						
1/2"	P1@.200	Ea	14.70	7.15	—	21.85
3/4"	P1@.249	Ea	18.40	8.90	—	27.30
1"	P1@.299	Ea	25.80	10.70	—	36.50
1¼"	P1@.398	Ea	32.70	14.20	—	46.90
1½"	P1@.448	Ea	43.90	16.00	—	59.90
2"	P1@.498	Ea	73.30	17.80	—	91.10
2½"	P1@.830	Ea	122.00	29.70	—	151.70
3"	P1@1.24	Ea	175.00	44.30	—	219.30

Description	Craft@Hrs	Unit	Material $	Labor $	Equipment $	Total $
Class 125 iron body gate valve, flanged ends						
2"	P1@.500	Ea	295.00	17.90	—	312.90
2½"	P1@.600	Ea	398.00	21.50	—	419.50
3"	P1@.750	Ea	432.00	26.80	—	458.80
4"	P1@1.35	Ea	634.00	48.30	—	682.30
5"	P1@2.00	Ea	1,210.00	71.50	—	1,281.50
6"	P1@2.50	Ea	1,210.00	89.40	—	1,299.40

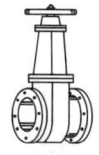

Description	Craft@Hrs	Unit	Material $	Labor $	Equipment $	Total $
Class 125 bronze body globe valve, with soft-soldered joints						
1/2"	P1@.200	Ea	28.60	7.15	—	35.75
3/4"	P1@.249	Ea	38.10	8.90	—	47.00
1"	P1@.299	Ea	65.60	10.70	—	76.30
1¼"	P1@.398	Ea	92.30	14.20	—	106.50
1½"	P1@.448	Ea	125.00	16.00	—	141.00
2"	P1@.498	Ea	202.00	17.80	—	219.80

Description	Craft@Hrs	Unit	Material $	Labor $	Equipment $	Total $
Class 125 iron body globe valve, flanged ends						
2½"	P1@.600	Ea	334.00	21.50	—	355.50
3"	P1@.750	Ea	401.00	26.80	—	427.80
4"	P1@1.35	Ea	538.00	48.30	—	586.30

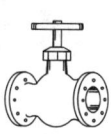

Description	Craft@Hrs	Unit	Material $	Labor $	Equipment $	Total $
200 PSIG iron body butterfly valve, lug type, lever operated						
2"	P1@.450	Ea	128.00	16.10	—	144.10
2½"	P1@.450	Ea	132.00	16.10	—	148.10
3"	P1@.550	Ea	139.00	19.70	—	158.70
4"	P1@.550	Ea	173.00	19.70	—	192.70

Description	Craft@Hrs	Unit	Material $	Labor $	Equipment $	Total $
200 PSIG iron body butterfly valve, wafer type, lever operated						
2"	P1@.450	Ea	116.00	16.10	—	132.10
2½"	P1@.450	Ea	119.00	16.10	—	135.10
3"	P1@.550	Ea	128.00	19.70	—	147.70
4"	P1@.550	Ea	155.00	19.70	—	174.70

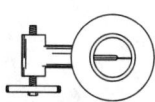

Copper, Type L with Soft-Soldered Joints

Description	Craft@Hrs	Unit	Material $	Labor $	Equipment $	Total $

Class 125 bronze body 2-piece ball valve, soft-soldered joints

Description	Craft@Hrs	Unit	Material $	Labor $	Equipment $	Total $
1/2"	P1@.200	Ea	7.53	7.15	—	14.68
3/4"	P1@.249	Ea	10.00	8.90	—	18.90
1"	P1@.299	Ea	17.30	10.70	—	28.00
1¼"	P1@.398	Ea	28.90	14.20	—	43.10
1½"	P1@.448	Ea	39.00	16.00	—	55.00
2"	P1@.498	Ea	49.40	17.80	—	67.20
3"	P1@1.24	Ea	338.00	44.30	—	382.30
4"	P1@1.45	Ea	443.00	51.90	—	494.90

Class 125 bronze body swing check valve, soft-soldered joints

Description	Craft@Hrs	Unit	Material $	Labor $	Equipment $	Total $
1/2"	P1@.200	Ea	16.90	7.15	—	24.05
3/4"	P1@.249	Ea	24.20	8.90	—	33.10
1"	P1@.299	Ea	31.70	10.70	—	42.40
1¼"	P1@.398	Ea	45.00	14.20	—	59.20
1½"	P1@.448	Ea	63.50	16.00	—	79.50
2"	P1@.498	Ea	106.00	17.80	—	123.80
2½"	P1@.830	Ea	212.00	29.70	—	241.70
3"	P1@1.24	Ea	305.00	44.30	—	349.30

Class 125 iron body swing check valve, flanged ends

Description	Craft@Hrs	Unit	Material $	Labor $	Equipment $	Total $
2"	P1@.500	Ea	148.00	17.90	—	165.90
2½"	P1@.600	Ea	188.00	21.50	—	209.50
3"	P1@.750	Ea	233.00	26.80	—	259.80
4"	P1@1.35	Ea	343.00	48.30	—	391.30

Class 125 iron body silent check valve, wafer type

Description	Craft@Hrs	Unit	Material $	Labor $	Equipment $	Total $
2"	P1@.500	Ea	108.00	17.90	—	125.90
2½"	P1@.600	Ea	120.00	21.50	—	141.50
3"	P1@.750	Ea	138.00	26.80	—	164.80
4"	P1@1.35	Ea	185.00	48.30	—	233.30

Class 125 bronze body strainer, threaded ends

Description	Craft@Hrs	Unit	Material $	Labor $	Equipment $	Total $
1/2"	P1@.230	Ea	26.40	8.22	—	34.62
3/4"	P1@.260	Ea	34.70	9.30	—	44.00
1"	P1@.330	Ea	42.50	11.80	—	54.30
1¼"	P1@.440	Ea	59.30	15.70	—	75.00
1½"	P1@.495	Ea	89.30	17.70	—	107.00
2"	P1@.550	Ea	154.00	19.70	—	173.70

Class 125 iron body strainer, flanged ends

Description	Craft@Hrs	Unit	Material $	Labor $	Equipment $	Total $
2"	P1@.500	Ea	117.00	17.90	—	134.90
2½"	P1@.600	Ea	131.00	21.50	—	152.50
3"	P1@.750	Ea	151.00	26.80	—	177.80
4"	P1@1.35	Ea	258.00	48.30	—	306.30

Description	Craft@Hrs	Unit	Material $	Labor $	Equipment $	Total $

Installation of 2-way control valve, threaded joints

Description	Craft@Hrs	Unit	Material $	Labor $	Equipment $	Total $
1/2"	P1@.210	Ea	—	7.51	—	7.51
3/4"	P1@.275	Ea	—	9.83	—	9.83
1"	P1@.350	Ea	—	12.50	—	12.50
1¼"	P1@.430	Ea	—	15.40	—	15.40
1½"	P1@.505	Ea	—	18.10	—	18.10
2"	P1@.675	Ea	—	24.10	—	24.10
2½"	P1@.830	Ea	—	29.70	—	29.70
3"	P1@.990	Ea	—	35.40	—	35.40

Installation of 3-way control valve, threaded joints

Description	Craft@Hrs	Unit	Material $	Labor $	Equipment $	Total $
1/2"	P1@.260	Ea	—	9.30	—	9.30
3/4"	P1@.365	Ea	—	13.10	—	13.10
1"	P1@.475	Ea	—	17.00	—	17.00
1¼"	P1@.575	Ea	—	20.60	—	20.60
1½"	P1@.680	Ea	—	24.30	—	24.30
2"	P1@.910	Ea	—	32.50	—	32.50
2½"	P1@1.12	Ea	—	40.10	—	40.10
3"	P1@1.33	Ea	—	47.60	—	47.60

Companion flange, 150 pound cast brass

Description	Craft@Hrs	Unit	Material $	Labor $	Equipment $	Total $
2"	P1@.290	Ea	343.00	10.40	—	353.40
2½"	P1@.380	Ea	424.00	13.60	—	437.60
3"	P1@.460	Ea	424.00	16.40	—	440.40
4"	P1@.600	Ea	674.00	21.50	—	695.50

Bolt and gasket sets

Description	Craft@Hrs	Unit	Material $	Labor $	Equipment $	Total $
2"	P1@.500	Ea	3.86	17.90	—	21.76
2½"	P1@.650	Ea	4.51	23.20	—	27.71
3"	P1@.750	Ea	7.67	26.80	—	34.47
4"	P1@1.00	Ea	12.80	35.80	—	48.60

Thermometer with well

Description	Craft@Hrs	Unit	Material $	Labor $	Equipment $	Total $
7"	P1@.250	Ea	159.00	8.94	—	167.94
9"	P1@.250	Ea	164.00	8.94	—	172.94

Dial type pressure gauge

Description	Craft@Hrs	Unit	Material $	Labor $	Equipment $	Total $
2½"	P1@.200	Ea	33.10	7.15	—	40.25
3½"	P1@.200	Ea	43.80	7.15	—	50.95

Pressure/temperature tap

Description	Craft@Hrs	Unit	Material $	Labor $	Equipment $	Total $
tap	P1@.150	Ea	15.70	5.36	—	21.06

Copper, Type L with Soft-Soldered Joints

Description	Craft@Hrs	Unit	Material $	Labor $	Equipment $	Total $

Hanger with swivel assembly

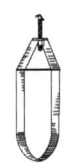

Description	Craft@Hrs	Unit	Material $	Labor $	Equipment $	Total $
1/2"	P1@.250	Ea	3.74	8.94	—	12.68
3/4"	P1@.250	Ea	4.13	8.94	—	13.07
1"	P1@.250	Ea	4.33	8.94	—	13.27
1¼"	P1@.300	Ea	4.44	10.70	—	15.14
1½"	P1@.300	Ea	4.77	10.70	—	15.47
2"	P1@.300	Ea	4.98	10.70	—	15.68
2½"	P1@.350	Ea	6.78	12.50	—	19.28
3"	P1@.350	Ea	8.37	12.50	—	20.87
4"	P1@.350	Ea	9.19	12.50	—	21.69

Riser clamp

Description	Craft@Hrs	Unit	Material $	Labor $	Equipment $	Total $
1/2"	P1@.100	Ea	2.41	3.58	—	5.99
3/4"	P1@.100	Ea	3.55	3.58	—	7.13
1"	P1@.100	Ea	3.59	3.58	—	7.17
1¼"	P1@.105	Ea	4.33	3.75	—	8.08
1½"	P1@.110	Ea	4.58	3.93	—	8.51
2"	P1@.115	Ea	4.84	4.11	—	8.95
2½"	P1@.120	Ea	5.09	4.29	—	9.38
3"	P1@.120	Ea	5.52	4.29	—	9.81
4"	P1@.125	Ea	7.03	4.47	—	11.50

Type M hard-drawn copper pipe with wrought copper fittings and brazed joints is used in a wide variety of plumbing and HVAC systems such as potable water, heating hot water and chilled water.

Brazed joints are those made with silver or other alloy filler metals having melting points at, or above, 1,000 degrees F. Maximum working pressure/temperature relationships for brazed joints are approximately as follows:

Maximum Working Pressure (PSIG)*			
Water Temperature (degrees F.)	Nominal Pipe Size (inches)		
	Up to 1	1¼ to 2	2½ to 4
Up to 350	270	190	150

For copper pipe and solder-type fittings using brazing alloys melting at, or above, 1,000 degrees F.

This section has been arranged to save the estimator's time by including all normally-used system components such as pipe, fittings, valves, hanger assemblies, riser clamps and miscellaneous items under one heading. Additional items can be found under "Plumbing and Piping Specialties." The cost estimates in this section are based on the conditions, limitations and wage rates described in the section "How to Use This Book" beginning on page 5.

Description	Craft@Hrs	Unit	Material $	Labor $	Equipment $	Total $

Type M copper pipe with brazed joints installed horizontally. Complete installation including six to twelve wrought copper tees and six to twelve wrought copper elbows every 100', and hangers spaced to meet plumbing code (*a tee & elbow every 8.3' for ½" dia., a tee and elbow every 16.6' for 4" dia.*).

1/2"	P1@.113	LF	1.89	4.04	—	5.93
3/4"	P1@.123	LF	2.92	4.40	—	7.32
1"	P1@.143	LF	6.08	5.11	—	11.19
1¼"	P1@.163	LF	10.30	5.83	—	16.13
1½"	P1@.183	LF	12.20	6.54	—	18.74
2"	P1@.213	LF	25.40	7.62	—	33.02
2½"	P1@.243	LF	37.20	8.69	—	45.89
3"	P1@.273	LF	46.80	9.76	—	56.56
4"	P1@.313	LF	68.90	11.20	—	80.10

Type M copper pipe with brazed joints, installed risers. Complete installation including a wrought copper reducing tee every floor and a riser clamp every other floor.

1/2"	P1@.062	LF	1.60	2.22	—	3.82
3/4"	P1@.072	LF	2.60	2.57	—	5.17
1"	P1@.077	LF	5.09	2.75	—	7.84
1¼"	P1@.082	LF	7.50	2.93	—	10.43
1½"	P1@.102	LF	9.90	3.65	—	13.55
2"	P1@.132	LF	13.10	4.72	—	17.82
2½"	P1@.152	LF	28.50	5.44	—	33.94
3"	P1@.172	LF	34.90	6.15	—	41.05
4"	P1@.192	LF	51.80	6.87	—	58.67

Copper, Type M with Brazed Joints

Description	Craft@Hrs	Unit	Material $	Labor $	Equipment $	Total $

Type M copper pipe with brazed joints, pipe only

Description	Craft@Hrs	Unit	Material $	Labor $	Equipment $	Total $
1/2"	P1@.032	LF	1.13	1.14	—	2.27
3/4"	P1@.035	LF	1.85	1.25	—	3.10
1"	P1@.038	LF	2.90	1.36	—	4.26
1¼"	P1@.042	LF	4.36	1.50	—	5.86
1½"	P1@.046	LF	6.11	1.64	—	7.75
2"	P1@.053	LF	11.20	1.90	—	13.10
2½"	P1@.060	LF	15.50	2.15	—	17.65
3"	P1@.066	LF	19.60	2.36	—	21.96
4"	P1@.080	LF	31.30	2.86	—	34.16

Type M copper 45 degree ell C x C with brazed joints

Description	Craft@Hrs	Unit	Material $	Labor $	Equipment $	Total $
1/2"	P1@.129	Ea	.99	4.61	—	5.60
3/4"	P1@.181	Ea	1.88	6.47	—	8.35
1"	P1@.232	Ea	5.54	8.30	—	13.84
1¼"	P1@.284	Ea	7.50	10.20	—	17.70
1½"	P1@.335	Ea	11.10	12.00	—	23.10
2"	P1@.447	Ea	18.30	16.00	—	34.30
2½"	P1@.550	Ea	46.70	19.70	—	66.40
3"	P1@.654	Ea	61.60	23.40	—	85.00
4"	P1@.860	Ea	119.00	30.80	—	149.80

Type M copper 90 degree ell C x C with brazed joints

Description	Craft@Hrs	Unit	Material $	Labor $	Equipment $	Total $
1/2"	P1@.129	Ea	.56	4.61	—	5.17
3/4"	P1@.181	Ea	1.18	6.47	—	7.65
1"	P1@.232	Ea	4.21	8.30	—	12.51
1¼"	P1@.284	Ea	6.83	10.20	—	17.03
1½"	P1@.335	Ea	10.40	12.00	—	22.40
2"	P1@.447	Ea	18.60	16.00	—	34.60
2½"	P1@.550	Ea	38.60	19.70	—	58.30
3"	P1@.654	Ea	48.80	23.40	—	72.20
4"	P1@.860	Ea	104.00	30.80	—	134.80

Type M copper 90 degree ell Ftg. x C with brazed joints

Description	Craft@Hrs	Unit	Material $	Labor $	Equipment $	Total $
1/2"	P1@.129	Ea	.91	4.61	—	5.52
3/4"	P1@.181	Ea	2.00	6.47	—	8.47
1"	P1@.232	Ea	5.98	8.30	—	14.28
1¼"	P1@.284	Ea	8.33	10.20	—	18.53
1½"	P1@.335	Ea	12.50	12.00	—	24.50
2"	P1@.447	Ea	25.00	16.00	—	41.00
2½"	P1@.550	Ea	52.50	19.70	—	72.20
3"	P1@.654	Ea	59.00	23.40	—	82.40
4"	P1@.860	Ea	119.00	30.80	—	149.80

Description	Craft@Hrs	Unit	Material $	Labor $	Equipment $	Total $

Type M copper tee C x C x C with brazed joints

Description	Craft@Hrs	Unit	Material $	Labor $	Equipment $	Total $
1/2"	P1@.156	Ea	.98	5.58	—	6.56
3/4"	P1@.204	Ea	2.64	7.30	—	9.94
1"	P1@.263	Ea	9.98	9.40	—	19.38
1¼"	P1@.322	Ea	15.80	11.50	—	27.30
1½"	P1@.380	Ea	23.00	13.60	—	36.60
2"	P1@.507	Ea	36.80	18.10	—	54.90
2½"	P1@.624	Ea	74.60	22.30	—	96.90
3"	P1@.741	Ea	115.00	26.50	—	141.50
4"	P1@.975	Ea	192.00	34.90	—	226.90

Type M copper reducing branch tee C x C x C with brazed joints

Description	Craft@Hrs	Unit	Material $	Labor $	Equipment $	Total $
1/2" x 3/8"	P1@.129	Ea	1.98	4.61	—	6.59
3/4" x 1/2"	P1@.182	Ea	4.20	6.51	—	10.71
1" x 1/2"	P1@.209	Ea	8.25	7.47	—	15.72
1" x 3/4"	P1@.234	Ea	8.25	8.37	—	16.62
1¼" x 1/2"	P1@.241	Ea	12.60	8.62	—	21.22
1¼" x 3/4"	P1@.264	Ea	12.60	9.44	—	22.04
1¼" x 1"	P1@.287	Ea	12.60	10.30	—	22.90
1½" x 1/2"	P1@.291	Ea	15.40	10.40	—	25.80
1½" x 3/4"	P1@.307	Ea	15.40	11.00	—	26.40
1½" x 1"	P1@.323	Ea	15.40	11.60	—	27.00
1½" x 1¼"	P1@.339	Ea	15.40	12.10	—	27.50
2" x 1/2"	P1@.348	Ea	20.60	12.40	—	33.00
2" x 3/4"	P1@.373	Ea	20.60	13.30	—	33.90
2" x 1"	P1@.399	Ea	20.60	14.30	—	34.90
2" x 1¼"	P1@.426	Ea	20.60	15.20	—	35.80
2" x 1½"	P1@.452	Ea	20.60	16.20	—	36.80
2½" x 1/2"	P1@.468	Ea	62.10	16.70	—	78.80
2½" x 3/4"	P1@.485	Ea	62.10	17.30	—	79.40
2½" x 1"	P1@.502	Ea	62.10	18.00	—	80.10
2½" x 1½"	P1@.537	Ea	62.10	19.20	—	81.30
2½" x 2"	P1@.555	Ea	62.10	19.80	—	81.90
3" x 1¼"	P1@.562	Ea	70.20	20.10	—	90.30
3" x 2"	P1@.626	Ea	72.90	22.40	—	95.30
3" x 2½"	P1@.660	Ea	72.90	23.60	—	96.50
4" x 1¼"	P1@.669	Ea	127.00	23.90	—	150.90
4" x 1½"	P1@.718	Ea	127.00	25.70	—	152.70
4" x 2"	P1@.768	Ea	127.00	27.50	—	154.50
4" x 2½"	P1@.819	Ea	127.00	29.30	—	156.30
4" x 3"	P1@.868	Ea	127.00	31.00	—	158.00

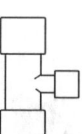

Description	Craft@Hrs	Unit	Material $	Labor $	Equipment $	Total $

Type M copper reducer with brazed joints

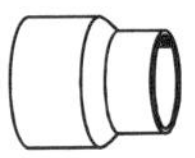

Description	Craft@Hrs	Unit	Material $	Labor $	Equipment $	Total $
1/2" x 3/8"	P1@.148	Ea	.71	5.29	—	6.00
3/4" x 1/2"	P1@.155	Ea	1.34	5.54	—	6.88
1" x 3/4"	P1@.206	Ea	2.41	7.37	—	9.78
1¼" x 1/2"	P1@.206	Ea	4.05	7.37	—	11.42
1¼" x 3/4"	P1@.232	Ea	4.05	8.30	—	12.35
1¼" x 1"	P1@.258	Ea	4.05	9.23	—	13.28
1½" x 1/2"	P1@.232	Ea	6.19	8.30	—	14.49
1½" x 3/4"	P1@.257	Ea	6.19	9.19	—	15.38
1½" x 1"	P1@.283	Ea	6.19	10.10	—	16.29
1½" x 1¼"	P1@.308	Ea	6.19	11.00	—	17.19
2" x 1/2"	P1@.287	Ea	9.60	10.30	—	19.90
2" x 3/4"	P1@.277	Ea	9.60	9.91	—	19.51
2" x 1"	P1@.338	Ea	9.60	12.10	—	21.70
2" x 1¼"	P1@.365	Ea	9.60	13.10	—	22.70
2" x 1½"	P1@.390	Ea	9.60	13.90	—	23.50
2½" x 1"	P1@.390	Ea	26.20	13.90	—	40.10
2½" x 1¼"	P1@.416	Ea	26.20	14.90	—	41.10
2½" x 1½"	P1@.442	Ea	26.20	15.80	—	42.00
2½" x 2"	P1@.497	Ea	26.20	17.80	—	44.00
3" x 1¼"	P1@.468	Ea	30.60	16.70	—	47.30
3" x 1½"	P1@.493	Ea	30.60	17.60	—	48.20
3" x 2"	P1@.548	Ea	30.60	19.60	—	50.20
3" x 2½"	P1@.600	Ea	30.60	21.50	—	52.10
4" x 2"	P1@.652	Ea	59.00	23.30	—	82.30
4" x 2½"	P1@.704	Ea	59.00	25.20	—	84.20
4" x 3"	P1@.755	Ea	59.00	27.00	—	86.00

Type M copper adapter C x MPT with brazed joint

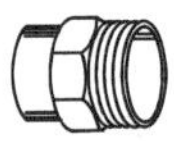

Description	Craft@Hrs	Unit	Material $	Labor $	Equipment $	Total $
1/2"	P1@.090	Ea	1.76	3.22	—	4.98
3/4"	P1@.126	Ea	2.75	4.51	—	7.26
1"	P1@.162	Ea	6.83	5.79	—	12.62
1¼"	P1@.198	Ea	12.80	7.08	—	19.88
1½"	P1@.234	Ea	14.30	8.37	—	22.67
2"	P1@.312	Ea	18.50	11.20	—	29.70
2½"	P1@.384	Ea	60.10	13.70	—	73.80
3"	P1@.456	Ea	99.80	16.30	—	116.10

Type M copper adapter C x FPT with brazed joint

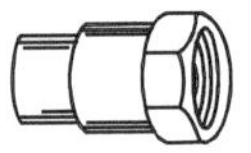

Description	Craft@Hrs	Unit	Material $	Labor $	Equipment $	Total $
1/2"	P1@.090	Ea	2.93	3.22	—	6.15
3/4"	P1@.126	Ea	5.30	4.51	—	9.81
1"	P1@.162	Ea	8.03	5.79	—	13.82
1¼"	P1@.198	Ea	16.80	7.08	—	23.88
1½"	P1@.234	Ea	20.40	8.37	—	28.77
2"	P1@.312	Ea	31.30	11.20	—	42.50
2½"	P1@.384	Ea	78.80	13.70	—	92.50
3"	P1@.456	Ea	109.00	16.30	—	125.30

Description	Craft@Hrs	Unit	Material $	Labor $	Equipment $	Total $

Type M copper flush bushing with brazed joints

Description	Craft@Hrs	Unit	Material $	Labor $	Equipment $	Total $
1/2" x 3/8"	P1@.148	Ea	1.79	5.29	—	7.08
3/4" x 1/2"	P1@.155	Ea	3.04	5.54	—	8.58
1" x 3/4"	P1@.206	Ea	7.88	7.37	—	15.25
1" x 1/2"	P1@.206	Ea	9.00	7.37	—	16.37
1¼" x 1"	P1@.258	Ea	9.38	9.23	—	18.61
1½" x 1¼"	P1@.308	Ea	11.90	11.00	—	22.90
2" x 1½"	P1@.390	Ea	22.40	13.90	—	36.30

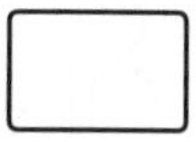

Type M copper union with brazed joints

Description	Craft@Hrs	Unit	Material $	Labor $	Equipment $	Total $
1/2"	P1@.146	Ea	7.14	5.22	—	12.36
3/4"	P1@.205	Ea	8.55	7.33	—	15.88
1"	P1@.263	Ea	19.10	9.40	—	28.50
1¼"	P1@.322	Ea	34.80	11.50	—	46.30
1½"	P1@.380	Ea	42.80	13.60	—	56.40
2"	P1@.507	Ea	63.80	18.10	—	81.90

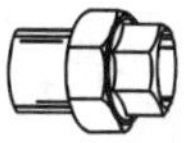

Type M copper dielectric union with brazed joints

Description	Craft@Hrs	Unit	Material $	Labor $	Equipment $	Total $
1/2"	P1@.146	Ea	6.70	5.22	—	11.02
3/4"	P1@.205	Ea	6.70	7.33	—	14.03
1"	P1@.263	Ea	13.80	9.40	—	23.20
1¼"	P1@.322	Ea	23.60	11.50	—	35.10
1½"	P1@.380	Ea	32.50	13.60	—	46.10
2"	P1@.507	Ea	44.90	18.10	—	63.00

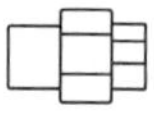

Type M copper cap with brazed joints

Description	Craft@Hrs	Unit	Material $	Labor $	Equipment $	Total $
1/2"	P1@.083	Ea	.58	2.97	—	3.55
3/4"	P1@.116	Ea	1.19	4.15	—	5.34
1"	P1@.149	Ea	2.69	5.33	—	8.02
1¼"	P1@.182	Ea	3.86	6.51	—	10.37
1½"	P1@.215	Ea	5.63	7.69	—	13.32
2"	P1@.286	Ea	10.20	10.20	—	20.40
2½"	P1@.352	Ea	25.70	12.60	—	38.30
3"	P1@.418	Ea	40.20	14.90	—	55.10
4"	P1@.550	Ea	61.70	19.70	—	81.40

Type M copper coupling with brazed joints

Description	Craft@Hrs	Unit	Material $	Labor $	Equipment $	Total $
1/2"	P1@.129	Ea	.37	4.61	—	4.98
3/4"	P1@.181	Ea	.93	6.47	—	7.40
1"	P1@.232	Ea	2.54	8.30	—	10.84
1¼"	P1@.284	Ea	4.45	10.20	—	14.65
1½"	P1@.335	Ea	6.44	12.00	—	18.44
2"	P1@.447	Ea	9.83	16.00	—	25.83
2½"	P1@.550	Ea	23.00	19.70	—	42.70
3"	P1@.654	Ea	35.10	23.40	—	58.50
4"	P1@.860	Ea	68.90	30.80	—	99.70

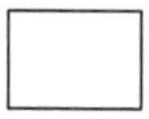

Copper, Type M with Brazed Joints

Description	Craft@Hrs	Unit	Material $	Labor $	Equipment $	Total $

Class 125 bronze body gate valve, brazed joints

Description	Craft@Hrs	Unit	Material $	Labor $	Equipment $	Total $
1/2"	P1@.240	Ea	15.40	8.58	—	23.98
3/4"	P1@.300	Ea	19.30	10.70	—	30.00
1"	P1@.360	Ea	27.10	12.90	—	40.00
1¼"	P1@.480	Ea	34.30	17.20	—	51.50
1½"	P1@.540	Ea	46.10	19.30	—	65.40
2"	P1@.600	Ea	77.00	21.50	—	98.50
2½"	P1@1.00	Ea	128.00	35.80	—	163.80
3"	P1@1.50	Ea	184.00	53.60	—	237.60

Class 125 iron body gate valve, flanged ends

Description	Craft@Hrs	Unit	Material $	Labor $	Equipment $	Total $
2"	P1@.500	Ea	295.00	17.90	—	312.90
2½"	P1@.600	Ea	398.00	21.50	—	419.50
3"	P1@.750	Ea	432.00	26.80	—	458.80
4"	P1@1.35	Ea	634.00	48.30	—	682.30
5"	P1@2.00	Ea	1,210.00	71.50	—	1,281.50
6"	P1@2.50	Ea	1,210.00	89.40	—	1,299.40

Class 125 bronze body globe valve, brazed joints

Description	Craft@Hrs	Unit	Material $	Labor $	Equipment $	Total $
1/2"	P1@.240	Ea	28.60	8.58	—	37.18
3/4"	P1@.300	Ea	38.10	10.70	—	48.80
1"	P1@.360	Ea	65.60	12.90	—	78.50
1¼"	P1@.480	Ea	92.30	17.20	—	109.50
1½"	P1@.540	Ea	125.00	19.30	—	144.30
2"	P1@.600	Ea	202.00	21.50	—	223.50

Class 125 iron body globe valve, flanged ends

Description	Craft@Hrs	Unit	Material $	Labor $	Equipment $	Total $
2½"	P1@.650	Ea	334.00	23.20	—	357.20
3"	P1@.750	Ea	401.00	26.80	—	427.80
4"	P1@1.35	Ea	541.00	48.30	—	589.30

200 PSIG iron body butterfly valve, lug type, lever operated

Description	Craft@Hrs	Unit	Material $	Labor $	Equipment $	Total $
2"	P1@.450	Ea	128.00	16.10	—	144.10
2½"	P1@.450	Ea	132.00	16.10	—	148.10
3"	P1@.550	Ea	139.00	19.70	—	158.70
4"	P1@.550	Ea	173.00	19.70	—	192.70

200 PSIG iron body butterfly valve, wafer type, lever operated

Description	Craft@Hrs	Unit	Material $	Labor $	Equipment $	Total $
2"	P1@.450	Ea	116.00	16.10	—	132.10
2½"	P1@.450	Ea	119.00	16.10	—	135.10
3"	P1@.550	Ea	128.00	19.70	—	147.70
4"	P1@.550	Ea	155.00	19.70	—	174.70

Description	Craft@Hrs	Unit	Material $	Labor $	Equipment $	Total $

Class 125 bronze body 2-piece ball valve, brazed joints

1/2"	P1@.240	Ea	7.53	8.58	—	16.11
3/4"	P1@.300	Ea	10.00	10.70	—	20.70
1"	P1@.360	Ea	17.30	12.90	—	30.20
1¼"	P1@.480	Ea	28.90	17.20	—	46.10
1½"	P1@.540	Ea	39.00	19.30	—	58.30
2"	P1@.600	Ea	49.40	21.50	—	70.90
3"	P1@1.50	Ea	338.00	53.60	—	391.60
4"	P1@1.80	Ea	443.00	64.40	—	507.40

Class 125 bronze body swing check valve, brazed joints

1/2"	P1@.240	Ea	16.90	8.58	—	25.48
3/4"	P1@.300	Ea	24.20	10.70	—	34.90
1"	P1@.360	Ea	31.70	12.90	—	44.60
1¼"	P1@.480	Ea	45.00	17.20	—	62.20
1½"	P1@.540	Ea	63.50	19.30	—	82.80
2"	P1@.600	Ea	106.00	21.50	—	127.50
2½"	P1@1.00	Ea	212.00	35.80	—	247.80
3"	P1@1.50	Ea	305.00	53.60	—	358.60

Class 125 iron body swing check valve, flanged ends

2"	P1@.500	Ea	140.00	17.90	—	157.90
2½"	P1@.600	Ea	188.00	21.50	—	209.50
3"	P1@.750	Ea	233.00	26.80	—	259.80
4"	P1@1.35	Ea	343.00	48.30	—	391.30

Class 125 iron body silent check valve, wafer type

2"	P1@.500	Ea	108.00	17.90	—	125.90
2½"	P1@.600	Ea	120.00	21.50	—	141.50
3"	P1@.750	Ea	138.00	26.80	—	164.80
4"	P1@1.35	Ea	185.00	48.30	—	233.30

Class 125 bronze body strainer, threaded ends

1/2"	P1@.230	Ea	26.40	8.22	—	34.62
3/4"	P1@.260	Ea	34.70	9.30	—	44.00
1"	P1@.330	Ea	42.50	11.80	—	54.30
1¼"	P1@.440	Ea	59.30	15.70	—	75.00
1½"	P1@.495	Ea	89.30	17.70	—	107.00
2"	P1@.550	Ea	154.00	19.70	—	173.70

Class 125 iron body strainer, flanged ends

2"	P1@.500	Ea	118.00	17.90	—	135.90
2½"	P1@.600	Ea	132.00	21.50	—	153.50
3"	P1@.750	Ea	152.00	26.80	—	178.80
4"	P1@1.35	Ea	260.00	48.30	—	308.30

Copper, Type M with Brazed Joints

Description	Craft@Hrs	Unit	Material $	Labor $	Equipment $	Total $

Installation of 2-way control valve, threaded joints

Description	Craft@Hrs	Unit	Material $	Labor $	Equipment $	Total $
1/2"	P1@.210	Ea	—	7.51	—	7.51
3/4"	P1@.275	Ea	—	9.83	—	9.83
1"	P1@.350	Ea	—	12.50	—	12.50
1¼"	P1@.430	Ea	—	15.40	—	15.40
1½"	P1@.505	Ea	—	18.10	—	18.10
2"	P1@.675	Ea	—	24.10	—	24.10
2½"	P1@.830	Ea	—	29.70	—	29.70
3"	P1@.990	Ea	—	35.40	—	35.40

Installation of 3-way control valve, threaded joints

Description	Craft@Hrs	Unit	Material $	Labor $	Equipment $	Total $
1/2"	P1@.260	Ea	—	9.30	—	9.30
3/4"	P1@.365	Ea	—	13.10	—	13.10
1"	P1@.475	Ea	—	17.00	—	17.00
1¼"	P1@.575	Ea	—	20.60	—	20.60
1½"	P1@.680	Ea	—	24.30	—	24.30
2"	P1@.910	Ea	—	32.50	—	32.50
2½"	P1@1.12	Ea	—	40.10	—	40.10
3"	P1@1.33	Ea	—	47.60	—	47.60

Companion flange, 150 pound cast brass

Description	Craft@Hrs	Unit	Material $	Labor $	Equipment $	Total $
2"	P1@.290	Ea	348.00	10.40	—	358.40
2½"	P1@.380	Ea	429.00	13.60	—	442.60
3"	P1@.460	Ea	429.00	16.40	—	445.40
4"	P1@.600	Ea	683.00	21.50	—	704.50

Bolt and gasket sets

Description	Craft@Hrs	Unit	Material $	Labor $	Equipment $	Total $
2"	P1@.500	Ea	3.86	17.90	—	21.76
2½"	P1@.650	Ea	4.51	23.20	—	27.71
3"	P1@.750	Ea	7.67	26.80	—	34.47
4"	P1@1.00	Ea	12.80	35.80	—	48.60

Thermometer with well

Description	Craft@Hrs	Unit	Material $	Labor $	Equipment $	Total $
7"	P1@.250	Ea	159.00	8.94	—	167.94
9"	P1@.250	Ea	164.00	8.94	—	172.94

Dial type pressure gauge

Description	Craft@Hrs	Unit	Material $	Labor $	Equipment $	Total $
2½"	P1@.200	Ea	33.10	7.15	—	40.25
3½"	P1@.200	Ea	43.80	7.15	—	50.95

Pressure/temperature tap

Description	Craft@Hrs	Unit	Material $	Labor $	Equipment $	Total $
tap	P1@.150	Ea	15.70	5.36	—	21.06

Description	Craft@Hrs	Unit	Material $	Labor $	Equipment $	Total $

Hanger with swivel assembly

Description	Craft@Hrs	Unit	Material $	Labor $	Equipment $	Total $
1/2"	P1@.250	Ea	3.74	8.94	—	12.68
3/4"	P1@.250	Ea	4.13	8.94	—	13.07
1"	P1@.250	Ea	4.33	8.94	—	13.27
1¼"	P1@.300	Ea	4.44	10.70	—	15.14
1½"	P1@.300	Ea	4.77	10.70	—	15.47
2"	P1@.300	Ea	4.98	10.70	—	15.68
2½"	P1@.350	Ea	6.78	12.50	—	19.28
3"	P1@.350	Ea	8.37	12.50	—	20.87
4"	P1@.350	Ea	9.19	12.50	—	21.69

Riser clamp

Description	Craft@Hrs	Unit	Material $	Labor $	Equipment $	Total $
1/2"	P1@.100	Ea	2.41	3.58	—	5.99
3/4"	P1@.100	Ea	3.55	3.58	—	7.13
1"	P1@.100	Ea	3.59	3.58	—	7.17
1¼"	P1@.105	Ea	4.33	3.75	—	8.08
1½"	P1@.110	Ea	4.58	3.93	—	8.51
2"	P1@.115	Ea	4.84	4.11	—	8.95
2½"	P1@.120	Ea	5.09	4.29	—	9.38
3"	P1@.120	Ea	5.52	4.29	—	0.01
4"	P1@.125	Ea	7.03	4.47	—	11.50

Copper, Type M with Soft-Soldered Joints

Type M hard-drawn copper pipe with wrought copper fittings and soft-soldered joints is used in a wide variety of plumbing and HVAC systems such as potable water, heating hot water, chilled water, and A.C. condensate.

Soft-soldered joints are those made with solders having melting points in the 350 degree F. to 500 degree F. range. Maximum working pressure/temperature relationships for soft-soldered joints are approximately as follows:

Maximum Working Pressures (PSIG)*				
Soft-solder Type	Water Temperature (degrees F.)	Nominal Pipe Size (inches)		
		Up to 1	1¼ to 2	2½ x 4
50-50 tin-lead**	100	200	175	150
	150	150	125	100
	200	100	90	75
	250	85	75	50
95-5 tin-antimony	100	500	400	300
	150	400	350	275
	200	300	250	200
	250	200	175	150

*For copper pipe and solder-type fittings using soft-solders melting at approximately 350 degrees F. to 500 degrees F.

**The use of any solder containing lead is not allowed in potable water systems.

This section has been arranged to save the estimator's time by including all normally-used system components such as pipe, fittings, valves, hanger assemblies, riser clamps and miscellaneous items under one heading. Additional items can be found under "Plumbing and Piping Specialties." The cost estimates in this section are based on the conditions, limitations and wage rates described in the section "How to Use This Book" beginning on page 5.

Description	Craft@Hrs	Unit	Material $	Labor $	Equipment $	Total $

Type M copper pipe with soft-soldered joints installed horizontally.
Complete installation including six to twelve wrought copper tees and six to twelve wrought copper elbows every 100', and hangers spaced to meet plumbing code (*a tee & elbow every 8.3' for ½" dia., a tee and elbow every 16.6' for 4" dia.*).

Description	Craft@Hrs	Unit	Material $	Labor $	Equipment $	Total $
1/2"	P1@.110	LF	1.89	3.93	—	5.82
3/4"	P1@.120	LF	2.92	4.29	—	7.21
1"	P1@.140	LF	6.08	5.01	—	11.09
1¼"	P1@.160	LF	10.30	5.72	—	16.02
1½"	P1@.180	LF	12.20	6.44	—	18.64
2"	P1@.210	LF	25.40	7.51	—	32.91
2½"	P1@.240	LF	37.20	8.58	—	45.78
3"	P1@.270	LF	46.80	9.66	—	56.46
4"	P1@.312	LF	68.90	11.20	—	80.10

Description	Craft@Hrs	Unit	Material $	Labor $	Equipment $	Total $

Type M copper pipe with soft-soldered joints, installed risers.
Complete installation including a wrought copper reducing tee every floor and a riser clamp every other floor.

Description	Craft@Hrs	Unit	Material $	Labor $	Equipment $	Total $
1/2"	P1@.060	LF	1.60	2.15	—	3.75
3/4"	P1@.070	LF	2.60	2.50	—	5.10
1"	P1@.075	LF	5.09	2.68	—	7.77
1¼"	P1@.080	LF	7.50	2.86	—	10.36
1½"	P1@.100	LF	9.90	3.58	—	13.48
2"	P1@.130	LF	13.10	4.65	—	17.75
2½"	P1@.150	LF	28.50	5.36	—	33.86
3"	P1@.170	LF	34.90	6.08	—	40.98
4"	P1@.190	LF	51.80	6.79	—	58.59

Type M copper pipe with soft-soldered joints, pipe only

Description	Craft@Hrs	Unit	Material $	Labor $	Equipment $	Total $
1/2"	P1@.032	LF	1.13	1.14	—	2.27
3/4"	P1@.035	LF	1.85	1.25	—	3.10
1"	P1@.038	LF	2.90	1.36	—	4.26
1¼"	P1@.042	LF	4.36	1.50	—	5.86
1½"	P1@.046	LF	6.11	1.64	—	7.75
2"	P1@.053	LF	11.20	1.90	—	13.10
2½"	P1@.060	LF	15.50	2.15	—	17.65
3"	P1@.066	LF	19.60	2.36	—	21.96
4"	P1@.080	LF	31.30	2.86	—	34.16

Type M copper 45 degree ell C x C with soft-soldered joints

Description	Craft@Hrs	Unit	Material $	Labor $	Equipment $	Total $
1/2"	P1@.107	Ea	.99	3.83	—	4.82
3/4"	P1@.150	Ea	1.69	5.36	—	7.05
1"	P1@.193	Ea	5.54	6.90	—	12.44
1¼"	P1@.236	Ea	6.25	8.44	—	14.69
1½"	P1@.278	Ea	11.10	9.94	—	21.04
2"	P1@.371	Ea	18.30	13.30	—	31.60
2½"	P1@.457	Ea	46.70	16.30	—	63.00
3"	P1@.543	Ea	61.60	19.40	—	81.00
4"	P1@.714	Ea	119.00	25.50	—	144.50

Type M copper 90 degree ell C x C with soft-soldered joints

Description	Craft@Hrs	Unit	Material $	Labor $	Equipment $	Total $
1/2"	P1@.107	Ea	.52	3.83	—	4.35
3/4"	P1@.150	Ea	1.10	5.36	—	6.46
1"	P1@.193	Ea	4.21	6.90	—	11.11
1¼"	P1@.236	Ea	6.83	8.44	—	15.27
1½"	P1@.278	Ea	10.40	9.94	—	20.34
2"	P1@.371	Ea	18.60	13.30	—	31.90
2½"	P1@.457	Ea	38.60	16.30	—	54.90
3"	P1@.543	Ea	48.80	19.40	—	68.20
4"	P1@.714	Ea	104.00	25.50	—	129.50

Copper, Type M with Soft-Soldered Joints

Description	Craft@Hrs	Unit	Material $	Labor $	Equipment $	Total $

Type M copper 90 degree ell Ftg. x C with soft-soldered joint

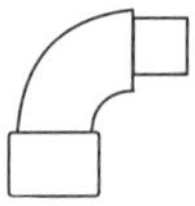

Description	Craft@Hrs	Unit	Material $	Labor $	Equipment $	Total $
1/2"	P1@.107	Ea	.91	3.83	—	4.74
3/4"	P1@.150	Ea	2.00	5.36	—	7.36
1"	P1@.193	Ea	5.98	6.90	—	12.88
1¼"	P1@.236	Ea	8.33	8.44	—	16.77
1½"	P1@.278	Ea	12.50	9.94	—	22.44
2"	P1@.371	Ea	25.00	13.30	—	38.30
2½"	P1@.457	Ea	52.50	16.30	—	68.80
3"	P1@.543	Ea	59.00	19.40	—	78.40
4"	P1@.714	Ea	119.00	25.50	—	144.50

Type M copper tee C x C x C with soft-soldered joints

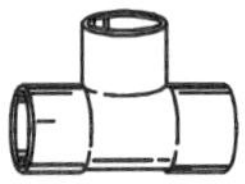

Description	Craft@Hrs	Unit	Material $	Labor $	Equipment $	Total $
1/2"	P1@.129	Ea	.89	4.61	—	5.50
3/4"	P1@.181	Ea	2.40	6.47	—	8.87
1"	P1@.233	Ea	9.98	8.33	—	18.31
1¼"	P1@.285	Ea	15.80	10.20	—	26.00
1½"	P1@.337	Ea	23.00	12.10	—	35.10
2"	P1@.449	Ea	36.80	16.10	—	52.90
2½"	P1@.552	Ea	74.60	19.70	—	94.30
3"	P1@.656	Ea	115.00	23.50	—	138.50
4"	P1@.863	Ea	144.00	30.90	—	174.90

Type M copper branch reducing tee C x C x C with soft-soldered joints

Description	Craft@Hrs	Unit	Material $	Labor $	Equipment $	Total $
1/2" x 3/8"	P1@.129	Ea	1.98	4.61	—	6.59
3/4" x 1/2"	P1@.182	Ea	4.20	6.51	—	10.71
1" x 1/2"	P1@.195	Ea	8.25	6.97	—	15.22
1" x 3/4"	P1@.219	Ea	8.25	7.83	—	16.08
1¼" x 1/2"	P1@.225	Ea	12.60	8.05	—	20.65
1¼" x 3/4"	P1@.247	Ea	12.60	8.83	—	21.43
1¼" x 1"	P1@.268	Ea	12.60	9.58	—	22.18
1½" x 1/2"	P1@.272	Ea	15.40	9.73	—	25.13
1½" x 3/4"	P1@.287	Ea	15.40	10.30	—	25.70
1½" x 1"	P1@.302	Ea	15.40	10.80	—	26.20
1½" x 1¼"	P1@.317	Ea	15.40	11.30	—	26.70
2" x 1/2"	P1@.325	Ea	20.60	11.60	—	32.20
2" x 3/4"	P1@.348	Ea	20.60	12.40	—	33.00
2" x 1"	P1@.373	Ea	20.60	13.30	—	33.90
2" x 1¼"	P1@.398	Ea	20.60	14.20	—	34.80
2" x 1½"	P1@.422	Ea	20.60	15.10	—	35.70
2½" x 1/2"	P1@.437	Ea	62.10	15.60	—	77.70
2½" x 3/4"	P1@.453	Ea	62.10	16.20	—	78.30
2½" x 1"	P1@.469	Ea	62.10	16.80	—	78.90
2½" x 1½"	P1@.502	Ea	62.10	18.00	—	80.10
2½" x 2"	P1@.519	Ea	62.10	18.60	—	80.70
3" x 1¼"	P1@.525	Ea	70.20	18.80	—	89.00
3" x 2"	P1@.585	Ea	72.90	20.90	—	93.80
3" x 2½"	P1@.617	Ea	72.90	22.10	—	95.00
4" x 1¼"	P1@.625	Ea	127.00	22.40	—	149.40
4" x 1½"	P1@.671	Ea	127.00	24.00	—	151.00
4" x 2"	P1@.718	Ea	127.00	25.70	—	152.70
4" x 2½"	P1@.765	Ea	127.00	27.40	—	154.40
4" x 3"	P1@.811	Ea	127.00	29.00	—	156.00

Description	Craft@Hrs	Unit	Material $	Labor $	Equipment $	Total $

Type M copper reducer with soft-soldered joints

Description	Craft@Hrs	Unit	Material $	Labor $	Equipment $	Total $
1/2" x 3/8"	P1@.124	Ea	.71	4.43	—	5.14
3/4" x 1/2"	P1@.129	Ea	1.34	4.61	—	5.95
1" x 3/4"	P1@.172	Ea	2.41	6.15	—	8.56
1¼" x 1/2"	P1@.172	Ea	4.05	6.15	—	10.20
1¼" x 3/4"	P1@.193	Ea	4.05	6.90	—	10.95
1¼" x 1"	P1@.215	Ea	4.05	7.69	—	11.74
1½" x 1/2"	P1@.193	Ea	6.19	6.90	—	13.09
1½" x 3/4"	P1@.214	Ea	6.19	7.65	—	13.84
1½" x 1"	P1@.236	Ea	6.19	8.44	—	14.63
1½" x 1¼"	P1@.257	Ea	6.19	9.19	—	15.38
2" x 1/2"	P1@.239	Ea	9.60	8.55	—	18.15
2" x 3/4"	P1@.261	Ea	9.60	9.33	—	18.93
2" x 1"	P1@.282	Ea	9.60	10.10	—	19.70
2" x 1¼"	P1@.304	Ea	9.60	10.90	—	20.50
2" x 1½"	P1@.325	Ea	9.60	11.60	—	21.20
2½" x 1"	P1@.325	Ea	26.20	11.60	—	37.80
2½" x 1¼"	P1@.347	Ea	26.20	12.40	—	38.60
2½" x 1½"	P1@.368	Ea	26.20	13.20	—	39.40
2½" x 2"	P1@.414	Ea	26.20	14.80	—	41.00
3" x 2"	P1@.457	Ea	30.60	16.30	—	46.90
3" x 2½"	P1@.500	Ea	30.60	17.90	—	48.50
4" x 2"	P1@.543	Ea	59.00	19.40	—	78.40
4" x 2½"	P1@.586	Ea	59.00	21.00	—	80.00
4" x 3"	P1@.629	Ea	59.00	22.50	—	81.50

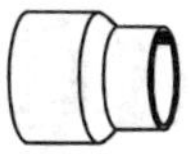

Type M copper adapter C x MPT with soft-soldered joint

Description	Craft@Hrs	Unit	Material $	Labor $	Equipment $	Total $
1/2"	P1@.075	Ea	1.76	2.68	—	4.44
3/4"	P1@.105	Ea	2.75	3.75	—	6.50
1"	P1@.134	Ea	6.83	4.79	—	11.62
1¼"	P1@.164	Ea	12.80	5.86	—	18.66
1½"	P1@.194	Ea	14.30	6.94	—	21.24
2"	P1@.259	Ea	18.50	9.26	—	27.76
2½"	P1@.319	Ea	60.10	11.40	—	71.50
3"	P1@.378	Ea	99.80	13.50	—	113.30

Type M copper adapter C x FPT with soft-soldered joint

Description	Craft@Hrs	Unit	Material $	Labor $	Equipment $	Total $
1/2"	P1@.075	Ea	2.93	2.68	—	5.61
3/4"	P1@.105	Ea	5.30	3.75	—	9.05
1"	P1@.134	Ea	8.03	4.79	—	12.82
1¼"	P1@.164	Ea	16.80	5.86	—	22.66
1½"	P1@.194	Ea	20.40	6.94	—	27.34
2"	P1@.259	Ea	31.30	9.26	—	40.56
2½"	P1@.319	Ea	78.80	11.40	—	90.20
3"	P1@.378	Ea	109.00	13.50	—	122.50

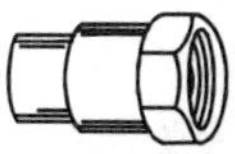

Copper, Type M with Soft-Soldered Joints

Description	Craft@Hrs	Unit	Material $	Labor $	Equipment $	Total $

Type M copper flush bushing with soft-soldered joints

Description	Craft@Hrs	Unit	Material $	Labor $	Equipment $	Total $
1/2" x 3/8"	P1@.124	Ea	1.79	4.43	—	6.22
3/4" x 1/2"	P1@.129	Ea	3.04	4.61	—	7.65
1" x 3/4"	P1@.172	Ea	7.88	6.15	—	14.03
1" x 1/2"	P1@.172	Ea	9.00	6.15	—	15.15
1¼" x 1"	P1@.215	Ea	9.38	7.69	—	17.07
1½" x 1¼"	P1@.257	Ea	11.90	9.19	—	21.09
2" x 1½"	P1@.325	Ea	22.40	11.60	—	34.00

Type M copper union with soft-soldered joint

Description	Craft@Hrs	Unit	Material $	Labor $	Equipment $	Total $
1/2"	P1@.121	Ea	7.14	4.33	—	11.47
3/4"	P1@.170	Ea	8.55	6.08	—	14.63
1"	P1@.218	Ea	19.10	7.80	—	26.90
1¼"	P1@.267	Ea	34.80	9.55	—	44.35
1½"	P1@.315	Ea	42.80	11.30	—	54.10
2"	P1@.421	Ea	63.80	15.10	—	78.90

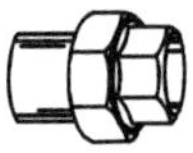

Type M copper dielectric union with soft-soldered joint

Description	Craft@Hrs	Unit	Material $	Labor $	Equipment $	Total $
1/2"	P1@.121	Ea	6.70	4.33	—	11.03
3/4"	P1@.170	Ea	6.70	6.08	—	12.78
1"	P1@.218	Ea	13.80	7.80	—	21.60
1¼"	P1@.267	Ea	23.60	9.55	—	33.15
1½"	P1@.315	Ea	32.50	11.30	—	43.80
2"	P1@.421	Ea	44.90	15.10	—	60.00

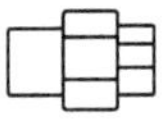

Type M copper cap with soft-soldered joint

Description	Craft@Hrs	Unit	Material $	Labor $	Equipment $	Total $
1/2"	P1@.069	Ea	.58	2.47	—	3.05
3/4"	P1@.096	Ea	1.19	3.43	—	4.62
1"	P1@.124	Ea	2.69	4.43	—	7.12
1¼"	P1@.151	Ea	3.86	5.40	—	9.26
1½"	P1@.178	Ea	5.63	6.37	—	12.00
2"	P1@.237	Ea	10.20	8.48	—	18.68
2½"	P1@.292	Ea	25.70	10.40	—	36.10
3"	P1@.347	Ea	40.20	12.40	—	52.60
4"	P1@.457	Ea	61.70	16.30	—	78.00

Type M copper coupling with soft-soldered joints

Description	Craft@Hrs	Unit	Material $	Labor $	Equipment $	Total $
1/2"	P1@.107	Ea	.37	3.83	—	4.20
3/4"	P1@.150	Ea	.93	5.36	—	6.29
1"	P1@.193	Ea	2.54	6.90	—	9.44
1¼"	P1@.236	Ea	4.45	8.44	—	12.89
1½"	P1@.278	Ea	6.44	9.94	—	16.38
2"	P1@.371	Ea	9.83	13.30	—	23.13
2½"	P1@.457	Ea	23.00	16.30	—	39.30
3"	P1@.543	Ea	35.10	19.40	—	54.50
4"	P1@.714	Ea	68.90	25.50	—	94.40

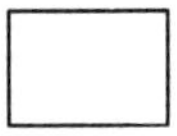

Description	Craft@Hrs	Unit	Material $	Labor $	Equipment $	Total $

Class 125 bronze body gate valve with soft-soldered joints

Description	Craft@Hrs	Unit	Material $	Labor $	Equipment $	Total $
1/2"	P1@.200	Ea	15.40	7.15	—	22.55
3/4"	P1@.249	Ea	19.30	8.90	—	28.20
1"	P1@.299	Ea	27.10	10.70	—	37.80
1¼"	P1@.398	Ea	34.30	14.20	—	48.50
1½"	P1@.448	Ea	46.10	16.00	—	62.10
2"	P1@.498	Ea	77.00	17.80	—	94.80
2½"	P1@.830	Ea	128.00	29.70	—	157.70
3"	P1@1.24	Ea	184.00	44.30	—	228.30

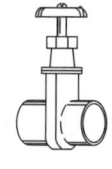

Class 125 iron body gate valve with flanged ends

Description	Craft@Hrs	Unit	Material $	Labor $	Equipment $	Total $
2"	P1@.500	Ea	295.00	17.90	—	312.90
2½"	P1@.600	Ea	398.00	21.50	—	419.50
3"	P1@.750	Ea	432.00	26.80	—	458.80
4"	P1@1.35	Ea	634.00	48.30	—	682.30
5"	P1@2.00	Ea	1,210.00	71.50	—	1,281.50
6"	P1@2.50	Ea	1,210.00	89.40	—	1,299.40

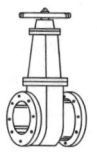

Class 125 bronze body globe valve with soft-soldered joints

Description	Craft@Hrs	Unit	Material $	Labor $	Equipment $	Total $
1/2"	P1@.200	Ea	28.60	7.15	—	35.75
3/4"	P1@.249	Ea	38.10	8.90	—	47.00
1"	P1@.299	Ea	65.60	10.70	—	76.30
1¼"	P1@.398	Ea	92.30	14.20	—	106.50
1½"	P1@.448	Ea	125.00	16.00	—	141.00
2"	P1@.498	Ea	202.00	17.80	—	219.80

Class 125 iron body globe valve with flanged ends

Description	Craft@Hrs	Unit	Material $	Labor $	Equipment $	Total $
2½"	P1@.600	Ea	334.00	21.50	—	355.50
3"	P1@.750	Ea	401.00	26.80	—	427.80
4"	P1@1.35	Ea	538.00	48.30	—	586.30

200 PSIG iron body butterfly valve, lug type, lever operated

Description	Craft@Hrs	Unit	Material $	Labor $	Equipment $	Total $
2"	P1@.450	Ea	128.00	16.10	—	144.10
2½"	P1@.450	Ea	132.00	16.10	—	148.10
3"	P1@.550	Ea	139.00	19.70	—	158.70
4"	P1@.550	Ea	173.00	19.70	—	192.70

200 PSIG iron body butterfly valve, wafer type, lever operated

Description	Craft@Hrs	Unit	Material $	Labor $	Equipment $	Total $
2"	P1@.450	Ea	116.00	16.10	—	132.10
2½"	P1@.450	Ea	119.00	16.10	—	135.10
3"	P1@.550	Ea	128.00	19.70	—	147.70
4"	P1@.550	Ea	155.00	19.70	—	174.70

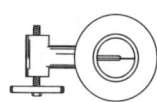

Copper, Type M with Soft-Soldered Joints

Description	Craft@Hrs	Unit	Material $	Labor $	Equipment $	Total $

Class 125 bronze body 2-piece ball valve with soft-soldered joints

Description	Craft@Hrs	Unit	Material $	Labor $	Equipment $	Total $
1/2"	P1@.200	Ea	7.53	7.15	—	14.68
3/4"	P1@.249	Ea	10.00	8.90	—	18.90
1"	P1@.299	Ea	17.30	10.70	—	28.00
1¼"	P1@.398	Ea	28.90	14.20	—	43.10
1½"	P1@.448	Ea	39.00	16.00	—	55.00
2"	P1@.498	Ea	49.40	17.80	—	67.20
3"	P1@1.24	Ea	338.00	44.30	—	382.30
4"	P1@1.45	Ea	443.00	51.90	—	494.90

Class 125 bronze body swing check valve with soft-soldered joints

Description	Craft@Hrs	Unit	Material $	Labor $	Equipment $	Total $
1/2"	P1@.200	Ea	16.90	7.15	—	24.05
3/4"	P1@.249	Ea	24.20	8.90	—	33.10
1"	P1@.299	Ea	31.70	10.70	—	42.40
1¼"	P1@.398	Ea	45.00	14.20	—	59.20
1½"	P1@.448	Ea	63.50	16.00	—	79.50
2"	P1@.498	Ea	106.00	17.80	—	123.80
2½"	P1@.830	Ea	212.00	29.70	—	241.70
3"	P1@1.24	Ea	305.00	44.30	—	349.30

Class 125 iron body swing check valve, flanged ends

Description	Craft@Hrs	Unit	Material $	Labor $	Equipment $	Total $
2"	P1@.500	Ea	148.00	17.90	—	165.90
2½"	P1@.600	Ea	188.00	21.50	—	209.50
3"	P1@.750	Ea	233.00	26.80	—	259.80
4"	P1@1.35	Ea	343.00	48.30	—	391.30

Class 125 iron body silent check valve, wafer type

Description	Craft@Hrs	Unit	Material $	Labor $	Equipment $	Total $
2"	P1@.500	Ea	108.00	17.90	—	125.90
2½"	P1@.600	Ea	120.00	21.50	—	141.50
3"	P1@.750	Ea	138.00	26.80	—	164.80
4"	P1@1.35	Ea	185.00	48.30	—	233.30

Class 125 bronze body strainer, threaded ends

Description	Craft@Hrs	Unit	Material $	Labor $	Equipment $	Total $
1/2"	P1@.230	Ea	26.40	8.22	—	34.62
3/4"	P1@.260	Ea	34.70	9.30	—	44.00
1"	P1@.330	Ea	42.50	11.80	—	54.30
1¼"	P1@.440	Ea	59.30	15.70	—	75.00
1½"	P1@.495	Ea	89.30	17.70	—	107.00
2"	P1@.550	Ea	154.00	19.70	—	173.70

Class 125 iron body strainer, flanged ends

Description	Craft@Hrs	Unit	Material $	Labor $	Equipment $	Total $
2"	P1@.500	Ea	118.00	17.90	—	135.90
2½"	P1@.600	Ea	132.00	21.50	—	153.50
3"	P1@.750	Ea	152.00	26.80	—	178.80
4"	P1@1.35	Ea	260.00	48.30	—	308.30

Description	Craft@Hrs	Unit	Material $	Labor $	Equipment $	Total $

Installation of 2-way control valve, threaded joints

Description	Craft@Hrs	Unit	Material $	Labor $	Equipment $	Total $
1/2"	P1@.210	Ea	—	7.51	—	7.51
3/4"	P1@.275	Ea	—	9.83	—	9.83
1"	P1@.350	Ea	—	12.50	—	12.50
1¼"	P1@.430	Ea	—	15.40	—	15.40
1½"	P1@.505	Ea	—	18.10	—	18.10
2"	P1@.675	Ea	—	24.10	—	24.10
2½"	P1@.830	Ea	—	29.70	—	29.70
3"	P1@.990	Ea	—	35.40	—	35.40

Installation of 3-way control valve, threaded joints

Description	Craft@Hrs	Unit	Material $	Labor $	Equipment $	Total $
1/2"	P1@.260	Ea	—	9.30	—	9.30
3/4"	P1@.365	Ea	—	13.10	—	13.10
1"	P1@.475	Ea	—	17.00	—	17.00
1¼"	P1@.575	Ea	—	20.60	—	20.60
1½"	P1@.680	Ea	—	24.30	—	24.30
2"	P1@.910	Ea	—	32.50	—	32.50
2½"	P1@1.12	Ea	—	40.10	—	40.10
3"	P1@1.33	Ea	—	47.60	—	47.60

Companion flange, 150 pound cast brass

Description	Craft@Hrs	Unit	Material $	Labor $	Equipment $	Total $
2"	P1@.290	Ea	359.00	10.40	—	369.40
2½"	P1@.380	Ea	439.00	13.60	—	452.60
3"	P1@.460	Ea	439.00	16.40	—	455.40
4"	P1@.600	Ea	702.00	21.50	—	723.50

Bolt and gasket sets

Description	Craft@Hrs	Unit	Material $	Labor $	Equipment $	Total $
2"	P1@.500	Ea	3.89	17.90	—	21.79
2½"	P1@.650	Ea	4.54	23.20	—	27.74
3"	P1@.750	Ea	7.73	26.80	—	34.53
4"	P1@1.00	Ea	12.80	35.80	—	48.60

Thermometer with well

Description	Craft@Hrs	Unit	Material $	Labor $	Equipment $	Total $
7"	P1@.250	Ea	159.00	8.94	—	167.94
9"	P1@.250	Ea	164.00	8.94	—	172.94

Dial type pressure gauge

Description	Craft@Hrs	Unit	Material $	Labor $	Equipment $	Total $
2½"	P1@.200	Ea	33.10	7.15	—	40.25
3½"	P1@.200	Ea	43.80	7.15	—	50.95

Pressure/temperature tap

Description	Craft@Hrs	Unit	Material $	Labor $	Equipment $	Total $
tap	P1@.150	Ea	15.70	5.36	—	21.06

Copper, Type M with Soft-Soldered Joints

Description	Craft@Hrs	Unit	Material $	Labor $	Equipment $	Total $

Hanger with swivel assembly

Description	Craft@Hrs	Unit	Material $	Labor $	Equipment $	Total $
1/2"	P1@.250	Ea	3.74	8.94	—	12.68
3/4"	P1@.250	Ea	4.13	8.94	—	13.07
1"	P1@.250	Ea	4.33	8.94	—	13.27
1¼"	P1@.300	Ea	4.44	10.70	—	15.14
1½"	P1@.300	Ea	4.77	10.70	—	15.47
2"	P1@.300	Ea	4.98	10.70	—	15.68
2½"	P1@.350	Ea	6.78	12.50	—	19.28
3"	P1@.350	Ea	8.37	12.50	—	20.87
4"	P1@.350	Ea	9.19	12.50	—	21.69

Riser clamp

Description	Craft@Hrs	Unit	Material $	Labor $	Equipment $	Total $
1/2"	P1@.100	Ea	2.41	3.58	—	5.99
3/4"	P1@.100	Ea	3.55	3.58	—	7.13
1"	P1@.100	Ea	3.59	3.58	—	7.17
1¼"	P1@.105	Ea	4.33	3.75	—	8.08
1½"	P1@.110	Ea	4.58	3.93	—	8.51
2"	P1@.115	Ea	4.84	4.11	—	8.95
2½"	P1@.120	Ea	5.09	4.29	—	9.38
3"	P1@.120	Ea	5.52	4.29	—	9.81
4"	P1@.125	Ea	7.03	4.47	—	11.50

Hard drawn rigid copper roll-grooved pipe with factory grooved copper fittings is commonly used for domestic or potable water distribution systems.

Maximum operating pressure for fittings is 300 psi (2065kPa). Operating temperature for fittings ranges from -30F to 250F. (-34C – 121C)

Consult manufacturer for pipe pressure and temperature ratings and for the various combinations of pipe, fittings and system applications.

The cost estimates in this section are based on the conditions, limitations and wage rates described in the section "How to Use This Book" beginning on page 5.

Description	Craft@Hrs	Unit	Material $	Labor $	Equipment $	Total $

Type K Copper pipe only, roll grooved, no hangers or fittings

2"	P1@.080	LF	14.70	2.86	—	17.56
2½"	P1@.090	LF	21.40	3.22	—	24.62
3"	P1@.105	LF	29.00	3.75	—	32.75
4"	P1@.145	LF	45.40	5.19	—	50.59
6"	P1@.195	LF	69.30	6.97	—	76.27

Type L Copper pipe only, roll grooved, no hangers or fittings

2"	P1@.070	LF	13.30	2.50	—	15.80
2½"	P1@.070	LF	19.30	2.50	—	21.80
3"	P1@.090	LF	26.10	3.22	—	29.32
4"	P1@.130	LF	40.80	4.65	—	45.45
6"	P1@.180	LF	62.40	6.44	—	68.84

90 degree copper elbow, roll grooved, style #610, (Victaulic)

2"	P1@.410	Ea	72.20	14.70	—	86.90
2½"	P1@.430	Ea	79.40	15.40	—	94.80
3"	P1@.500	Ea	112.00	17.90	—	129.90
4"	P1@.690	Ea	239.00	24.70	—	263.70
6"	P1@1.19	Ea	1,410.00	42.60	—	1,452.60

45 degree copper elbow, roll grooved, style #611, (Victaulic)

2"	P1@.410	Ea	63.50	14.70	—	78.20
2½"	P1@.430	Ea	69.20	15.40	—	84.60
3"	P1@.500	Ea	94.90	17.90	—	112.80
4"	P1@.690	Ea	218.00	24.70	—	242.70
6"	P1@1.19	Ea	1,200.00	42.60	—	1,242.60

Tee, copper, roll grooved, style #620, (Victaulic)

2"	P1@.510	Ea	123.00	18.20	—	141.20
2½"	P1@.540	Ea	130.00	19.30	—	149.30
3"	P1@.630	Ea	196.00	22.50	—	218.50
4"	P1@.860	Ea	433.00	30.80	—	463.80
6"	P1@1.49	Ea	1,610.00	53.30	—	1,663.30

Copper, Type K & L with Roll Grooved Joints

Description	Craft@Hrs	Unit	Material $	Labor $	Equipment $	Total $

Reducing tee, copper, roll grooved, style #625, (Victaulic)

Description	Craft@Hrs	Unit	Material $	Labor $	Equipment $	Total $
2½" x 2"	P1@.540	Ea	202.00	19.30	—	221.30
3" x 2½"	P1@.630	Ea	215.00	22.50	—	237.50
4" x 3"	P1@.860	Ea	504.00	30.80	—	534.80

Reducing tee, copper, roll grooved, style #626, (Victaulic)

Description	Craft@Hrs	Unit	Material $	Labor $	Equipment $	Total $
2½" x 1"	P1@.500	Ea	255.00	17.90	—	272.90
2½" x 1½"	P1@.520	Ea	255.00	18.60	—	273.60
3" x 1½"	P1@.600	Ea	269.00	21.50	—	290.50

Reducer, copper, roll grooved, style #650, (Victaulic)

Description	Craft@Hrs	Unit	Material $	Labor $	Equipment $	Total $
2½" x 2"	P1@.430	Ea	108.00	15.40	—	123.40
3" x 2"	P1@.500	Ea	108.00	17.90	—	125.90
3" x 2½"	P1@.520	Ea	108.00	18.60	—	126.60
4" x 2"	P1@.690	Ea	210.00	24.70	—	234.70
4" x 3"	P1@.720	Ea	210.00	25.70	—	235.70
6" x 3"	P1@1.20	Ea	731.00	42.90	—	773.90

Roll-grooved copper coupling with gasket, style #650, (Victaulic)

Description	Craft@Hrs	Unit	Material $	Labor $	Equipment $	Total $
2"	P1@.300	Ea	35.50	10.70	—	46.20
2 1/2"	P1@.350	Ea	40.00	12.50	—	52.50
3"	P1@.400	Ea	43.30	14.30	—	57.60
4"	P1@.500	Ea	55.70	17.90	—	73.60
5"	P1@.600	Ea	113.00	21.50	—	134.50
6"	P1@.700	Ea	128.00	25.00	—	153.00
8"	P1@.900	Ea	200.00	32.20	—	232.20
10"	P1@1.10	Ea	294.00	39.30	—	333.30
12"	P1@1.35	Ea	441.00	48.30	—	489.30

Flange Adapter, copper, roll grooved, style #641, (Victaulic)

Description	Craft@Hrs	Unit	Material $	Labor $	Equipment $	Total $
2½"	P1@.300	Ea	212.00	10.70	—	222.70
3"	P1@.400	Ea	227.00	14.30	—	241.30
4"	P1@.550	Ea	311.00	19.70	—	330.70

Butterfly Valve, brass, roll grooved, lever handle, style #608, (Victaulic)

Description	Craft@Hrs	Unit	Material $	Labor $	Equipment $	Total $
2½"	P1@.450	Ea	659.00	16.10	—	675.10
3"	P1@.450	Ea	845.00	16.10	—	861.10
4"	P1@.550	Ea	1,150.00	19.70	—	1,169.70

PVC (Polyvinyl Chloride) Schedule 40 pipe with Schedule 40 socket-type fittings and solvent-welded joints is widely used in process, chemical, A.C. condensate, potable water and irrigation piping systems. PVC is available in Type I (normal impact) and Type II (high impact). Type I has a maximum temperature rating of 150 degrees F., and Type II is rated at 140 degrees F.

Consult the manufacturers for maximum pressure ratings and recommended joint solvents for specific applications.

Because of the current unreliability of many plastic valves, most engineers specify standard bronze-body and iron-body valves for use in PVC water piping systems. Plastic valves, however, have to be used in systems conveying liquids that may be injurious to metallic valves.

This section has been arranged to save the estimator's time by including all normally-used system components such as pipe, fittings, valves, hanger assemblies, riser clamps and miscellaneous items under one heading. Additional items can be found under "Plumbing and Piping Specialties." The cost estimates in this section are based on the conditions, limitations and wage rates described in the section "How to Use This Book" beginning on page 5.

Equipment cost, where shown, is $110 per hour for a 10-ton hydraulic truck-mounted crane.

Description	Craft@Hrs	Unit	Material $	Labor $	Equipment $	Total $

Schedule 40 PVC pipe assembly with solvent-weld joints installed horizontally. Complete installation including three to eight tees and three to eight elbows every 100' and hangers spaced to meet plumbing code (a tee and elbow every 12' for 1/2" dia., a tee and elbow every 40' for 8" dia.).

Description	Craft@Hrs	Unit	Material $	Labor $	Equipment $	Total $
½"	P1@.084	LF	1.50	3.00	—	4.50
3/4"	P1@.091	LF	1.81	3.25	—	5.06
1"	P1@.099	LF	1.87	3.54	—	5.41
1¼"	P1@.114	LF	1.96	4.08	—	6.04
1½"	P1@.121	LF	2.23	4.33	—	6.56
2"	P1@.130	LF	2.75	4.65	—	7.40
2½"	P1@.134	LF	4.45	4.79	—	9.24
3"	P1@.146	LF	6.97	5.22	—	12.19
4"	P1@.177	LF	8.82	6.33	—	15.15
6"	P1@.231	LF	14.80	8.26	—	23.06
8"	P1@.236	LF	26.20	8.44	—	34.64

Schedule 40 PVC pipe assembly with solvent-weld joints, installed risers. Complete installation including a reducing tee every floor and a riser clamp every other.

Description	Craft@Hrs	Unit	Material $	Labor $	Equipment $	Total $
1/2"	P1@.040	LF	.74	1.43	—	2.17
3/4"	P1@.047	LF	.97	1.68	—	2.65
1"	P1@.052	LF	1.24	1.86	—	3.10
1¼"	P1@.061	LF	1.56	2.18	—	3.74
1½"	P1@.069	LF	1.91	2.47	—	4.38
2"	P1@.077	LF	2.51	2.75	—	5.26
2½"	P1@.088	LF	3.99	3.15	—	7.14
3"	P1@.099	LF	6.07	3.54	—	9.61
4"	P1@.125	LF	7.79	4.47	—	12.26
6"	P1@.202	LF	16.30	7.22	—	23.52
8"	P1@.266	LF	30.10	9.51	—	39.61

PVC, Schedule 40, with Solvent-Weld Joints

Description	Craft@Hrs	Unit	Material $	Labor $	Equipment $	Total $

Schedule 40 PVC pipe with solvent-weld joints

Description	Craft@Hrs	Unit	Material $	Labor $	Equipment $	Total $
1/2"	P1@.020	LF	.39	.72	—	1.11
3/4"	P1@.025	LF	.60	.89	—	1.49
1"	P1@.030	LF	.80	1.07	—	1.87
1¼"	P1@.035	LF	.98	1.25	—	2.23
1½"	P1@.040	LF	1.16	1.43	—	2.59
2"	P1@.045	LF	1.56	1.61	—	3.17
2½"	P1@.050	LF	2.46	1.79	—	4.25
3"	P1@.055	LF	3.95	1.97	—	5.92
4"	P1@.070	LF	4.61	2.50	—	7.11
5"	P1@.095	LF	7.88	3.40	—	11.28
6"	P1@.120	LF	7.88	4.29	—	12.17
8"	P1@.160	LF	12.30	5.72	—	18.02

Schedule 40 PVC 45 degree ell S x S with solvent-weld joints

Description	Craft@Hrs	Unit	Material $	Labor $	Equipment $	Total $
1/2"	P1@.100	Ea	.66	3.58	—	4.24
3/4"	P1@.115	Ea	1.02	4.11	—	5.13
1"	P1@.120	Ea	1.24	4.29	—	5.53
1¼"	P1@.160	Ea	1.73	5.72	—	7.45
1½"	P1@.180	Ea	2.17	6.44	—	8.61
2"	P1@.200	Ea	2.93	7.15	—	10.08
2½"	P1@.250	Ea	7.32	8.94	—	16.26
3"	P1@.300	Ea	11.40	10.70	—	22.10
4"	P1@.500	Ea	20.40	17.90	—	38.30
5"	P1@.550	Ea	40.40	19.70	—	60.10
6"	P1@.600	Ea	50.40	21.50	—	71.90
8"	P1@.800	Ea	121.00	28.60	—	149.60

Schedule 40 PVC 90 degree ell S x S with solvent-weld joints

Description	Craft@Hrs	Unit	Material $	Labor $	Equipment $	Total $
1/2"	P1@.100	Ea	.42	3.58	—	4.00
3/4"	P1@.115	Ea	.46	4.11	—	4.57
1"	P1@.120	Ea	.79	4.29	—	5.08
1¼"	P1@.160	Ea	1.44	5.72	—	7.16
1½"	P1@.180	Ea	1.52	6.44	—	7.96
2"	P1@.200	Ea	2.40	7.15	—	9.55
2½"	P1@.250	Ea	7.30	8.94	—	16.24
3"	P1@.300	Ea	8.72	10.70	—	19.42
4"	P1@.500	Ea	15.60	17.90	—	33.50
5"	P1@.550	Ea	40.40	19.70	—	60.10
6"	P1@.600	Ea	49.90	21.50	—	71.40
8"	P1@.800	Ea	128.00	28.60	—	156.60

Description	Craft@Hrs	Unit	Material $	Labor $	Equipment $	Total $

Schedule 40 PVC tee S x S x S with solvent-weld joints

Description	Craft@Hrs	Unit	Material $	Labor $	Equipment $	Total $
1/2"	P1@.130	Ea	.52	4.65	—	5.17
3/4"	P1@.140	Ea	.56	5.01	—	5.57
1"	P1@.170	Ea	1.06	6.08	—	7.14
1¼"	P1@.220	Ea	1.67	7.87	—	9.54
1½"	P1@.250	Ea	1.87	8.94	—	10.81
2"	P1@.280	Ea	2.40	10.00	—	12.40
2½"	P1@.350	Ea	9.79	12.50	—	22.29
3"	P1@.420	Ea	12.70	15.00	—	27.70
4"	P1@.560	Ea	23.40	20.00	—	43.40
5"	P1@.710	Ea	56.20	25.40	—	81.60
6"	P1@.840	Ea	81.60	30.00	—	111.60
8"	P1@1.12	Ea	181.00	40.10	—	221.10

Schedule 40 PVC reducing tee S x S x S with solvent-weld joints

Description	Craft@Hrs	Unit	Material $	Labor $	Equipment $	Total $
3/4" x 1/2"	P1@.140	Ea	.66	5.01	—	5.67
1" x 1/2"	P1@.170	Ea	1.24	6.08	—	7.32
1" x 3/4"	P1@.170	Ea	1.84	6.08	—	7.92
1¼" x 1/2"	P1@.220	Ea	1.84	7.87	—	9.71
1¼" x 3/4"	P1@.220	Ea	1.84	7.87	—	9.71
1¼" x 1"	P1@.220	Ea	1.84	7.87	—	9.71
1½" x 1/2"	P1@.250	Ea	3.16	8.94	—	12.10
1½" x 3/4"	P1@.250	Ea	3.16	8.94	—	12.10
1½" x 1"	P1@.250	Ea	3.16	8.94	—	12.10
1½" x 1¼"	P1@.250	Ea	3.16	8.94	—	12.10
2" x 1/2"	P1@.280	Ea	4.52	10.00	—	14.52
2" x 3/4"	P1@.280	Ea	4.52	10.00	—	14.52
2" x 1"	P1@.280	Ea	4.52	10.00	—	14.52
2" x 1¼"	P1@.280	Ea	4.52	10.00	—	14.52
2" x 1½"	P1@.280	Ea	4.52	10.00	—	14.52
2½" x 1"	P1@.350	Ea	9.35	12.50	—	21.85
2½" x 1¼"	P1@.350	Ea	9.35	12.50	—	21.85
2½" x 1½"	P1@.350	Ea	9.35	12.50	—	21.85
2½" x 2"	P1@.350	Ea	9.35	12.50	—	21.85
3" x 2"	P1@.420	Ea	13.90	15.00	—	28.90
3" x 2½"	P1@.420	Ea	13.90	15.00	—	28.90
4" x 2"	P1@.560	Ea	23.40	20.00	—	43.40
4" x 2½"	P1@.560	Ea	23.40	20.00	—	43.40
4" x 3"	P1@.560	Ea	23.40	20.00	—	43.40
5" x 3"	P1@.710	Ea	41.60	25.40	—	67.00
5" x 4"	P1@.710	Ea	41.60	25.40	—	67.00
6" x 3"	P1@.840	Ea	68.20	30.00	—	98.20
6" x 4"	P1@.840	Ea	68.20	30.00	—	98.20
6" x 5"	P1@.840	Ea	68.20	30.00	—	98.20
8" x 4"	P1@1.12	Ea	151.00	40.10	—	191.10
8" x 5"	P1@1.12	Ea	151.00	40.10	—	191.10
8" x 6"	P1@1.12	Ea	151.00	40.10	—	191.10

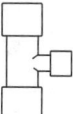

PVC, Schedule 40, with Solvent-Weld Joints

Description	Craft@Hrs	Unit	Material $	Labor $	Equipment $	Total $

Schedule 40 PVC reducing bushing SPIG x S with solvent-weld joints

Description	Craft@Hrs	Unit	Material $	Labor $	Equipment $	Total $
3/4" x 1/2"	P1@.050	Ea	.43	1.79	—	2.22
1" x 1/2"	P1@.060	Ea	.77	2.15	—	2.92
1" x 3/4"	P1@.060	Ea	.77	2.15	—	2.92
1¼" x 1/2"	P1@.080	Ea	1.02	2.86	—	3.88
1¼" x 3/4"	P1@.080	Ea	1.02	2.86	—	3.88
1¼" x 1"	P1@.080	Ea	1.02	2.86	—	3.88
1½" x 1/2"	P1@.100	Ea	1.06	3.58	—	4.64
1½" x 3/4"	P1@.100	Ea	1.06	3.58	—	4.64
1½" x 1"	P1@.100	Ea	1.06	3.58	—	4.64
1½" x 1¼"	P1@.100	Ea	1.06	3.58	—	4.64
2" x 1"	P1@.110	Ea	1.78	3.93	—	5.71
2" x 1¼"	P1@.110	Ea	1.78	3.93	—	5.71
2" x 1½"	P1@.110	Ea	1.78	3.93	—	5.71
2½" x 1¼"	P1@.130	Ea	2.86	4.65	—	7.51
2½" x 1½"	P1@.130	Ea	2.86	4.65	—	7.51
2½" x 2"	P1@.130	Ea	2.86	4.65	—	7.51
3" x 2"	P1@.150	Ea	4.26	5.36	—	9.62
3" x 2½"	P1@.150	Ea	4.26	5.36	—	9.62
4" x 3"	P1@.200	Ea	9.52	7.15	—	16.67
5" x 3"	P1@.250	Ea	18.80	8.94	—	27.74
5" x 4"	P1@.250	Ea	18.80	8.94	—	27.74
6" x 4"	P1@.300	Ea	27.40	10.70	—	38.10
6" x 5"	P1@.300	Ea	27.40	10.70	—	38.10
8" x 4"	P1@.400	Ea	82.00	14.30	—	96.30
8" x 5"	P1@.400	Ea	82.00	14.30	—	96.30

Schedule 40 PVC adapter MPT x S with solvent-weld joints

Description	Craft@Hrs	Unit	Material $	Labor $	Equipment $	Total $
1/2"	P1@.060	Ea	.37	2.15	—	2.52
3/4"	P1@.065	Ea	.42	2.32	—	2.74
1"	P1@.070	Ea	.72	2.50	—	3.22
1¼"	P1@.100	Ea	.89	3.58	—	4.47
1½"	P1@.110	Ea	1.20	3.93	—	5.13
2"	P1@.120	Ea	3.91	4.29	—	8.20
2½"	P1@.150	Ea	4.57	5.36	—	9.93
3"	P1@.180	Ea	6.66	6.44	—	13.10
4"	P1@.240	Ea	8.48	8.58	—	17.06
5"	P1@.370	Ea	14.40	13.20	—	27.60
6"	P1@.420	Ea	22.20	15.00	—	37.20

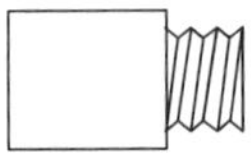

Schedule 40 PVC adapter FPT x S with solvent-weld joints

Description	Craft@Hrs	Unit	Material $	Labor $	Equipment $	Total $
1/2"	P1@.110	Ea	.46	3.93	—	4.39
3/4"	P1@.120	Ea	.56	4.29	—	4.85
1"	P1@.130	Ea	.66	4.65	—	5.31
1¼"	P1@.170	Ea	1.03	6.08	—	7.11
1½"	P1@.210	Ea	1.20	7.51	—	8.71
2"	P1@.290	Ea	1.61	10.40	—	12.01
2½"	P1@.330	Ea	4.01	11.80	—	15.81
3"	P1@.470	Ea	8.00	16.80	—	24.80
4"	P1@.630	Ea	14.90	22.50	—	37.40
5"	P1@.890	Ea	23.30	31.80	—	55.10
6"	P1@.970	Ea	32.80	34.70	—	67.50

Description	Craft@Hrs	Unit	Material $	Labor $	Equipment $	Total $

Schedule 40 PVC cap S with solvent-weld joints

Description	Craft@Hrs	Unit	Material $	Labor $	Equipment $	Total $
1/2"	P1@.060	Ea	.37	2.15	—	2.52
3/4"	P1@.065	Ea	.43	2.32	—	2.75
1"	P1@.070	Ea	.66	2.50	—	3.16
1¼"	P1@.100	Ea	1.02	3.58	—	4.60
1½"	P1@.110	Ea	1.03	3.93	—	4.96
2"	P1@.120	Ea	1.25	4.29	—	5.54
2½"	P1@.150	Ea	3.91	5.36	—	9.27
3"	P1@.180	Ea	4.30	6.44	—	10.74
4"	P1@.240	Ea	9.79	8.58	—	18.37
5"	P1@.370	Ea	16.40	13.20	—	29.60
6"	P1@.420	Ea	23.50	15.00	—	38.50

Schedule 40 PVC plug MPT with solvent-weld joints

Description	Craft@Hrs	Unit	Material $	Labor $	Equipment $	Total $
1/2"	P1@.150	Ea	1.09	5.36	—	6.45
3/4"	P1@.160	Ea	1.18	5.72	—	6.90
1"	P1@.180	Ea	1.91	6.44	—	8.35
1¼"	P1@.210	Ea	1.99	7.51	—	9.50
1½"	P1@.230	Ea	2.14	8.22	—	10.36
2"	P1@.240	Ea	2.80	8.58	—	11.38
2½"	P1@.260	Ea	3.85	9.30	—	13.15
3"	P1@.300	Ea	5.82	10.70	—	16.52
4"	P1@.400	Ea	13.20	14.30	—	27.50

Schedule 40 PVC coupling S x S with solvent-weld joints

Description	Craft@Hrs	Unit	Material $	Labor $	Equipment $	Total $
1/2"	P1@.100	Ea	.25	3.58	—	3.83
3/4"	P1@.115	Ea	.37	4.11	—	4.48
1"	P1@.120	Ea	.64	4.29	—	4.93
1¼"	P1@.160	Ea	.89	5.72	—	6.61
1½"	P1@.180	Ea	.92	6.44	—	7.36
2"	P1@.200	Ea	1.45	7.15	—	8.60
2½"	P1@.250	Ea	3.16	8.94	—	12.10
3"	P1@.300	Ea	5.00	10.70	—	15.70
4"	P1@.500	Ea	7.18	17.90	—	25.08
5"	P1@.550	Ea	13.20	19.70	—	32.90
6"	P1@.600	Ea	22.70	21.50	—	44.20
8"	P1@.800	Ea	42.50	28.60	—	71.10

Schedule 40 PVC union S x S with solvent-weld joints

Description	Craft@Hrs	Unit	Material $	Labor $	Equipment $	Total $
1/2"	P1@.125	Ea	3.32	4.47	—	7.79
3/4"	P1@.140	Ea	4.15	5.01	—	9.16
1"	P1@.160	Ea	4.30	5.72	—	10.02
1¼"	P1@.200	Ea	8.88	7.15	—	16.03
1½"	P1@.225	Ea	10.10	8.05	—	18.15
2"	P1@.250	Ea	15.70	8.94	—	24.64
2½"	P1@.275	Ea	23.30	9.83	—	33.13
3"	P1@.300	Ea	28.70	10.70	—	39.40

PVC, Schedule 40, with Solvent-Weld Joints

Description	Craft@Hrs	Unit	Material $	Labor $	Equipment $	Total $

Class 125 bronze body gate valve, threaded ends

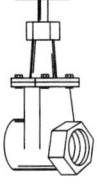

Description	Craft@Hrs	Unit	Material $	Labor $	Equipment $	Total $
1/2"	P1@.210	Ea	15.40	7.51	—	22.91
3/4"	P1@.250	Ea	19.30	8.94	—	28.24
1"	P1@.300	Ea	27.10	10.70	—	37.80
1¼"	P1@.400	Ea	34.30	14.30	—	48.60
1½"	P1@.450	Ea	46.10	16.10	—	62.20
2"	P1@.500	Ea	77.00	17.90	—	94.90
2½"	P1@.750	Ea	128.00	26.80	—	154.80
3"	P1@.950	Ea	184.00	34.00	—	218.00

Class 125 iron body gate valve, flanged ends

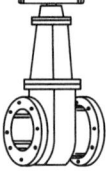

Description	Craft@Hrs	Unit	Material $	Labor $	Equipment $	Total $
2"	P1@.500	Ea	295.00	17.90	—	312.90
2½"	P1@.600	Ea	398.00	21.50	—	419.50
3"	P1@.750	Ea	432.00	26.80	—	458.80
4"	P1@1.35	Ea	634.00	48.30	—	682.30
5"	ER@2.00	Ea	1,210.00	76.20	31.40	1,317.60
6"	ER@2.50	Ea	1,210.00	95.30	39.30	1,344.60
8"	ER@3.00	Ea	1,990.00	114.00	47.10	2,151.10

Class 125 bronze body globe valve, threaded ends

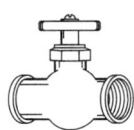

Description	Craft@Hrs	Unit	Material $	Labor $	Equipment $	Total $
1/2"	P1@.210	Ea	28.60	7.51	—	36.11
3/4"	P1@.250	Ea	38.10	8.94	—	47.04
1"	P1@.300	Ea	54.70	10.70	—	65.40
1¼"	P1@.400	Ea	76.90	14.30	—	91.20
1½"	P1@.450	Ea	104.00	16.10	—	120.10
2"	P1@.500	Ea	168.00	17.90	—	185.90

200 PSIG iron body butterfly valve, lug type, lever operated

Description	Craft@Hrs	Unit	Material $	Labor $	Equipment $	Total $
2"	P1@.450	Ea	128.00	16.10	—	144.10
2½"	P1@.450	Ea	132.00	16.10	—	148.10
3"	P1@.550	Ea	139.00	19.70	—	158.70
4"	P1@.550	Ea	173.00	19.70	—	192.70
5"	ER@.800	Ea	229.00	30.50	12.70	272.20
6"	ER@.800	Ea	279.00	30.50	12.70	322.20
8"	ER@.900	Ea	382.00	34.30	14.10	430.40

200 PSIG iron body butterfly valve, wafer type, lever operated

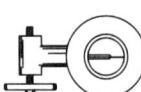

Description	Craft@Hrs	Unit	Material $	Labor $	Equipment $	Total $
2"	P1@.450	Ea	116.00	16.10	—	132.10
2½"	P1@.450	Ea	119.00	16.10	—	135.10
3"	P1@.550	Ea	128.00	19.70	—	147.70
4"	P1@.550	Ea	155.00	19.70	—	174.70
5"	ER@.800	Ea	207.00	30.50	12.70	250.20
6"	ER@.800	Ea	259.00	30.50	12.70	302.20
8"	ER@.900	Ea	359.00	34.30	14.10	407.40

Description	Craft@Hrs	Unit	Material $	Labor $	Equipment $	Total $

Class 125 bronze body 2-piece ball valve, threaded ends

Description	Craft@Hrs	Unit	Material $	Labor $	Equipment $	Total $
1/2"	P1@.210	Ea	7.53	7.51	—	15.04
3/4"	P1@.250	Ea	10.00	8.94	—	18.94
1"	P1@.300	Ea	17.30	10.70	—	28.00
1¼"	P1@.400	Ea	28.90	14.30	—	43.20
1½"	P1@.450	Ea	39.00	16.10	—	55.10
2"	P1@.500	Ea	49.40	17.90	—	67.30
3"	P1@.625	Ea	338.00	22.40	—	360.40
4"	P1@.690	Ea	443.00	24.70	—	467.70

PVC Ball Valve, Solid body, Solvent weld joints, EDPM (female socket)

Description	Craft@Hrs	Unit	Material $	Labor $	Equipment $	Total $
1/2"	P1@.170	Ea	3.00	6.08	—	9.08
3/4"	P1@.200	Ea	3.40	7.15	—	10.55
1"	P1@.220	Ea	4.95	7.87	—	12.82
1¼"	P1@.280	Ea	5.70	10.00	—	15.70
1½"	P1@.315	Ea	9.50	11.30	—	20.80
2"	P1@.385	Ea	11.90	13.80	—	25.70

PVC Ball Valve, Solid body, Threaded joints, EDPM (female pipe thread)

Description	Craft@Hrs	Unit	Material $	Labor $	Equipment $	Total $
1/2"	P1@.200	Ea	3.00	7.15	—	10.15
3/4"	P1@.240	Ea	3.40	8.58	—	11.98
1"	P1@.290	Ea	4.95	10.40	—	15.35
1¼"	P1@.390	Ea	5.70	13.90	—	19.60
1½"	P1@.420	Ea	9.50	15.00	—	24.50
2"	P1@.490	Ea	11.90	17.50	—	29.40

PVC Ball Valve, Tru-union, Threaded joints, EDPM (female pipe thread)

Description	Craft@Hrs	Unit	Material $	Labor $	Equipment $	Total $
1/2"	P1@.200	Ea	3.00	7.15	—	10.15
3/4"	P1@.240	Ea	3.40	8.58	—	11.98
1"	P1@.290	Ea	4.95	10.40	—	15.35
1¼"	P1@.390	Ea	5.70	13.90	—	19.60
1½"	P1@.420	Ea	9.50	15.00	—	24.50
2"	P1@.490	Ea	11.90	17.50	—	29.40
2½"	P1@.520	Ea	73.90	18.60	—	92.50
3"	P1@.590	Ea	170.00	21.10	—	191.10
4"	P1@.630	Ea	315.00	22.50	—	337.50

PVC, Schedule 40, with Solvent-Weld Joints

Description	Craft@Hrs	Unit	Material $	Labor $	Equipment $	Total $

PVC Ball Valve, Union type body, Solvent-weld joints, (female socket)

Description	Craft@Hrs	Unit	Material $	Labor $	Equipment $	Total $
1/2"	P1@.185	Ea	3.00	6.62	—	9.62
3/4"	P1@.225	Ea	3.40	8.05	—	11.45
1"	P1@.270	Ea	4.95	9.66	—	14.61
1¼"	P1@.360	Ea	5.70	12.90	—	18.60
1½"	P1@.400	Ea	9.50	14.30	—	23.80
2"	P1@.465	Ea	11.90	16.60	—	28.50
2½"	P1@.490	Ea	73.90	17.50	—	91.40
3"	P1@.540	Ea	170.00	19.30	—	189.30
4"	P1@.600	Ea	315.00	21.50	—	336.50

Class 125 bronze body swing check valve, threaded

Description	Craft@Hrs	Unit	Material $	Labor $	Equipment $	Total $
1/2"	P1@.210	Ea	16.90	7.51	—	24.41
3/4"	P1@.250	Ea	24.20	8.94	—	33.14
1"	P1@.300	Ea	31.70	10.70	—	42.40
1¼"	P1@.400	Ea	45.00	14.30	—	59.30
1½"	P1@.450	Ea	63.50	16.10	—	79.60
2"	P1@.500	Ea	106.00	17.90	—	123.90

Class 125 iron body swing check valve, flanged ends

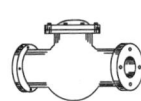

Description	Craft@Hrs	Unit	Material $	Labor $	Equipment $	Total $
2"	P1@.500	Ea	148.00	17.90	—	165.90
2½"	P1@.600	Ea	188.00	21.50	—	209.50
3"	P1@.750	Ea	233.00	26.80	—	259.80
4"	P1@1.35	Ea	343.00	48.30	—	391.30
5"	ER@2.00	Ea	660.00	76.20	31.40	767.60
6"	ER@2.50	Ea	660.00	95.30	39.30	794.60
8"	ER@3.00	Ea	1,190.00	114.00	47.10	1,351.10

Class 125 iron body silent check valve, wafer type

Description	Craft@Hrs	Unit	Material $	Labor $	Equipment $	Total $
2"	P1@.500	Ea	108.00	17.90	—	125.90
2½"	P1@.600	Ea	120.00	21.50	—	141.50
3"	P1@.750	Ea	138.00	26.80	—	164.80
4"	P1@1.35	Ea	185.00	48.30	—	233.30
5"	ER@2.00	Ea	284.00	76.20	31.40	391.60
6"	ER@2.50	Ea	385.00	95.30	39.30	519.60
8"	ER@3.00	Ea	660.00	114.00	47.10	821.10

Class 125 bronze body strainer, threaded ends

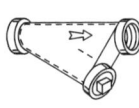

Description	Craft@Hrs	Unit	Material $	Labor $	Equipment $	Total $
1/2"	P1@.210	Ea	26.40	7.51	—	33.91
3/4"	P1@.250	Ea	34.70	8.94	—	43.64
1"	P1@.300	Ea	42.50	10.70	—	53.20
1¼"	P1@.400	Ea	59.30	14.30	—	73.60
1½"	P1@.450	Ea	89.30	16.10	—	105.40
2"	P1@.500	Ea	154.00	17.90	—	171.90

Description	Craft@Hrs	Unit	Material $	Labor $	Equipment $	Total $

Class 125 iron body strainer, flanged ends

Description	Craft@Hrs	Unit	Material $	Labor $	Equipment $	Total $
2"	P1@.500	Ea	118.00	17.90	—	135.90
2½"	P1@.600	Ea	132.00	21.50	—	153.50
3"	P1@.750	Ea	152.00	26.80	—	178.80
4"	P1@1.35	Ea	260.00	48.30	—	308.30
5"	ER@2.00	Ea	528.00	76.20	31.40	635.60
6"	ER@2.50	Ea	528.00	95.30	39.30	662.60
8"	ER@3.00	Ea	892.00	114.00	47.10	1,053.10

Installation of 2-way control valve, threaded joints

Description	Craft@Hrs	Unit	Material $	Labor $	Equipment $	Total $
1/2"	P1@.210	Ea	—	7.51	—	7.51
3/4"	P1@.275	Ea	—	9.83	—	9.83
1"	P1@.350	Ea	—	12.50	—	12.50
1¼"	P1@.430	Ea	—	15.40	—	15.40
1½"	P1@.505	Ea	—	18.10	—	18.10
2"	P1@.675	Ea	—	24.10	—	24.10
2½"	P1@.830	Ea	—	29.70	—	29.70
3"	P1@.990	Ea	—	35.40	—	35.40

Installation of 3-way control valve, threaded joints

Description	Craft@Hrs	Unit	Material $	Labor $	Equipment $	Total $
1/2"	P1@.260	Ea	—	9.30	—	9.30
3/4"	P1@.365	Ea	—	13.10	—	13.10
1"	P1@.475	Ea	—	17.00	—	17.00
1¼"	P1@.575	Ea	—	20.60	—	20.60
1½"	P1@.680	Ea	—	24.30	—	24.30
2"	P1@.910	Ea	—	32.50	—	32.50
2½"	P1@1.12	Ea	—	40.10	—	40.10
3"	P1@1.33	Ea	—	47.60	—	47.60

PVC companion flange

Description	Craft@Hrs	Unit	Material $	Labor $	Equipment $	Total $
2"	P1@.140	Ea	5.21	5.01	—	10.22
2½"	P1@.180	Ea	15.60	6.44	—	22.04
3"	P1@.210	Ea	16.20	7.51	—	23.71
4"	P1@.280	Ea	16.70	10.00	—	26.70
5"	P1@.310	Ea	37.40	11.10	—	48.50
6"	P1@.420	Ea	37.40	15.00	—	52.40
8"	P1@.560	Ea	55.90	20.00	—	75.90

Bolt and gasket sets

Description	Craft@Hrs	Unit	Material $	Labor $	Equipment $	Total $
2"	P1@.500	Ea	3.92	17.90	—	21.82
2½"	P1@.650	Ea	4.60	23.20	—	27.80
3"	P1@.750	Ea	7.84	26.80	—	34.64
4"	P1@1.00	Ea	13.10	35.80	—	48.90
5"	P1@1.10	Ea	21.60	39.30	—	60.90
6"	P1@1.20	Ea	21.60	42.90	—	64.50
8"	P1@1.25	Ea	24.10	44.70	—	68.80

PVC, Schedule 40, with Solvent-Weld Joints

Description	Craft@Hrs	Unit	Material $	Labor $	Equipment $	Total $
Thermometer with well						
7"	P1@.250	Ea	159.00	8.94	—	167.94
9"	P1@.250	Ea	164.00	8.94	—	172.94
Dial type pressure gauge						
2½"	P1@.200	Ea	33.10	7.15	—	40.25
3½"	P1@.200	Ea	43.80	7.15	—	50.95
Pressure/temperature tap						
tap	P1@.150	Ea	15.70	5.36	—	21.06
Hanger with swivel assembly						
1/2"	P1@.250	Ea	3.74	8.94	—	12.68
3/4"	P1@.250	Ea	4.13	8.94	—	13.07
1"	P1@.250	Ea	4.33	8.94	—	13.27
1¼"	P1@.300	Ea	4.44	10.70	—	15.14
1½"	P1@.300	Ea	4.77	10.70	—	15.47
2"	P1@.300	Ea	4.98	10.70	—	15.68
2½"	P1@.350	Ea	6.78	12.50	—	19.28
3"	P1@.350	Ea	8.37	12.50	—	20.87
4"	P1@.350	Ea	9.19	12.50	—	21.69
5"	P1@.450	Ea	13.40	16.10	—	29.50
6"	P1@.450	Ea	15.10	16.10	—	31.20
8"	P1@.450	Ea	19.00	16.10	—	35.10
Riser clamp						
1/2"	P1@.100	Ea	2.41	3.58	—	5.99
3/4"	P1@.100	Ea	3.55	3.58	—	7.13
1"	P1@.100	Ea	3.59	3.58	—	7.17
1¼"	P1@.105	Ea	4.33	3.75	—	8.08
1½"	P1@.110	Ea	4.58	3.93	—	8.51
2"	P1@.115	Ea	4.84	4.11	—	8.95
2½"	P1@.120	Ea	5.09	4.29	—	9.38
3"	P1@.120	Ea	5.52	4.29	—	9.81
4"	P1@.125	Ea	7.03	4.47	—	11.50
5"	P1@.180	Ea	10.10	6.44	—	16.54
6"	P1@.200	Ea	12.20	7.15	—	19.35
8"	P1@.200	Ea	19.80	7.15	—	26.95

Description	Craft@Hrs	Unit	Material $	Labor $	Equipment $	Total $

PVC (Polyvinyl Chloride) Schedule 80 pipe with Schedule 80 socket-type fittings and solvent-welded joints is widely used in process, chemical, A.C. condensate, potable water and irrigation piping systems. PVC is available in Type I (normal impact) and Type II (high impact). Type I has a maximum temperature rating of 150 degrees F., and Type II is rated at 140 degrees F.

Schedule 80 PVC can be threaded, but it is more economical to use solvent-welded joints.

Manufacturers should be consulted for maximum pressure ratings and recommended joints solvents for specific applications.

Because of the current unreliability of many plastic valves, most engineers specify standard bronze-body and iron-body valves for use in PVC water piping systems. Plastic valves, however, have to be used in systems conveying liquids that may be injurious to metallic valves.

This section has been arranged to save the estimator's time by including all normally-used system components such as pipe, fittings, valves, hanger assemblies, riser clamps and miscellaneous items under one heading. Additional items can be found under "Plumbing and Piping Specialties." The cost estimates in this section are based on the conditions, limitations and wage rates described in the section "How to Use This Book" beginning on page 5.

Equipment cost, where shown, is $110 per hour for a 10-ton hydraulic truck-mounted crane.

Description	Craft@Hrs	Unit	Material $	Labor $	Equipment $	Total $

Schedule 80 PVC pipe assembly with solvent-weld joints installed horizontally.
Complete installation including three to eight tees and three to eight elbows every 100' and hangers spaced to meet plumbing code (*a tee and elbow every 12' for 1/2" dia., a tee and elbow every 40' for 8"dia.*).

1/2"	P1@.085	LF	2.03	3.04	—	5.07
3/4"	P1@.093	LF	2.54	3.33	—	5.87
1"	P1@.099	LF	2.80	3.54	—	6.34
1¼"	P1@.115	LF	3.28	4.11	—	7.39
1½"	P1@.117	LF	3.49	4.18	—	7.67
2"	P1@.120	LF	4.44	4.29	—	8.73
2½"	P1@.125	LF	7.31	4.47	—	11.78
3"	P1@.129	LF	9.88	4.61	—	14.49
4"	P1@.161	LF	14.30	5.76	—	20.06
6"	P1@.253	LF	21.60	9.05	—	30.65
8"	P1@.299	LF	36.00	10.70	—	46.70

PVC, Schedule 80, with Solvent-Weld Joints

Description	Craft@Hrs	Unit	Material $	Labor $	Equipment $	Total $

Schedule 80 PVC pipe assembly with solvent-weld joints, installed risers. Complete installation including a reducing tee every floor and a riser clamp every other.

Description	Craft@Hrs	Unit	Material $	Labor $	Equipment $	Total $
1/2"	P1@.041	LF	1.00	1.47	—	2.47
3/4"	P1@.048	LF	1.45	1.72	—	3.17
1"	P1@.054	LF	1.75	1.93	—	3.68
1¼"	P1@.063	LF	2.06	2.25	—	4.31
1½"	P1@.071	LF	2.69	2.54	—	5.23
2"	P1@.079	LF	3.60	2.83	—	6.43
2½"	P1@.090	LF	5.91	3.22	—	9.13
3"	P1@.101	LF	7.63	3.61	—	11.24
4"	P1@.129	LF	10.90	4.61	—	15.51
6"	P1@.242	LF	19.40	8.65	—	28.05
8"	P1@.296	LF	32.90	10.60	—	43.50

Schedule 80 PVC pipe with solvent-weld joints

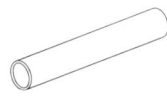

Description	Craft@Hrs	Unit	Material $	Labor $	Equipment $	Total $
1/2"	P1@.021	LF	.66	.75	—	1.41
3/4"	P1@.026	LF	.88	.93	—	1.81
1"	P1@.032	LF	1.16	1.14	—	2.30
1¼"	P1@.037	LF	1.36	1.32	—	2.68
1½"	P1@.042	LF	1.84	1.50	—	3.34
2"	P1@.047	LF	2.48	1.68	—	4.16
2½"	P1@.052	LF	3.95	1.86	—	5.81
3"	P1@.057	LF	5.29	2.04	—	7.33
4"	P1@.074	LF	7.80	2.65	—	10.45
6"	P1@.160	LF	14.80	5.72	—	20.52
8"	P1@.190	LF	22.60	6.79	—	29.39

Schedule 80 PVC 45 degree ell S x S with solvent-weld joints

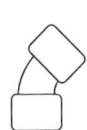

Description	Craft@Hrs	Unit	Material $	Labor $	Equipment $	Total $
1/2"	P1@.105	Ea	3.23	3.75	—	6.98
3/4"	P1@.120	Ea	4.88	4.29	—	9.17
1"	P1@.125	Ea	7.33	4.47	—	11.80
1¼"	P1@.170	Ea	9.36	6.08	—	15.44
1½"	P1@.190	Ea	11.10	6.79	—	17.89
2"	P1@.210	Ea	14.30	7.51	—	21.81
2½"	P1@.262	Ea	30.00	9.37	—	39.37
3"	P1@.315	Ea	36.60	11.30	—	47.90
4"	P1@.520	Ea	65.90	18.60	—	84.50
6"	P1@.630	Ea	82.80	22.50	—	105.30
8"	P1@.840	Ea	179.00	30.00	—	209.00

Description	Craft@Hrs	Unit	Material $	Labor $	Equipment $	Total $

Schedule 80 PVC 90 degree ell S x S with solvent-weld joints

Description	Craft@Hrs	Unit	Material $	Labor $	Equipment $	Total $
1/2"	P1@.105	Ea	1.75	3.75	—	5.50
3/4"	P1@.120	Ea	2.21	4.29	—	6.50
1"	P1@.125	Ea	3.49	4.47	—	7.96
1¼"	P1@.170	Ea	4.73	6.08	—	10.81
1½"	P1@.190	Ea	5.05	6.79	—	11.84
2"	P1@.210	Ea	6.06	7.51	—	13.57
2½"	P1@.262	Ea	14.10	9.37	—	23.47
3"	P1@.315	Ea	16.00	11.30	—	27.30
4"	P1@.520	Ea	24.30	18.60	—	42.90
6"	P1@.630	Ea	69.30	22.50	—	91.80
8"	P1@.840	Ea	189.00	30.00	—	219.00

Schedule 80 PVC tee S x S x S with solvent-weld joints

Description	Craft@Hrs	Unit	Material $	Labor $	Equipment $	Total $
1/2"	P1@.135	Ea	4.83	4.83	—	9.66
3/4"	P1@.145	Ea	5.05	5.19	—	10.24
1"	P1@.180	Ea	6.28	6.44	—	12.72
1¼"	P1@.230	Ea	17.30	8.22	—	25.52
1½"	P1@.265	Ea	17.30	9.48	—	26.78
2"	P1@.295	Ea	21.60	10.50	—	32.10
2½"	P1@.368	Ea	23.50	13.20	—	36.70
3"	P1@.440	Ea	29.40	15.70	—	45.10
4"	P1@.585	Ea	34.10	20.90	—	55.00
6"	P1@.880	Ea	115.00	31.50	—	146.50
8"	P1@1.18	Ea	269.00	42.20	—	311.20

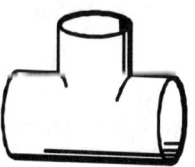

Schedule 80 PVC reducing tee S x S x S with solvent-weld joints

Description	Craft@Hrs	Unit	Material $	Labor $	Equipment $	Total $
3/4" x 1/2"	P1@.145	Ea	5.46	5.19	—	10.65
1" x 1/2"	P1@.175	Ea	5.88	6.26	—	12.14
1" x 3/4"	P1@.175	Ea	5.88	6.26	—	12.14
1¼" x 1/2"	P1@.230	Ea	19.30	8.22	—	27.52
1¼" x 3/4"	P1@.230	Ea	19.30	8.22	—	27.52
1¼" x 1"	P1@.230	Ea	19.30	8.22	—	27.52
1½" x 3/4"	P1@.260	Ea	20.80	9.30	—	30.10
1½" x 1"	P1@.260	Ea	20.80	9.30	—	30.10
1½" x 1¼"	P1@.260	Ea	20.80	9.30	—	30.10
2" x 1/2"	P1@.290	Ea	25.40	10.40	—	35.80
2" x 3/4"	P1@.290	Ea	25.40	10.40	—	35.80
2" x 1"	P1@.290	Ea	25.40	10.40	—	35.80
2" x 1½"	P1@.290	Ea	25.40	10.40	—	35.80
3" x 2"	P1@.430	Ea	36.00	15.40	—	51.40
4" x 2"	P1@.570	Ea	40.90	20.40	—	61.30
4" x 3"	P1@.570	Ea	40.90	20.40	—	61.30
6" x 4"	P1@.850	Ea	138.00	30.40	—	168.40
8" x 4"	P1@1.16	Ea	308.00	41.50	—	349.50
8" x 5"	P1@1.16	Ea	308.00	41.50	—	349.50
8" x 6"	P1@1.16	Ea	308.00	41.50	—	349.50

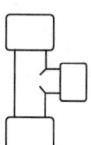

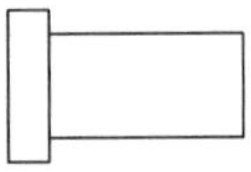

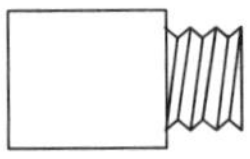

PVC, Schedule 80, with Solvent-Weld Joints

Description	Craft@Hrs	Unit	Material $	Labor $	Equipment $	Total $

Schedule 80 PVC reducing bushing SPIG x S with solvent-weld joints

Description	Craft@Hrs	Unit	Material $	Labor $	Equipment $	Total $
3/4" x 1/2"	P1@.053	Ea	1.00	1.90	—	2.90
1" x 1/2"	P1@.063	Ea	2.85	2.25	—	5.10
1" x 3/4"	P1@.063	Ea	2.85	2.25	—	5.10
1¼" x 1/2"	P1@.084	Ea	4.56	3.00	—	7.56
1¼" x 3/4"	P1@.084	Ea	4.56	3.00	—	7.56
1¼" x 1"	P1@.084	Ea	4.56	3.00	—	7.56
1½" x 1/2"	P1@.105	Ea	6.06	3.75	—	9.81
1½" x 3/4"	P1@.105	Ea	6.06	3.75	—	9.81
1½" x 1"	P1@.105	Ea	6.06	3.75	—	9.81
1½" x 1¼"	P1@.105	Ea	6.06	3.75	—	9.81
2" x 1"	P1@.116	Ea	8.69	4.15	—	12.84
2" x 1¼"	P1@.116	Ea	8.69	4.15	—	12.84
2" x 1½"	P1@.116	Ea	8.69	4.15	—	12.84
3" x 2"	P1@.158	Ea	24.00	5.65	—	29.65
4" x 2½"	P1@.210	Ea	32.90	7.51	—	40.41
4" x 3"	P1@.210	Ea	32.90	7.51	—	40.41
6" x 3"	P1@.315	Ea	45.90	11.30	—	57.20
6" x 4"	P1@.315	Ea	45.90	11.30	—	57.20
8" x 4"	P1@.420	Ea	111.00	15.00	—	126.00
8" x 6"	P1@.420	Ea	111.00	15.00	—	126.00

Schedule 80 PVC adapter MPT x S with solvent-weld joints

Description	Craft@Hrs	Unit	Material $	Labor $	Equipment $	Total $
1/2"	P1@.063	Ea	3.63	2.25	—	5.88
3/4"	P1@.068	Ea	3.95	2.43	—	6.38
1"	P1@.074	Ea	6.81	2.65	—	9.46
1¼"	P1@.105	Ea	8.04	3.75	—	11.79
1½"	P1@.116	Ea	11.50	4.15	—	15.65
2"	P1@.126	Ea	16.60	4.51	—	21.11
2½"	P1@.158	Ea	19.10	5.65	—	24.75
3"	P1@.189	Ea	21.10	6.76	—	27.86
4"	P1@.252	Ea	37.30	9.01	—	46.31
5"	P1@.385	Ea	52.00	13.80	—	65.80
6"	P1@.435	Ea	120.00	15.60	—	135.60

Schedule 80 PVC adapter FPT x S with solvent-weld joints

Description	Craft@Hrs	Unit	Material $	Labor $	Equipment $	Total $
1/2"	P1@.116	Ea	2.90	4.15	—	7.05
3/4"	P1@.126	Ea	4.30	4.51	—	8.81
1"	P1@.137	Ea	6.30	4.90	—	11.20
1¼"	P1@.179	Ea	10.20	6.40	—	16.60
1½"	P1@.221	Ea	12.50	7.90	—	20.40
2"	P1@.305	Ea	21.80	10.90	—	32.70
2½"	P1@.345	Ea	34.40	12.30	—	46.70
3"	P1@.490	Ea	38.60	17.50	—	56.10
4"	P1@.655	Ea	66.60	23.40	—	90.00
5"	P1@.920	Ea	108.00	32.90	—	140.90
6"	P1@1.10	Ea	131.00	39.30	—	170.30

PVC, Schedule 80, with Solvent-Weld Joints

Description	Craft@Hrs	Unit	Material $	Labor $	Equipment $	Total $

Schedule 80 PVC cap S with solvent-weld joints

Description	Craft@Hrs	Unit	Material $	Labor $	Equipment $	Total $
1/2"	P1@.063	Ea	3.03	2.25	—	5.28
3/4"	P1@.070	Ea	3.14	2.50	—	5.64
1"	P1@.075	Ea	5.60	2.68	—	8.28
1¼"	P1@.105	Ea	6.80	3.75	—	10.55
1½"	P1@.115	Ea	6.80	4.11	—	10.91
2"	P1@.125	Ea	13.40	4.47	—	17.87
2½"	P1@.158	Ea	27.10	5.65	—	32.75
3"	P1@.189	Ea	32.30	6.76	—	39.06
4"	P1@.250	Ea	54.40	8.94	—	63.34
5"	P1@.390	Ea	112.00	13.90	—	125.90
6"	P1@.450	Ea	135.00	16.10	—	151.10

Schedule 80 PVC plug MPT with solvent-weld joints

Description	Craft@Hrs	Unit	Material $	Labor $	Equipment $	Total $
1/2"	P1@.155	Ea	2.93	5.54	—	8.47
3/4"	P1@.165	Ea	3.03	5.90	—	8.93
1"	P1@.190	Ea	3.71	6.79	—	10.50
1¼"	P1@.220	Ea	5.46	7.87	—	13.33
1½"	P1@.240	Ea	6.60	8.58	—	15.18
2"	P1@.255	Ea	6.79	9.12	—	15.91
2½"	P1@.275	Ea	17.00	9.83	—	26.83
3"	P1@.320	Ea	22.10	11.40	—	33.50
4"	P1@.425	Ea	43.00	15.20	—	58.20

Schedule 80 PVC coupling S x S with solvent-weld joints

Description	Craft@Hrs	Unit	Material $	Labor $	Equipment $	Total $
1/2"	P1@.105	Ea	3.08	3.75	—	6.83
3/4"	P1@.120	Ea	4.15	4.29	—	8.44
1"	P1@.125	Ea	4.30	4.47	—	8.77
1¼"	P1@.170	Ea	6.56	6.08	—	12.64
1½"	P1@.190	Ea	7.04	6.79	—	13.83
2"	P1@.210	Ea	7.54	7.51	—	15.05
2½"	P1@.262	Ea	18.60	9.37	—	27.97
3"	P1@.315	Ea	21.30	11.30	—	32.60
4"	P1@.520	Ea	26.90	18.60	—	45.50
6"	P1@.630	Ea	57.30	22.50	—	79.80
8"	P1@.840	Ea	79.00	30.00	—	109.00

Schedule 80 PVC union S x S with solvent-weld joints

Description	Craft@Hrs	Unit	Material $	Labor $	Equipment $	Total $
1/2"	P1@.125	Ea	6.21	4.47	—	10.68
3/4"	P1@.140	Ea	8.55	5.01	—	13.56
1"	P1@.160	Ea	9.01	5.72	—	14.73
1¼"	P1@.200	Ea	17.90	7.15	—	25.05
1½"	P1@.225	Ea	20.30	8.05	—	28.35
2"	P1@.250	Ea	27.50	8.94	—	36.44
2½"	P1@.275	Ea	43.40	9.83	—	53.23
3"	P1@.300	Ea	51.30	10.70	—	62.00

PVC, Schedule 80, with Solvent-Weld Joints

Description	Craft@Hrs	Unit	Material $	Labor $	Equipment $	Total $

Class 125 bronze body gate valve, threaded ends

Description	Craft@Hrs	Unit	Material $	Labor $	Equipment $	Total $
1/2"	P1@.210	Ea	15.40	7.51	—	22.91
3/4"	P1@.250	Ea	19.30	8.94	—	28.24
1"	P1@.300	Ea	27.10	10.70	—	37.80
1¼"	P1@.400	Ea	34.30	14.30	—	48.60
1½"	P1@.450	Ea	46.10	16.10	—	62.20
2"	P1@.500	Ea	77.00	17.90	—	94.90
2½"	P1@.750	Ea	128.00	26.80	—	154.80
3"	P1@.950	Ea	184.00	34.00	—	218.00

Class 125 iron body gate valve, flanged ends

Description	Craft@Hrs	Unit	Material $	Labor $	Equipment $	Total $
2"	P1@.500	Ea	295.00	17.90	—	312.90
2½"	P1@.600	Ea	398.00	21.50	—	419.50
3"	P1@.750	Ea	432.00	26.80	—	458.80
4"	P1@1.35	Ea	634.00	48.30	—	682.30
5"	ER@2.00	Ea	1,210.00	76.20	31.40	1,317.60
6"	ER@2.50	Ea	1,210.00	95.30	39.30	1,344.60
8"	ER@3.00	Ea	1,990.00	114.00	47.10	2,151.10

Class 125 bronze body globe valve, threaded ends

Description	Craft@Hrs	Unit	Material $	Labor $	Equipment $	Total $
1/2"	P1@.210	Ea	28.60	7.51	—	36.11
3/4"	P1@.250	Ea	38.10	8.94	—	47.04
1"	P1@.300	Ea	54.70	10.70	—	65.40
1¼"	P1@.400	Ea	76.90	14.30	—	91.20
1½"	P1@.450	Ea	104.00	16.10	—	120.10
2"	P1@.500	Ea	168.00	17.90	—	185.90

200 PSIG iron body butterfly valve, lug type, lever operated

Description	Craft@Hrs	Unit	Material $	Labor $	Equipment $	Total $
2"	P1@.450	Ea	128.00	16.10	—	144.10
2½"	P1@.450	Ea	132.00	16.10	—	148.10
3"	P1@.550	Ea	139.00	19.70	—	158.70
4"	P1@.550	Ea	173.00	19.70	—	192.70
5"	ER@.800	Ea	229.00	30.50	12.70	272.20
6"	ER@.800	Ea	279.00	30.50	12.70	322.20
8"	ER@.900	Ea	382.00	34.30	14.10	430.40

200 PSIG iron body butterfly valve, wafer type, lever operated

Description	Craft@Hrs	Unit	Material $	Labor $	Equipment $	Total $
2"	P1@.450	Ea	116.00	16.10	—	132.10
2½"	P1@.450	Ea	119.00	16.10	—	135.10
3"	P1@.550	Ea	128.00	19.70	—	147.70
4"	P1@.550	Ea	155.00	19.70	—	174.70
5"	ER@.800	Ea	207.00	30.50	12.70	250.20
6"	ER@.800	Ea	259.00	30.50	12.70	302.20
8"	ER@.900	Ea	359.00	34.30	14.10	407.40

Description	Craft@Hrs	Unit	Material $	Labor $	Equipment $	Total $

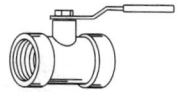

Class 125 bronze body 2-piece ball valve, threaded ends

Description	Craft@Hrs	Unit	Material $	Labor $	Equipment $	Total $
1/2"	P1@.210	Ea	7.53	7.51	—	15.04
3/4"	P1@.250	Ea	10.00	8.94	—	18.94
1"	P1@.300	Ea	17.30	10.70	—	28.00
1¼"	P1@.400	Ea	28.90	14.30	—	43.20
1½"	P1@.450	Ea	39.00	16.10	—	55.10
2"	P1@.500	Ea	49.40	17.90	—	67.30
3"	P1@.625	Ea	338.00	22.40	—	360.40
4"	P1@.690	Ea	443.00	24.70	—	467.70

PVC Ball Valve, Solid body, Solvent weld joints, EDPM (female socket)

Description	Craft@Hrs	Unit	Material $	Labor $	Equipment $	Total $
1/2"	P1@.170	Ea	3.00	6.08	—	9.08
3/4"	P1@.200	Ea	3.40	7.15	—	10.55
1"	P1@.220	Ea	4.95	7.87	—	12.82
1¼"	P1@.280	Ea	5.70	10.00	—	15.70
1½"	P1@.315	Ea	9.50	11.30	—	20.80
2"	P1@.385	Ea	11.90	13.80	—	25.70

PVC Ball Valve, Solid body, Threaded joints, EDPM (female pipe thread)

Description	Craft@Hrs	Unit	Material $	Labor $	Equipment $	Total $
1/2"	P1@.200	Ea	3.00	7.15	—	10.15
3/4"	P1@.240	Ea	3.40	8.58	—	11.98
1"	P1@.290	Ea	4.95	10.40	—	15.35
1¼"	P1@.390	Ea	5.70	13.90	—	19.60
1½"	P1@.420	Ea	9.50	15.00	—	24.50
2"	P1@.490	Ea	11.90	17.50	—	29.40

PVC Ball Valve, Tru-union, Threaded joints, EDPM (female pipe thread)

Description	Craft@Hrs	Unit	Material $	Labor $	Equipment $	Total $
1/2"	P1@.200	Ea	3.00	7.15	—	10.15
3/4"	P1@.240	Ea	3.40	8.58	—	11.98
1"	P1@.290	Ea	4.95	10.40	—	15.35
1¼"	P1@.390	Ea	5.70	13.90	—	19.60
1½"	P1@.420	Ea	9.50	15.00	—	24.50
2"	P1@.490	Ea	11.90	17.50	—	29.40
2½"	P1@.520	Ea	73.90	18.60	—	92.50
3"	P1@.590	Ea	170.00	21.10	—	191.10
4"	P1@.630	Ea	315.00	22.50	—	337.50

PVC Ball Valve, Union type body, Solvent-weld joints, (female socket)

Description	Craft@Hrs	Unit	Material $	Labor $	Equipment $	Total $
1/2"	P1@.185	Ea	3.00	6.62	—	9.62
3/4"	P1@.225	Ea	3.40	8.05	—	11.45
1"	P1@.270	Ea	4.95	9.66	—	14.61
1¼"	P1@.360	Ea	5.70	12.90	—	18.60
1½"	P1@.400	Ea	9.50	14.30	—	23.80
2"	P1@.465	Ea	11.90	16.60	—	28.50
2½"	P1@.490	Ea	73.90	17.50	—	91.40
3"	P1@.540	Ea	170.00	19.30	—	189.30
4"	P1@.600	Ea	315.00	21.50	—	336.50

PVC, Schedule 80, with Solvent-Weld Joints

Description	Craft@Hrs	Unit	Material $	Labor $	Equipment $	Total $

Class 125 bronze body swing check valve, threaded ends

Description	Craft@Hrs	Unit	Material $	Labor $	Equipment $	Total $
1/2"	P1@.210	Ea	16.90	7.51	—	24.41
3/4"	P1@.250	Ea	24.20	8.94	—	33.14
1"	P1@.300	Ea	31.70	10.70	—	42.40
1¼"	P1@.400	Ea	45.00	14.30	—	59.30
1½"	P1@.450	Ea	63.50	16.10	—	79.60
2"	P1@.500	Ea	106.00	17.90	—	123.90

Class 125 iron body swing check valve, flanged ends

Description	Craft@Hrs	Unit	Material $	Labor $	Equipment $	Total $
2"	P1@.500	Ea	148.00	17.90	—	165.90
2½"	P1@.600	Ea	188.00	21.50	—	209.50
3"	P1@.750	Ea	233.00	26.80	—	259.80
4"	P1@1.35	Ea	343.00	48.30	—	391.30
5"	ER@2.00	Ea	660.00	76.20	31.40	767.60
6"	ER@2.50	Ea	660.00	95.30	39.30	794.60
8"	ER@3.00	Ea	1,190.00	114.00	47.10	1,351.10

Class 125 iron body silent check valve, wafer type

Description	Craft@Hrs	Unit	Material $	Labor $	Equipment $	Total $
2"	P1@.500	Ea	108.00	17.90	—	125.90
2½"	P1@.600	Ea	120.00	21.50	—	141.50
3"	P1@.750	Ea	138.00	26.80	—	164.80
4"	P1@1.35	Ea	185.00	48.30	—	233.30
5"	ER@2.00	Ea	284.00	76.20	31.40	391.60
6"	ER@2.50	Ea	385.00	95.30	39.30	519.60
8"	ER@3.00	Ea	660.00	114.00	47.10	821.10

Class 125 bronze body strainer, threaded ends

Description	Craft@Hrs	Unit	Material $	Labor $	Equipment $	Total $
1/2"	P1@.210	Ea	26.40	7.51	—	33.91
3/4"	P1@.250	Ea	34.70	8.94	—	43.64
1"	P1@.300	Ea	42.50	10.70	—	53.20
1¼"	P1@.400	Ea	59.30	14.30	—	73.60
1½"	P1@.450	Ea	89.30	16.10	—	105.40
2"	P1@.500	Ea	154.00	17.90	—	171.90

Class 125 iron body strainer, flanged ends

Description	Craft@Hrs	Unit	Material $	Labor $	Equipment $	Total $
2"	P1@.500	Ea	157.00	17.90	—	174.90
2½"	P1@.600	Ea	175.00	21.50	—	196.50
3"	P1@.750	Ea	203.00	26.80	—	229.80
4"	P1@1.35	Ea	347.00	48.30	—	395.30
5"	ER@2.00	Ea	703.00	76.20	31.40	810.60
6"	ER@2.50	Ea	703.00	95.30	39.30	837.60
8"	ER@3.00	Ea	1,190.00	114.00	47.10	1,351.10

Description	Craft@Hrs	Unit	Material $	Labor $	Equipment $	Total $

Installation of 2-way control valve, threaded joints

Description	Craft@Hrs	Unit	Material $	Labor $	Equipment $	Total $
1/2"	P1@.210	Ea	—	7.51	—	7.51
3/4"	P1@.275	Ea	—	9.83	—	9.83
1"	P1@.350	Ea	—	12.50	—	12.50
1¼"	P1@.430	Ea	—	15.40	—	15.40
1½"	P1@.505	Ea	—	18.10	—	18.10
2"	P1@.675	Ea	—	24.10	—	24.10
2½"	P1@.830	Ea	—	29.70	—	29.70
3"	P1@.990	Ea	—	35.40	—	35.40

Installation of 3-way control valve, threaded joints

Description	Craft@Hrs	Unit	Material $	Labor $	Equipment $	Total $
1/2"	P1@.260	Ea	—	9.30	—	9.30
3/4"	P1@.365	Ea	—	13.10	—	13.10
1"	P1@.475	Ea	—	17.00	—	17.00
1¼"	P1@.575	Ea	—	20.60	—	20.60
1½"	P1@.680	Ea	—	24.30	—	24.30
2"	P1@.910	Ea	—	32.50	—	32.50
2½"	P1@1.12	Ea	—	40.10	—	40.10
3"	P1@1.33	Ea	—	47.60	—	47.60

PVC companion flange

Description	Craft@Hrs	Unit	Material $	Labor $	Equipment $	Total $
2"	P1@.140	Ea	5.67	5.01	—	10.68
2½"	P1@.180	Ea	16.90	6.44	—	23.34
3"	P1@.210	Ea	17.60	7.51	—	25.11
4"	P1@.280	Ea	18.30	10.00	—	28.30
5"	P1@.310	Ea	40.50	11.10	—	51.60
6"	P1@.420	Ea	40.50	15.00	—	55.50
8"	P1@.560	Ea	60.50	20.00	—	80.50

Bolt and gasket set

Description	Craft@Hrs	Unit	Material $	Labor $	Equipment $	Total $
2"	P1@.500	Ea	3.86	17.90	—	21.76
2½"	P1@.650	Ea	4.51	23.20	—	27.71
3"	P1@.750	Ea	7.64	26.80	—	34.44
4"	P1@1.00	Ea	12.70	35.80	—	48.50
5"	P1@1.10	Ea	21.20	39.30	—	60.50
6"	P1@1.20	Ea	21.20	42.90	—	64.10
8"	P1@1.25	Ea	23.80	44.70	—	68.50

Thermometer with well

Description	Craft@Hrs	Unit	Material $	Labor $	Equipment $	Total $
7"	P1@.250	Ea	159.00	8.94	—	167.94
9"	P1@.250	Ea	164.00	8.94	—	172.94

Dial type pressure gauge

Description	Craft@Hrs	Unit	Material $	Labor $	Equipment $	Total $
2½"	P1@.200	Ea	33.10	7.15	—	40.25
3½"	P1@.200	Ea	43.80	7.15	—	50.95

PVC, Schedule 80, with Solvent-Weld Joints

Description	Craft@Hrs	Unit	Material $	Labor $	Equipment $	Total $
Pressure/temperature tap						
tap	P1@.150	Ea	15.70	5.36	—	21.06
Hanger with swivel assembly						
1/2"	P1@.250	Ea	3.74	8.94	—	12.68
3/4"	P1@.250	Ea	4.13	8.94	—	13.07
1"	P1@.250	Ea	4.33	8.94	—	13.27
1¼"	P1@.300	Ea	4.44	10.70	—	15.14
1½"	P1@.300	Ea	4.77	10.70	—	15.47
2"	P1@.300	Ea	4.98	10.70	—	15.68
2½"	P1@.350	Ea	6.78	12.50	—	19.28
3"	P1@.350	Ea	8.37	12.50	—	20.87
4"	P1@.350	Ea	9.19	12.50	—	21.69
5"	P1@.450	Ea	13.40	16.10	—	29.50
6"	P1@.450	Ea	15.10	16.10	—	31.20
8"	P1@.450	Ea	19.00	16.10	—	35.10
Riser clamp						
1/2"	P1@.100	Ea	2.41	3.58	—	5.99
3/4"	P1@.100	Ea	3.55	3.58	—	7.13
1"	P1@.100	Ea	3.59	3.58	—	7.17
1¼"	P1@.105	Ea	4.33	3.75	—	8.08
1½"	P1@.110	Ea	4.58	3.93	—	8.51
2"	P1@.115	Ea	4.84	4.11	—	8.95
2½"	P1@.120	Ea	5.09	4.29	—	9.38
3"	P1@.120	Ea	5.52	4.29	—	9.81
4"	P1@.125	Ea	7.03	4.47	—	11.50
5"	P1@.180	Ea	10.10	6.44	—	16.54
6"	P1@.200	Ea	12.20	7.15	—	19.35
8"	P1@.200	Ea	19.80	7.15	—	26.95

Composite pressure pipe is aluminum tube laminated between 2 layers of plastic pipe. It is manufactured and sold in various coiled lengths (200' to 1,000' rolls). The jointing method can be either crimped or compression connections. It will not rust or corrode and because of its flexibility, installations require approximately 40% less fittings. Expansion rates are similar to copper, but its thermal coefficient (heat loss) is over 800 times less than copper. It is also chemically resistant to most acids, salt solutions, alkalis, fats and oils.

Common applications include ice or snow melting systems, radiant floor heating systems, water service tubing, hot & cold domestic water service, chilled water systems, compressed air systems, solar and process piping applications.

Maximum temperature	Maximum pressure	Pipe selection
73 degrees F	200 psi	PE-AL or PEX-AL
140 degrees F	160 psi	PE-AL or PEX-AL
180 degrees F	125 psi	PEX-AL
210 degrees F	115 psi	PEX-AL

PE-AL is the designation for polyethylene-aluminum composite pipe
PEX-AL is the designation for cross linked polyethylene-aluminum composite pipe

Description	Craft@Hrs	Unit	Material $	Labor $	Equipment $	Total $

Cross linked Polyethylene-Aluminum pipe with crimped joints (PEX-AL), pipe only

Description	Craft@Hrs	Unit	Material $	Labor $	Equipment $	Total $
3/8"	P1@.028	LF	.42	1.00	—	1.42
1/2"	P1@.028	LF	.62	1.00	—	1.62
5/8"	P1@.030	LF	.86	1.07	—	1.93
3/4"	P1@.030	LF	1.22	1.07	—	2.29
1"	P1@.034	LF	1.97	1.22	—	3.19

Polyethylene-Aluminum pipe with crimped joints (PE-AL), pipe only

Description	Craft@Hrs	Unit	Material $	Labor $	Equipment $	Total $
1/2"	P1@.028	LF	.47	1.00	—	1.47
5/8"	P1@.030	LF	.65	1.07	—	1.72
3/4"	P1@.030	LF	.86	1.07	—	1.93
1"	P1@.034	LF	1.39	1.22	—	2.61

90 degree brass ell with crimped joints (PEX-AL/PE-AL)

Description	Craft@Hrs	Unit	Material $	Labor $	Equipment $	Total $
1/2"	P1@.080	Ea	1.29	2.86	—	4.15
5/8"	P1@.085	Ea	1.55	3.04	—	4.59
3/4"	P1@.090	Ea	3.00	3.22	—	6.22
1"	P1@.105	Ea	3.27	3.75	—	7.02

90 degree brass ell with crimped joints x male pipe thread (PEX-AL x MPT/PE-AL x MPT)

Description	Craft@Hrs	Unit	Material $	Labor $	Equipment $	Total $
1/2" x 1/2"	P1@.085	Ea	1.41	3.04	—	4.45
1/2" x 3/8"	P1@.095	Ea	1.78	3.40	—	5.18

Polyethylene-Aluminum Pipe with Crimped Joints

Description	Craft@Hrs	Unit	Material $	Labor $	Equipment $	Total $

90 degree brass ell with crimped joints x copper socket (PEX-AL x C/PE-AL x C)

Description	Craft@Hrs	Unit	Material $	Labor $	Equipment $	Total $
1/2"	P1@.085	Ea	1.41	3.04	—	4.45
1/2" x 3/4"	P1@.090	Ea	1.78	3.22	—	5.00
5/8" x 3/4"	P1@.095	Ea	1.94	3.40	—	5.34
3/4"	P1@.095	Ea	1.58	3.40	—	4.98

90 degree brass wingback ell with crimped joints x female pipe thread (PEX-AL x FPT/PE-AL x FPT)

Description	Craft@Hrs	Unit	Material $	Labor $	Equipment $	Total $
1/2"	P1@.085	Ea	2.75	3.04	—	5.79
3/4"	P1@.095	Ea	3.27	3.40	—	6.67

Tee (brass) with crimped joints (PEX-AL/PE-AL)

Description	Craft@Hrs	Unit	Material $	Labor $	Equipment $	Total $
1/2"	P1@.095	Ea	1.37	3.40	—	4.77
5/8"	P1@.105	Ea	1.62	3.75	—	5.37
3/4"	P1@.110	Ea	3.27	3.93	—	7.20
1"	P1@.120	Ea	4.01	4.29	—	8.30

Reducing tee (brass) with crimped joints (PEX-AL/PE-AL)

Description	Craft@Hrs	Unit	Material $	Labor $	Equipment $	Total $
1/2" x 1/2" x 5/8"	P1@.110	Ea	1.62	3.93	—	5.55
1/2" x 1/2" x 3/4"	P1@.115	Ea	1.89	4.11	—	6.00
5/8" x 1/2" x 1/2"	P1@.115	Ea	1.62	4.11	—	5.73
5/8" x 1/2" x 5/8"	P1@.115	Ea	1.62	4.11	—	5.73
5/8" x 5/8" x 1/2"	P1@.115	Ea	1.62	4.11	—	5.73
3/4" x 1/2" x 1/2"	P1@.115	Ea	1.89	4.11	—	6.00
3/4" x 1/2" x 3/4"	P1@.115	Ea	2.02	4.11	—	6.13
3/4" x 3/4" x 1/2"	P1@.115	Ea	2.02	4.11	—	6.13
3/4" x 3/4" x 5/8"	P1@.115	Ea	2.02	4.11	—	6.13
3/4" x 3/4" x 1"	P1@.120	Ea	3.87	4.29	—	8.16
1" x 1/2" x 1/2"	P1@.120	Ea	3.76	4.29	—	8.05
1" x 1/2" x 3/4"	P1@.120	Ea	3.51	4.29	—	7.80
1" x 1/2" x 1"	P1@.120	Ea	3.87	4.29	—	8.16
1" x 3/4" x 1/2"	P1@.120	Ea	3.87	4.29	—	8.16
1" x 3/4" x 3/4"	P1@.120	Ea	3.87	4.29	—	8.16
1" x 3/4" x 1"	P1@.120	Ea	3.87	4.29	—	8.16
1" x 1" x 1/2"	P1@.120	Ea	3.49	4.29	—	7.78
1" x 1" x 3/4"	P1@.120	Ea	4.01	4.29	—	8.30

Coupling (brass) with crimped joints (PEX-AL/PE-AL)

Description	Craft@Hrs	Unit	Material $	Labor $	Equipment $	Total $
1/2"	P1@.080	Ea	.63	2.86	—	3.49
5/8"	P1@.085	Ea	.80	3.04	—	3.84
3/4"	P1@.090	Ea	1.76	3.22	—	4.98
1"	P1@.105	Ea	1.80	3.75	—	5.55

Polyethylene-Aluminum Pipe with Crimped Joints

Description	Craft@Hrs	Unit	Material $	Labor $	Equipment $	Total $

Reducing coupling (brass) with crimped joints (PEX-AL/PE-AL)

Description	Craft@Hrs	Unit	Material $	Labor $	Equipment $	Total $
1/2"	P1@.080	Ea	.80	2.86	—	3.66
5/8"	P1@.085	Ea	1.14	3.04	—	4.18
3/4"	P1@.090	Ea	1.20	3.22	—	4.42
1"	P1@.105	Ea	1.48	3.75	—	5.23

Adapter (brass) with crimped joints x male pipe thread (PEX-AL x MPT/PE-AL x MPT)

Description	Craft@Hrs	Unit	Material $	Labor $	Equipment $	Total $
1/2" x 1/2"	P1@.080	Ea	.95	2.86	—	3.81
1/2" x 3/4"	P1@.085	Ea	1.20	3.04	—	4.24
5/8" x 3/4"	P1@.085	Ea	1.18	3.04	—	4.22
3/4" x 1/2"	P1@.090	Ea	1.62	3.22	—	4.84
3/4" x 3/4"	P1@.090	Ea	1.72	3.22	—	4.94
3/4" x 1"	P1@.105	Ea	2.75	3.75	—	6.50
1" x 1"	P1@.105	Ea	3.00	3.75	—	6.75

Adapter (brass) with crimped joints x female pipe thread (PEX-AL x FPT/PE-AL x FPT)

Description	Craft@Hrs	Unit	Material $	Labor $	Equipment $	Total $
1/2" x 1/2"	P1@.080	Ea	.99	2.86	—	3.85
1/2" x 3/4"	P1@.085	Ea	1.37	3.04	—	4.41
5/8" x 3/4"	P1@.085	Ea	1.37	3.04	—	4.41
3/4" x 1/2"	P1@.090	Ea	1.62	3.22	—	4.84
3/4" x 3/4"	P1@.090	Ea	1.78	3.22	—	5.00
3/4" x 1"	P1@.105	Ea	2.97	3.75	—	6.72
1" x 1"	P1@.105	Ea	3.27	3.75	—	7.02

Adapter (brass) with crimped joints x copper fitting (spigot) (PEX-AL x C ftg/PE-AL x C ftg)

Description	Craft@Hrs	Unit	Material $	Labor $	Equipment $	Total $
1/2" x 1/2"	P1@.080	Ea	1.14	2.86	—	4.00
1/2" x 3/4"	P1@.085	Ea	1.39	3.04	—	4.43
5/8" x 3/4"	P1@.085	Ea	1.37	3.04	—	4.41
3/4" x 3/4"	P1@.090	Ea	2.75	3.22	—	5.97
1" x 1"	P1@.105	Ea	3.27	3.75	—	7.02

Adapter (brass) with crimped joints x copper socket (PEX-AL x C/PE-AL x C)

Description	Craft@Hrs	Unit	Material $	Labor $	Equipment $	Total $
1/2" x 1/2"	P1@.080	Ea	.95	2.86	—	3.81
1/2" x 3/4"	P1@.085	Ea	1.40	3.04	—	4.44
5/8" x 3/4"	P1@.085	Ea	1.45	3.04	—	4.49
3/4" x 1/2"	P1@.090	Ea	1.56	3.22	—	4.78
3/4" x 3/4"	P1@.090	Ea	2.25	3.22	—	5.47
3/4" x 1"	P1@.105	Ea	3.27	3.75	—	7.02
1" x 1"	P1@.105	Ea	3.76	3.75	—	7.51

Polyethylene-Aluminum Pipe with Crimped Joints

Description	Craft@Hrs	Unit	Material $	Labor $	Equipment $	Total $

Cap (brass) with crimped joints (PEX-AL/PE-AL)

Description	Craft@Hrs	Unit	Material $	Labor $	Equipment $	Total $
1/2"	P1@.060	Ea	.43	2.15	—	2.58
5/8"	P1@.065	Ea	.60	2.32	—	2.92
3/4"	P1@.070	Ea	.80	2.50	—	3.30
1"	P1@.085	Ea	.99	3.04	—	4.03

Mini ball valve (brass) with crimped joints (PEX-AL/PE-AL)

Description	Craft@Hrs	Unit	Material $	Labor $	Equipment $	Total $
1/2" x 3/8" straight	P1@.150	Ea	5.03	5.36	—	10.39
1/2" x 3/8" angle	P1@.150	Ea	4.63	5.36	—	9.99
1/2" x copper	P1@.150	Ea	5.30	5.36	—	10.66
1/2" x MPT	P1@.150	Ea	5.58	5.36	—	10.94
1/2"	P1@.150	Ea	5.90	5.36	—	11.26
1/2" x comp	P1@.150	Ea	5.98	5.36	—	11.34
5/8"	P1@.165	Ea	8.40	5.90	—	14.30
5/8" x 3/4" copper	P1@.170	Ea	6.70	6.08	—	12.78
5/8" x 3/4" MPT	P1@.170	Ea	7.35	6.08	—	13.43
3/4"	P1@.170	Ea	9.03	6.08	—	15.11
3/4" x copper	P1@.170	Ea	8.50	6.08	—	14.58

Balancing valve (brass) with crimped joints (PEX-AL/PE-AL)

Description	Craft@Hrs	Unit	Material $	Labor $	Equipment $	Total $
1/2" x copper	P1@.150	Ea	34.50	5.36	—	39.86
1/2" x MPT	P1@.150	Ea	39.80	5.36	—	45.16

Manifolds (headers) copper with crimped joints (PEX-AL)

Description	Craft@Hrs	Unit	Material $	Labor $	Equipment $	Total $
3 - 1/2" outlets	P1@.750	Ea	29.00	26.80	—	55.80
3 - 5/8" outlets	P1@.750	Ea	32.80	26.80	—	59.60
4 - 1/2" outlets	P1@.850	Ea	37.20	30.40	—	67.60
4 - 5/8" outlets	P1@.850	Ea	43.00	30.40	—	73.40
5 - 1/2" outlets	P1@1.00	Ea	47.60	35.80	—	83.40
6 - 1/2" outlets	P1@1.20	Ea	54.00	42.90	—	96.90
8 - 1/2" outlets	P1@1.45	Ea	77.00	51.90	—	128.90
10 - 1/2" outlets	P1@1.65	Ea	99.00	59.00	—	158.00
12 - 1/2" outlets	P1@1.85	Ea	120.00	66.20	—	186.20

Manifolds (headers) copper with mini ball valves and crimped joints (PEX-AL)

Description	Craft@Hrs	Unit	Material $	Labor $	Equipment $	Total $
3 - 1/2" outlets	P1@.750	Ea	107.00	26.80	—	133.80
3 - 5/8" outlets	P1@.750	Ea	129.00	26.80	—	155.80
4 - 1/2" outlets	P1@.850	Ea	140.00	30.40	—	170.40
4 - 5/8" outlets	P1@.850	Ea	170.00	30.40	—	200.40
5 - 1/2" outlets	P1@1.00	Ea	177.00	35.80	—	212.80
6 - 1/2" outlets	P1@1.20	Ea	203.00	42.90	—	245.90

Description	Craft@Hrs	Unit	Material $	Labor $	Equipment $	Total $

Crimp rings, nickel plated soft copper

Description	Craft@Hrs	Unit	Material $	Labor $	Equipment $	Total $
1/2"	P1@.000	Ea	.16	—	—	.16
5/8"	P1@.000	Ea	.17	—	—	.17
3/4"	P1@.000	Ea	.20	—	—	.20
1"	P1@.000	Ea	.30	—	—	.30

Replacement O-rings

Description	Craft@Hrs	Unit	Material $	Labor $	Equipment $	Total $
1/2"	P1@.000	Ea	.11	—	—	.11
5/8"	P1@.000	Ea	.11	—	—	.11
3/4"	P1@.000	Ea	.16	—	—	.16
1"	P1@.000	Ea	.16	—	—	.16

Pipe hangers (polypropylene)

Description	Craft@Hrs	Unit	Material $	Labor $	Equipment $	Total $
1/2"	P1@.050	Ea	.05	1.79	—	1.84
5/8"	P1@.050	Ea	.05	1.79	—	1.84
3/4"	P1@.050	Ea	.05	1.79	—	1.84
1"	P1@.050	Ea	.07	1.79	—	1.86

Pipe clips (nail clips)

Description	Craft@Hrs	Unit	Material $	Labor $	Equipment $	Total $
1/2"	P1@.030	Ea	.12	1.07	—	1.19
5/8"	P1@.030	Ea	.18	1.07	—	1.25
1"	P1@.030	Ea	.42	1.07	—	1.49

Miscellaneous tools

Description	Craft@Hrs	Unit	Material $	Labor $	Equipment $	Total $
Crimp tool	P1@.000	Ea	242.00	—	—	242.00
Beveling tool	P1@.000	Ea	4.91	—	—	4.91
Reaming tool	P1@.000	Ea	29.40	—	—	29.40
Pipe bender kit	P1@.000	Ea	603.00	—	—	603.00
Bending spring	P1@.000	Ea	26.10	—	—	26.10
Pipe cutter	P1@.000	Ea	17.60	—	—	17.60

Polyethylene-Aluminum Pipe with Compression Joints

Description	Craft@Hrs	Unit	Material $	Labor $	Equipment $	Total $

Composite pressure pipe is aluminum tube laminated between 2 layers of plastic pipe. It is manufactured and sold in various coiled lengths (200' to 1000' rolls). The jointing method can be either crimped or compression connections. It will not rust or corrode and because of its flexibility, installations require approximately 40% less fittings. Installed like soft copper, it can be supported every 8.2 feet where applicable. Expansion rates are similar to copper, but its thermal coefficient (heat loss) is over 800 times less than copper. It is also chemically resistant to most acids, salt solutions, alkalis, fats and oils.

Common applications include ice or snow melting systems, radiant floor heating systems, water service tubing, hot & cold domestic water service, chilled water systems, compressed air systems, solar and process piping applications.

Maximum temperature	Maximum pressure	Pipe selection
73 degrees F	200 psi	PE-AL or PEX-AL
140 degrees F	160 psi	PE-AL or PEX-AL
180 degrees F	125 psi	PEX-AL
210 degrees F	115 psi	PEX-AL

PE-AL is the designation for polyethylene-aluminum composite pipe
PEX-AL is the designation for cross linked polyethylene-aluminum composite pipe

Description	Craft@Hrs	Unit	Material $	Labor $	Equipment $	Total $

Cross linked Polyethylene-Aluminum pipe with compression joints (PEX-AL), pipe only

3/8"	P1@.028	LF	.38	1.00	—	1.38
1/2"	P1@.028	LF	.57	1.00	—	1.57
5/8"	P1@.030	LF	.77	1.07	—	1.84
3/4"	P1@.030	LF	1.10	1.07	—	2.17
1"	P1@.034	LF	1.76	1.22	—	2.98

Polyethylene-Aluminum pipe with compression joints (PE-AL), pipe only

1/2"	P1@.028	LF	.46	1.00	—	1.46
5/8"	P1@.030	LF	.62	1.07	—	1.69
3/4"	P1@.030	LF	.84	1.07	—	1.91
1"	P1@.034	LF	1.34	1.22	—	2.56

90 degree brass ell with compression joints (PEX-AL/PE-AL)

1/2"	P1@.092	Ea	2.70	3.29	—	5.99
5/8"	P1@.098	Ea	3.75	3.50	—	7.25
3/4"	P1@.104	Ea	5.48	3.72	—	9.20
1"	P1@.121	Ea	9.60	4.33	—	13.93

90 degree brass ell with compression joints x male pipe thread (PEX-AL x MPT/PE-AL x MPT)

1/2" x 1/2"	P1@.098	Ea	4.23	3.50	—	7.73
1/2" x 3/4"	P1@.105	Ea	5.13	3.75	—	8.88
3/4" x 3/4"	P1@.109	Ea	4.73	3.90	—	8.63

Description	Craft@Hrs	Unit	Material $	Labor $	Equipment $	Total $

90 degree brass ell with compression joints x copper socket (PEX-AL x C/PE-AL x C)

Description	Craft@Hrs	Unit	Material $	Labor $	Equipment $	Total $
1/2"	P1@.098	Ea	2.31	3.50	—	5.81
1/2" x 3/4"	P1@.109	Ea	2.45	3.90	—	6.35
5/8" x 3/4"	P1@.112	Ea	2.93	4.01	—	6.94
3/4"	P1@.115	Ea	4.43	4.11	—	8.54

90 degree brass wingback ell with compression joints x female pipe thread (PEX-AL x FPT/PE-AL x FPT)

Description	Craft@Hrs	Unit	Material $	Labor $	Equipment $	Total $
1/2"	P1@.098	Ea	2.00	3.50	—	5.50
5/8"	P1@.105	Ea	3.43	3.75	—	7.18
3/4"	P1@.115	Ea	5.25	4.11	—	9.36

Tee (brass) with compression joints (PEX-AL/PE-AL)

Description	Craft@Hrs	Unit	Material $	Labor $	Equipment $	Total $
1/2"	P1@.109	Ea	3.30	3.90	—	7.20
5/8"	P1@.121	Ea	5.30	4.33	—	9.63
3/4"	P1@.127	Ea	7.85	4.54	—	12.39
1"	P1@.141	Ea	13.00	5.04	—	18.04

Reducing tee (brass) with compression joints (PEX-AL/PE-AL)

Description	Craft@Hrs	Unit	Material $	Labor $	Equipment $	Total $
5/8" x 5/8" x 1/2"	P1@.138	Ea	4.85	4.93	—	9.78
3/4" x 1/2" x 3/4"	P1@.138	Ea	6.90	4.93	—	11.83
3/4" x 3/4" x 1/2"	P1@.138	Ea	7.05	4.93	—	11.98
3/4" x 1/2" x 3/4"	P1@.138	Ea	6.25	4.93	—	11.18
3/4" x 3/4" x 5/8"	P1@.138	Ea	7.13	4.93	—	12.06
1" x 1/2" x 1/2"	P1@.142	Ea	10.60	5.08	—	15.68
1" x 1" x 1/2"	P1@.142	Ea	14.30	5.08	—	19.38
1" x 3/4" x 3/4"	P1@.142	Ea	8.03	5.08	—	13.11
1" x 1" x 3/4"	P1@.142	Ea	12.40	5.08	—	17.48

Coupling (brass) with compression joints (PEX-AL/PE-AL)

Description	Craft@Hrs	Unit	Material $	Labor $	Equipment $	Total $
3/8"	P1@.090	Ea	2.28	3.22	—	5.50
1/2"	P1@.092	Ea	2.49	3.29	—	5.78
5/8"	P1@.098	Ea	3.38	3.50	—	6.88
3/4"	P1@.104	Ea	5.03	3.72	—	8.75
1"	P1@.121	Ea	9.03	4.33	—	13.36

Reducing coupling (brass) with compression joints (PEX-AL/PE-AL)

Description	Craft@Hrs	Unit	Material $	Labor $	Equipment $	Total $
5/8" x 1/2"	P1@.092	Ea	2.98	3.29	—	6.27
3/4" x 1/2"	P1@.098	Ea	4.50	3.50	—	8.00
3/4" x 5/8"	P1@.104	Ea	4.35	3.72	—	8.07
1" x 3/4"	P1@.121	Ea	6.05	4.33	—	10.38

Cap (brass) with compression joints (PEX-AL/PE-AL)

Description	Craft@Hrs	Unit	Material $	Labor $	Equipment $	Total $
1/2"	P1@.069	Ea	1.67	2.47	—	4.14
5/8"	P1@.075	Ea	2.31	2.68	—	4.99
3/4"	P1@.081	Ea	3.03	2.90	—	5.93
1"	P1@.098	Ea	4.50	3.50	—	8.00

Polyethylene-Aluminum Pipe with Compression Joints

Description	Craft@Hrs	Unit	Material $	Labor $	Equipment $	Total $
Mini ball valve (brass) with compression joints (PEX-AL/PE-AL)						
1/2" x 3/8" straight	P1@.173	Ea	6.08	6.19	—	12.27
1/2" x 3/8" angle	P1@.173	Ea	5.73	6.19	—	11.92
1/2" x copper	P1@.173	Ea	5.70	6.19	—	11.89
1/2" x MPT	P1@.173	Ea	6.00	6.19	—	12.19
1/2"	P1@.173	Ea	6.70	6.19	—	12.89
5/8"	P1@.173	Ea	8.85	6.19	—	15.04
5/8" x 3/4" copper	P1@.196	Ea	6.90	7.01	—	13.91
5/8" x 3/4" MPT	P1@.196	Ea	7.40	7.01	—	14.41
3/4"	P1@.196	Ea	8.23	7.01	—	15.24
3/4" x copper	P1@.196	Ea	8.55	7.01	—	15.56
Balancing valve (brass) with compression joints (PEX-AL/PE-AL)						
1/2" x copper	P1@.173	Ea	38.20	6.19	—	44.39
1/2" x MPT	P1@.173	Ea	42.40	6.19	—	48.59
Crimp rings, nickel plated soft copper						
1/2"	P1@.000	Ea	.16	—	—	.16
5/8"	P1@.000	Ea	.18	—	—	.18
3/4"	P1@.000	Ea	.21	—	—	.21
1"	P1@.000	Ea	.31	—	—	.31
Replacement O-rings						
1/2"	P1@.000	Ea	.11	—	—	.11
5/8"	P1@.000	Ea	.11	—	—	.11
3/4"	P1@.000	Ea	.16	—	—	.16
1"	P1@.000	Ea	.16	—	—	.16
Pipe hangers (polypropylene)						
1/2"	P1@.050	Ea	.05	1.79	—	1.84
5/8"	P1@.050	Ea	.05	1.79	—	1.84
3/4"	P1@.050	Ea	.05	1.79	—	1.84
1"	P1@.050	Ea	.07	1.79	—	1.86
Pipe clips (nail clips)						
1/2"	P1@.030	Ea	.12	1.07	—	1.19
5/8"	P1@.030	Ea	.18	1.07	—	1.25
1"	P1@.030	Ea	.42	1.07	—	1.49
Miscellaneous tools						
Crimp tool	P1@.000	Ea	242.00	—	—	242.00
Beveling tool	P1@.000	Ea	4.91	—	—	4.91
Reaming tool	P1@.000	Ea	29.40	—	—	29.40
Pipe bender kit	P1@.000	Ea	434.00	—	—	434.00
Bending spring	P1@.000	Ea	36.30	—	—	36.30
Pipe cutter	P1@.000	Ea	17.60	—	—	17.60

Description	Craft@Hrs	Unit	Material $	Labor $	Equipment $	Total $

Water meters, turbine type. Including brass connection unions

Description	Craft@Hrs	Unit	Material $	Labor $	Equipment $	Total $
1/2"	P1@.500	Ea	115.00	17.90	—	132.90
3/4"	P1@.500	Ea	186.00	17.90	—	203.90
1"	P1@.650	Ea	250.00	23.20	—	273.20
1½"	P1@.700	Ea	672.00	25.00	—	697.00
2"	P1@.800	Ea	991.00	28.60	—	1,019.60
3"	P1@.900	Ea	1,540.00	32.20	—	1,572.20
4"	P1@.975	Ea	2,880.00	34.90	—	2,914.90

Water meters, compound type. Including brass connection unions

Description	Craft@Hrs	Unit	Material $	Labor $	Equipment $	Total $
2"	P1@.800	Ea	2,440.00	28.60	—	2,468.60
3"	P1@.900	Ea	3,170.00	32.20	—	3,202.20
4"	P1@.975	Ea	5,210.00	34.90	—	5,244.90

Water meter by-pass and connection assembly. Includes three isolation ball valves, two tees, two 90 degree elbows, 10' of Type L copper pipe and standard support devices. Make additional allowances for the cost of the water meter.

Description	Craft@Hrs	Unit	Material $	Labor $	Equipment $	Total $
1/2"	P1@1.00	Ea	79.10	35.80	—	114.90
3/4"	P1@1.50	Ea	120.00	53.60	—	173.60
1"	P1@1.95	Ea	207.00	69.70	—	276.70
1½"	P1@2.45	Ea	381.00	87.60	—	468.60
2"	P1@3.25	Ea	560.00	116.00	—	676.00
3"	P1@5.50	Ea	1,670.00	197.00	—	1,867.00
4"	P1@6.75	Ea	2,840.00	241.00	—	3,081.00

Backflow preventers-reduced pressure. Including valves and test ports

Description	Craft@Hrs	Unit	Material $	Labor $	Equipment $	Total $
3/4"	P1@1.00	Ea	290.00	35.80	—	325.80
1"	P1@1.25	Ea	359.00	44.70	—	403.70
1½"	P1@2.00	Ea	577.00	71.50	—	648.50
2"	P1@2.15	Ea	718.00	76.90	—	794.90
2½"	P1@4.00	Ea	2,600.00	143.00	—	2,743.00
3"	P1@4.75	Ea	3,440.00	170.00	—	3,610.00
4"	P1@6.00	Ea	3,910.00	215.00	—	4,125.00
6"	P1@7.95	Ea	4,280.00	284.00	—	4,564.00

Backflow preventers-reduced pressure. Including integral ball valves and test ports

Description	Craft@Hrs	Unit	Material $	Labor $	Equipment $	Total $
1/2"	P1@.900	Ea	176.00	32.20	—	208.20
3/4"	P1@1.00	Ea	209.00	35.80	—	244.80
1"	P1@1.25	Ea	261.00	44.70	—	305.70
1¼"	P1@1.50	Ea	383.00	53.60	—	436.60
1½"	P1@2.00	Ea	416.00	71.50	—	487.50
2"	P1@2.15	Ea	469.00	76.90	—	545.90

Plumbing and Piping Specialties

Description	Craft@Hrs	Unit	Material $	Labor $	Equipment $	Total $

Backflow preventers, double check valve assembly. Including integral ball valves and test ports

1/2"	P1@1.00	Ea	132.00	35.80	—	167.80
3/4"	P1@1.00	Ea	146.00	35.80	—	181.80
1"	P1@1.25	Ea	176.00	44.70	—	220.70
1¼"	P1@1.95	Ea	297.00	69.70	—	366.70
1½"	P1@2.00	Ea	307.00	71.50	—	378.50
2"	P1@2.15	Ea	373.00	76.90	—	449.90
2½"	P1@2.95	Ea	1,530.00	105.00	—	1,635.00
3"	P1@3.45	Ea	1,950.00	123.00	—	2,073.00
4"	P1@3.95	Ea	2,650.00	141.00	—	2,791.00
6"	P1@5.75	Ea	4,460.00	206.00	—	4,666.00

Vacuum breakers, atmospheric. Female pipe thread

1/4"	P1@.350	Ea	46.20	12.50	—	58.70
1/2"	P1@.350	Ea	64.40	12.50	—	76.90
3/4"	P1@.350	Ea	78.80	12.50	—	91.30

Vacuum breakers, hose connection. Male/female hose thread, polished brass

1/2"	P1@.150	Ea	19.60	5.36	—	24.96
3/4"	P1@.150	Ea	23.90	5.36	—	29.26

Suction diffusers

2"	P1@1.25	Ea	270.00	44.70	—	314.70
3"	P1@2.50	Ea	454.00	89.40	—	543.40
4"	P1@3.25	Ea	580.00	116.00	—	696.00
6"	P1@4.80	Ea	819.00	172.00	—	991.00
8"	P1@5.95	Ea	1,530.00	213.00	—	1,743.00
10"	P1@6.75	Ea	2,090.00	241.00	—	2,331.00

Triple duty valves

2"	P1@1.25	Ea	359.00	44.70	—	403.70
3"	P1@2.50	Ea	486.00	89.40	—	575.40
4"	P1@3.25	Ea	878.00	116.00	—	994.00
6"	P1@4.80	Ea	1,440.00	172.00	—	1,612.00
8"	P1@5.95	Ea	2,120.00	213.00	—	2,333.00
10"	P1@6.75	Ea	3,120.00	241.00	—	3,361.00

In-line circulating pump, all bronze. 115 volt, including flange kit

1/25 HP	P1@1.50	Ea	344.00	53.60	—	397.60
1/16 HP	P1@1.55	Ea	433.00	55.40	—	488.40
1/12 HP	P1@1.60	Ea	598.00	57.20	—	655.20
1/6 HP	P1@1.75	Ea	1,020.00	62.60	—	1,082.60
1/4 HP	P1@1.95	Ea	1,520.00	69.70	—	1,589.70
1 HP	P1@2.25	Ea	1,760.00	80.50	—	1,840.50
1½ HP	P1@2.85	Ea	2,130.00	102.00	—	2,232.00

Description	Craft@Hrs	Unit	Material $	Labor $	Equipment $	Total $

In-line circulating pump, iron body. 115 volt, including flange kit

Description	Craft@Hrs	Unit	Material $	Labor $	Equipment $	Total $
1/25 HP	P1@1.50	Ea	169.00	53.60	—	222.60
1/16 HP	P1@1.55	Ea	189.00	55.40	—	244.40
1/12 HP	P1@1.60	Ea	354.00	57.20	—	411.20
1/6 HP	P1@1.75	Ea	602.00	62.60	—	664.60
1/4 HP	P1@1.95	Ea	697.00	69.70	—	766.70
1 HP	P1@2.25	Ea	817.00	80.50	—	897.50
1½ HP	P1@2.85	Ea	1,080.00	102.00	—	1,182.00

Primed steel cam-lock non-fire rated access doors

Description	Craft@Hrs	Unit	Material $	Labor $	Equipment $	Total $
8" x 8"	P1@.400	Ea	62.10	14.30	—	76.40
12" x 12"	P1@.500	Ea	71.40	17.90	—	89.30
16" x 16"	P1@.600	Ea	87.90	21.50	—	109.40
18" x 18"	P1@.800	Ea	101.00	28.60	—	129.60
24" x 24"	P1@1.25	Ea	149.00	44.70	—	193.70
36" x 36"	P1@1.60	Ea	275.00	57.20	—	332.20
Add for Allen key lock		Ea	4.50	—	—	4.50
Add for cylinder lock		Ea	11.10	—	—	11.10

Primed steel cam-lock 1 1/2 hour fire rated access doors

Description	Craft@Hrs	Unit	Material $	Labor $	Equipment $	Total $
8" x 8"	P1@.400	Ea	229.00	14.30	—	243.30
12" x 12"	P1@.500	Ea	240.00	17.90	—	257.90
16" x 16"	P1@.600	Ea	284.00	21.50	—	305.50
18" x 18"	P1@.800	Ea	289.00	28.60	—	317.60
24" x 24"	P1@1.25	Ea	404.00	44.70	—	448.70
24" x 36"	P1@1.50	Ea	523.00	53.60	—	576.60
48" x 48"	P1@2.00	Ea	809.00	71.50	—	880.50
Add for cylinder lock		Ea	11.40	—	—	11.40

Threaded automatic air vent

Description	Craft@Hrs	Unit	Material $	Labor $	Equipment $	Total $
Float type	P1@.150	Ea	11.60	5.36	—	16.96
35 PSIG	P1@.150	Ea	76.10	5.36	—	81.46
150 PSIG	P1@.150	Ea	125.00	5.36	—	130.36

Threaded manual air vent

Description	Craft@Hrs	Unit	Material $	Labor $	Equipment $	Total $
175 PSIG	P1@.150	Ea	7.21	5.36	—	12.57

150 pound forged steel slip-on companion flange, welding type

Description	Craft@Hrs	Unit	Material $	Labor $	Equipment $	Total $
2½"	P1@.610	Ea	38.20	21.80	—	60.00
3"	P1@.730	Ea	38.20	26.10	—	64.30
4"	P1@.980	Ea	52.70	35.00	—	87.70
6"	ER@1.47	Ea	69.50	56.00	23.10	148.60
8"	ER@1.77	Ea	129.00	67.50	27.90	224.40
10"	ER@2.20	Ea	214.00	83.80	34.70	332.50
12"	ER@2.64	Ea	320.00	101.00	41.40	462.40

Plumbing and Piping Specialties

Description	Craft@Hrs	Unit	Material $	Labor $	Equipment $	Total $

150 pound forged steel companion flange, threaded

2"	P1@.290	Ea	41.00	10.40	—	51.40
2½"	P1@.380	Ea	61.60	13.60	—	75.20
3"	P1@.460	Ea	68.60	16.40	—	85.00
4"	P1@.600	Ea	91.40	21.50	—	112.90
6"	ER@.680	Ea	135.00	25.90	10.60	171.50
8"	ER@.760	Ea	187.00	29.00	11.90	227.90

300 pound forged steel slip-on companion flange, welding type

2½"	P1@.810	Ea	43.80	29.00	—	72.80
3"	P1@.980	Ea	43.80	35.00	—	78.80
4"	P1@1.30	Ea	64.20	46.50	—	110.70
6"	ER@1.95	Ea	117.00	74.30	30.60	221.90
8"	ER@2.35	Ea	165.00	89.60	37.10	291.70
10"	ER@2.90	Ea	271.00	111.00	45.70	427.70
12"	ER@3.50	Ea	434.00	133.00	55.00	622.00

PVC companion flange

2"	P1@.140	Ea	4.48	5.01	—	9.49
2½"	P1@.180	Ea	13.50	6.44	—	19.94
3"	P1@.210	Ea	14.00	7.51	—	21.51
4"	P1@.280	Ea	14.50	10.00	—	24.50
5"	P1@.310	Ea	32.40	11.10	—	43.50
6"	P1@.420	Ea	32.40	15.00	—	47.40
8"	P1@.560	Ea	48.00	20.00	—	68.00

Bolt and gasket set

2"	P1@.500	Ea	4.03	17.90	—	21.93
2½"	P1@.650	Ea	4.70	23.20	—	27.90
3"	P1@.750	Ea	7.99	26.80	—	34.79
4"	P1@1.00	Ea	13.40	35.80	—	49.20
5"	P1@1.10	Ea	15.40	39.30	—	54.70
6"	P1@1.20	Ea	15.40	42.90	—	58.30
8"	P1@1.25	Ea	20.00	44.70	—	64.70
10"	P1@1.70	Ea	23.90	60.80	—	84.70
12"	P1@2.20	Ea	32.10	78.70	—	110.80

Dielectric union, brazed joints

1/2"	P1@.146	Ea	15.60	5.22	—	20.82
3/4"	P1@.205	Ea	15.60	7.33	—	22.93
1"	P1@.263	Ea	32.40	9.40	—	41.80
1¼"	P1@.322	Ea	55.40	11.50	—	66.90
1½"	P1@.380	Ea	75.60	13.60	—	89.20
2"	P1@.507	Ea	104.00	18.10	—	122.10

Description	Craft@Hrs	Unit	Material $	Labor $	Equipment $	Total $

125/150# galvanized ASME expansion tank

Description	Craft@Hrs	Unit	Material $	Labor $	Equipment $	Total $
15 gallon	P1@1.75	Ea	273.00	62.60	—	335.60
30 gallon	P1@2.00	Ea	491.00	71.50	—	562.50
40 gallon	P1@2.35	Ea	546.00	84.00	—	630.00
60 gallon	P1@2.65	Ea	740.00	94.80	—	834.80
80 gallon	P1@2.95	Ea	1,370.00	105.00	—	1,475.00
100 gallon	P1@3.25	Ea	1,640.00	116.00	—	1,756.00
120 gallon	P1@3.50	Ea	1,770.00	125.00	—	1,895.00

Expansion tank fitting

Description	Craft@Hrs	Unit	Material $	Labor $	Equipment $	Total $
3/4"	P1@.300	Ea	67.20	10.70	—	77.90

Lead roof pipe flashing

Description	Craft@Hrs	Unit	Material $	Labor $	Equipment $	Total $
1½"	P1@.250	Ea	31.60	8.94	—	40.54
2"	P1@.250	Ea	41.60	8.94	—	50.54
3"	P1@.250	Ea	48.30	8.94	—	57.24
4"	P1@.250	Ea	53.10	8.94	—	62.04
6"	P1@.350	Ea	68.10	12.50	—	80.60
8"	P1@.400	Ea	86.10	14.30	—	100.40

Roof flashing, 20" x 20"

Description	Craft@Hrs	Unit	Material $	Labor $	Equipment $	Total $
Pitched roof, neoprene						
all pipe	P1@.300	Ea	20.90	10.70	—	31.60
Pitched roof, aluminum/neoprene						
all pipe	P1@.300	Ea	22.20	10.70	—	32.90
Flat roof, aluminum/neoprene						
all pipe	P1@.300	Ea	42.10	10.70	—	52.80

Braided stainless steel pipe connector, solder ends, brazed joints

Description	Craft@Hrs	Unit	Material $	Labor $	Equipment $	Total $
1/2"	P1@.240	Ea	53.20	8.58	—	61.78
3/4"	P1@.300	Ea	70.50	10.70	—	81.20
1"	P1@.360	Ea	91.40	12.90	—	104.30
1¼"	P1@.480	Ea	106.00	17.20	—	123.20
1½"	P1@.540	Ea	121.00	19.30	—	140.30
2"	P1@.600	Ea	162.00	21.50	—	183.50

Braided stainless steel pipe connector, threaded joints

Description	Craft@Hrs	Unit	Material $	Labor $	Equipment $	Total $
1/2"	P1@.210	Ea	60.00	7.51	—	67.51
3/4"	P1@.250	Ea	64.90	8.94	—	73.84
1"	P1@.300	Ea	70.00	10.70	—	80.70
1¼"	P1@.400	Ea	71.40	14.30	—	85.70
1½"	P1@.450	Ea	96.80	16.10	—	112.90
2"	P1@.500	Ea	125.00	17.90	—	142.90

Plumbing and Piping Specialties

Description	Craft@Hrs	Unit	Material $	Labor $	Equipment $	Total $

Braided stainless steel pipe connector, flanged joints

Description	Craft@Hrs	Unit	Material $	Labor $	Equipment $	Total $
2"	P1@.500	Ea	143.00	17.90	—	160.90
2½"	P1@.630	Ea	177.00	22.50	—	199.50
3"	P1@.750	Ea	229.00	26.80	—	255.80
4"	P1@1.50	Ea	299.00	53.60	—	352.60
6"	P1@2.50	Ea	544.00	89.40	—	633.40
8"	P1@3.00	Ea	1,210.00	107.00	—	1,317.00

Galvanized steel band pipe hanger

Description	Craft@Hrs	Unit	Material $	Labor $	Equipment $	Total $
1/2"	P1@.250	Ea	1.87	8.94	—	10.81
3/4"	P1@.250	Ea	1.87	8.94	—	10.81
1"	P1@.250	Ea	1.87	8.94	—	10.81
1¼"	P1@.300	Ea	1.98	10.70	—	12.68
1½"	P1@.300	Ea	2.10	10.70	—	12.80
2"	P1@.300	Ea	2.20	10.70	—	12.90
2½"	P1@.350	Ea	2.70	12.50	—	15.20
3"	P1@.350	Ea	3.10	12.50	—	15.60
4"	P1@.350	Ea	3.93	12.50	—	16.43
6"	P1@.450	Ea	6.18	16.10	—	22.28
8"	P1@.450	Ea	9.19	16.10	—	25.29

Trapeze pipe hanger

Description	Craft@Hrs	Unit	Material $	Labor $	Equipment $	Total $
24"	P1@1.00	Ea	43.20	35.80	—	79.00
36"	P1@1.25	Ea	55.10	44.70	—	99.80
48"	P1@1.50	Ea	83.70	53.60	—	137.30

Wall bracket support with anchors and bolts

Description	Craft@Hrs	Unit	Material $	Labor $	Equipment $	Total $
6" x 6"	P1@.600	Ea	20.30	21.50	—	41.80
8" x 8"	P1@.650	Ea	23.30	23.20	—	46.50
10" x 10"	P1@.700	Ea	26.30	25.00	—	51.30
12" x 12"	P1@.800	Ea	31.70	28.60	—	60.30
15" x 15"	P1@.950	Ea	36.60	34.00	—	70.60
18" x 18"	P1@1.10	Ea	51.60	39.30	—	90.90
24" x 24"	P1@1.60	Ea	74.70	57.20	—	131.90

Galvanized U-bolts with nuts

Description	Craft@Hrs	Unit	Material $	Labor $	Equipment $	Total $
1/2"	P1@.160	Ea	1.59	5.72	—	7.31
3/4"	P1@.165	Ea	1.59	5.90	—	7.49
1"	P1@.170	Ea	1.75	6.08	—	7.83
1¼"	P1@.175	Ea	2.00	6.26	—	8.26
1½"	P1@.180	Ea	2.40	6.44	—	8.84
2"	P1@.190	Ea	2.57	6.79	—	9.36
2½"	P1@.210	Ea	3.97	7.51	—	11.48
3"	P1@.250	Ea	4.59	8.94	—	13.53
4"	P1@.300	Ea	5.20	10.70	—	15.90
6"	P1@.400	Ea	8.12	14.30	—	22.42

Description	Craft@Hrs	Unit	Material $	Labor $	Equipment $	Total $
Nail-on wire pipe hooks						
1/2"	P1@.100	Ea	.09	3.58	—	3.67
3/4"	P1@.100	Ea	.11	3.58	—	3.69
1"	P1@.100	Ea	.12	3.58	—	3.70
Galvanized steel pipe sleeves						
1"	P1@.120	Ea	3.97	4.29	—	8.26
1¼"	P1@.125	Ea	4.41	4.47	—	8.88
1½"	P1@.125	Ea	4.99	4.47	—	9.46
2"	P1@.130	Ea	5.37	4.65	—	10.02
2½"	P1@.150	Ea	5.46	5.36	—	10.82
3"	P1@.180	Ea	5.58	6.44	—	12.02
4"	P1@.220	Ea	6.34	7.87	—	14.21
5"	P1@.250	Ea	7.91	8.94	—	16.85
6"	P1@.270	Ea	8.54	9.66	—	18.20
8"	P1@.270	Ea	9.82	9.66	—	19.48
10"	P1@.290	Ea	11.40	10.40	—	21.80
12"	P1@.310	Ea	15.50	11.10	—	26.60
14"	P1@.330	Ea	28.90	11.80	—	40.70
Dial-type pressure gauge						
2½"	P1@.200	Ea	32.70	7.15	—	39.85
3½"	P1@.200	Ea	43.40	7.15	—	50.55
Riser clamp						
1/2"	P1@.100	Ea	2.41	3.58	—	5.99
3/4"	P1@.100	Ea	3.55	3.58	—	7.13
1"	P1@.100	Ea	3.59	3.58	—	7.17
1¼"	P1@.105	Ea	4.33	3.75	—	8.08
1½"	P1@.110	Ea	4.58	3.93	—	8.51
2"	P1@.115	Ea	4.84	4.11	—	8.95
2½"	P1@.120	Ea	5.09	4.29	—	9.38
3"	P1@.120	Ea	5.52	4.29	—	9.81
4"	P1@.125	Ea	7.03	4.47	—	11.50
5"	P1@.180	Ea	10.10	6.44	—	16.54
6"	P1@.200	Ea	12.20	7.15	—	19.35
8"	P1@.200	Ea	19.80	7.15	—	26.95
10"	P1@.250	Ea	29.40	8.94	—	38.34
12"	P1@.250	Ea	34.90	8.94	—	43.84
15 PSIG float and thermostatic steam trap, threaded						
3/4"	P1@.250	Ea	285.00	8.94	—	293.94
1"	P1@.300	Ea	285.00	10.70	—	295.70
1¼"	P1@.400	Ea	343.00	14.30	—	357.30
1½"	P1@.450	Ea	644.00	16.10	—	660.10
2"	P1@.500	Ea	826.00	17.90	—	843.90

Description	Craft@Hrs	Unit	Material $	Labor $	Equipment $	Total $

15 PSIG inverted bucket steam trap, threaded joints

Description	Craft@Hrs	Unit	Material $	Labor $	Equipment $	Total $
1/2"	P1@.210	Ea	165.00	7.51	—	172.51
3/4"	P1@.250	Ea	406.00	8.94	—	414.94
1"	P1@.300	Ea	581.00	10.70	—	591.70
1¼"	P1@.400	Ea	969.00	14.30	—	983.30
1½"	P1@.450	Ea	969.00	16.10	—	985.10
2"	P1@.500	Ea	1,450.00	17.90	—	1,467.90

Thermometer with well

Description	Craft@Hrs	Unit	Material $	Labor $	Equipment $	Total $
7"	P1@.250	Ea	159.00	8.94	—	167.94
9"	P1@.250	Ea	164.00	8.94	—	172.94

Class 125 bronze body 2-piece ball valve, brazed joints

Description	Craft@Hrs	Unit	Material $	Labor $	Equipment $	Total $
1½"	P1@.240	Ea	7.53	8.58	—	16.11
3/4"	P1@.300	Ea	10.00	10.70	—	20.70
1"	P1@.360	Ea	17.30	12.90	—	30.20
1¼"	P1@.480	Ea	28.90	17.20	—	46.10
1½"	P1@.540	Ea	39.00	19.30	—	58.30
2"	P1@.600	Ea	49.40	21.50	—	70.90
3"	P1@1.50	Ea	338.00	53.60	—	391.60
4"	P1@1.80	Ea	443.00	64.40	—	507.40

Class 125 bronze body 2-piece ball valve, soft-soldered joints

Description	Craft@Hrs	Unit	Material $	Labor $	Equipment $	Total $
1/2"	P1@.200	Ea	7.53	7.15	—	14.68
3/4"	P1@.249	Ea	10.00	8.90	—	18.90
1"	P1@.299	Ea	17.30	10.70	—	28.00
1¼"	P1@.308	Ea	28.90	11.00	—	39.90
1½"	P1@.498	Ea	39.00	17.80	—	56.80
2"	P1@.498	Ea	49.40	17.80	—	67.20
3"	P1@1.24	Ea	338.00	44.30	—	382.30
4"	P1@1.45	Ea	443.00	51.90	—	494.90

Class 125 bronze body 2-piece ball valve, threaded joints

Description	Craft@Hrs	Unit	Material $	Labor $	Equipment $	Total $
1/2"	P1@.210	Ea	7.53	7.51	—	15.04
3/4"	P1@.250	Ea	10.00	8.94	—	18.94
1"	P1@.300	Ea	17.30	10.70	—	28.00
1¼"	P1@.400	Ea	28.90	14.30	—	43.20
1½"	P1@.450	Ea	39.00	16.10	—	55.10
2"	P1@.500	Ea	49.40	17.90	—	67.30
3"	P1@.625	Ea	338.00	22.40	—	360.40
4"	P1@.690	Ea	443.00	24.70	—	467.70

Description	Craft@Hrs	Unit	Material $	Labor $	Equipment $	Total $

Class 150, 600 pound W.O.G. bronze body 2-piece ball valve, threaded joints

1/2"	P1@.210	Ea	12.30	7.51	—	19.81
3/4"	P1@.250	Ea	20.00	8.94	—	28.94
1"	P1@.300	Ea	25.40	10.70	—	36.10
1¼"	P1@.400	Ea	31.80	14.30	—	46.10
1½"	P1@.450	Ea	42.90	16.10	—	59.00
2"	P1@.500	Ea	54.30	17.90	—	72.20

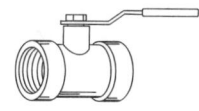

Bronze body circuit balance valve

1/2"	P1@.210	Ea	60.60	7.51	—	68.11
3/4"	P1@.250	Ea	64.40	8.94	—	73.34
1"	P1@.300	Ea	74.70	10.70	—	85.40
1¼"	P1@.400	Ea	92.90	14.30	—	107.20
1½"	P1@.400	Ea	117.00	14.30	—	131.30
2"	P1@.500	Ea	165.00	17.90	—	182.90
2½"	P1@1.75	Ea	623.00	62.60	—	685.60
3"	P1@2.50	Ea	959.00	89.40	—	1,048.40
4"	P1@3.15	Ea	1,640.00	113.00	—	1,753.00

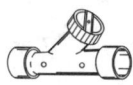

200 PSIG iron body butterfly valve, lug type, lever operated

2"	P1@.450	Ea	128.00	16.10	—	144.10
2½"	P1@.450	Ea	132.00	16.10	—	148.10
3"	P1@.550	Ea	139.00	19.70	—	158.70
4"	P1@.505	Ea	173.00	18.10	—	191.10
6"	ER@.800	Ea	279.00	30.50	12.70	322.20
8"	ER@.800	Ea	382.00	30.50	12.70	425.20
10"	ER@.900	Ea	530.00	34.30	14.10	578.40
12"	ER@1.00	Ea	694.00	38.10	15.70	747.80

200 PSIG iron body butterfly valve, wafer type, lever operated

2"	P1@.450	Ea	116.00	16.10	—	132.10
2½"	P1@.450	Ea	119.00	16.10	—	135.10
3"	P1@.550	Ea	128.00	19.70	—	147.70
4"	P1@.550	Ea	155.00	19.70	—	174.70
6"	ER@.800	Ea	259.00	30.50	12.70	302.20
8"	ER@.800	Ea	359.00	30.50	12.70	402.20
10"	ER@.900	Ea	502.00	34.30	14.10	550.40
12"	ER@1.00	Ea	659.00	38.10	15.70	712.80

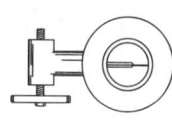

Class 125 bronze body swing check valve, with brazed joints

1/2"	P1@.240	Ea	16.90	8.58	—	25.48
3/4"	P1@.300	Ea	24.20	10.70	—	34.90
1"	P1@.360	Ea	31.70	12.90	—	44.60
1¼"	P1@.480	Ea	45.00	17.20	—	62.20
1½"	P1@.540	Ea	63.50	19.30	—	82.80
2"	P1@.600	Ea	106.00	21.50	—	127.50
2½"	P1@1.00	Ea	212.00	35.80	—	247.80
3"	P1@1.50	Ea	305.00	53.60	—	358.60

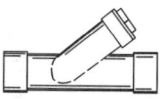

Description	Craft@Hrs	Unit	Material $	Labor $	Equipment $	Total $

Class 125 bronze body swing check valve with soft-soldered joints

Description	Craft@Hrs	Unit	Material $	Labor $	Equipment $	Total $
1/2"	P1@.200	Ea	16.90	7.15	—	24.05
3/4"	P1@.249	Ea	24.20	8.90	—	33.10
1"	P1@.299	Ea	31.70	10.70	—	42.40
1¼"	P1@.398	Ea	45.00	14.20	—	59.20
1½"	P1@.448	Ea	63.50	16.00	—	79.50
2"	P1@.498	Ea	106.00	17.80	—	123.80
2½"	P1@.830	Ea	212.00	29.70	—	241.70
3"	P1@1.24	Ea	305.00	44.30	—	349.30

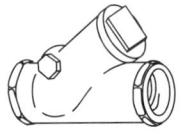

Class 125 bronze body swing check valve, threaded joints

Description	Craft@Hrs	Unit	Material $	Labor $	Equipment $	Total $
1/2"	P1@.210	Ea	16.90	7.51	—	24.41
3/4"	P1@.250	Ea	24.20	8.94	—	33.14
1"	P1@.300	Ea	31.70	10.70	—	42.40
1¼"	P1@.400	Ea	45.00	14.30	—	59.30
1½"	P1@.450	Ea	63.50	16.10	—	79.60
2"	P1@.500	Ea	106.00	17.90	—	123.90

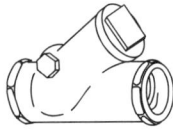

Class 300 bronze body swing check valve, threaded joints

Description	Craft@Hrs	Unit	Material $	Labor $	Equipment $	Total $
1/2"	P1@.210	Ea	57.00	7.51	—	64.51
3/4"	P1@.250	Ea	67.50	8.94	—	76.44
1"	P1@.300	Ea	82.50	10.70	—	93.20
1¼"	P1@.400	Ea	119.00	14.30	—	133.30
1½"	P1@.450	Ea	171.00	16.10	—	187.10
2"	P1@.500	Ea	238.00	17.90	—	255.90

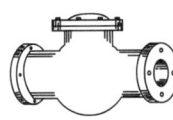

Class 125 iron body swing check valve, flanged joints

Description	Craft@Hrs	Unit	Material $	Labor $	Equipment $	Total $
2"	P1@.500	Ea	148.00	17.90	—	165.90
2½"	P1@.600	Ea	188.00	21.50	—	209.50
3"	P1@.750	Ea	233.00	26.80	—	259.80
4"	P1@1.35	Ea	343.00	48.30	—	391.30
6"	ER@2.50	Ea	666.00	95.30	39.30	800.60
8"	ER@3.00	Ea	1,190.00	114.00	47.10	1,351.10
10"	ER@4.00	Ea	1,960.00	152.00	62.90	2,174.90
12"	ER@4.50	Ea	2,660.00	171.00	70.80	2,901.80

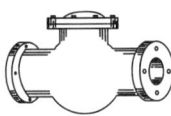

Class 250 iron body swing check valve, flanged joints

Description	Craft@Hrs	Unit	Material $	Labor $	Equipment $	Total $
2"	P1@.500	Ea	382.00	17.90	—	399.90
2½"	P1@.600	Ea	445.00	21.50	—	466.50
3"	P1@.750	Ea	553.00	26.80	—	579.80
4"	P1@1.35	Ea	678.00	48.30	—	726.30
6"	ER@2.50	Ea	1,310.00	95.30	39.30	1,444.60
8"	ER@3.00	Ea	2,420.00	114.00	47.10	2,581.10

Description	Craft@Hrs	Unit	Material $	Labor $	Equipment $	Total $

Class 125 iron body silent check valve, wafer type

Description	Craft@Hrs	Unit	Material $	Labor $	Equipment $	Total $
2"	P1@.500	Ea	108.00	17.90	—	125.90
2½"	P1@.600	Ea	120.00	21.50	—	141.50
3"	P1@.750	Ea	138.00	26.80	—	164.80
4"	P1@1.35	Ea	185.00	48.30	—	233.30
5"	P1@2.00	Ea	284.00	71.50	—	355.50
6"	P1@2.50	Ea	385.00	89.40	—	474.40
8"	P1@3.00	Ea	660.00	107.00	—	767.00

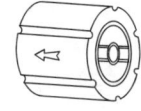

Class 125 iron body silent check valve, flanged joints

Description	Craft@Hrs	Unit	Material $	Labor $	Equipment $	Total $
2"	P1@.500	Ea	108.00	17.90	—	125.90
2½"	P1@.600	Ea	125.00	21.50	—	146.50
3"	P1@.750	Ea	140.00	26.80	—	166.80
4"	P1@1.35	Ea	184.00	48.30	—	232.30
6"	ER@2.50	Ea	330.00	95.30	39.30	464.60
8"	ER@3.00	Ea	595.00	114.00	47.10	756.10
10"	ER@4.00	Ea	924.00	152.00	62.90	1,138.90

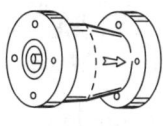

Class 250 iron body silent check valve, flanged joints

Description	Craft@Hrs	Unit	Material $	Labor $	Equipment $	Total $
2"	P1@.500	Ea	160.00	17.90	—	177.90
2½"	P1@.600	Ea	184.00	21.50	—	205.50
3"	P1@.750	Ea	205.00	26.80	—	231.80
4"	P1@1.35	Ea	278.00	48.30	—	326.30
6"	ER@2.50	Ea	488.00	95.30	39.30	622.60
8"	ER@3.00	Ea	888.00	114.00	47.10	1,049.10
10"	ER@4.00	Ea	1,400.00	152.00	62.90	1,614.90

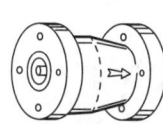

Class 125 bronze body gate valve, solder ends

Description	Craft@Hrs	Unit	Material $	Labor $	Equipment $	Total $
1/2"	P1@.240	Ea	15.40	8.58	—	23.98
3/4"	P1@.300	Ea	19.30	10.70	—	30.00
1"	P1@.360	Ea	27.10	12.90	—	40.00
1¼"	P1@.480	Ea	34.30	17.20	—	51.50
1½"	P1@.540	Ea	46.10	19.30	—	65.40
2"	P1@.600	Ea	77.00	21.50	—	98.50
2½"	P1@1.00	Ea	128.00	35.80	—	163.80
3"	P1@1.50	Ea	184.00	53.60	—	237.60

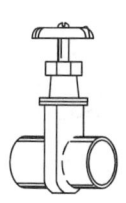

Class 125 bronze body gate valve, threaded joints

Description	Craft@Hrs	Unit	Material $	Labor $	Equipment $	Total $
1/2"	P1@.210	Ea	15.40	7.51	—	22.91
3/4"	P1@.250	Ea	19.30	8.94	—	28.24
1"	P1@.300	Ea	27.10	10.70	—	37.80
1¼"	P1@.400	Ea	34.30	14.30	—	48.60
1½"	P1@.450	Ea	46.10	16.10	—	62.20
2"	P1@.500	Ea	77.00	17.90	—	94.90
2½"	P1@.750	Ea	128.00	26.80	—	154.80
3"	P1@.950	Ea	184.00	34.00	—	218.00

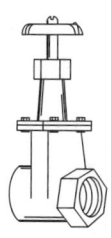

Plumbing and Piping Specialties

Description	Craft@Hrs	Unit	Material $	Labor $	Equipment $	Total $

Class 300 bronze body gate valve, threaded joints

Description	Craft@Hrs	Unit	Material $	Labor $	Equipment $	Total $
1/2"	P1@.210	Ea	31.40	7.51	—	38.91
3/4"	P1@.250	Ea	41.00	8.94	—	49.94
1"	P1@.300	Ea	49.80	10.70	—	60.50
1¼"	P1@.400	Ea	67.80	14.30	—	82.10
1½"	P1@.450	Ea	85.50	16.10	—	101.60
2"	P1@.500	Ea	126.00	17.90	—	143.90
2½"	P1@.750	Ea	257.00	26.80	—	283.80
3"	P1@.950	Ea	365.00	34.00	—	399.00

Class 125 iron body gate valve, flanged joints

Description	Craft@Hrs	Unit	Material $	Labor $	Equipment $	Total $
2"	P1@.500	Ea	246.00	17.90	—	263.90
2½"	P1@.600	Ea	332.00	21.50	—	353.50
3"	P1@.750	Ea	360.00	26.80	—	386.80
4"	P1@1.35	Ea	528.00	48.30	—	576.30
6"	ER@2.50	Ea	1,010.00	95.30	39.30	1,144.60
8"	ER@3.00	Ea	1,660.00	114.00	47.10	1,821.10
10"	ER@4.00	Ea	2,710.00	152.00	62.90	2,924.90
12"	ER@4.50	Ea	3,750.00	171.00	70.80	3,991.80

Class 250 iron body gate valve, flanged joints

Description	Craft@Hrs	Unit	Material $	Labor $	Equipment $	Total $
2"	P1@.500	Ea	430.00	17.90	—	447.90
2½"	P1@.600	Ea	310.00	21.50	—	331.50
3"	P1@.750	Ea	393.00	26.80	—	419.80
4"	P1@1.35	Ea	666.00	48.30	—	714.30
6"	ER@2.50	Ea	1,240.00	95.30	39.30	1,374.60
8"	ER@3.00	Ea	1,910.00	114.00	47.10	2,071.10
10"	ER@4.00	Ea	3,420.00	152.00	62.90	3,634.90
12"	ER@4.50	Ea	5,560.00	171.00	70.80	5,801.80

Class 125 bronze body globe valve, brazed joints

Description	Craft@Hrs	Unit	Material $	Labor $	Equipment $	Total $
1/2"	P1@.240	Ea	28.60	8.58	—	37.18
3/4"	P1@.300	Ea	38.10	10.70	—	48.80
1"	P1@.360	Ea	54.70	12.90	—	67.60
1¼"	P1@.480	Ea	76.90	17.20	—	94.10
1½"	P1@.540	Ea	104.00	19.30	—	123.30
2"	P1@.600	Ea	168.00	21.50	—	189.50

Class 125 bronze body globe valve, soft-soldered joints

Description	Craft@Hrs	Unit	Material $	Labor $	Equipment $	Total $
1/2"	P1@.200	Ea	28.60	7.15	—	35.75
3/4"	P1@.249	Ea	38.10	8.90	—	47.00
1"	P1@.299	Ea	54.70	10.70	—	65.40
1¼"	P1@.398	Ea	76.90	14.20	—	91.10
1½"	P1@.448	Ea	104.00	16.00	—	120.00
2"	P1@.498	Ea	168.00	17.80	—	185.80

Description	Craft@Hrs	Unit	Material $	Labor $	Equipment $	Total $

Class 125 bronze body globe valve, threaded ends

1/2"	P1@.210	Ea	28.60	7.51	—	36.11
3/4"	P1@.250	Ea	38.10	8.94	—	47.04
1"	P1@.300	Ea	54.70	10.70	—	65.40
1¼"	P1@.400	Ea	76.90	14.30	—	91.20
1½"	P1@.450	Ea	104.00	16.10	—	120.10
2"	P1@.500	Ea	168.00	17.90	—	185.90

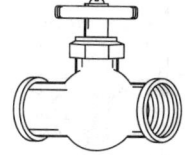

Class 300 bronze body globe valve, threaded joints

1/2"	P1@.210	Ea	53.80	7.51	—	61.31
3/4"	P1@.250	Ea	68.40	8.94	—	77.34
1"	P1@.300	Ea	97.20	10.70	—	107.90
1¼"	P1@.400	Ea	141.00	14.30	—	155.30
1½"	P1@.450	Ea	188.00	16.10	—	204.10
2"	P1@.500	Ea	260.00	17.90	—	277.90

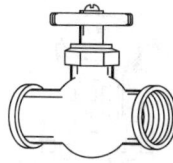

Class 125 iron body globe valve, flanged joints

2"	P1@.500	Ea	197.00	17.90	—	214.90
2½"	P1@.600	Ea	306.00	21.50	—	327.50
3"	P1@.750	Ea	307.00	26.80	—	393.80
4"	P1@1.35	Ea	493.00	48.30	—	541.30
6"	ER@2.50	Ea	865.00	95.30	39.30	999.60
8"	ER@3.00	Ea	1,330.00	114.00	47.10	1,491.10
10"	ER@4.00	Ea	2,020.00	152.00	62.90	2,234.90

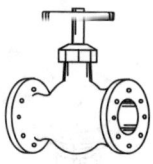

Class 250 iron body globe valve, flanged joints

2"	P1@.500	Ea	433.00	17.90	—	450.90
2½"	P1@.600	Ea	539.00	21.50	—	560.50
3"	P1@.750	Ea	641.00	26.80	—	667.80
4"	P1@1.35	Ea	884.00	48.30	—	932.30
6"	ER@2.50	Ea	1,740.00	95.30	39.30	1,874.60
8"	ER@3.00	Ea	3,880.00	114.00	47.10	4,041.10

Class 125 bronze body strainer, threaded joints

1/2"	P1@.230	Ea	26.40	8.22	—	34.62
3/4"	P1@.260	Ea	34.70	9.30	—	44.00
1"	P1@.330	Ea	42.50	11.80	—	54.30
1¼"	P1@.440	Ea	59.30	15.70	—	75.00
1½"	P1@.495	Ea	89.30	17.70	—	107.00
2"	P1@.550	Ea	154.00	19.70	—	173.70

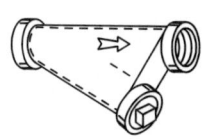
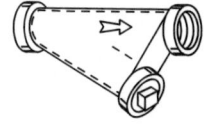

Class 125 iron body strainer, flanged joints

2"	P1@.500	Ea	108.00	17.90	—	125.90
2½"	P1@.600	Ea	121.00	21.50	—	142.50
3"	P1@.750	Ea	140.00	26.80	—	166.80
4"	P1@1.35	Ea	239.00	48.30	—	287.30
6"	ER@2.50	Ea	484.00	95.30	39.30	618.60
8"	ER@3.00	Ea	817.00	114.00	47.10	978.10

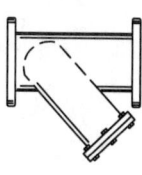

Description	Craft@Hrs	Unit	Material $	Labor $	Equipment $	Total $

Class 250 iron body strainer, flanged joints

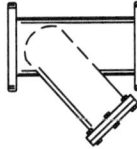

Description	Craft@Hrs	Unit	Material $	Labor $	Equipment $	Total $
2"	P1@.500	Ea	185.00	17.90	—	202.90
2½"	P1@.600	Ea	202.00	21.50	—	223.50
3"	P1@.750	Ea	237.00	26.80	—	263.80
4"	P1@1.35	Ea	420.00	48.30	—	468.30
6"	ER@2.50	Ea	672.00	95.30	39.30	806.60
8"	ER@3.00	Ea	1,230.00	114.00	47.10	1,391.10

Installation of 2-way control valve, threaded joints

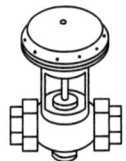

Description	Craft@Hrs	Unit	Material $	Labor $	Equipment $	Total $
1/2"	P1@.210	Ea	—	7.51	—	7.51
3/4"	P1@.250	Ea	—	8.94	—	8.94
1"	P1@.300	Ea	—	10.70	—	10.70
1¼"	P1@.400	Ea	—	14.30	—	14.30
1½"	P1@.450	Ea	—	16.10	—	16.10
2"	P1@.500	Ea	—	17.90	—	17.90
2½"	P1@.830	Ea	—	29.70	—	29.70
3"	P1@.990	Ea	—	35.40	—	35.40

Installation of 2-way control valve, flanged joints

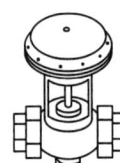

Description	Craft@Hrs	Unit	Material $	Labor $	Equipment $	Total $
2"	P1@.500	Ea	—	17.90	—	17.90
2½"	P1@.600	Ea	—	21.50	—	21.50
3"	P1@.750	Ea	—	26.80	—	26.80
4"	P1@1.35	Ea	—	48.30	—	48.30
6"	P1@2.50	Ea	—	89.40	—	89.40
8"	P1@3.00	Ea	—	107.00	—	107.00
10"	P1@4.00	Ea	—	143.00	—	143.00
12"	P1@4.50	Ea	—	161.00	—	161.00

Installation of 3-way control valve, threaded joints

Description	Craft@Hrs	Unit	Material $	Labor $	Equipment $	Total $
1/2"	P1@.260	Ea	—	9.30	—	9.30
3/4"	P1@.365	Ea	—	13.10	—	13.10
1"	P1@.475	Ea	—	17.00	—	17.00
1¼"	P1@.575	Ea	—	20.60	—	20.60
1½"	P1@.680	Ea	—	24.30	—	24.30
2"	P1@.910	Ea	—	32.50	—	32.50
2½"	P1@1.12	Ea	—	40.10	—	40.10
3"	P1@1.33	Ea	—	47.60	—	47.60

Description	Craft@Hrs	Unit	Material $	Labor $	Equipment $	Total $

Installation of 3-way control valve, flanged joints

Description	Craft@Hrs	Unit	Material $	Labor $	Equipment $	Total $
2"	P1@.910	Ea	—	32.50	—	32.50
2½"	P1@1.12	Ea	—	40.10	—	40.10
3"	P1@1.33	Ea	—	47.60	—	47.60
4"	P1@2.00	Ea	—	71.50	—	71.50
6"	P1@3.70	Ea	—	132.00	—	132.00
8"	P1@4.40	Ea	—	157.00	—	157.00
10"	P1@5.90	Ea	—	211.00	—	211.00
12"	P1@6.50	Ea	—	232.00	—	232.00

Lever handled gas valve, threaded joints

Description	Craft@Hrs	Unit	Material $	Labor $	Equipment $	Total $
1/2"	P1@.210	Ea	24.70	7.51	—	32.21
3/4"	P1@.250	Ea	27.30	8.94	—	36.24
1"	P1@.300	Ea	34.10	10.70	—	44.80
1¼"	P1@.400	Ea	45.50	14.30	—	59.80
1½"	P1@.450	Ea	61.20	16.10	—	77.30
2"	P1@.500	Ea	89.30	17.90	—	107.20

Plug-type gas valve, flanged joints, lubricated

Description	Craft@Hrs	Unit	Material $	Labor $	Equipment $	Total $
2"	P1@.500	Ea	194.00	17.90	—	211.90
2½"	P1@.600	Ea	315.00	21.50	—	336.50
3"	P1@.750	Ea	441.00	26.80	—	467.80
4"	P1@1.35	Ea	605.00	48.30	—	653.30
6"	P1@2.50	Ea	1,080.00	89.40	—	1,169.40

3/4" hose bibb

Description	Craft@Hrs	Unit	Material $	Labor $	Equipment $	Total $
Low quality	P1@.350	Ea	10.10	12.50	—	22.60
Med. quality	P1@.350	Ea	14.60	12.50	—	27.10
High quality	P1@.350	Ea	19.30	12.50	—	31.80

Water pressure reducing valve, threaded joints, 25-75 PSI

Description	Craft@Hrs	Unit	Material $	Labor $	Equipment $	Total $
1/2"	P1@.210	Ea	88.90	7.51	—	96.41
3/4"	P1@.250	Ea	114.00	8.94	—	122.94
1"	P1@.300	Ea	189.00	10.70	—	199.70
1¼"	P1@.400	Ea	229.00	14.30	—	243.30
1½"	P1@.450	Ea	327.00	16.10	—	343.10
2"	P1@.500	Ea	518.00	17.90	—	535.90

Water hammer arrester, threaded joints

Description	Craft@Hrs	Unit	Material $	Labor $	Equipment $	Total $
1/2"	P1@.160	Ea	42.90	5.72	—	48.62
3/4"	P1@.190	Ea	49.70	6.79	—	56.49
1"	P1@.230	Ea	127.00	8.22	—	135.22
1¼"	P1@.300	Ea	147.00	10.70	—	157.70
1½"	P1@.340	Ea	210.00	12.20	—	222.20
2"	P1@.380	Ea	320.00	13.60	—	333.60

Description	Craft@Hrs	Unit	Material $	Labor $	Equipment $	Total $

Carbon steel weldolet, Schedule 40

1/2"	P1@.330	Ea	13.80	11.80	—	25.60
3/4"	P1@.500	Ea	14.20	17.90	—	32.10
1"	P1@.670	Ea	14.90	24.00	—	38.90
1¼"	P1@.830	Ea	18.00	29.70	—	47.70
1½"	P1@1.00	Ea	18.00	35.80	—	53.80
2"	P1@1.33	Ea	19.80	47.60	—	67.40
2½"	P1@1.67	Ea	43.60	59.70	—	103.30
3"	P1@2.00	Ea	49.50	71.50	—	121.00
4"	P1@2.67	Ea	63.90	95.50	—	159.40
6"	P1@4.00	Ea	264.00	143.00	—	407.00
8"	P1@4.80	Ea	389.00	172.00	—	561.00
10"	P1@6.00	Ea	537.00	215.00	—	752.00
12"	P1@7.20	Ea	660.00	257.00	—	917.00

Carbon steel weldolet, Schedule 80

1/2"	P1@.440	Ea	15.20	15.70	—	30.90
3/4"	P1@.670	Ea	15.60	24.00	—	39.60
1"	P1@.890	Ea	17.00	31.80	—	48.80
1¼"	P1@1.11	Ea	21.40	39.70	—	61.10
1½"	P1@1.33	Ea	21.40	47.60	—	69.00
2"	P1@1.77	Ea	23.10	63.30	—	86.40
2½"	P1@2.22	Ea	51.80	79.40	—	131.20
3"	P1@2.66	Ea	53.20	95.10	—	148.30
4"	P1@3.55	Ea	118.00	127.00	—	245.00
6"	P1@5.32	Ea	371.00	190.00	—	561.00
8"	P1@6.38	Ea	495.00	228.00	—	723.00

Carbon steel threadolet, Schedule 40

3/4"	P1@.330	Ea	8.56	11.80	—	20.36
1"	P1@.440	Ea	9.50	15.70	—	25.20
1¼"	P1@.560	Ea	13.20	20.00	—	33.20
1½"	P1@.670	Ea	14.90	24.00	—	38.90
2"	P1@.890	Ea	17.30	31.80	—	49.10
2½"	P1@1.11	Ea	53.40	39.70	—	93.10

Carbon steel threadolet, Schedule 80

1/2	P1@.300	Ea	12.80	10.70	—	23.50
3/4"	P1@.440	Ea	14.10	15.70	—	29.80
1"	P1@.590	Ea	17.50	21.10	—	38.60
1¼"	P1@.740	Ea	78.60	26.50	—	105.10
1½"	P1@.890	Ea	78.60	31.80	—	110.40
2"	P1@1.18	Ea	103.00	42.20	—	145.20

Cast Iron, DWV, Service Weight, No-Hub with Coupled Joints

Service weight cast iron soil pipe with hub-less coupled joints is widely used for non-pressurized underground and in-building drain, waste and vent (DWV) systems.

This section has been arranged to save the estimator's time by including all normally-used system components such as pipe, fittings, hanger assemblies and riser clamps under one heading. Additional items can be found under "Plumbing and Piping Specialties." The cost estimates in this section are based on the conditions, limitations and wage rates described in the section "How to Use This Book" beginning on page 5.

Equipment cost, where shown, is $110 per hour for a 10-ton hydraulic truck-mounted crane.

Description	Craft@Hrs	Unit	Material $	Labor $	Equipment $	Total $

DWV cast iron no-hub (MJ) horizontal pipe assembly.
Horizontally hung in a building. Assembly includes fittings, couplings, hangers and rod. Based on a reducing wye and a 45 degree elbow every 15 feet and a hanger every 6 feet for 2" pipe. A reducing wye and a 45 degree elbow every 40 feet and a hanger every 8 feet for 10" pipe. *(Distance between fittings in the pipe assembly increases as pipe diameter increases.)* Use these figures for preliminary estimates.

Description	Craft@Hrs	Unit	Material $	Labor $	Equipment $	Total $
2"	P1@.130	LF	9.69	4.65	—	14.34
3"	P1@.160	LF	10.60	5.72	—	16.32
4"	P1@.190	LF	12.60	6.79	—	19.39
6"	FR@.220	LF	21.40	0.38	3.50	36.28
8"	ER@.260	LF	46.70	9.91	4.10	60.71
10"	ER@.310	LF	72.50	11.80	5.00	89.30

DWV cast iron no-hub (MJ), pipe only

Description	Craft@Hrs	Unit	Material $	Labor $	Equipment $	Total $
2"	P1@.050	LF	5.32	1.79	—	7.11
3"	P1@.060	LF	6.35	2.15	—	8.50
4"	P1@.080	LF	8.28	2.86	—	11.14
6"	ER@.110	LF	15.40	4.19	—	19.59
8"	ER@.130	LF	28.70	4.95	—	33.65
10"	ER@.140	LF	49.90	5.34	—	55.24

DWV cast iron no-hub 1/8 bend

Description	Craft@Hrs	Unit	Material $	Labor $	Equipment $	Total $
2"	P1@.250	Ea	6.16	8.94	—	15.10
3"	P1@.350	Ea	8.06	12.50	—	20.56
4"	P1@.500	Ea	10.20	17.90	—	28.10
6"	ER@.650	Ea	32.80	24.80	10.30	67.90
8"	ER@.710	Ea	87.50	27.10	11.20	125.80
10"	ER@.800	Ea	130.00	30.50	12.70	173.20

Cast Iron, DWV, Service Weight, No-Hub with Coupled Joints

Description	Craft@Hrs	Unit	Material $	Labor $	Equipment $	Total $
DWV cast iron no-hub 1/4 bend						
2"	P1@.250	Ea	7.54	8.94	—	16.48
3"	P1@.350	Ea	9.37	12.50	—	21.87
4"	P1@.500	Ea	14.30	17.90	—	32.20
6"	ER@.650	Ea	34.70	24.80	10.30	69.80
8"	ER@.710	Ea	134.00	27.10	11.20	172.30
10"	ER@.710	Ea	147.00	27.10	11.20	185.30
DWV cast iron no-hub long sweep 1/4 bend						
2"	P1@.250	Ea	13.10	8.94	—	22.04
3"	P1@.350	Ea	15.80	12.50	—	28.30
4"	P1@.500	Ea	25.00	17.90	—	42.90
5"	ER@.580	Ea	46.10	22.10	9.20	77.40
6"	ER@.650	Ea	56.10	24.80	10.30	91.20
DWV cast iron no-hub low heel outlet 1/4 bend						
3"	P1@.470	Ea	13.30	16.80	—	30.10
4"	P1@.650	Ea	16.20	23.20	—	39.40
DWV cast iron no-hub closet bend						
3"	P1@.350	Ea	30.80	12.50	—	43.30
4"	P1@.500	Ea	31.50	17.90	—	49.40
DWV cast iron no-hub closet flange						
4" x 3"	P1@.500	Ea	45.00	17.90	—	62.90
DWV cast iron no-hub P-trap						
2"	P1@.250	Ea	12.20	8.94	—	21.14
3"	P1@.350	Ea	38.00	12.50	—	50.50
4"	P1@.500	Ea	46.90	17.90	—	64.80
6"	ER@.650	Ea	167.00	24.80	10.30	202.10
DWV cast iron no-hub wye						
2"	P1@.380	Ea	9.64	13.60	—	23.24
3"	P1@.520	Ea	13.40	18.60	—	32.00
4"	P1@.700	Ea	21.00	25.00	—	46.00
6"	ER@.900	Ea	56.90	34.30	14.10	105.30
8"	ER@1.15	Ea	145.00	43.80	18.20	207.00
10"	ER@1.50	Ea	347.00	57.20	23.70	427.90

Cast Iron, DWV, Service Weight, No-Hub with Coupled Joints

Description	Craft@Hrs	Unit	Material $	Labor $	Equipment $	Total $

DWV cast iron no-hub reducing wye

Description	Craft@Hrs	Unit	Material $	Labor $	Equipment $	Total $
3" x 2"	P1@.470	Ea	12.90	16.80	—	29.70
4" x 2"	P1@.600	Ea	17.20	21.50	—	38.70
4" x 3"	P1@.650	Ea	18.20	23.20	—	41.40
6" x 2"	ER@.770	Ea	46.20	29.30	12.10	87.60
6" x 3"	ER@.770	Ea	50.10	29.30	12.10	91.50
6" x 4"	ER@.800	Ea	49.20	30.50	12.50	92.20
8" x 2"	ER@1.00	Ea	117.00	38.10	15.70	170.80
8" x 3"	ER@1.00	Ea	119.00	38.10	15.70	172.80
8" x 4"	ER@1.00	Ea	121.00	38.10	15.70	174.80
8" x 6"	ER@1.10	Ea	126.00	41.90	17.40	185.30

DWV cast iron no-hub combination upright wye and 1/8 bend

Description	Craft@Hrs	Unit	Material $	Labor $	Equipment $	Total $
2"	P1@.380	Ea	23.90	13.60	—	37.50
3"	P1@.520	Ea	38.40	18.60	—	57.00
4"	P1@.700	Ea	50.70	25.00	—	75.70

DWV cast iron no-hub combination upright reducing wye and 1/8 bend

Description	Craft@Hrs	Unit	Material $	Labor $	Equipment $	Total $
3" x 2"	P1@.470	Ea	41.70	16.80	—	58.50
4" x 2"	P1@.600	Ea	49.60	21.50	—	71.10
4" x 3"	P1@.650	Ea	52.00	23.20	—	75.20

DWV cast iron no-hub double wye

Description	Craft@Hrs	Unit	Material $	Labor $	Equipment $	Total $
2"	P1@.500	Ea	29.60	17.90	—	47.50
3"	P1@.700	Ea	30.00	25.00	—	55.00
4"	P1@.950	Ea	46.60	34.00	—	80.60
6"	ER@1.10	Ea	161.00	41.90	17.40	220.30
8"	ER@1.45	Ea	295.00	55.30	22.80	373.10

DWV cast iron no-hub double reducing wye

Description	Craft@Hrs	Unit	Material $	Labor $	Equipment $	Total $
3" x 2"	P1@.520	Ea	29.60	18.60	—	48.20
4" x 2"	P1@.740	Ea	41.20	26.50	—	67.70
4" x 3"	P1@1.00	Ea	42.60	35.80	—	78.40
6" x 4"	ER@1.75	Ea	159.00	66.70	27.50	253.20

DWV cast iron no-hub combination wye and 1/8 bend

Description	Craft@Hrs	Unit	Material $	Labor $	Equipment $	Total $
2"	P1@.380	Ea	18.50	13.60	—	32.10
3"	P1@.520	Ea	26.40	18.60	—	45.00
4"	P1@.700	Ea	42.10	25.00	—	67.10
6"	ER@.900	Ea	144.00	34.30	14.10	192.40
8"	ER@1.15	Ea	289.00	43.80	17.60	350.40

Cast Iron, DWV, Service Weight, No-Hub with Coupled Joints

Description	Craft@Hrs	Unit	Material $	Labor $	Equipment $	Total $

DWV cast iron no-hub combination reducing wye and 1/8 bend

Description	Craft@Hrs	Unit	Material $	Labor $	Equipment $	Total $
3" x 2"	P1@.470	Ea	22.40	16.80	—	39.20
4" x 2"	P1@.600	Ea	34.10	21.50	—	55.60
4" x 3"	P1@.650	Ea	36.50	23.20	—	59.70
6" x 2"	ER@.770	Ea	162.00	29.30	12.10	203.40
6" x 3"	ER@.770	Ea	171.00	29.30	12.10	212.40
6" x 4"	ER@.800	Ea	175.00	30.50	12.70	218.20
8" x 4"	ER@1.20	Ea	319.00	45.70	18.90	383.60

DWV cast iron no-hub double combination wye and 1/8 bend

Description	Craft@Hrs	Unit	Material $	Labor $	Equipment $	Total $
2"	P1@.500	Ea	32.00	17.90	—	49.90
3"	P1@.700	Ea	51.50	25.00	—	76.50
4"	P1@.950	Ea	63.40	34.00	—	97.40

DWV cast iron no-hub double combination reducing wye and 1/8 bend

Description	Craft@Hrs	Unit	Material $	Labor $	Equipment $	Total $
3" x 2"	P1@.600	Ea	36.30	21.50	—	57.80
4" x 2"	P1@1.25	Ea	57.20	44.70	—	101.90
4" x 3"	P1@1.35	Ea	59.90	48.30	—	108.20

DWV cast iron no-hub sanitary tee (TY)

Description	Craft@Hrs	Unit	Material $	Labor $	Equipment $	Total $
2"	P1@.380	Ea	11.10	13.60	—	24.70
3"	P1@.520	Ea	13.40	18.60	—	32.00
4"	P1@.700	Ea	22.10	25.00	—	47.10
6"	ER@.900	Ea	67.40	34.30	14.10	115.80
8"	ER@1.15	Ea	177.00	43.80	18.20	239.00

DWV cast iron no-hub reducing sanitary tee (TY)

Description	Craft@Hrs	Unit	Material $	Labor $	Equipment $	Total $
3" x 2"	P1@.470	Ea	12.80	16.80	—	29.60
4" x 2"	P1@.600	Ea	17.00	21.50	—	38.50
4" x 3"	P1@.650	Ea	19.50	23.20	—	42.70
6" x 2"	ER@.770	Ea	53.40	29.30	12.10	94.80
6" x 3"	ER@.770	Ea	50.50	29.30	12.10	91.90
6" x 4"	ER@.800	Ea	54.80	30.50	12.70	98.00
8" x 3"	ER@1.00	Ea	140.00	38.10	15.70	193.80
8" x 4"	ER@1.00	Ea	137.00	38.10	15.70	190.80
8" x 6"	ER@1.10	Ea	145.00	41.90	17.40	204.30

DWV cast iron no-hub tapped sanitary tee (tapped TY)

Description	Craft@Hrs	Unit	Material $	Labor $	Equipment $	Total $
2" x 1½"	P1@.200	Ea	31.30	7.15	—	38.45
2" x 2"	P1@.250	Ea	31.30	8.94	—	40.24
3" x 2"	P1@.350	Ea	39.90	12.50	—	52.40
4" x 2"	P1@.425	Ea	47.30	15.20	—	62.50
6" x 2"	ER@.700	Ea	122.00	26.70	11.00	159.70

Cast Iron, DWV, Service Weight, No-Hub with Coupled Joints

Description	Craft@Hrs	Unit	Material $	Labor $	Equipment $	Total $

DWV cast iron no-hub test tee

Description	Craft@Hrs	Unit	Material $	Labor $	Equipment $	Total $
2"	P1@.370	Ea	39.60	13.20	—	52.80
3"	P1@.520	Ea	56.10	18.60	—	74.70
4"	P1@.750	Ea	71.40	26.80	—	98.20
6"	ER@1.12	Ea	256.00	42.70	17.60	316.30
8"	ER@1.25	Ea	464.00	47.60	19.60	531.20

DWV cast iron no-hub cross (double TY)

Description	Craft@Hrs	Unit	Material $	Labor $	Equipment $	Total $
2"	P1@.500	Ea	31.50	17.90	—	49.40
3"	P1@.700	Ea	33.60	25.00	—	58.60
4"	P1@.950	Ea	48.30	34.00	—	82.30
6"	ER@1.30	Ea	180.00	49.50	20.40	249.90

DWV cast iron no-hub reducing cross (double TY)

Description	Craft@Hrs	Unit	Material $	Labor $	Equipment $	Total $
3" x 2"	P1@.600	Ea	29.70	21.50	—	51.20
4" x 2"	P1@.900	Ea	41.00	32.20	—	73.20
4" x 3"	P1@.900	Ea	42.10	32.20	—	74.30
6" x 4"	ER@1.20	Ea	164.00	45.70	18.90	228.60
8" x 4"	ER@1.50	Ea	414.00	57.20	23.70	494.90

DWV cast iron no-hub reducer

Description	Craft@Hrs	Unit	Material $	Labor $	Equipment $	Total $
2"	P1@.120	Ea	6.20	4.29	—	10.49
3"	P1@.130	Ea	6.67	4.65	—	11.32
4"	P1@.150	Ea	9.91	5.36	—	15.27
6"	ER@.200	Ea	26.50	7.62	3.10	37.22
8"	ER@.400	Ea	68.10	15.20	6.30	89.60
10"	ER@.650	Ea	105.00	24.80	10.30	140.10

DWV cast iron no-hub cap

Description	Craft@Hrs	Unit	Material $	Labor $	Equipment $	Total $
2"	P1@.125	Ea	3.79	4.47	—	8.26
3"	P1@.175	Ea	6.49	6.26	—	12.75
4"	P1@.250	Ea	8.46	8.94	—	17.40
6"	ER@.375	Ea	16.00	14.30	5.90	36.20
8"	ER@.500	Ea	29.60	19.10	7.90	56.60
10"	ER@.750	Ea	45.70	28.60	11.90	86.20

MJ (mechanical joint) coupling with gasket (labor included with fittings)

Description	Craft@Hrs	Unit	Material $	Labor $	Equipment $	Total $
2"	--@--	Ea	7.33	—	—	7.33
3"	--@--	Ea	8.02	—	—	8.02
4"	--@--	Ea	8.62	—	—	8.62
6"	--@--	Ea	23.40	—	—	23.40
8"	--@--	Ea	43.60	—	—	43.60
10"	--@--	Ea	60.50	—	—	60.50

Cast Iron, DWV, Service Weight, No-Hub with Coupled Joints

Description	Craft@Hrs	Unit	Material $	Labor $	Equipment $	Total $
Hanger with swivel assembly						
2"	P1@.400	Ea	6.03	14.30	—	20.33
3"	P1@.450	Ea	7.78	16.10	—	23.88
4"	P1@.450	Ea	9.63	16.10	—	25.73
5"	P1@.500	Ea	14.90	17.90	—	32.80
6"	P1@.500	Ea	16.60	17.90	—	34.50
8"	P1@.550	Ea	21.80	19.70	—	41.50
10"	P1@.650	Ea	28.50	23.20	—	51.70

Description	Craft@Hrs	Unit	Material $	Labor $	Equipment $	Total $
Riser clamp						
2"	P1@.115	Ea	4.84	4.11	—	8.95
3"	P1@.120	Ea	5.52	4.29	—	9.81
4"	P1@.125	Ea	7.03	4.47	—	11.50
5"	P1@.180	Ea	10.10	6.44	—	16.54
6"	P1@.200	Ea	12.20	7.15	—	19.35
8"	P1@.200	Ea	19.80	7.15	—	26.95
10"	P1@.250	Ea	29.40	8.94	—	38.34

Cast Iron, DWV, Service Weight, Hub & Spigot with Gasketed Joints

Service weight hub and spigot cast iron soil pipe with gasketed joints is widely used for non-pressurized underground and in-building drain, waste and vent (DWV) systems.

This section has been arranged to save the estimator's time by including all normally-used system components such as pipe, fittings, valves, hanger assemblies and riser clamps under one heading. Additional items can be found under "Plumbing and Piping Specialties." The cost estimates in this section are based on the conditions, limitations and wage rates described in the section "How to Use This Book" beginning on page 5.

Equipment cost, where shown, is $110 per hour for a 10-ton hydraulic truck-mounted crane.

Description	Craft@Hrs	Unit	Material $	Labor $	Equipment $	Total $

DWV cast iron pipe, hub and spigot, gasketed joints

Description	Craft@Hrs	Unit	Material $	Labor $	Equipment $	Total $
2"	P1@.060	LF	9.13	2.15	—	11.28
3"	P1@.070	LF	10.10	2.50	—	12.60
4"	P1@.090	LF	13.20	3.22	—	16.42
6"	ER@.110	LF	25.20	4.19	1.70	31.09
8"	ER@.140	LF	42.40	5.34	2.20	49.94
10"	ER@.180	LF	67.70	6.86	2.80	77.36
12"	ER@.230	LF	87.40	8.77	3.70	99.87
15"	ER@.280	LF	131.00	10.70	4.40	146.10

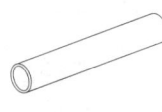

DWV cast iron 1/8 bend, hub and spigot, gasketed joints

Description	Craft@Hrs	Unit	Material $	Labor $	Equipment $	Total $
2"	P1@.300	Ea	10.60	10.70	—	21.30
3"	P1@.400	Ea	14.60	14.30	—	28.90
4"	P1@.550	Ea	20.40	19.70	—	40.10
6"	ER@.700	Ea	51.30	26.70	11.00	89.00
8"	ER@1.15	Ea	121.00	43.80	18.20	183.00
10"	ER@1.25	Ea	164.00	47.60	19.60	231.20
12"	ER@2.25	Ea	380.00	85.70	35.50	501.20
15"	ER@2.50	Ea	636.00	95.30	39.30	770.60

DWV cast iron 1/4 bend, hub and spigot, gasketed joints

Description	Craft@Hrs	Unit	Material $	Labor $	Equipment $	Total $
2"	P1@.300	Ea	11.80	10.70	—	22.50
3"	P1@.400	Ea	18.30	14.30	—	32.60
4"	P1@.550	Ea	32.00	19.70	—	51.70
6"	ER@.900	Ea	60.80	34.30	14.10	109.20
8"	ER@1.15	Ea	162.00	43.80	18.20	224.00
10"	ER@1.25	Ea	227.00	47.60	19.60	294.20
12"	ER@2.25	Ea	474.00	85.70	35.50	595.20
15"	ER@2.50	Ea	955.00	95.30	39.30	1,089.60

DWV cast iron heel inlet 1/4 bend, hub and spigot, gasketed joints

Description	Craft@Hrs	Unit	Material $	Labor $	Equipment $	Total $
3" x 2"	P1@.550	Ea	22.20	19.70	—	41.90
4" x 2"	P1@.700	Ea	27.00	25.00	—	52.00

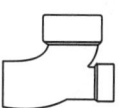

Cast Iron, DWV, Service Weight, Hub & Spigot with Gasketed Joints

Description	Craft@Hrs	Unit	Material $	Labor $	Equipment $	Total $

DWV cast iron 1/16 bend, hub and spigot, gasketed joints

Description	Craft@Hrs	Unit	Material $	Labor $	Equipment $	Total $
2"	P1@.300	Ea	6.73	10.70	—	17.43
3"	P1@.400	Ea	11.20	14.30	—	25.50
4"	P1@.550	Ea	15.60	19.70	—	35.30
6"	ER@.700	Ea	26.20	26.70	11.00	63.90
8"	ER@1.15	Ea	92.20	43.80	18.20	154.20

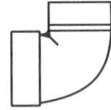

DWV cast iron closet bend, hub and spigot, gasketed joints

Description	Craft@Hrs	Unit	Material $	Labor $	Equipment $	Total $
4" x 4"	P1@.400	Ea	39.00	14.30	—	53.30

DWV cast iron closet flange, hub and spigot, gasketed joints

Description	Craft@Hrs	Unit	Material $	Labor $	Equipment $	Total $
4" x 2"	P1@.075	Ea	12.70	2.68	—	15.38
4" x 3"	P1@.100	Ea	17.40	3.58	—	20.98
4" x 4"	P1@.125	Ea	22.70	4.47	—	27.17

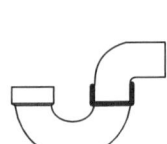

DWV cast iron regular P-trap, hub and spigot, gasketed joints

Description	Craft@Hrs	Unit	Material $	Labor $	Equipment $	Total $
2"	P1@.300	Ea	45.10	10.70	—	55.80
3"	P1@.400	Ea	62.80	14.30	—	77.10
4"	P1@.550	Ea	97.10	19.70	—	116.80
6"	ER@.700	Ea	212.00	26.70	11.00	249.70
8"	ER@1.15	Ea	628.00	43.80	18.20	690.00
10"	ER@2.00	Ea	861.00	76.20	31.40	968.60

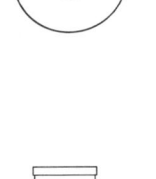

DWV cast iron sanitary tee, (TY) hub and spigot, gasketed joints

Description	Craft@Hrs	Unit	Material $	Labor $	Equipment $	Total $
2"	P1@.400	Ea	21.70	14.30	—	36.00
3"	P1@.600	Ea	38.00	21.50	—	59.50
4"	P1@.750	Ea	76.10	26.80	—	102.90
6"	ER@.900	Ea	174.00	34.30	14.10	222.40
8"	ER@1.50	Ea	253.00	57.20	21.50	331.70

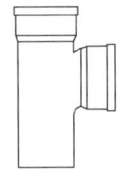

DWV cast iron sanitary reducing tee, (TY) hub and spigot, gasketed joints

Description	Craft@Hrs	Unit	Material $	Labor $	Equipment $	Total $
3" x 2"	P1@.400	Ea	30.80	14.30	—	45.10
4" x 2"	P1@.650	Ea	80.40	23.20	—	103.60
4" x 3"	P1@.700	Ea	80.40	25.00	—	105.40
6" x 2"	ER@.750	Ea	108.00	28.60	11.80	148.40
6" x 3"	ER@.800	Ea	108.00	30.50	12.70	151.20
6" x 4"	ER@.850	Ea	108.00	32.40	13.40	153.80

DWV cast iron sanitary tapped tee, (tapped TY) hub and spigot, gasketed joints

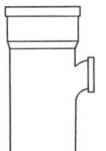

Description	Craft@Hrs	Unit	Material $	Labor $	Equipment $	Total $
2"	P1@.300	Ea	18.60	10.70	—	29.30
3"	P1@.400	Ea	27.30	14.30	—	41.60
4"	P1@.550	Ea	27.00	19.70	—	46.70

Cast Iron, DWV, Service Weight, Hub & Spigot with Gasketed Joints

Description	Craft@Hrs	Unit	Material $	Labor $	Equipment $	Total $

DWV cast iron combination wye and 1/8 bend, hub and spigot, gasketed joints

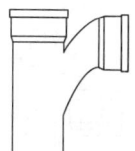

Description	Craft@Hrs	Unit	Material $	Labor $	Equipment $	Total $
2"	P1@.400	Ea	19.20	14.30	—	33.50
3"	P1@.600	Ea	29.00	21.50	—	50.50
4"	P1@.750	Ea	40.30	26.80	—	67.10
5"	ER@.850	Ea	75.80	32.40	13.40	121.60
6"	ER@.900	Ea	96.10	34.30	14.10	144.50

DWV cast iron combination reducing wye and 1/8 bend, hub and spigot, gasketed joints

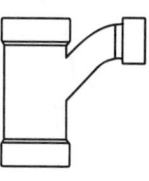

Description	Craft@Hrs	Unit	Material $	Labor $	Equipment $	Total $
3" x 2"	P1@.550	Ea	21.70	19.70	—	41.40
4" x 2"	P1@.650	Ea	29.70	23.20	—	52.90
4" x 3"	P1@.700	Ea	34.60	25.00	—	59.60
5" x 2"	ER@.700	Ea	61.10	26.70	11.00	98.80
5" x 4"	ER@.800	Ea	72.60	30.50	12.70	115.80
6" x 2"	ER@.750	Ea	62.00	28.60	11.80	102.40
6" x 3"	ER@.750	Ea	65.90	28.60	11.80	106.30
6" x 4"	ER@.800	Ea	69.10	30.50	12.70	112.30

DWV cast iron combination double wye and 1/8 bend, hub and spigot, gasketed joints

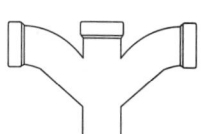

Description	Craft@Hrs	Unit	Material $	Labor $	Equipment $	Total $
2"	P1@.500	Ea	36.80	17.90	—	54.70
3"	P1@.750	Ea	50.30	26.80	—	77.10
4"	P1@.950	Ea	71.50	34.00	—	105.50
6"	ER@1.10	Ea	246.00	41.90	17.30	305.20

DWV cast iron combination double reducing wye and 1/8 bend, hub and spigot, gasketed joints

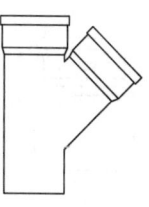

Description	Craft@Hrs	Unit	Material $	Labor $	Equipment $	Total $
3" x 2"	P1@.550	Ea	40.50	19.70	—	60.20
4" x 2"	P1@.650	Ea	50.80	23.20	—	74.00
4" x 3"	P1@.700	Ea	57.60	25.00	—	82.60
6" x 4"	ER@.800	Ea	161.00	30.50	12.70	204.20

DWV cast iron wye, hub and spigot, gasketed joints

Description	Craft@Hrs	Unit	Material $	Labor $	Equipment $	Total $
2"	P1@.400	Ea	17.70	14.30	—	32.00
3"	P1@.600	Ea	30.70	21.50	—	52.20
4"	P1@.750	Ea	40.80	26.80	—	67.60
6"	ER@.900	Ea	97.90	34.30	14.10	146.30
8"	ER@1.50	Ea	219.00	57.20	23.70	299.90
10"	ER@2.00	Ea	368.00	76.20	31.40	475.60
12"	ER@2.50	Ea	828.00	95.30	39.30	962.60
15"	ER@3.00	Ea	1,640.00	114.00	47.10	1,801.10

Cast Iron, DWV, Service Weight, Hub & Spigot with Gasketed Joints

Description	Craft@Hrs	Unit	Material $	Labor $	Equipment $	Total $

DWV cast iron reducing wye, hub and spigot, gasketed joints

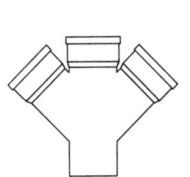

Description	Craft@Hrs	Unit	Material $	Labor $	Equipment $	Total $
4" x 2"	P1@.650	Ea	30.80	23.20	—	54.00
4" x 3"	P1@.700	Ea	37.80	25.00	—	62.80
6" x 2"	ER@.750	Ea	78.50	28.60	11.80	118.90
6" x 3"	ER@.750	Ea	78.50	28.60	11.80	118.90
6" x 4"	ER@.800	Ea	79.40	30.50	12.70	122.60
8" x 4"	ER@1.35	Ea	183.00	51.40	21.20	255.60
8" x 6"	ER@1.45	Ea	194.00	55.30	22.80	272.10
10" x 4"	ER@1.85	Ea	314.00	70.50	29.20	413.70
10" x 6"	ER@1.90	Ea	324.00	72.40	29.90	426.30
10" x 8"	ER@1.95	Ea	346.00	74.30	30.80	451.10
12" x 6"	ER@2.35	Ea	585.00	89.60	37.10	711.70
12" x 8"	ER@2.40	Ea	641.00	91.50	37.70	770.20
12" x 10"	ER@2.45	Ea	691.00	93.40	38.70	823.10
15" x 6"	ER@2.85	Ea	945.00	109.00	44.90	1,098.90
15" x 8"	ER@2.85	Ea	955.00	109.00	44.90	1,108.90
15" x 10"	ER@2.90	Ea	1,070.00	111.00	45.70	1,226.70
15" x 12"	ER@2.95	Ea	1,080.00	112.00	46.50	1,238.50

DWV cast iron double wye, hub and spigot, gasketed joints

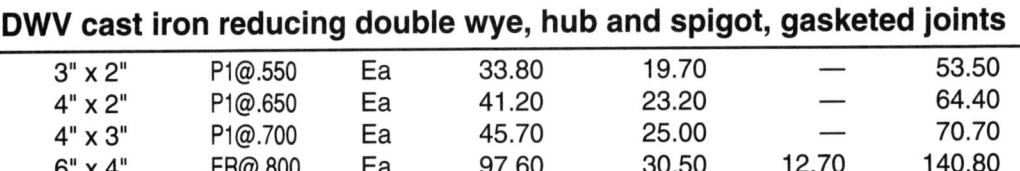

Description	Craft@Hrs	Unit	Material $	Labor $	Equipment $	Total $
2"	P1@.500	Ea	23.00	17.90	—	40.90
3"	P1@.750	Ea	40.00	26.80	—	66.80
4"	P1@.950	Ea	55.40	34.00	—	89.40
6"	ER@1.10	Ea	129.00	41.90	17.30	188.20
8"	ER@1.85	Ea	317.00	70.50	29.20	416.70
10"	ER@2.50	Ea	444.00	95.30	39.30	578.60
12"	ER@2.75	Ea	642.00	105.00	43.50	790.50
15"	ER@3.50	Ea	1,940.00	133.00	55.00	2,128.00

DWV cast iron reducing double wye, hub and spigot, gasketed joints

Description	Craft@Hrs	Unit	Material $	Labor $	Equipment $	Total $
3" x 2"	P1@.550	Ea	33.80	19.70	—	53.50
4" x 2"	P1@.650	Ea	41.20	23.20	—	64.40
4" x 3"	P1@.700	Ea	45.70	25.00	—	70.70
6" x 4"	ER@.800	Ea	97.60	30.50	12.70	140.80

DWV cast iron reducer, hub and spigot, gasketed joints

Description	Craft@Hrs	Unit	Material $	Labor $	Equipment $	Total $
3"	P1@.150	Ea	11.80	5.36	—	17.16
4"	P1@.200	Ea	30.80	7.15	—	37.95
6"	ER@.320	Ea	48.00	12.20	5.00	65.20
8"	ER@.350	Ea	102.00	13.30	5.50	120.80
10"	ER@.570	Ea	214.00	21.70	9.00	244.70
12"	ER@.620	Ea	285.00	23.60	9.70	318.30
15"	ER@1.12	Ea	527.00	42.70	17.60	587.30

Cast Iron, DWV, Service Weight, Hub & Spigot with Gasketed Joints

Description	Craft@Hrs	Unit	Material $	Labor $	Equipment $	Total $

Gaskets for hub and spigot joints

Description	Craft@Hrs	Unit	Material $	Labor $	Equipment $	Total $
2"	--@--	Ea	2.77	—	—	2.77
3"	--@--	Ea	3.67	—	—	3.67
4"	--@--	Ea	4.67	—	—	4.67
5"	--@--	Ea	7.16	—	—	7.16
6"	--@--	Ea	7.54	—	—	7.54
8"	--@--	Ea	16.40	—	—	16.40
10"	--@--	Ea	25.40	—	—	25.40
12"	--@--	Ea	32.30	—	—	32.30
15"	--@--	Ea	38.60	—	—	38.60

Hanger with swivel assembly

Description	Craft@Hrs	Unit	Material $	Labor $	Equipment $	Total $
2"	P1@.400	Ea	6.05	14.30	—	20.35
3"	P1@.450	Ea	7.82	16.10	—	23.92
4"	P1@.450	Ea	9.67	16.10	—	25.77
5"	P1@.500	Ea	14.90	17.90	—	32.80
6"	P1@.500	Ea	16.70	17.90	—	34.60
8"	P1@.550	Ea	21.80	19.70	—	41.50
10"	P1@.650	Ea	28.90	23.20	—	52.10
12"	P1@.750	Ea	39.70	26.80	—	66.50
16"	P1@.850	Ea	52.80	30.40	—	83.20

Riser clamp

Description	Craft@Hrs	Unit	Material $	Labor $	Equipment $	Total $
2"	P1@.115	Ea	4.84	4.11	—	8.95
3"	P1@.120	Ea	5.52	4.29	—	9.81
4"	P1@.125	Ea	7.03	4.47	—	11.50
5"	P1@.180	Ea	10.10	6.44	—	16.54
6"	P1@.200	Ea	12.20	7.15	—	19.35
8"	P1@.200	Ea	19.80	7.15	—	26.95
10"	P1@.250	Ea	29.40	8.94	—	38.34
12"	P1@.250	Ea	34.90	8.94	—	43.84

Copper, DWV with Soft-Soldered Joints

Copper drainage tubing is used for non-pressurized above grade, interior building drain, waste and vent (DWV) systems.

Because of its thin walls, it must be handled carefully and protected from damage during building construction.

This section has been arranged to save the estimator's time by including all normally-used system components such as pipe, fittings, hanger assemblies and riser clamps under one heading. Additional items can be found under "Plumbing and Piping Specialties." The cost estimates in this section are based on the conditions, limitations and wage rates described in the section "How to Use This Book" beginning on page 5.

Description	Craft@Hrs	Unit	Material $	Labor $	Equipment $	Total $

DWV copper pipe assembly, installed horizontally. Assembly includes fittings, couplings, hangers and rod. Based on a reducing wye and a 45 degree elbow every 12 feet for 1¼" pipe. A reducing wye and a 45 degree elbow every 25 feet for 4" pipe, and hangers spaced to meet plumbing code.

Description	Craft@Hrs	Unit	Material $	Labor $	Equipment $	Total $
1¼"	P1@.130	LF	12.00	4.65	—	16.65
1½"	P1@.132	LF	12.20	4.72	—	16.92
2"	P1@.137	LF	15.50	4.90	—	20.40
3"	P1@.179	LF	27.90	6.40	—	34.30
4"	P1@.202	LF	52.10	7.22	—	59.32

DWV copper pipe assembly, installed risers. Complete installation including a reducing tee every floor and a riser clamp every other.

Description	Craft@Hrs	Unit	Material $	Labor $	Equipment $	Total $
1¼"	P1@.069	LF	9.28	2.47	—	11.75
1½"	P1@.074	LF	10.50	2.65	—	13.15
2"	P1@.086	LF	14.30	3.08	—	17.38
3"	P1@.110	LF	27.70	3.93	—	31.63
4"	P1@.134	LF	46.40	4.79	—	51.19

DWV copper pipe, soft-soldered joints

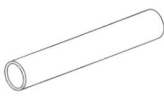

Description	Craft@Hrs	Unit	Material $	Labor $	Equipment $	Total $
1¼"	P1@.040	LF	4.78	1.43	—	6.21
1½"	P1@.045	LF	6.06	1.61	—	7.67
2"	P1@.050	LF	7.96	1.79	—	9.75
3"	P1@.060	LF	12.60	2.15	—	14.75
4"	P1@.070	LF	16.10	2.50	—	18.60

DWV copper 1/8 bend, soft-soldered joints

Description	Craft@Hrs	Unit	Material $	Labor $	Equipment $	Total $
1¼"	P1@.170	Ea	5.98	6.08	—	12.06
1½"	P1@.200	Ea	5.51	7.15	—	12.66
2"	P1@.270	Ea	10.00	9.66	—	19.66
3"	P1@.400	Ea	24.70	14.30	—	39.00
4"	P1@.540	Ea	102.00	19.30	—	121.30

Description	Craft@Hrs	Unit	Material $	Labor $	Equipment $	Total $

DWV copper 1/4 bend, soft-soldered joints

1¼"	P1@.170	Ea	6.73	6.08	—	12.81
1½"	P1@.200	Ea	7.69	7.15	—	14.84
2"	P1@.270	Ea	12.40	9.66	—	22.06
3"	P1@.400	Ea	32.60	14.30	—	46.90
4"	P1@.540	Ea	157.00	19.30	—	176.30

DWV copper closet flange, soft-soldered joints

4" x 3"	P1@.680	Ea	31.90	24.30	—	56.20
4" x 4"	P1@.980	Ea	39.40	35.00	—	74.40

DWV copper P-trap, soft-soldered joints

1¼"	P1@.240	Ea	45.40	8.58	—	53.98
1½"	P1@.240	Ea	39.50	8.58	—	48.08
2"	P1@.240	Ea	60.30	8.58	—	68.88
3"	P1@.320	Ea	173.00	11.40	—	184.40

DWV copper sanitary tee, (TY) soft-soldered joints

1½"	P1@.240	Ea	13.90	8.58	—	22.48
2"	P1@.320	Ea	19.00	11.40	—	30.40
3"	P1@.490	Ea	66.00	17.50	—	83.50
4"	P1@.650	Ea	166.00	23.20	—	189.20

DWV copper sanitary cross, soft-soldered joints

1½"	P1@.640	Ea	92.80	22.90	—	115.70
2"	P1@.900	Ea	134.00	32.20	—	166.20
3"	P1@1.36	Ea	379.00	48.60	—	427.60

DWV copper combination wye and 1/8 bend, soft-soldered joints

1½"	P1@.240	Ea	29.40	8.58	—	37.98
2"	P1@.320	Ea	50.30	11.40	—	61.70
3"	P1@.490	Ea	166.00	17.50	—	183.50
4"	P1@.650	Ea	373.00	23.20	—	396.20

DWV copper wye, soft-soldered joints

1½"	P1@.240	Ea	22.90	8.58	—	31.48
2"	P1@.320	Ea	32.90	11.40	—	44.30
3"	P1@.490	Ea	82.40	17.50	—	99.90
4"	P1@.650	Ea	166.00	23.20	—	189.20

DWV copper cleanout with threaded plug

1¼"	P1@.100	Ea	11.60	3.58	—	15.18
1½"	P1@.160	Ea	17.30	5.72	—	23.02
2"	P1@.230	Ea	22.60	8.22	—	30.82

Copper, DWV with Soft-Soldered Joints

Description	Craft@Hrs	Unit	Material $	Labor $	Equipment $	Total $
DWV copper female adapter						
1½"	P1@.130	Ea	8.96	4.65	—	13.61
2"	P1@.170	Ea	17.00	6.08	—	23.08
3"	P1@.230	Ea	68.00	8.22	—	76.22
4"	P1@.300	Ea	116.00	10.70	—	126.70
DWV copper male adapter						
1½"	P1@.130	Ea	10.20	4.65	—	14.85
2"	P1@.170	Ea	12.20	6.08	—	18.28
3"	P1@.230	Ea	59.50	8.22	—	67.72
4"	P1@.300	Ea	100.00	10.70	—	110.70
DWV copper soil adapter (copper x MJ)						
2"	P1@.170	Ea	14.60	6.08	—	20.68
3"	P1@.250	Ea	29.80	8.94	—	38.74
4"	P1@.320	Ea	55.80	11.40	—	67.20
DWV copper test cap, soft-soldered joint						
1½"	P1@.100	Ea	.34	3.58	—	3.92
2"	P1@.130	Ea	.49	4.65	—	5.14
3"	P1@.200	Ea	.62	7.15	—	7.77
4"	P1@.250	Ea	1.30	8.94	—	10.24
DWV copper test tee, soft-soldered joints						
1½"	P1@.200	Ea	33.40	7.15	—	40.55
2"	P1@.270	Ea	43.60	9.66	—	53.26
3"	P1@.400	Ea	117.00	14.30	—	131.30
DWV copper fitting reducer, soft-soldered joints						
1½"	P1@.100	Ea	5.51	3.58	—	9.09
2"	P1@.140	Ea	7.07	5.01	—	12.08
3"	P1@.200	Ea	24.70	7.15	—	31.85
4"	P1@.270	Ea	39.20	9.66	—	48.86
DWV copper coupling, soft-soldered joints						
1¼"	P1@.170	Ea	2.93	6.08	—	9.01
1½"	P1@.200	Ea	3.38	7.15	—	10.53
2"	P1@.270	Ea	4.84	9.66	—	14.50
3"	P1@.400	Ea	8.96	14.30	—	23.26
4"	P1@.540	Ea	35.20	19.30	—	54.50

Description	Craft@Hrs	Unit	Material $	Labor $	Equipment $	Total $

Hanger with swivel assembly

Description	Craft@Hrs	Unit	Material $	Labor $	Equipment $	Total $
1¼"	P1@.300	Ea	4.44	10.70	—	15.14
1½"	P1@.300	Ea	4.77	10.70	—	15.47
2"	P1@.300	Ea	4.98	10.70	—	15.68
3"	P1@.350	Ea	8.37	12.50	—	20.87
4"	P1@.350	Ea	9.19	12.50	—	21.69

Riser clamp

Description	Craft@Hrs	Unit	Material $	Labor $	Equipment $	Total $
1¼"	P1@.105	Ea	4.33	3.75	—	8.08
1½"	P1@.110	Ea	4.58	3.93	—	8.51
2"	P1@.115	Ea	4.84	4.11	—	8.95
3"	P1@.120	Ea	5.52	4.29	—	9.81
4"	P1@.125	Ea	7.03	4.47	—	11.50

ABS, DWV with Solvent-Weld Joints

ABS (Acrylonitrile Butadiene Styrene) ASTM D 2661-73 is made from various grades of rubber-based resins and is primarily used for non-pressurized drain, waste and vent (DWV) systems.

ABS can also be used for low-temperature (100 degree F. maximum) water distribution systems, but consult the manufacturers for the proper ASTM type, maximum pressure limitations and recommended joint solvent.

The current Uniform Plumbing Code (UPC) states in Sections 401 and 503 that "ABS piping systems shall be limited to those structures where combustible construction is allowed." This does not preclude its use for underground DWV systems.

This section has been arranged to save the estimator's time by including all normally-used system components such as pipe, fittings, hanger assemblies and riser clamps under one heading. The cost estimates in this section are based on the conditions, limitations and wage rates described in the section "How to Use This Book" beginning on page 5.

Description	Craft@Hrs	Unit	Material $	Labor $	Equipment $	Total $
ABS DWV pipe with solvent-weld joints						
1½"	P1@.020	LF	1.00	.72	—	1.72
2"	P1@.025	LF	1.58	.89	—	2.47
3"	P1@.030	LF	2.76	1.07	—	3.83
4"	P1@.040	LF	4.48	1.43	—	5.91
6"	P1@.055	LF	8.28	1.97	—	10.25
ABS DWV 1/8 bend with solvent-weld joints						
1½"	P1@.120	Ea	.95	4.29	—	5.24
2"	P1@.125	Ea	1.71	4.47	—	6.18
3"	P1@.175	Ea	4.14	6.26	—	10.40
4"	P1@.250	Ea	7.82	8.94	—	16.76
6"	P1@.325	Ea	61.90	11.60	—	73.50
ABS DWV 1/8 street bend with solvent-weld joints						
1½"	P1@.060	Ea	1.05	2.15	—	3.20
2"	P1@.065	Ea	1.99	2.32	—	4.31
3"	P1@.088	Ea	4.40	3.15	—	7.55
4"	P1@.125	Ea	12.10	4.47	—	16.57
6"	P1@.163	Ea	60.80	5.83	—	66.63
ABS DWV 1/4 bend with solvent-weld joints						
1½"	P1@.120	Ea	1.16	4.29	—	5.45
2"	P1@.125	Ea	1.96	4.47	—	6.43
3"	P1@.175	Ea	4.94	6.26	—	11.20
4"	P1@.250	Ea	11.30	8.94	—	20.24
6"	P1@.325	Ea	63.90	11.60	—	75.50
ABS DWV closet bend with solvent-weld joints						
3" x 4"	P1@.250	Ea	16.70	8.94	—	25.64
ABS DWV closet flange with solvent-weld joints						
4"	P1@.125	Ea	7.54	4.47	—	12.01

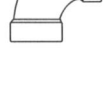
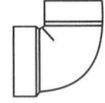

Description	Craft@Hrs	Unit	Material $	Labor $	Equipment $	Total $

ABS DWV P-trap with solvent-weld joints

Description	Craft@Hrs	Unit	Material $	Labor $	Equipment $	Total $
1½"	P1@.150	Ea	3.47	5.36	—	8.83
2"	P1@.190	Ea	8.58	6.79	—	15.37
3"	P1@.260	Ea	20.40	9.30	—	29.70
4"	P1@.350	Ea	39.50	12.50	—	52.00

ABS DWV sanitary tee (TY) with solvent-weld joints

Description	Craft@Hrs	Unit	Material $	Labor $	Equipment $	Total $
1½"	P1@.150	Ea	1.66	5.36	—	7.02
2"	P1@.190	Ea	3.84	6.79	—	10.63
3"	P1@.260	Ea	7.40	9.30	—	16.70
4"	P1@.350	Ea	15.60	12.50	—	28.10
6"	P1@.450	Ea	71.40	16.10	—	87.50

ABS DWV reducing sanitary tee (TY) with solvent-weld joints

Description	Craft@Hrs	Unit	Material $	Labor $	Equipment $	Total $
2" x 1½"	P1@.160	Ea	2.79	5.72	—	8.51
3" x 1½"	P1@.210	Ea	4.77	7.51	—	12.28
3" x 2"	P1@.235	Ea	5.30	8.40	—	13.70
4" x 2"	P1@.300	Ea	15.30	10.70	—	26.00
4" x 3"	P1@.325	Ea	18.20	11.60	—	29.80

ABS DWV double sanitary tee (TY) with solvent-weld joints

Description	Craft@Hrs	Unit	Material $	Labor $	Equipment $	Total $
1½"	P1@.210	Ea	5.10	7.51	—	12.61
2"	P1@.250	Ea	7.67	8.94	—	16.61
3"	P1@.350	Ea	12.10	12.50	—	24.60
4"	P1@.475	Ea	47.90	17.00	—	64.90

ABS DWV combination wye and 1/8 bend with solvent-weld joints

Description	Craft@Hrs	Unit	Material $	Labor $	Equipment $	Total $
1½"	P1@.150	Ea	10.10	5.36	—	15.46
2"	P1@.190	Ea	13.10	6.79	—	19.89
3"	P1@.260	Ea	17.60	9.30	—	26.90
4"	P1@.350	Ea	30.70	12.50	—	43.20

ABS DWV combination reducing wye and 1/8 bend with solvent-weld joints

Description	Craft@Hrs	Unit	Material $	Labor $	Equipment $	Total $
2" x 1½"	P1@.160	Ea	14.50	5.72	—	20.22
3" x 1½"	P1@.210	Ea	22.20	7.51	—	29.71
3" x 2"	P1@.235	Ea	15.50	8.40	—	23.90
4" x 2"	P1@.300	Ea	28.10	10.70	—	38.80
4" x 3"	P1@.325	Ea	33.10	11.60	—	44.70

ABS, DWV with Solvent-Weld Joints

Description	Craft@Hrs	Unit	Material $	Labor $	Equipment $	Total $
ABS DWV bushing with solvent-weld joints						
2" x 1½"	P1@.062	Ea	.87	2.22	—	3.09
3" x 1½"	P1@.088	Ea	4.31	3.15	—	7.46
3" x 2"	P1@.088	Ea	3.08	3.15	—	6.23
4" x 2"	P1@.125	Ea	6.23	4.47	—	10.70
4" x 3"	P1@.125	Ea	6.23	4.47	—	10.70
6" x 4"	P1@.163	Ea	31.10	5.83	—	36.93
ABS DWV FIP adapter						
1½"	P1@.080	Ea	1.31	2.86	—	4.17
2"	P1@.085	Ea	2.42	3.04	—	5.46
3"	P1@.155	Ea	5.73	5.54	—	11.27
4"	P1@.220	Ea	10.90	7.87	—	18.77
6"	P1@.285	Ea	42.00	10.20	—	52.20
ABS DWV test cap						
1½"	P1@.030	Ea	.66	1.07	—	1.73
2"	P1@.040	Ea	1.06	1.43	—	2.49
3"	P1@.050	Ea	1.67	1.79	—	3.46
4"	P1@.090	Ea	2.64	3.22	—	5.86
ABS DWV cleanout plug						
1½"	P1@.090	Ea	3.06	3.22	—	6.28
2"	P1@.120	Ea	4.48	4.29	—	8.77
3"	P1@.240	Ea	7.98	8.58	—	16.56
4"	P1@.320	Ea	17.30	11.40	—	28.70
6"	P1@.450	Ea	55.10	16.10	—	71.20
ABS DWV cleanout tee with solvent-weld joints						
1½"	P1@.120	Ea	4.74	4.29	—	9.03
2"	P1@.125	Ea	7.67	4.47	—	12.14
3"	P1@.175	Ea	12.70	6.26	—	18.96
4"	P1@.250	Ea	26.00	8.94	—	34.94
ABS DWV wye with solvent-weld joints						
1½"	P1@.150	Ea	2.22	5.36	—	7.58
2"	P1@.190	Ea	4.14	6.79	—	10.93
3"	P1@.260	Ea	7.67	9.30	—	16.97
4"	P1@.350	Ea	17.70	12.50	—	30.20
6"	P1@.450	Ea	74.80	16.10	—	90.90
ABS DWV reducing wye with solvent-weld joints						
2" x 1½"	P1@.160	Ea	3.84	5.72	—	9.56
3" x 1½"	P1@.210	Ea	5.36	7.51	—	12.87
3" x 2"	P1@.235	Ea	6.00	8.40	—	14.40
4" x 2"	P1@.300	Ea	12.90	10.70	—	23.60
4" x 3"	P1@.325	Ea	16.10	11.60	—	27.70
6" x 4"	P1@.400	Ea	61.90	14.30	—	76.20

Description	Craft@Hrs	Unit	Material $	Labor $	Equipment $	Total $

ABS DWV reducer with solvent-weld joints

Description	Craft@Hrs	Unit	Material $	Labor $	Equipment $	Total $
2" x 1½"	P1@.125	Ea	.87	4.47	—	5.34
3" x 1½"	P1@.150	Ea	4.31	5.36	—	9.67
3" x 2"	P1@.150	Ea	3.08	5.36	—	8.44
4" x 2"	P1@.185	Ea	6.23	6.62	—	12.85
4" x 3"	P1@.210	Ea	6.16	7.51	—	13.67

ABS DWV coupling with solvent-weld joints

Description	Craft@Hrs	Unit	Material $	Labor $	Equipment $	Total $
1½"	P1@.120	Ea	.74	4.29	—	5.03
2"	P1@.125	Ea	1.09	4.47	—	5.56
3"	P1@.175	Ea	2.53	6.26	—	8.79
4"	P1@.250	Ea	4.60	8.94	—	13.54
6"	P1@.325	Ea	27.20	11.60	—	38.80

Hanger with swivel assembly

Description	Craft@Hrs	Unit	Material $	Labor $	Equipment $	Total $
1½"	P1@.300	Ea	4.77	10.70	—	15.47
2"	P1@.300	Ea	4.98	10.70	—	15.68
3"	P1@.350	Ea	8.37	12.50	—	20.87
4"	P1@.350	Ea	9.19	12.50	—	21.69
0"	P1@.450	Ea	15.10	16.10	—	31.20

Riser clamp

Description	Craft@Hrs	Unit	Material $	Labor $	Equipment $	Total $
1½"	P1@.110	Ea	4.58	3.93	—	8.51
2"	P1@.115	Ea	4.84	4.11	—	8.95
3"	P1@.120	Ea	5.52	4.29	—	9.81
4"	P1@.125	Ea	7.03	4.47	—	11.50
6"	P1@.130	Ea	12.20	4.65	—	16.85

Galvanized steel pipe sleeves

Description	Craft@Hrs	Unit	Material $	Labor $	Equipment $	Total $
2"	P1@.130	Ea	6.34	4.65	—	10.99
2½"	P1@.150	Ea	6.47	5.36	—	11.83
3"	P1@.180	Ea	6.62	6.44	—	13.06
4"	P1@.220	Ea	7.52	7.87	—	15.39
5"	P1@.250	Ea	9.37	8.94	—	18.31
6"	P1@.270	Ea	10.10	9.66	—	19.76

Lead roof pipe flashing

Description	Craft@Hrs	Unit	Material $	Labor $	Equipment $	Total $
1½"	P1@.250	Ea	31.60	8.94	—	40.54
2"	P1@.250	Ea	41.60	8.94	—	50.54
3"	P1@.250	Ea	48.30	8.94	—	57.24
4"	P1@.250	Ea	53.10	8.94	—	62.04
6"	P1@.350	Ea	68.10	12.50	—	80.60

PVC, DWV with Solvent-Weld Joints

PVC (Polyvinyl Chloride) meets ASTM – D3034, D3212, F1336, F679 & CSA 182.2 (CSA - Canadian Standards Association – similar to UL – Underwriters Laboratories). The applications for PVC DWV have increased in number due to the unique flame spread characteristics of their compounds, as well as the development of new fire stop devices for use with combustible piping systems.

PVC will not rust, pit or degrade when exposed to moisture and is extremely resistant to a broad range of corrosive agents.

PVC DWV is suitable for sewer, vent and storm water drainage applications in both below and above grade applications.

When penetrating a vertical or horizontal fire separation (fire rated wall or floor) with combustible piping, certified fire stop devices must be used in compliance with local and national codes.
PVC DWV is not permitted in a ceiling space used as a return air plenum and is not permitted in a vertical shaft. PVC DWV has a flame spread rating of 15.

This section has been arranged to save the estimator's time by including all normally-used system components such as pipe, fittings, hanger assemblies and riser clamps under one heading. The cost estimates in this section are based on the conditions, limitations and wage rates described in the section "How to Use This Book" beginning on page 5.

Description	Craft@Hrs	Unit	Material $	Labor $	Equipment $	Total $

PVC DWV pipe with solvent-weld joints

Description	Craft@Hrs	Unit	Material $	Labor $	Equipment $	Total $
1½"	P1@.021	LF	1.01	.75	—	1.76
2"	P1@.026	LF	1.83	.93	—	2.76
3"	P1@.032	LF	3.88	1.14	—	5.02
4"	P1@.042	LF	4.75	1.50	—	6.25
6"	P1@.058	LF	8.54	2.07	—	10.61

PVC DWV 90 degree elbow with solvent-weld joints

Description	Craft@Hrs	Unit	Material $	Labor $	Equipment $	Total $
1½"	P1@.126	Ea	1.02	4.51	—	5.53
2"	P1@.131	Ea	2.38	4.68	—	7.06
3"	P1@.184	Ea	6.27	6.58	—	12.85
4"	P1@.263	Ea	12.40	9.40	—	21.80
6"	P1@.341	Ea	52.00	12.20	—	64.20

PVC DWV 90 degree street elbow with solvent-weld joints

Description	Craft@Hrs	Unit	Material $	Labor $	Equipment $	Total $
1½"	P1@.063	Ea	1.68	2.25	—	3.93
2"	P1@.068	Ea	2.85	2.43	—	5.28
3"	P1@.092	Ea	8.31	3.29	—	11.60
4"	P1@.131	Ea	20.10	4.68	—	24.78
6"	P1@.171	Ea	73.50	6.11	—	79.61

PVC DWV 45 degree elbow with solvent-weld joints

Description	Craft@Hrs	Unit	Material $	Labor $	Equipment $	Total $
1½"	P1@.126	Ea	1.31	4.51	—	5.82
2"	P1@.131	Ea	1.59	4.68	—	6.27
3"	P1@.184	Ea	3.99	6.58	—	10.57
4"	P1@.263	Ea	10.10	9.40	—	19.50
6"	P1@.341	Ea	61.20	12.20	—	73.40

PVC, DWV with Solvent-Weld Joints

Description	Craft@Hrs	Unit	Material $	Labor $	Equipment $	Total $
PVC DWV wye with solvent-weld joints						
1½"	P1@.158	Ea	2.47	5.65	—	8.12
2"	P1@.200	Ea	5.47	7.15	—	12.62
3"	P1@.273	Ea	9.04	9.76	—	18.80
4"	P1@.368	Ea	14.80	13.20	—	28.00
6"	P1@.473	Ea	58.10	16.90	—	75.00
PVC DWV reducing wye with solvent-weld joints						
2" x 1½"	P1@.168	Ea	4.74	6.01	—	10.75
3" x 1½"	P1@.221	Ea	5.44	7.90	—	13.34
3" x 2"	P1@.247	Ea	7.82	8.83	—	16.65
4" x 2"	P1@.315	Ea	15.10	11.30	—	26.40
4" x 3"	P1@.341	Ea	17.90	12.20	—	30.10
6" x 3"	P1@.473	Ea	66.10	16.90	—	83.00
6" x 4"	P1@.473	Ea	70.40	16.90	—	87.30
PVC DWV sanitary tee (TY) with solvent-weld joints						
1½"	P1@.158	Ea	1.75	5.65	—	7.40
2"	P1@.200	Ea	3.23	7.15	—	10.38
3"	P1@.273	Ea	6.95	9.76	—	16.71
4"	P1@.368	Ea	13.20	13.20	—	26.40
6"	P1@.473	Ea	72.90	16.90	—	89.80
PVC DWV reducing sanitary tee (TY) with solvent-weld joints						
2" x 1½"	P1@.168	Ea	2.52	6.01	—	8.53
3" x 1½"	P1@.221	Ea	4.34	7.90	—	12.24
3" x 2"	P1@.247	Ea	5.83	8.83	—	14.66
4" x 2"	P1@.315	Ea	12.80	11.30	—	24.10
4" x 3"	P1@.341	Ea	14.70	12.20	—	26.90
6" x 3"	P1@.473	Ea	55.90	16.90	—	72.80
6" x 4"	P1@.473	Ea	55.90	16.90	—	72.80
PVC DWV double wye with solvent-weld joints						
2" x 1½"	P1@.263	Ea	7.13	9.40	—	16.53
3" x 2"	P1@.368	Ea	14.50	13.20	—	27.70
4" x 3"	P1@.473	Ea	47.20	16.90	—	64.10
6" x 4"	P1@.578	Ea	57.00	20.70	—	77.70
PVC DWV closet flange with solvent-weld joints, adjustable with test plate						
4" x 3"	P1@.131	Ea	12.40	4.68	—	17.08
PVC DWV closet flange with solvent-weld joints, non-adjustable with test plate						
4" x 3"	P1@.131	Ea	9.32	4.68	—	14.00

PVC, DWV with Solvent-Weld Joints

Description	Craft@Hrs	Unit	Material $	Labor $	Equipment $	Total $

PVC DWV closet flange with solvent-weld joints, non-adjustable without test plate

Description	Craft@Hrs	Unit	Material $	Labor $	Equipment $	Total $
4" x 3"	P1@.131	Ea	8.49	4.68	—	13.17

PVC DWV closet flange with solvent-weld joints, adjustable, offset without test plate

Description	Craft@Hrs	Unit	Material $	Labor $	Equipment $	Total $
4" x 3"	P1@.170	Ea	18.20	6.08	—	24.28

PVC DWV P-trap with solvent-weld joints

Description	Craft@Hrs	Unit	Material $	Labor $	Equipment $	Total $
1½"	P1@.160	Ea	3.97	5.72	—	9.69
2"	P1@.200	Ea	9.68	7.15	—	16.83
3"	P1@.285	Ea	22.90	10.20	—	33.10
4"	P1@.370	Ea	44.90	13.20	—	58.10
6"	P1@.580	Ea	171.00	20.70	—	191.70

PVC DWV bushing with solvent-weld joints

Description	Craft@Hrs	Unit	Material $	Labor $	Equipment $	Total $
2" x 1½"	P1@.062	Ea	.82	2.22	—	3.04
3" x 2"	P1@.088	Ea	2.40	3.15	—	5.55
4" x 3"	P1@.125	Ea	5.04	4.47	—	9.51
6" x 4"	P1@.163	Ea	26.00	5.83	—	31.83

PVC DWV FIP adapter

Description	Craft@Hrs	Unit	Material $	Labor $	Equipment $	Total $
1½"	P1@.080	Ea	1.10	2.86	—	3.96
2"	P1@.085	Ea	1.93	3.04	—	4.97
3"	P1@.155	Ea	4.69	5.54	—	10.23
4"	P1@.220	Ea	9.12	7.87	—	16.99
6"	P1@.285	Ea	41.20	10.20	—	51.40

PVC DWV MIP adapter

Description	Craft@Hrs	Unit	Material $	Labor $	Equipment $	Total $
1½"	P1@.080	Ea	1.22	2.86	—	4.08
2"	P1@.085	Ea	2.02	3.04	—	5.06
3"	P1@.155	Ea	4.16	5.54	—	9.70
4"	P1@.220	Ea	11.10	7.87	—	18.97
6"	P1@.285	Ea	20.50	10.20	—	30.70

PVC DWV coupling with solvent-weld joints

Description	Craft@Hrs	Unit	Material $	Labor $	Equipment $	Total $
1½"	P1@.120	Ea	.58	4.29	—	4.87
2"	P1@.125	Ea	.91	4.47	—	5.38
3"	P1@.175	Ea	2.13	6.26	—	8.39
4"	P1@.250	Ea	4.49	8.94	—	13.43
6"	P1@.325	Ea	21.90	11.60	—	33.50

Description	Craft@Hrs	Unit	Material $	Labor $	Equipment $	Total $

PVC DWV reducer with solvent-weld joints

Description	Craft@Hrs	Unit	Material $	Labor $	Equipment $	Total $
2" x 1½"	P1@.125	Ea	1.35	4.47	—	5.82
3" x 2"	P1@.150	Ea	4.09	5.36	—	9.45
4" x 3"	P1@.185	Ea	5.32	6.62	—	11.94
6" x 4"	P1@.210	Ea	53.80	7.51	—	61.31

PVC DWV in-line cleanout with solvent-weld joints

Description	Craft@Hrs	Unit	Material $	Labor $	Equipment $	Total $
1½"	P1@.120	Ea	5.08	4.29	—	9.37
2"	P1@.125	Ea	8.69	4.47	—	13.16
3"	P1@.175	Ea	12.80	6.26	—	19.06
4"	P1@.250	Ea	26.60	8.94	—	35.54
6" x 4"	P1@.250	Ea	90.70	8.94	—	99.64
6"	P1@.250	Ea	90.70	8.94	—	99.64

PVC DWV end cleanout

Description	Craft@Hrs	Unit	Material $	Labor $	Equipment $	Total $
1½"	P1@.090	Ea	2.23	3.22	—	5.45
2"	P1@.120	Ea	3.63	4.29	—	7.92
3"	P1@.240	Ea	7.36	8.58	—	15.94
4"	P1@.320	Ea	13.70	11.40	—	25.10
6"	P1@.450	Ea	47.70	16.10	—	63.80

PVC DWV fitting end cleanout

Description	Craft@Hrs	Unit	Material $	Labor $	Equipment $	Total $
1½"	P1@.090	Ea	2.12	3.22	—	5.34
2"	P1@.120	Ea	4.20	4.29	—	8.49
3"	P1@.240	Ea	4.90	8.58	—	13.48
4"	P1@.320	Ea	10.50	11.40	—	21.90
6"	P1@.450	Ea	54.40	16.10	—	70.50

Hanger with swivel assembly

Description	Craft@Hrs	Unit	Material $	Labor $	Equipment $	Total $
1½"	P1@.300	Ea	6.08	10.70	—	16.78
2"	P1@.300	Ea	6.36	10.70	—	17.06
3"	P1@.350	Ea	10.70	12.50	—	23.20
4"	P1@.350	Ea	11.70	12.50	—	24.20
6"	P1@.450	Ea	19.10	16.10	—	35.20

Riser clamp

Description	Craft@Hrs	Unit	Material $	Labor $	Equipment $	Total $
1½"	P1@.110	Ea	4.58	3.93	—	8.51
2"	P1@.115	Ea	4.84	4.11	—	8.95
3"	P1@.120	Ea	5.52	4.29	—	9.81
4"	P1@.125	Ea	7.03	4.47	—	11.50
6"	P1@.130	Ea	12.20	4.65	—	16.85

PVC, DWV with Solvent-Weld Joints

Description	Craft@Hrs	Unit	Material $	Labor $	Equipment $	Total $
Galvanized steel pipe sleeves						
2"	P1@.130	Ea	6.34	4.65	—	10.99
2½"	P1@.150	Ea	6.47	5.36	—	11.83
3"	P1@.180	Ea	6.62	6.44	—	13.06
4"	P1@.220	Ea	7.52	7.87	—	15.39
5"	P1@.250	Ea	9.37	8.94	—	18.31
6"	P1@.270	Ea	10.10	9.66	—	19.76
Lead roof pipe flashing						
1½"	P1@.400	Ea	33.10	14.30	—	47.40
2"	P1@.450	Ea	43.70	16.10	—	59.80
3"	P1@.500	Ea	50.50	17.90	—	68.40
4"	P1@.550	Ea	55.90	19.70	—	75.60

PVC, DWV with Gasketed Bell and Spigot Joints

PVC (Polyvinyl Chloride) is polymerized vinyl chloride produced from acetylene and anhydrous hydrochloric acid.

PVC sewer pipe (ASTM D-3034) 10' lay length, is for underground installation and is used primarily to convey sewage and rain water. It is of the bell & spigot type with joints made water-tight by means of an elastomeric gasket (o-ring) factory-installed in the hub end of the pipe and fittings.

This section includes pipe and all commonly-used fittings. Additional items can be found under "Plumbing and Piping Specialties." The cost estimates in this section are based on the conditions, limitations, and wage rates described in the section "How to Use This Book" beginning on page 5.

Description	Craft@Hrs	Unit	Material $	Labor $	Equipment $	Total $

PVC SDR35 sewer pipe with bell and spigot gasketed joints

Description	Craft@Hrs	Unit	Material $	Labor $	Equipment $	Total $
4"	P1@.050	LF	3.59	1.79	—	5.38
6"	P1@.070	LF	7.88	2.50	—	10.38
8"	P1@.090	LF	12.50	3.22	—	15.72
10"	P1@.110	LF	17.50	3.93	—	21.43
12"	P1@.140	LF	22.10	5.01	—	27.11
15"	P1@.180	LF	32.60	6.44	—	39.04
18"	P1@.230	LF	50.70	8.22	—	58.92

PVC sewer pipe 1/16 bend B x B with bell and spigot gasketed joints

Description	Craft@Hrs	Unit	Material $	Labor $	Equipment $	Total $
4"	P1@.320	Ea	16.60	11.40	—	28.00
6"	P1@.400	Ea	32.50	14.30	—	46.80
8"	P1@.480	Ea	95.10	17.20	—	112.30
10"	P1@.800	Ea	260.00	28.60	—	288.60
12"	P1@1.06	Ea	340.00	37.90	—	377.90
15"	P1@1.34	Ea	902.00	47.90	—	949.90
18"	P1@1.60	Ea	1,310.00	57.20	—	1,367.20

PVC sewer pipe 1/16 bend B x S with bell and spigot gasketed joints

Description	Craft@Hrs	Unit	Material $	Labor $	Equipment $	Total $
4"	P1@.320	Ea	15.80	11.40	—	27.20
6"	P1@.400	Ea	31.10	14.30	—	45.40
8"	P1@.480	Ea	100.00	17.20	—	117.20
10"	P1@.800	Ea	255.00	28.60	—	283.60
12"	P1@1.06	Ea	330.00	37.90	—	367.90
15"	P1@1.34	Ea	721.00	47.90	—	768.90
18"	P1@1.60	Ea	1,100.00	57.20	—	1,157.20

PVC sewer pipe 1/8 bend B x B with bell and spigot gasketed joints

Description	Craft@Hrs	Unit	Material $	Labor $	Equipment $	Total $
4"	P1@.320	Ea	16.70	11.40	—	28.10
6"	P1@.400	Ea	33.60	14.30	—	47.90
8"	P1@.480	Ea	90.20	17.20	—	107.40
10"	P1@.800	Ea	243.00	28.60	—	271.60
12"	P1@1.06	Ea	358.00	37.90	—	395.90
15"	P1@1.34	Ea	794.00	47.90	—	841.90
18"	P1@1.60	Ea	1,270.00	57.20	—	1,327.20

PVC, DWV with Gasketed Bell and Spigot Joints

Description	Craft@Hrs	Unit	Material $	Labor $	Equipment $	Total $

PVC sewer pipe 1/8 bend B x S with bell and spigot gasketed joints

Description	Craft@Hrs	Unit	Material $	Labor $	Equipment $	Total $
4"	P1@.320	Ea	15.00	11.40	—	26.40
6"	P1@.400	Ea	30.00	14.30	—	44.30
8"	P1@.480	Ea	90.20	17.20	—	107.40
10"	P1@.800	Ea	240.00	28.60	—	268.60
12"	P1@1.06	Ea	347.00	37.90	—	384.90
15"	P1@1.34	Ea	623.00	47.90	—	670.90
18"	P1@1.60	Ea	1,050.00	57.20	—	1,107.20

PVC sewer pipe 1/4 bend B x B with bell and spigot gasketed joints

Description	Craft@Hrs	Unit	Material $	Labor $	Equipment $	Total $
4"	P1@.320	Ea	23.80	11.40	—	35.20
6"	P1@.400	Ea	44.30	14.30	—	58.60
8"	P1@.480	Ea	121.00	17.20	—	138.20
10"	P1@.800	Ea	411.00	28.60	—	439.60
12"	P1@1.06	Ea	528.00	37.90	—	565.90
15"	P1@1.34	Ea	1,120.00	47.90	—	1,167.90
18"	P1@1.60	Ea	1,890.00	57.20	—	1,947.20

PVC sewer pipe 1/4 bend B x S with bell and spigot gasketed joints

Description	Craft@Hrs	Unit	Material $	Labor $	Equipment $	Total $
4"	P1@.320	Ea	22.00	11.40	—	33.40
6"	P1@.400	Ea	46.60	14.30	—	60.90
8"	P1@.480	Ea	127.00	17.20	—	144.20
10"	P1@.800	Ea	406.00	28.60	—	434.60
12"	P1@1.06	Ea	513.00	37.90	—	550.90
15"	P1@1.34	Ea	1,010.00	47.90	—	1,057.90
18"	P1@1.60	Ea	1,750.00	57.20	—	1,807.20

PVC sewer pipe coupling B x B with bell and spigot gasketed joints

Description	Craft@Hrs	Unit	Material $	Labor $	Equipment $	Total $
4"	P1@.320	Ea	20.50	11.40	—	31.90
6"	P1@.400	Ea	41.10	14.30	—	55.40
8"	P1@.480	Ea	70.00	17.20	—	87.20
10"	P1@.800	Ea	154.00	28.60	—	182.60
12"	P1@1.06	Ea	225.00	37.90	—	262.90
15"	P1@1.34	Ea	466.00	47.90	—	513.90
18"	P1@1.60	Ea	909.00	57.20	—	966.20

PVC sewer pipe wye B x B x B with bell and spigot gasketed joints

Description	Craft@Hrs	Unit	Material $	Labor $	Equipment $	Total $
4"	P1@.480	Ea	27.80	17.20	—	45.00
6"	P1@.600	Ea	63.80	21.50	—	85.30
8"	P1@.720	Ea	176.00	25.70	—	201.70
10"	P1@1.20	Ea	513.00	42.90	—	555.90
12"	P1@1.59	Ea	728.00	56.90	—	784.90
15"	P1@2.01	Ea	1,270.00	71.90	—	1,341.90
18"	P1@2.40	Ea	2,230.00	85.80	—	2,315.80

Description	Craft@Hrs	Unit	Material $	Labor $	Equipment $	Total $

PVC sewer pipe wye B x S x B with bell and spigot gasketed joints

Description	Craft@Hrs	Unit	Material $	Labor $	Equipment $	Total $
10"	P1@1.20	Ea	506.00	42.90	—	548.90
12"	P1@1.59	Ea	721.00	56.90	—	777.90
15"	P1@2.01	Ea	1,170.00	71.90	—	1,241.90
18"	P1@2.40	Ea	2,050.00	85.80	—	2,135.80

PVC sewer pipe reducing wye B x B x B with bell and spigot gasketed joints

Description	Craft@Hrs	Unit	Material $	Labor $	Equipment $	Total $
6" x 4"	P1@.560	Ea	56.80	20.00	—	76.80
8" x 4"	P1@.640	Ea	84.30	22.90	—	107.20
8" x 6"	P1@.680	Ea	101.00	24.30	—	125.30
10" x 4"	P1@.960	Ea	255.00	34.30	—	289.30
10" x 6"	P1@1.00	Ea	260.00	35.80	—	295.80
10" x 8"	P1@1.04	Ea	411.00	37.20	—	448.20
12" x 4"	P1@1.22	Ea	370.00	43.60	—	413.60
12" x 6"	P1@1.26	Ea	381.00	45.10	—	426.10
12" x 8"	P1@1.30	Ea	568.00	46.50	—	614.50
12" x 10"	P1@1.46	Ea	691.00	52.20	—	743.20
15" x 4"	P1@1.50	Ea	558.00	53.60	—	611.60
15" x 6"	P1@1.54	Ea	661.00	55.10	—	716.10
15" x 8"	P1@1.58	Ea	746.00	56.50	—	802.50
15" x 10"	P1@1.74	Ea	891.00	62.20	—	953.20
15" x 12"	P1@1.87	Ea	998.00	66.90	—	1,064.90
18" x 4"	P1@1.76	Ea	1,270.00	62.90	—	1,332.90
18" x 6"	P1@1.80	Ea	1,290.00	64.40	—	1,354.40
18" x 8"	P1@1.84	Ea	1,350.00	65.80	—	1,415.80
18" x 10"	P1@2.00	Ea	1,500.00	71.50	—	1,571.50
18" x 12"	P1@2.13	Ea	1,620.00	76.20	—	1,696.20
18" x 15"	P1@2.27	Ea	1,730.00	81.20	—	1,811.20

PVC sewer pipe reducing wye B x S x B with bell and spigot gasketed joints

Description	Craft@Hrs	Unit	Material $	Labor $	Equipment $	Total $
10" x 4"	P1@.960	Ea	248.00	34.30	—	282.30
10" x 6"	P1@1.00	Ea	262.00	35.80	—	297.80
10" x 8"	P1@1.04	Ea	423.00	37.20	—	460.20
12" x 4"	P1@1.22	Ea	375.00	43.60	—	418.60
12" x 6"	P1@1.26	Ea	381.00	45.10	—	426.10
12" x 8"	P1@1.30	Ea	560.00	46.50	—	606.50
12" x 10"	P1@1.46	Ea	681.00	52.20	—	733.20
15" x 4"	P1@1.50	Ea	570.00	53.60	—	623.60
15" x 6"	P1@1.54	Ea	646.00	55.10	—	701.10
15" x 8"	P1@1.58	Ea	756.00	56.50	—	812.50
15" x 10"	P1@1.74	Ea	866.00	62.20	—	928.20
15" x 12"	P1@1.87	Ea	1,010.00	66.90	—	1,076.90
18" x 4"	P1@1.76	Ea	1,200.00	62.90	—	1,262.90
18" x 6"	P1@1.80	Ea	1,290.00	64.40	—	1,354.40
18" x 8"	P1@1.84	Ea	1,350.00	65.80	—	1,415.80
18" x 10"	P1@2.00	Ea	1,500.00	71.50	—	1,571.50
18" x 12"	P1@2.13	Ea	1,620.00	76.20	—	1,696.20
18" x 15"	P1@2.27	Ea	1,730.00	81.20	—	1,811.20

PVC, DWV with Gasketed Bell and Spigot Joints

Description	Craft@Hrs	Unit	Material $	Labor $	Equipment $	Total $

PVC sewer pipe tee-wye B x B x B with bell and spigot gasketed joints

Description	Craft@Hrs	Unit	Material $	Labor $	Equipment $	Total $
4"	P1@.480	Ea	31.10	17.20	—	48.30
6"	P1@.600	Ea	78.50	21.50	—	100.00
6" x 4"	P1@.560	Ea	67.00	20.00	—	87.00
8" x 4"	P1@.640	Ea	85.10	22.90	—	108.00
8" x 6"	P1@.680	Ea	97.90	24.30	—	122.20

PVC sewer pipe tee B x S x B with bell and spigot gasketed joints

Description	Craft@Hrs	Unit	Material $	Labor $	Equipment $	Total $
10"	P1@1.20	Ea	428.00	42.90	—	470.90
12"	P1@1.59	Ea	593.00	56.90	—	649.90
15"	P1@2.01	Ea	944.00	71.90	—	1,015.90
18"	P1@2.40	Ea	1,770.00	85.80	—	1,855.80

PVC sewer pipe reducing tee B x S x B with bell and spigot gasketed joints

Description	Craft@Hrs	Unit	Material $	Labor $	Equipment $	Total $
10" x 4"	P1@.960	Ea	273.00	34.30	—	307.30
10" x 6"	P1@1.00	Ea	283.00	35.80	—	318.80
10" x 8"	P1@1.04	Ea	443.00	37.20	—	480.20
12" x 4"	P1@1.22	Ea	334.00	43.60	—	377.60
12" x 6"	P1@1.26	Ea	340.00	45.10	—	385.10
12" x 8"	P1@1.30	Ea	468.00	46.50	—	514.50
12" x 10"	P1@1.46	Ea	560.00	52.20	—	612.20
15" x 4"	P1@1.50	Ea	541.00	53.60	—	594.60
15" x 6"	P1@1.54	Ea	568.00	55.10	—	623.10
15" x 8"	P1@1.58	Ea	581.00	56.50	—	637.50
15" x 10"	P1@1.74	Ea	681.00	62.20	—	743.20
15" x 12"	P1@1.87	Ea	746.00	66.90	—	812.90
18" x 4"	P1@1.76	Ea	1,360.00	62.90	—	1,422.90
18" x 6"	P1@1.80	Ea	1,400.00	64.40	—	1,464.40
18" x 8"	P1@1.84	Ea	1,470.00	65.80	—	1,535.80
18" x 10"	P1@2.00	Ea	1,520.00	71.50	—	1,591.50
18" x 12"	P1@2.13	Ea	1,580.00	76.20	—	1,656.20
18" x 15"	P1@2.27	Ea	1,680.00	81.20	—	1,761.20

PVC, DWV with Gasketed Bell and Spigot Joints

Description	Craft@Hrs	Unit	Material $	Labor $	Equipment $	Total $

PVC sewer pipe reducer S x B with bell and spigot gasketed joints

Description	Craft@Hrs	Unit	Material $	Labor $	Equipment $	Total $
6" x 4"	P1@.360	Ea	30.00	12.90	—	42.90
8" x 4"	P1@.400	Ea	81.70	14.30	—	96.00
8" x 6"	P1@.440	Ea	91.90	15.70	—	107.60
10" x 4"	P1@.560	Ea	215.00	20.00	—	235.00
10" x 6"	P1@.600	Ea	221.00	21.50	—	242.50
10" x 8"	P1@.640	Ea	263.00	22.90	—	285.90
12" x 4"	P1@.690	Ea	283.00	24.70	—	307.70
12" x 6"	P1@.730	Ea	290.00	26.10	—	316.10
12" x 8"	P1@.770	Ea	334.00	27.50	—	361.50
12" x 10"	P1@.930	Ea	370.00	33.30	—	403.30
15" x 4"	P1@.830	Ea	476.00	29.70	—	505.70
15" x 6"	P1@.870	Ea	493.00	31.10	—	524.10
15" x 8"	P1@.910	Ea	560.00	32.50	—	592.50
15" x 10"	P1@1.07	Ea	560.00	38.30	—	598.30
15" x 12"	P1@1.20	Ea	645.00	42.90	—	687.90
18" x 8"	P1@1.04	Ea	691.00	37.20	—	728.20
18" x 10"	P1@1.20	Ea	700.00	42.90	—	742.90
18" x 12"	P1@1.33	Ea	704.00	47.60	—	751.60
18" x 15"	P1@1.47	Ea	721.00	52.60	—	773.60

PVC sewer pipe adapter B x S with bell and spigot gasketed joints

Description	Craft@Hrs	Unit	Material $	Labor $	Equipment $	Total $
4"	P1@.320	Ea	17.50	11.40	—	28.90
6"	P1@.400	Ea	32.50	14.30	—	46.80
8"	P1@.480	Ea	83.40	17.20	—	100.60
10"	P1@.800	Ea	104.00	28.60	—	132.60
12"	P1@1.06	Ea	158.00	37.90	—	195.90
15"	P1@1.34	Ea	238.00	47.90	—	285.90
18"	P1@1.60	Ea	523.00	57.20	—	580.20

PVC sewer pipe cap with bell and spigot gasketed joints

Description	Craft@Hrs	Unit	Material $	Labor $	Equipment $	Total $
4"	P1@.160	Ea	10.70	5.72	—	16.42
6"	P1@.200	Ea	20.00	7.15	—	27.15
8"	P1@.240	Ea	54.10	8.58	—	62.68
10"	P1@.400	Ea	169.00	14.30	—	183.30
12"	P1@.530	Ea	255.00	19.00	—	274.00
15"	P1@.670	Ea	413.00	24.00	—	437.00
18"	P1@.800	Ea	593.00	28.60	—	621.60

PVC sewer pipe test plug with bell and spigot gasketed joints

Description	Craft@Hrs	Unit	Material $	Labor $	Equipment $	Total $
4"	P1@.160	Ea	8.08	5.72	—	13.80
6"	P1@.200	Ea	12.40	7.15	—	19.55
8"	P1@.240	Ea	44.30	8.58	—	52.88
10"	P1@.400	Ea	146.00	14.30	—	160.30
12"	P1@.530	Ea	175.00	19.00	—	194.00
15"	P1@.670	Ea	347.00	24.00	—	371.00
18"	P1@.800	Ea	481.00	28.60	—	509.60

Polypropylene, Schedule 40, with Heat-Fusioned Joints

Polypropylene is used almost exclusively for acid and laboratory drain, waste and vent (DWV) systems.

Joints are made by applying low-voltage current to electrical resistance coils imbedded in propylene collars, which are slipped over the ends of the pipe, then inserted into the hubs of the fittings. Compression clamps are placed around the assemblies and tightened to compress the joints prior to applying electrical current from the power unit to fuse the joint.

This section has been arranged to save the estimator's time by including all normally-used system components such as pipe, fittings, hanger assemblies, and riser clamps under one heading. The cost estimates in this section are based on the conditions, limitations and wage rates described in the section "How to Use This Book" beginning on page 5.

Description	Craft@Hrs	Unit	Material $	Labor $	Equipment $	Total $

Schedule 40 DWV polypropylene pipe with heat-fusioned joints

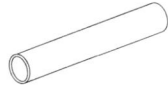

Description	Craft@Hrs	Unit	Material $	Labor $	Equipment $	Total $
1½"	P1@.050	LF	5.09	1.79	—	6.88
2"	P1@.055	LF	6.91	1.97	—	8.88
3"	P1@.060	LF	12.40	2.15	—	14.55
4"	P1@.080	LF	17.60	2.86	—	20.46
6"	P1@.110	LF	31.80	3.93	—	35.73

Schedule 40 DWV polypropylene 1/8 bend with heat-fusioned joints

Description	Craft@Hrs	Unit	Material $	Labor $	Equipment $	Total $
1½"	P1@.350	Ea	18.50	12.50	—	31.00
2"	P1@.400	Ea	22.60	14.30	—	36.90
3"	P1@.450	Ea	41.30	16.10	—	57.40
4"	P1@.500	Ea	46.80	17.90	—	64.70
6"	P1@.750	Ea	129.00	26.80	—	155.80

Schedule 40 DWV polypropylene 1/4 bend with heat-fusioned joints

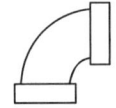

Description	Craft@Hrs	Unit	Material $	Labor $	Equipment $	Total $
1½"	P1@.350	Ea	18.70	12.50	—	31.20
2"	P1@.400	Ea	23.10	14.30	—	37.40
3"	P1@.450	Ea	39.50	16.10	—	55.60
4"	P1@.500	Ea	62.90	17.90	—	80.80
6"	P1@.750	Ea	155.00	26.80	—	181.80

Schedule 40 DWV polypropylene long sweep 1/4 bend with heat-fusioned joints

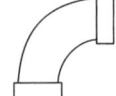

Description	Craft@Hrs	Unit	Material $	Labor $	Equipment $	Total $
1½"	P1@.350	Ea	19.10	12.50	—	31.60
2"	P1@.400	Ea	26.70	14.30	—	41.00
3"	P1@.450	Ea	45.40	16.10	—	61.50
4"	P1@.500	Ea	64.90	17.90	—	82.80

Schedule 40 DWV polypropylene P-trap with heat-fusioned joints

Description	Craft@Hrs	Unit	Material $	Labor $	Equipment $	Total $
1½"	P1@.500	Ea	36.70	17.90	—	54.60
2"	P1@.560	Ea	51.10	20.00	—	71.10
3"	P1@.630	Ea	92.70	22.50	—	115.20
4"	P1@.700	Ea	165.00	25.00	—	190.00

Polypropylene, Schedule 40, with Heat-Fusioned Joints

Description	Craft@Hrs	Unit	Material $	Labor $	Equipment $	Total $

Schedule 40 DWV polypropylene sanitary tee (TY) with heat-fusioned joints

Description	Craft@Hrs	Unit	Material $	Labor $	Equipment $	Total $
1½"	P1@.500	Ea	23.10	17.90	—	41.00
2"	P1@.560	Ea	28.00	20.00	—	48.00
3"	P1@.630	Ea	46.50	22.50	—	69.00
4"	P1@.700	Ea	83.70	25.00	—	108.70

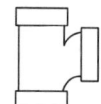

Schedule 40 DWV polypropylene reducing sanitary tee (TY) with heat-fusioned joints

Description	Craft@Hrs	Unit	Material $	Labor $	Equipment $	Total $
2" x 1½"	P1@.510	Ea	28.00	18.20	—	46.20
3" x 1½"	P1@.530	Ea	48.50	19.00	—	67.50
3" x 2"	P1@.580	Ea	52.00	20.70	—	72.70
4" x 2"	P1@.650	Ea	74.80	23.20	—	98.00
4" x 3"	P1@.700	Ea	77.70	25.00	—	102.70
6" x 4"	P1@.850	Ea	142.00	30.40	—	172.40

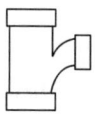

Schedule 40 DWV polypropylene combination wye and 1/8 bend with heat-fusioned joints

Description	Craft@Hrs	Unit	Material $	Labor $	Equipment $	Total $
1½"	P1@.500	Ea	31.10	17.90		49.00
2"	P1@.560	Ea	41.60	20.00	—	61.60
3"	P1@.630	Ea	66.00	22.50	—	88.50
4"	P1@.700	Ea	93.90	25.00	—	118.90

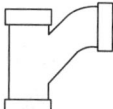

Schedule 40 DWV polypropylene combination reducing wye and 1/8 bend with heat-fusioned joints

Description	Craft@Hrs	Unit	Material $	Labor $	Equipment $	Total $
2" x 1½"	P1@.500	Ea	37.10	17.90	—	55.00
3" x 1½"	P1@.520	Ea	54.90	18.60	—	73.50
3" x 2"	P1@.570	Ea	59.50	20.40	—	79.90
4" x 2"	P1@.580	Ea	82.40	20.70	—	103.10
4" x 3"	P1@.630	Ea	86.30	22.50	—	108.80

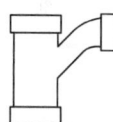

Schedule 40 DWV polypropylene wye with heat-fusioned joints

Description	Craft@Hrs	Unit	Material $	Labor $	Equipment $	Total $
1½"	P1@.500	Ea	23.80	17.90	—	41.70
2"	P1@.560	Ea	33.90	20.00	—	53.90
3"	P1@.630	Ea	59.50	22.50	—	82.00
4"	P1@.700	Ea	86.30	25.00	—	111.30
6"	P1@1.05	Ea	218.00	37.50	—	255.50

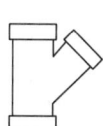

Polypropylene, Schedule 40, with Heat-Fusioned Joints

Description	Craft@Hrs	Unit	Material $	Labor $	Equipment $	Total $

Schedule 40 DWV polypropylene reducing wye with heat-fusioned joints

Description	Craft@Hrs	Unit	Material $	Labor $	Equipment $	Total $
2" x 1½"	P1@.510	Ea	33.20	18.20	—	51.40
3" x 2"	P1@.580	Ea	55.40	20.70	—	76.10
4" x 2"	P1@.600	Ea	79.80	21.50	—	101.30
4" x 3"	P1@.650	Ea	83.10	23.20	—	106.30
6" x 2"	P1@.900	Ea	140.00	32.20	—	172.20
6" x 3"	P1@.900	Ea	142.00	32.20	—	174.20
6" x 4"	P1@.950	Ea	143.00	34.00	—	177.00

Schedule 40 DWV polypropylene female adapter with heat-fusioned joints

Description	Craft@Hrs	Unit	Material $	Labor $	Equipment $	Total $
1½"	P1@.280	Ea	11.50	10.00	—	21.50
2"	P1@.330	Ea	15.60	11.80	—	27.40
3"	P1@.350	Ea	26.00	12.50	—	38.50
4"	P1@.400	Ea	49.80	14.30	—	64.10

Schedule 40 DWV polypropylene male adapter with heat-fusioned joints

Description	Craft@Hrs	Unit	Material $	Labor $	Equipment $	Total $
1½"	P1@.280	Ea	11.40	10.00	—	21.40
2"	P1@.330	Ea	13.80	11.80	—	25.60
3"	P1@.350	Ea	22.80	12.50	—	35.30
4"	P1@.400	Ea	47.30	14.30	—	61.60

Schedule 40 DWV polypropylene cleanout adapter with heat-fusioned joints

Description	Craft@Hrs	Unit	Material $	Labor $	Equipment $	Total $
1½"	P1@.210	Ea	12.10	7.51	—	19.61
2"	P1@.280	Ea	14.30	10.00	—	24.30
3"	P1@.430	Ea	26.50	15.40	—	41.90
4"	P1@.640	Ea	45.50	22.90	—	68.40
6"	P1@.850	Ea	78.50	30.40	—	108.90

Schedule 40 DWV polypropylene plug with heat-fusioned joints

Description	Craft@Hrs	Unit	Material $	Labor $	Equipment $	Total $
1½"	P1@.060	Ea	4.20	2.15	—	6.35
2"	P1@.065	Ea	4.58	2.32	—	6.90
3"	P1@.070	Ea	8.02	2.50	—	10.52
4"	P1@.090	Ea	9.00	3.22	—	12.22

Schedule 40 DWV polypropylene coupling with heat-fusioned joints

Description	Craft@Hrs	Unit	Material $	Labor $	Equipment $	Total $
1½"	P1@.350	Ea	14.70	12.50	—	27.20
2"	P1@.400	Ea	18.20	14.30	—	32.50
3"	P1@.450	Ea	23.30	16.10	—	39.40
4"	P1@.500	Ea	33.20	17.90	—	51.10
6"	P1@.750	Ea	52.90	26.80	—	79.70

Polypropylene, Schedule 40, with Heat-Fusioned Joints

Description	Craft@Hrs	Unit	Material $	Labor $	Equipment $	Total $

Schedule 40 DWV polypropylene reducer with heat-fusioned joints

Description	Craft@Hrs	Unit	Material $	Labor $	Equipment $	Total $
2" x 1½"	P1@.360	Ea	15.60	12.90	—	28.50
3" x 2"	P1@.410	Ea	17.90	14.70	—	32.60
4" x 2"	P1@.430	Ea	39.70	15.40	—	55.10
4" x 3"	P1@.450	Ea	40.80	16.10	—	56.90

Schedule 40 DWV polypropylene flange with heat-fusioned joints

Description	Craft@Hrs	Unit	Material $	Labor $	Equipment $	Total $
2"	P1@.350	Ea	51.20	12.50	—	63.70
3"	P1@.380	Ea	106.00	13.60	—	119.60
4"	P1@.430	Ea	138.00	15.40	—	153.40
6"	P1@.630	Ea	181.00	22.50	—	203.50

Schedule 40 DWV polypropylene bolt and gasket set

Description	Craft@Hrs	Unit	Material $	Labor $	Equipment $	Total $
2"	P1@.500	Ea	5.04	17.90	—	22.94
2½"	P1@.650	Ea	5.34	23.20	—	28.54
3"	P1@.750	Ea	5.60	26.80	—	32.40
4"	P1@1.00	Ea	10.20	35.80	—	46.00

Hanger with swivel assembly

Description	Craft@Hrs	Unit	Material $	Labor $	Equipment $	Total $
1½"	P1@.300	Ea	4.77	10.70	—	15.47
2"	P1@.300	Ea	4.98	10.70	—	15.68
3"	P1@.350	Ea	8.37	12.50	—	20.87
4"	P1@.350	Ea	9.19	12.50	—	21.69
6"	P1@.450	Ea	15.10	16.10	—	31.20

Riser clamp

Description	Craft@Hrs	Unit	Material $	Labor $	Equipment $	Total $
1½"	P1@.110	Ea	4.58	3.93	—	8.51
2"	P1@.115	Ea	4.84	4.11	—	8.95
3"	P1@.120	Ea	5.52	4.29	—	9.81
4"	P1@.125	Ea	7.03	4.47	—	11.50
6"	P1@.200	Ea	12.20	7.15	—	19.35

Pipe sleeve

Description	Craft@Hrs	Unit	Material $	Labor $	Equipment $	Total $
2"	P1@.130	Ea	6.49	4.65	—	11.14
2½"	P1@.150	Ea	6.59	5.36	—	11.95
3"	P1@.180	Ea	6.75	6.44	—	13.19
4"	P1@.220	Ea	7.67	7.87	—	15.54
5"	P1@.250	Ea	9.55	8.94	—	18.49
6"	P1@.270	Ea	10.30	9.66	—	19.96
8"	P1@.270	Ea	11.90	9.66	—	21.56

Lead roof pipe flashing

Description	Craft@Hrs	Unit	Material $	Labor $	Equipment $	Total $
1½"	P1@.250	Ea	33.30	8.94	—	42.24
2"	P1@.250	Ea	44.30	8.94	—	53.24
3"	P1@.250	Ea	50.80	8.94	—	59.74
4"	P1@.250	Ea	56.40	8.94	—	65.34

Floor, Area, Roof and Planter Drains

Description	Craft@Hrs	Unit	Material $	Labor $	Equipment $	Total $

Floor drain, cast iron or plastic body, nickel-bronze grate. Add for rough-in (P trap, fittings, hangers, etc.).

Description	Craft@Hrs	Unit	Material $	Labor $	Equipment $	Total $
3" x 2"	P1@.500	Ea	141.00	17.90	—	158.90
5" x 3"	P1@.500	Ea	176.00	17.90	—	193.90
6" x 4"	P1@.500	Ea	227.00	17.90	—	244.90
8" x 6"	P1@.500	Ea	302.00	17.90	—	319.90

Area drain, cast iron or plastic body, cast iron grate. Add for rough-in (P trap, fittings, hangers, etc.).

Description	Craft@Hrs	Unit	Material $	Labor $	Equipment $	Total $
10" x 3"	P1@.650	Ea	302.00	23.20	—	325.20
12" x 4"	P1@.650	Ea	329.00	23.20	—	352.20
14" x 6"	P1@.700	Ea	354.00	25.00	—	379.00

Roof and overflow drain, cast iron or plastic body, with plastic dome, flow control weir and deck clamp. Add for rough-in (pipe, fittings, hangers, etc.).

Description	Craft@Hrs	Unit	Material $	Labor $	Equipment $	Total $
8" x 3"	P1@1.50	Ea	404.00	53.60	—	457.60
10" x 4"	P1@1.50	Ea	427.00	53.60	—	480.60
12" x 6"	P1@1.65	Ea	427.00	59.00	—	486.00

Planter drain, cast iron. Add for rough-in (pipe, fittings, hangers, etc.).

Description	Craft@Hrs	Unit	Material $	Labor $	Equipment $	Total $
2"	P1@1.00	Ea	77.70	35.80	—	113.50
3"	P1@1.10	Ea	89.60	39.30	—	128.90
4"	P1@1.15	Ea	109.00	41.10	—	150.10
6"	P1@1.25	Ea	158.00	44.70	—	202.70

Planter drain, plastic. Add for rough-in (pipe, fittings, hangers, etc.).

Description	Craft@Hrs	Unit	Material $	Labor $	Equipment $	Total $
4"	P1@.800	Ea	23.70	28.60	—	52.30
5"	P1@1.00	Ea	25.10	35.80	—	60.90
6"	P1@1.05	Ea	41.20	37.50	—	78.70
8"	P1@1.10	Ea	62.90	39.30	—	102.20

Description	Craft@Hrs	Unit	Material $	Labor $	Equipment $	Total $

Cleanout, in-line, (Barrett) cast iron, (MJ) brass plug. Add for MJ couplings.

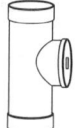

Description	Craft@Hrs	Unit	Material $	Labor $	Equipment $	Total $
2"	P1@.350	Ea	66.00	12.50	—	78.50
3"	P1@.350	Ea	84.40	12.50	—	96.90
4"	P1@.400	Ea	119.00	14.30	—	133.30
6"	P1@.400	Ea	420.00	14.30	—	434.30

Cleanout, in-line, (Barrett) ABS/PVC

Description	Craft@Hrs	Unit	Material $	Labor $	Equipment $	Total $
2"	P1@.300	Ea	18.40	10.70	—	29.10
3"	P1@.300	Ea	29.80	10.70	—	40.50
4"	P1@.350	Ea	54.00	12.50	—	66.50
6"	P1@.350	Ea	210.00	12.50	—	222.50
8"	P1@.400	Ea	407.00	14.30	—	421.30
10"	P1@.550	Ea	502.00	19.70	—	521.70

Cleanout, end of line (Malcolm), cast iron body, brass top. Add for MJ coupling.

Description	Craft@Hrs	Unit	Material $	Labor $	Equipment $	Total $
2" raised	P1@.250	Ea	23.70	8.94	—	32.64
2" recessed	P1@.250	Ea	59.30	8.94	—	68.24
3" raised	P1@.250	Ea	27.70	8.94	—	36.64
3" recessed	P1@.300	Ea	86.90	10.70	—	97.60
4" raised	P1@.300	Ea	37.00	10.70	—	47.70
4" recessed	P1@.350	Ea	113.00	12.50	—	125.50
6" raised	P1@.350	Ea	117.00	12.50	—	129.50
6" recessed	P1@.400	Ea	216.00	14.30	—	230.30
8" raised	P1@.450	Ea	252.00	16.10	—	268.10

Cleanout, end-of-line (Malcolm), ABS/PVC

Description	Craft@Hrs	Unit	Material $	Labor $	Equipment $	Total $
2"	P1@.250	Ea	9.82	8.94	—	18.76
3"	P1@.250	Ea	16.40	8.94	—	25.34
4"	P1@.300	Ea	29.80	10.70	—	40.50
6"	P1@.300	Ea	125.00	10.70	—	135.70
8"	P1@.400	Ea	225.00	14.30	—	239.30
10"	P1@.450	Ea	258.00	16.10	—	274.10
12"	P1@.550	Ea	302.00	19.70	—	321.70

Fire Protection Sprinklers

Description	Craft@Hrs	Unit	Material $	Labor $	Equipment $	Total $

Complete wet sprinkler system (square foot costs). System includes subcontractors' overhead and profit. Cost includes design drawings, zone and supervision valves, sprinkler heads, pipe, and connection to water supply service. Make additional allowances if a fire pump is required. Costs are based on ordinary hazard coverage (110 sf per head). Coverage density can vary depending on room sizes and/or hazard rating. Use these costs for preliminary estimates.

Description	Craft@Hrs	Unit	Material $	Labor $	Equipment $	Total $
Exposed piping to 5,000		SF	—	—	—	4.31
Exposed piping 5,000-15,000		SF	—	—	—	3.63
Exposed piping over 15,000		SF	—	—	—	3.23
Concealed piping to 5,000		SF	—	—	—	4.18
Concealed piping 5,000-15,000		SF	—	—	—	3.36
Concealed piping over 15,000		SF	—	—	—	2.63
Add for dry system		%	—	—	—	20.0

Complete wet sprinkler system (per head costs). Cost includes sprinkler heads, sprinkler mains, branch piping, and supports. Make additional allowances for zone and supervision valves, alarms, connection to water service and fire pump when required. Costs are based on ordinary hazard 155 degree heads. Use these costs for preliminary estimates.

Description	Craft@Hrs	Unit	Material $	Labor $	Equipment $	Total $
Upright heads to 5,000 SF		Ea	—	—	—	300.00
Upright heads 5,000 to 15,000 SF		Ea	—	—	—	286.00
Upright heads over 15,000 SF		Ea	—	—	—	272.00
Pendent heads to 5,000 SF		Ea	—	—	—	455.00
Pendent heads 5000 to 15,000 SF		Ea	—	—	—	379.00
Pendent heads over 15,000 SF		Ea	—	—	—	320.00

Supervision (zone) valves, flanged. Add for alarm trim if required.

Description	Craft@Hrs	Unit	Material $	Labor $	Equipment $	Total $
2½" OS&Y gate valve	SL@1.70	Ea	289.00	60.40	—	349.40
3" OS&Y gate vlv	SL@1.80	Ea	306.00	63.90	—	369.90
4" OS&Y gate vlv	SL@2.20	Ea	352.00	78.10	—	430.10
6" OS&Y gate vlv	SL@2.80	Ea	542.00	99.50	—	641.50
8" OS&Y gate vlv	SL@3.50	Ea	862.00	124.00	—	986.00
2½" butterfly valve (lug)	SL@.600	Ea	440.00	21.30	—	461.30
3" butterfly vlv	SL@.700	Ea	444.00	24.90	—	468.90
4" butterfly vlv	SL@.850	Ea	454.00	30.20	—	484.20
6" butterfly vlv	SL@1.15	Ea	630.00	40.80	—	670.80
8" butterfly vlv	SL@1.85	Ea	940.00	65.70	—	1,005.70

Description	Craft@Hrs	Unit	Material $	Labor $	Equipment $	Total $

Supervision (zone) valves, grooved. Add for alarm trim if required

4" butterfly valve	SL@.550	Ea	340.00	19.50	—	359.50
6" butterfly valve	SL@.850	Ea	442.00	30.20	—	472.20
8" butterfly valve	SL@1.25	Ea	667.00	44.40	—	711.40

Alarm valves (flanged or grooved)

4" alarm valve	SL@3.50	Ea	464.00	124.00	—	588.00
6" alarm valve	SL@4.00	Ea	570.00	142.00	—	712.00
8" alarm valve	SL@4.75	Ea	831.00	169.00	—	1,000.00
4" alarm vlv trim	SL@1.25	Ea	312.00	44.40	—	356.40
6" alarm vlv trim	SL@1.25	Ea	312.00	44.40	—	356.40
8" alarm vlv trim	SL@1.25	Ea	312.00	44.40	—	356.40
4" alarm vlv pkg	SL@4.75	Ea	1,460.00	169.00	—	1,629.00
6" alarm vlv pkg	SL@5.50	Ea	1,610.00	195.00	—	1,805.00
8" alarm vlv pkg	SL@6.00	Ea	1,970.00	213.00	—	2,183.00

Dry valves and trim (deluge/pre-action system)

2" dry valve	SL@2.25	Ea	868.00	79.90	—	947.90
3" dry valve	SL@2.75	Ea	890.00	97.70	—	995.70
4" dry valve	SL@3.05	Ea	1,020.00	108.00	—	1,128.00
6" dry valve	SL@3.65	Ea	1,300.00	130.00	—	1,430.00
2" dry valve trim	SL@1.25	Ea	423.00	44.40	—	467.40
3" dry valve trim	SL@1.25	Ea	423.00	44.40	—	467.40
4" dry valve trim	SL@1.25	Ea	423.00	44.40	—	467.40
6" dry valve trim	SL@1.25	Ea	423.00	44.40	—	467.40

Check valves, grooved

2½"	SL@.450	Ea	168.00	16.00	—	184.00
3"	SL@.500	Ea	184.00	17.80	—	201.80
4"	SL@.550	Ea	166.00	19.50	—	185.50
6"	SL@.850	Ea	326.00	30.20	—	356.20

Check valves, wafer type, flanged

2½"	SL@1.65	Ea	208.00	58.60	—	266.60
3"	SL@1.70	Ea	217.00	60.40	—	277.40
4"	SL@2.10	Ea	234.00	74.60	—	308.60
6"	SL@2.60	Ea	384.00	92.40	—	476.40
8"	SL@3.20	Ea	545.00	114.00	—	659.00

Double check detector valve assembly (flanged)

2½"	SL@4.00	Ea	2,180.00	142.00	—	2,322.00
3"	SL@4.35	Ea	2,330.00	155.00	—	2,485.00
4"	SL@4.95	Ea	2,540.00	176.00	—	2,716.00
6"	SL@6.25	Ea	3,980.00	222.00	—	4,202.00
8"	SL@6.95	Ea	7,190.00	247.00	—	7,437.00

Fire Protection Sprinklers

Description	Craft@Hrs	Unit	Material $	Labor $	Equipment $	Total $
Switches, flow, supervisory, monitor, pressure						
Flow switch	SL@1.00	Ea	151.00	35.50	—	186.50
Pressure switch	SL@1.00	Ea	114.00	35.50	—	149.50
Monitor switch	SL@1.00	Ea	110.00	35.50	—	145.50
Supervisory sw.	SL@1.00	Ea	101.00	35.50	—	136.50
Sprinkler heads, 155 degree						
Pendent, brass	SL@.350	Ea	6.64	12.40	—	19.04
Pendent, chrome	SL@.350	Ea	7.11	12.40	—	19.51
Upright, brass	SL@.350	Ea	6.64	12.40	—	19.04
Upright, chrome	SL@.350	Ea	16.70	12.40	—	29.10
Sidewall, brass	SL@.350	Ea	8.62	12.40	—	21.02
Sidewall, chrome	SL@.350	Ea	9.23	12.40	—	21.63
Sprinkler heads, dry, 165 degree						
Dry pendent	SL@.450	Ea	66.20	16.00	—	82.20
Dry sidewall	SL@.450	Ea	93.00	16.00	—	109.00
Sprinkler heads, 200 degree						
Pendent, brass	SL@.350	Ea	6.64	12.40	—	19.04
Pendent, chrome	SL@.350	Ea	7.11	12.40	—	19.51
Upright, brass	SL@.350	Ea	6.64	12.40	—	19.04
Upright, chrome	SL@.350	Ea	16.70	12.40	—	29.10
Sidewall, brass	SL@.350	Ea	8.62	12.40	—	21.02
Sidewall, chrome	SL@.350	Ea	9.23	12.40	—	21.63
Sprinkler heads, 286 degree						
Pendent, brass	SL@.350	Ea	6.64	12.40	—	19.04
Pendent, chrome	SL@.350	Ea	7.11	12.40	—	19.51
Upright, brass	SL@.350	Ea	6.64	12.40	—	19.04
Upright, chrome	SL@.350	Ea	16.70	12.40	—	29.10
Sidewall, brass	SL@.350	Ea	8.62	12.40	—	21.02
Sidewall, chrome	SL@.350	Ea	9.23	12.40	—	21.63
Sprinkler heads, 360 degree						
Pendent, brass	SL@.350	Ea	4.47	12.40	—	16.87
Upright, brass	SL@.350	Ea	4.47	12.40	—	16.87
Sprinkler heads, 400 degree						
Pendent, brass	SL@.350	Ea	19.80	12.40	—	32.20
Upright, brass	SL@.350	Ea	19.80	12.40	—	32.20

Description	Craft@Hrs	Unit	Material $	Labor $	Equipment $	Total $

Fire hose cabinet complete assembly. Includes recessed steel cabinet, 2½" and 1½" angle valves, 100' fire hose, hose rack and fog nozzle.

Description	Craft@Hrs	Unit	Material $	Labor $	Equipment $	Total $
FHC, chrome trim	SL@1.95	Ea	399.00	69.30	—	468.30
FHC, brass trim	SL@1.95	Ea	389.00	69.30	—	458.30
FHC, plastic trim	SL@1.95	Ea	378.00	69.30	—	447.30

Fire hose cabinet components

Description	Craft@Hrs	Unit	Material $	Labor $	Equipment $	Total $
Recessed cabinet	SL@1.00	Ea	179.00	35.50	—	214.50
Surface cabinet	SL@.850	Ea	208.00	30.20	—	238.20
Cabinet glass	SL@.350	Ea	13.30	12.40	—	25.70
100' fire hose	SL@.250	Ea	169.00	8.88	—	177.88
75' hose rack	SL@.350	Ea	52.00	12.40	—	64.40
100' hose rack	SL@.350	Ea	70.50	12.40	—	82.90
1½" brass fire hose angle valve	SL@.350	Ea	80.90	12.40	—	93.30
1½" chrome fire hose angle valve	SL@.350	Ea	135.00	12.40	—	147.40
2½" brass fire hose angle valve	SL@.450	Ea	155.00	16.00	—	171.00
2½" chrome fire hose angle valve	SL@.450	Ea	61.30	16.00	—	77.30
Brass fog nozzle	SL@.150	Ea	63.00	5.33	—	68.33
Plastic fog nozzle	SL@.150	Ea	17.40	5.33	—	22.73
Chrome fog nozzle	SL@.150	Ea	64.70	5.33	—	70.03

Fire Protection Equipment

Description	Craft@Hrs	Unit	Material $	Labor $	Equipment $	Total $

Electric fire pump, skid mounted with controller and jockey pump

Description	Craft@Hrs	Unit	Material $	Labor $	Equipment $	Total $
15 KW, 500 gpm		Ea	—	—	—	40,000.00
22 KW, 500 gpm		Ea	—	—	—	42,840.00
30 KW, 750 gpm		Ea	—	—	—	45,260.00
38 KW, 1000 gpm		Ea	—	—	—	50,190.00
56 KW, 1000 gpm		Ea	—	—	—	50,900.00

Diesel fire pump, skid mounted with controller and jockey pump

Description	Craft@Hrs	Unit	Material $	Labor $	Equipment $	Total $
250 gpm		Ea	—	—	—	59,600.00
500 gpm		Ea	—	—	—	62,300.00
750 gpm		Ea	—	—	—	63,100.00
1,000 gpm		Ea	—	—	—	64,500.00

Excess pressure pump

Description	Craft@Hrs	Unit	Material $	Labor $	Equipment $	Total $
1/4 hp	SL@1.35	Ea	430.00	48.00	—	478.00
1/3 hp	SL@1.35	Ea	445.00	48.00	—	493.00
1/2 hp	SL@1.50	Ea	462.00	53.30	—	515.30

Siamese connection, flush, fire dept. connection

Description	Craft@Hrs	Unit	Material $	Labor $	Equipment $	Total $
4" brass	SL@4.00	Ea	661.00	142.00	—	803.00
4" chrome	SL@4.00	Ea	713.00	142.00	—	855.00
6" brass	SL@4.50	Ea	768.00	160.00	—	928.00
6" chrome	SL@4.50	Ea	782.00	160.00	—	942.00

Siamese connection, surface, fire dept. connection

Description	Craft@Hrs	Unit	Material $	Labor $	Equipment $	Total $
4" brass	SL@2.85	Ea	156.00	101.00	—	257.00
6" brass	SL@3.25	Ea	296.00	115.00	—	411.00

Siamese connection, sidewalk, fire dept. connection

Description	Craft@Hrs	Unit	Material $	Labor $	Equipment $	Total $
4" brass	SL@3.25	Ea	598.00	115.00	—	713.00
6" brass	SL@3.75	Ea	845.00	133.00	—	978.00

Description	Craft@Hrs	Unit	Material $	Labor $	Equipment $	Total $

Water motor gong

Description	Craft@Hrs	Unit	Material $	Labor $	Equipment $	Total $
gong	SL@1.50	Ea	275.00	53.30	—	328.30

Multi-purpose fire extinguisher

Description	Craft@Hrs	Unit	Material $	Labor $	Equipment $	Total $
2 lb	SL@.500	Ea	32.60	17.80	—	50.40
5 lb	SL@.500	Ea	45.70	17.80	—	63.50
10 lb	SL@.500	Ea	65.20	17.80	—	83.00
20 lb	SL@.500	Ea	138.00	17.80	—	155.80

Fire hydrant with spool

Description	Craft@Hrs	Unit	Material $	Labor $	Equipment $	Total $
6"	P1@5.00	Ea	1,250.00	179.00	—	1,429.00

Indicator post

Description	Craft@Hrs	Unit	Material $	Labor $	Equipment $	Total $
6"	P1@6.50	Ea	726.00	232.00	—	958.00

Fire Protection Sprinkler Pipe and Fittings (Roll Grooved)

Description	Craft@Hrs	Unit	Material $	Labor $	Equipment $	Total $

Black steel pipe, (A53) installed horizontal, roll grooved, schedule 40.
Standard weight, including hangers every 10' and a fitting every 33'. Use these figures for preliminary estimates.

Description	Craft@Hrs	Unit	Material $	Labor $	Equipment $	Total $
2" (50mm)	SL@.160	LF	6.75	5.68	—	12.43
2½" (65mm)	SL@.180	LF	10.90	6.39	—	17.29
3" (75mm)	SL@.210	LF	13.90	7.46	—	21.36
4" (10cm)	SL@.300	LF	18.90	10.70	—	29.60
6" (15cm)	SK@.450	LF	39.00	17.00	4.20	60.20
8" (20cm)	SK@.630	LF	57.90	23.80	5.90	87.60
10" (25cm)	SK@.750	LF	112.00	28.40	7.10	147.50
12" (30cm)	SK@.800	LF	148.00	30.30	7.50	185.80

Black steel pipe, (A53) installed riser, roll grooved, schedule 40.
Standard weight, including a riser clamp every other floor and a tee at every floor. Make additional allowances for sleeving or coring as required. Use these figures for preliminary estimates.

Description	Craft@Hrs	Unit	Material $	Labor $	Equipment $	Total $
2" (50mm)	SL@.110	LF	6.97	3.91	—	10.88
2½" (65mm)	SL@.120	LF	11.20	4.26	—	15.46
3" (75mm)	SL@.150	LF	13.30	5.33	—	18.63
4" (10cm)	SL@.190	LF	19.10	6.75	—	25.85
6" (15cm)	SK@.300	LF	37.70	11.30	2.90	51.90
8" (20cm)	SK@.400	LF	58.70	15.10	3.80	77.60
10" (25cm)	SK@.510	LF	91.60	19.30	4.80	115.70
12" (30cm)	SK@.560	LF	132.00	21.20	5.30	158.50

Black steel pipe, (A53) installed horizontal, roll grooved, thin wall, schedule 10.
Light weight, including hangers every 10' and a fitting every 33'. Use these figures for preliminary estimates.

Description	Craft@Hrs	Unit	Material $	Labor $	Equipment $	Total $
2" (50mm)	SL@.160	LF	5.61	5.68	—	11.29
2½" (65mm)	SL@.180	LF	8.17	6.39	—	14.56
3" (75mm)	SL@.210	LF	9.61	7.46	—	17.07
4" (10cm)	SL@.300	LF	13.80	10.70	—	24.50
6" (15cm)	SK@.450	LF	27.70	17.00	4.20	48.90
8" (20cm)	SK@.630	LF	51.30	23.80	5.90	81.00

Black steel pipe, (A53) installed riser, roll grooved, thin wall, schedule 10.
Light weight, including a riser clamp every other floor and a tee at every floor. Make additional allowances for sleeving or coring as required. Use these figures for preliminary estimates.

Description	Craft@Hrs	Unit	Material $	Labor $	Equipment $	Total $
2" (50mm)	SL@.110	LF	6.31	3.91	—	10.22
2½" (65mm)	SL@.120	LF	9.12	4.26	—	13.38
3" (75mm)	SL@.150	LF	10.60	5.33	—	15.93
4" (10cm)	SL@.190	LF	14.90	6.75	—	21.65
6" (15cm)	SK@.300	LF	29.80	11.30	2.90	44.00
8" (20cm)	SK@.400	LF	57.90	15.10	3.80	76.80

Fire Protection Sprinkler Pipe and Fittings (Roll Grooved)

Description	Craft@Hrs	Unit	Material $	Labor $	Equipment $	Total $

Black steel pipe only, (A53) roll grooved, schedule 40. Standard weight, no hangers or fittings.

Description	Craft@Hrs	Unit	Material $	Labor $	Equipment $	Total $
2" (50mm)	SL@.070	LF	2.99	2.49	—	5.48
2½" (65mm)	SL@.070	LF	6.37	2.49	—	8.86
3" (75mm)	SL@.090	LF	7.09	3.20	—	10.29
4" (10cm)	SL@.130	LF	10.10	4.62	—	14.72
6" (15cm)	SK@.160	LF	15.20	6.05	1.50	22.75
8" (20cm)	SK@.210	LF	23.00	7.94	2.00	32.94
10" (25cm)	SK@.280	LF	33.40	10.60	2.60	46.60
12" (30cm)	SK@.340	LF	56.60	12.90	3.20	72.70
14" (36cm)	SK@.400	LF	65.20	15.10	3.80	84.10
16" (41cm)	SK@.550	LF	75.60	20.80	5.20	101.60

Black steel pipe only, (A53) roll grooved, thin wall, schedule 10. Light weight, no hangers or fittings.

Description	Craft@Hrs	Unit	Material $	Labor $	Equipment $	Total $
2" (50mm)	SL@.070	LF	2.26	2.49	—	4.75
2½" (65mm)	SL@.070	LF	4.76	2.49	—	7.25
3" (75mm)	SL@.090	LF	5.29	3.20	—	8.49
4" (10cm)	SL@.130	LF	7.57	4.62	—	12.19
6" (15cm)	SK@.160	LF	11.40	6.05	1.50	18.95
8" (20cm)	SK@.210	LF	19.50	7.94	2.00	29.44
10" (25cm)	SK@.280	LF	28.40	10.60	2.60	41.60
12" (30cm)	SK@.340	LF	48.00	12.90	3.20	64.10
14" (36cm)	SK@.400	LF	55.50	15.10	3.80	74.40
16" (41cm)	SK@.550	LF	64.60	20.80	5.20	90.60

90 degree elbow, roll grooved, style #10, (Victaulic)

Description	Craft@Hrs	Unit	Material $	Labor $	Equipment $	Total $
2"	SL@.410	Ea	28.20	14.60	—	42.80
2½"	SL@.430	Ea	28.20	15.30	—	43.50
3"	SL@.500	Ea	37.30	17.80	—	55.10
4"	SL@.690	Ea	58.20	24.50	—	82.70
6"	SL@1.19	Ea	145.00	42.30	—	187.30
8"	SL@1.69	Ea	248.00	60.00	—	308.00
10"	SL@2.23	Ea	477.00	79.20	—	556.20
12"	SL@2.48	Ea	755.00	88.10	—	843.10

45 degree elbow, roll grooved, style #11, (Victaulic)

Description	Craft@Hrs	Unit	Material $	Labor $	Equipment $	Total $
2"	SL@.410	Ea	28.20	14.60	—	42.80
2½"	SL@.430	Ea	28.20	15.30	—	43.50
3"	SL@.500	Ea	37.30	17.80	—	55.10
4"	SL@.690	Ea	58.20	24.50	—	82.70
6"	SL@1.19	Ea	145.00	42.30	—	187.30
8"	SL@1.69	Ea	248.00	60.00	—	308.00
10"	SL@2.23	Ea	477.00	79.20	—	556.20
12"	SL@2.48	Ea	755.00	88.10	—	843.10

Fire Protection Sprinkler Pipe and Fittings (Roll Grooved)

Description	Craft@Hrs	Unit	Material $	Labor $	Equipment $	Total $

22-1/2 degree elbow, roll grooved, style #12, (Victaulic)

Description	Craft@Hrs	Unit	Material $	Labor $	Equipment $	Total $
2"	SL@.410	Ea	33.60	14.60	—	48.20
2½"	SL@.430	Ea	33.60	15.30	—	48.90
3"	SL@.500	Ea	35.20	17.80	—	53.00
4"	SL@.690	Ea	66.30	24.50	—	90.80
6"	SL@1.19	Ea	166.00	42.30	—	208.30
8"	SL@1.69	Ea	341.00	60.00	—	401.00
10"	SL@2.23	Ea	480.00	79.20	—	559.20
12"	SL@2.48	Ea	515.00	88.10	—	603.10

Tee, roll grooved, style #20, (Victaulic)

Description	Craft@Hrs	Unit	Material $	Labor $	Equipment $	Total $
2"	SL@.510	Ea	54.60	18.10	—	72.70
2½"	SL@.540	Ea	54.60	19.20	—	73.80
3"	SL@.630	Ea	65.70	22.40	—	88.10
4"	SL@.860	Ea	92.10	30.50	—	122.60
6"	SL@1.49	Ea	245.00	52.90	—	297.90
8"	SL@2.11	Ea	411.00	74.90	—	485.90
10"	SL@2.79	Ea	729.00	99.10	—	828.10
12"	SL@3.10	Ea	852.00	110.00	—	962.00

Reducing tee, roll grooved, style #25, (Victaulic)

Description	Craft@Hrs	Unit	Material $	Labor $	Equipment $	Total $
2"	SL@.510	Ea	54.60	18.10	—	72.70
2½"	SL@.540	Ea	71.70	19.20	—	90.90
3"	SL@.630	Ea	64.80	22.40	—	87.20
4"	SL@.860	Ea	111.00	30.50	—	141.50
6"	SL@1.49	Ea	235.00	52.90	—	287.90
8"	SL@2.11	Ea	428.00	74.90	—	502.90
10"	SL@2.79	Ea	566.00	99.10	—	665.10
12"	SL@3.10	Ea	625.00	110.00	—	735.00

Reducer, roll grooved, style #50, (Victaulic)

Description	Craft@Hrs	Unit	Material $	Labor $	Equipment $	Total $
2"	SL@.410	Ea	18.10	14.60	—	32.70
2½"	SL@.430	Ea	18.10	15.30	—	33.40
3"	SL@.500	Ea	25.30	17.80	—	43.10
4"	SL@.690	Ea	42.30	24.50	—	66.80
6"	SL@1.19	Ea	70.70	42.30	—	113.00
8"	SL@1.69	Ea	125.00	60.00	—	185.00
10"	SL@2.23	Ea	260.00	79.20	—	339.20

Cap, roll grooved, style #60, (Victaulic)

Description	Craft@Hrs	Unit	Material $	Labor $	Equipment $	Total $
2"	SL@.270	Ea	15.90	9.59	—	25.49
2½"	SL@.280	Ea	15.90	9.95	—	25.85
3"	SL@.330	Ea	20.00	11.70	—	31.70
4"	SL@.450	Ea	23.20	16.00	—	39.20
6"	SL@.770	Ea	52.00	27.40	—	79.40
8"	SL@1.10	Ea	83.90	39.10	—	123.00
10"	SL@1.61	Ea	359.00	57.20	—	416.20

Description	Craft@Hrs	Unit	Material $	Labor $	Equipment $	Total $

Mechanical tee (saddle tee), roll grooved, style #920, (Victaulic)

3"	SL@.750	Ea	55.00	26.60	—	81.60
4"	SL@1.00	Ea	62.50	35.50	—	98.00
6"	SL@1.30	Ea	85.50	46.20	—	131.70
8"	SL@1.50	Ea	135.00	53.30	—	188.30

Coupling, roll grooved, 800 PSI, style #77, (Victaulic). Labor included with fittings.

3"	SL@.000	Ea	27.70	—	—	27.70
4"	SL@.000	Ea	46.10	—	—	46.10
6"	SL@.000	Ea	79.80	—	—	79.80
8"	SL@.000	Ea	119.00	—	—	119.00
10"	SL@.000	Ea	159.00	—	—	159.00
12"	SL@.000	Ea	212.00	—	—	212.00

Coupling, roll grooved, light weight, style #75, (Victaulic). Labor included with fittings.

3"	SL@.000	Ea	22.50	—	—	22.50
4"	SL@.000	Ea	24.30	—	—	24.30
6"	SL@.000	Ea	49.80	—	—	49.80
8"	SL@.000	Ea	76.40	—	—	76.40

Coupling, roll grooved, zeroflex, style #07, (Victaulic). Labor included with fittings.

3"	SL@.000	Ea	26.00	—	—	26.00
4"	SL@.000	Ea	34.10	—	—	34.10
6"	SL@.000	Ea	57.60	—	—	57.60
8"	SL@.000	Ea	87.80	—	—	87.80
10"	SL@.000	Ea	129.00	—	—	129.00
12"	SL@.000	Ea	177.00	—	—	177.00

Flange, roll grooved, style #741, (Victaulic)

3"	SL@.500	Ea	84.40	17.80	—	102.20
4"	SL@.690	Ea	119.00	24.50	—	143.50
6"	SL@1.19	Ea	167.00	42.30	—	209.30
8"	SL@1.69	Ea	193.00	60.00	—	253.00
10"	SL@2.23	Ea	288.00	79.20	—	367.20
12"	SL@2.48	Ea	391.00	88.10	—	479.10

Description	Craft@Hrs	Unit	Material $	Labor $	Equipment $	Total $

Schedule 40 carbon steel pipe, threaded and coupled, pipe only. Add for hangers, fittings etc.

Description	Craft@Hrs	Unit	Material $	Labor $	Equipment $	Total $
1"	P1@.090	LF	2.48	3.22	—	5.70
1¼"	P1@.100	LF	3.23	3.58	—	6.81
1½"	P1@.110	LF	3.85	3.93	—	7.78
2"	P1@.120	LF	5.20	4.29	—	9.49

125# cast iron 90 degree ell, threaded joints

Description	Craft@Hrs	Unit	Material $	Labor $	Equipment $	Total $
1" x 3/4"	P1@.130	Ea	4.26	4.65	—	8.91
1"	P1@.180	Ea	3.95	6.44	—	10.39
1¼"	P1@.240	Ea	5.57	8.58	—	14.15
1½"	P1@.300	Ea	7.78	10.70	—	18.48
2"	P1@.380	Ea	11.80	13.60	—	25.40

125# cast iron 45 degree ell, threaded joints

Description	Craft@Hrs	Unit	Material $	Labor $	Equipment $	Total $
1"	P1@.180	Ea	5.18	6.44	—	11.62
1¼"	P1@.240	Ea	7.37	8.58	—	15.95
1½"	P1@.300	Ea	9.49	10.70	—	20.19
2"	P1@.380	Ea	12.90	13.60	—	26.50

125# cast iron tee, threaded joints

Description	Craft@Hrs	Unit	Material $	Labor $	Equipment $	Total $
1"	P1@.230	Ea	5.43	8.22	—	13.65
1¼"	P1@.310	Ea	9.87	11.10	—	20.97
1½"	P1@.390	Ea	12.60	13.90	—	26.50
2"	P1@.490	Ea	17.40	17.50	—	34.90

125# cast iron reducing tee, threaded joints

Description	Craft@Hrs	Unit	Material $	Labor $	Equipment $	Total $
1" x 3/4"	P1@.220	Ea	6.94	7.87	—	14.81
1¼" x 3/4"	P1@.290	Ea	8.84	10.40	—	19.24
1¼" x 1"	P1@.290	Ea	9.67	10.40	—	20.07
1½" x 3/4"	P1@.370	Ea	12.20	13.20	—	25.40
1½" x 1"	P1@.370	Ea	13.40	13.20	—	26.60
1½" x 1¼"	P1@.370	Ea	14.80	13.20	—	28.00
2" x 1"	P1@.460	Ea	19.30	16.40	—	35.70
2" x 1¼"	P1@.460	Ea	20.60	16.40	—	37.00
2" x 1½"	P1@.460	Ea	22.20	16.40	—	38.60

125# cast iron reducer, threaded joints

Description	Craft@Hrs	Unit	Material $	Labor $	Equipment $	Total $
1" x 3/4"	P1@.200	Ea	5.43	7.15	—	12.58
1¼" x 1"	P1@.210	Ea	6.53	7.51	—	14.04
1½" x 1"	P1@.270	Ea	8.03	9.66	—	17.69
1½" x 1¼"	P1@.270	Ea	8.49	9.66	—	18.15
2" x 1¼"	P1@.340	Ea	12.70	12.20	—	24.90
2" x 1½"	P1@.340	Ea	13.30	12.20	—	25.50

Description	Craft@Hrs	Unit	Material $	Labor $	Equipment $	Total $

125# cast iron cross, threaded joints

Description	Craft@Hrs	Unit	Material $	Labor $	Equipment $	Total $
1"	P1@.360	Ea	12.70	12.90	—	25.60
1¼"	P1@.480	Ea	20.90	17.20	—	38.10
1½"	P1@.600	Ea	25.30	21.50	—	46.80
2"	P1@.700	Ea	42.00	25.00	—	67.00

125# cast iron cap, threaded joint

Description	Craft@Hrs	Unit	Material $	Labor $	Equipment $	Total $
3/4"	P1@.100	Ea	2.68	3.58	—	6.26
1"	P1@.140	Ea	3.24	5.01	—	8.25
1¼"	P1@.180	Ea	4.13	6.44	—	10.57
1½"	P1@.230	Ea	5.25	8.22	—	13.47
2"	P1@.290	Ea	8.23	10.40	—	18.63

125# cast iron plug, threaded joint

Description	Craft@Hrs	Unit	Material $	Labor $	Equipment $	Total $
3/4"	P1@.100	Ea	1.75	3.58	—	5.33
1"	P1@.140	Ea	2.30	5.01	—	7.31
1¼"	P1@.180	Ea	3.59	6.44	—	10.03
1½"	P1@.230	Ea	4.56	8.22	—	12.78
2"	P1@.290	Ea	5.64	10.40	—	16.04

125# cast iron coupling, threaded joints

Description	Craft@Hrs	Unit	Material $	Labor $	Equipment $	Total $
1"	P1@.180	Ea	4.71	6.44	—	11.15
1¼"	P1@.240	Ea	5.81	8.58	—	14.39
1½"	P1@.300	Ea	7.30	10.70	—	18.00
2"	P1@.380	Ea	9.87	13.60	—	23.47

Fire Protection Sprinkler Branch Fittings (Threaded)

Description	Craft@Hrs	Unit	Material $	Labor $	Equipment $	Total $
Steel pipe nipples — 3" long, standard right hand thread						
1"	P1@.025	Ea	1.82	.89	—	2.71
1¼"	P1@.025	Ea	2.35	.89	—	3.24
1½"	P1@.025	Ea	2.78	.89	—	3.67
2"	P1@.030	Ea	3.47	1.07	—	4.54
Steel pipe nipples — 4" long, standard right hand thread						
1"	P1@.025	Ea	2.17	.89	—	3.06
1¼"	P1@.030	Ea	2.71	1.07	—	3.78
1½"	P1@.030	Ea	3.36	1.07	—	4.43
2"	P1@.030	Ea	4.32	1.07	—	5.39
Steel pipe nipples — 6" long, standard right hand thread						
1"	P1@.030	Ea	2.78	1.07	—	3.85
1¼"	P1@.030	Ea	3.49	1.07	—	4.56
1½"	P1@.030	Ea	4.26	1.07	—	5.33
2"	P1@.035	Ea	5.71	1.25	—	6.96

Fire Protection Sprinkler Pipe and Fittings (CPVC)

Description	Craft@Hrs	Unit	Material $	Labor $	Equipment $	Total $

Chlorinated Polyvinyl Chloride pipe (CPVC), CPVC pipe only, solvent weld. No hangers or fittings.

Description	Craft@Hrs	Unit	Material $	Labor $	Equipment $	Total $
3/4"	SL@.055	LF	.97	1.95	—	2.92
1"	SL@.055	LF	1.41	1.95	—	3.36
1¼"	SL@.065	LF	2.23	2.31	—	4.54
1½"	SL@.075	LF	3.10	2.66	—	5.76
2"	SL@.085	LF	4.61	3.02	—	7.63
2½"	SL@.110	LF	7.13	3.91	—	11.04
3"	SL@.125	LF	10.90	4.44	—	15.34

CPVC 90 degree elbow, solvent weld

Description	Craft@Hrs	Unit	Material $	Labor $	Equipment $	Total $
3/4"	SL@.100	Ea	2.21	3.55	—	5.76
1"	SL@.120	Ea	3.32	4.26	—	7.58
1¼"	SL@.170	Ea	4.25	6.04	—	10.29
1½"	SL@.180	Ea	6.40	6.39	—	12.79
2"	SL@.200	Ea	7.44	7.10	—	14.54

CPVC 45 degree elbow, solvent weld

Description	Craft@Hrs	Unit	Material $	Labor $	Equipment $	Total $
3/4"	SL@.100	Ea	2.29	3.55	—	5.84
1"	SL@.120	Ea	2.98	4.26	—	7.24
1¼"	SL@.170	Ea	3.92	6.04	—	9.96
1½"	SL@.180	Ea	5.29	6.39	—	11.68
2"	SL@.200	Ea	6.58	7.10	—	13.68

CPVC tee, solvent weld

Description	Craft@Hrs	Unit	Material $	Labor $	Equipment $	Total $
3/4"	SL@.150	Ea	2.43	5.33	—	7.76
1"	SL@.170	Ea	4.15	6.04	—	10.19
1¼"	SL@.230	Ea	6.53	8.17	—	14.70
1½"	SL@.250	Ea	8.65	8.88	—	17.53
2"	SL@.280	Ea	13.40	9.95	—	23.35

CPVC reducing tee, solvent weld

Description	Craft@Hrs	Unit	Material $	Labor $	Equipment $	Total $
1"	SL@.170	Ea	3.56	6.04	—	9.60
1¼"	SL@.230	Ea	6.49	8.17	—	14.66
1½"	SL@.250	Ea	8.66	8.88	—	17.54
2"	SL@.280	Ea	13.80	9.95	—	23.75
3"	SL@.350	Ea	21.80	12.40	—	34.20

CPVC coupling, solvent weld

Description	Craft@Hrs	Unit	Material $	Labor $	Equipment $	Total $
3/4"	SL@.100	Ea	1.61	3.55	—	5.16
1"	SL@.120	Ea	2.16	4.26	—	6.42
1¼"	SL@.170	Ea	3.43	6.04	—	9.47
1½"	SL@.180	Ea	4.33	6.39	—	10.72
2"	SL@.200	Ea	5.45	7.10	—	12.55
3"	SL@.280	Ea	11.10	9.95	—	21.05

Description	Craft@Hrs	Unit	Material $	Labor $	Equipment $	Total $

CPVC head adapter, solvent weld

Description	Craft@Hrs	Unit	Material $	Labor $	Equipment $	Total $
3/4" x 1/2"	SL@.150	Ea	3.62	5.33	—	8.95
1" x 1/2"	SL@.165	Ea	3.87	5.86	—	9.73

CPVC cap, solvent weld

Description	Craft@Hrs	Unit	Material $	Labor $	Equipment $	Total $
3/4"	SL@.060	Ea	1.19	2.13	—	3.32
1"	SL@.065	Ea	1.66	2.31	—	3.97
1¼"	SL@.075	Ea	2.84	2.66	—	5.50
1½"	SL@.085	Ea	3.65	3.02	—	6.67
2"	SL@.095	Ea	4.24	3.37	—	7.61

CPVC flange, solvent weld

Description	Craft@Hrs	Unit	Material $	Labor $	Equipment $	Total $
1"	SL@.165	Ea	4.24	5.86	—	10.10
1¼"	SL@.175	Ea	5.45	6.22	—	11.67
1½"	SL@.185	Ea	8.44	6.57	—	15.01
2"	SL@.200	Ea	12.80	7.10	—	19.90
3"	SL@.225	Ea	21.10	7.99	—	29.09

CPVC female adapter, solvent weld

Description	Craft@Hrs	Unit	Material $	Labor $	Equipment $	Total $
1"	SL@.065	Ea	10.50	2.31	—	12.81
1¼"	SL@.075	Ea	17.30	2.66	—	19.96
1½"	SL@.085	Ea	22.40	3.02	—	25.42
2"	SL@.095	Ea	28.90	3.37	—	32.27

Boilers for industrial, commercial and institutional applications. ASME "H" or "U" stamped, LFUE (Levelized Fuel Utilization Efficiency) certified 90% or better. US Green Building council ocrtified. Cleaver-Brooks or equal. Shipped factory assembled and skid mounted. Includes the boiler shell and structure, I-beam skids, combustion chamber, fire tubes, crown sheet, tube sheets (for fire tube), water tubes, mud drum(s), steam drum (for water tube boiler), finned water tubes and check valves (for hot water generators), refractory brick lining, explosion doors, combustion air inlets and stack outlet. Add the cost of the fuel train piping, electrical wiring, refractory, stack, feedwater system, combustion controls, water treatment, expansion tanks (if hot water) or condensate return (if steam) and system start-up. See the sections following packaged boilers. Costs are based on boiler horsepower rating (BHP). Add the cost of commissioning, insurance inspection, permit, permit inspection, hydraulic test, calibration and lockout sealing of emissions monitor, and final run test. Equipment cost assumes a 10-ton hydraulic truck-mounted crane (at $110 per hour). LEED (Leadership in Energy & Environmental Design) certification requires 88% boiler efficiency, recording controls, compliance with SCAQMD 1146.2 emission standards and zone recording thermostats. A LEED certified boiler with a U.S. Green Building Council rating will add 4% to 15% to the cost of the boiler, as identified in the cost tables that follow.

Description	Craft@Hrs	Unit	Material $	Labor $	Equipment $	Total $

Hot water or steam cast iron gas boiler, 30 PSIG.
Costs shown are based on boiler horsepower rating (BHP). On site assembly. Includes allowances for associated appurtenances and pipe connections.

Description	Craft@Hrs	Unit	Material $	Labor $	Equipment $	Total $
4.5 BHP	SN@8.50	Ea	4,340.00	320.00	134.00	4,794.00
5.9 BHP	SN@9.50	Ea	5,420.00	358.00	149.00	5,927.00
9.0 BHP	SN@10.0	Ea	8,080.00	377.00	157.00	8,614.00
15.0 BHP	SN@12.0	Ea	12,500.00	452.00	189.00	13,141.00
20.0 BHP	SN@14.0	Ea	16,700.00	527.00	220.00	17,447.00
30.0 BHP	SN@16.0	Ea	21,500.00	602.00	251.00	22,353.00
45.0 BHP	SN@18.0	Ea	31,200.00	678.00	283.00	32,161.00
60.0 BHP	SN@20.0	Ea	40,500.00	753.00	314.00	41,567.00
Add for LEED certified boiler	—	Ea	6%	—	—	—
Add for LEED inspection	—	Ea	—	—	—	2,000.00
Add for central monitor/ recorder	—	Ea	1,500.00	—	—	1,500.00
Add for recording Thermostat	—	Ea	350.00	—	—	350.00

Commercial Boilers

Description	Craft@Hrs	Unit	Material $	Labor $	Equipment $	Total $

Hot water Scotch marine firetube gas boiler, 30 PSIG. Costs shown are based on boiler horsepower rating (BHP) Set in place only. Make additional allowances for associated appurtenances and pipe connections.

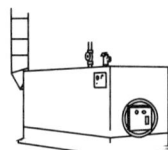

Description	Craft@Hrs	Unit	Material $	Labor $	Equipment $	Total $
30 BHP	SN@10.0	Ea	31,000.00	377.00	157.00	31,534.00
40 BHP	SN@10.0	Ea	32,900.00	377.00	157.00	33,434.00
50 BHP	SN@10.0	Ea	34,100.00	377.00	157.00	34,634.00
60 BHP	SN@12.0	Ea	38,100.00	452.00	189.00	38,741.00
70 BHP	SN@12.0	Ea	41,100.00	452.00	189.00	41,741.00
80 BHP	SN@14.0	Ea	43,700.00	527.00	220.00	44,447.00
100 BHP	SN@14.0	Ea	52,200.00	527.00	220.00	52,947.00
125 BHP	SN@16.0	Ea	56,000.00	602.00	251.00	56,853.00
150 BHP	SN@18.0	Ea	59,900.00	678.00	283.00	60,861.00
200 BHP	SN@18.0	Ea	70,700.00	678.00	283.00	71,661.00
250 BHP	SN@26.0	Ea	77,800.00	979.00	409.00	79,188.00
300 BHP	SN@26.0	Ea	83,600.00	979.00	409.00	84,988.00
350 BHP	SN@26.0	Ea	94,900.00	979.00	409.00	96,288.00
400 BHP	SN@32.0	Ea	97,500.00	1,200.00	503.00	99,203.00
500 BHP	SN@32.0	Ea	116,000.00	1,200.00	503.00	117,703.00
600 BHP	SN@32.0	Ea	131,000.00	1,200.00	503.00	132,703.00
750 BHP	SN@40.0	Ea	137,000.00	1,510.00	629.00	139,139.00
Add for LEED certified boiler	—	Ea	4%	—	—	—
Add for LEED inspection	—	Ea	—	—	—	2,000.00
Add for central monitor/ recorder	—	Ea	3,500.00	—	—	3,500.00
Add for recording Thermostat	—	Ea	350.00	—	—	350.00

Description	Craft@Hrs	Unit	Material $	Labor $	Equipment $	Total $

Packaged Scotch marine firetube oil or gas fired boiler, 150 PSIG.
Costs shown are based on boiler horsepower rating (BHP). Set in place only.
Make additional allowances for associated appurtenances and pipe connections.

Description	Craft@Hrs	Unit	Material $	Labor $	Equipment $	Total $
60 BHP	SN@12.0	Ea	43,700.00	452.00	189.00	44,341.00
70 BHP	SN@12.0	Ea	47,100.00	452.00	189.00	47,741.00
80 BHP	SN@14.0	Ea	50,500.00	527.00	220.00	51,247.00
100 BHP	SN@14.0	Ea	60,400.00	527.00	220.00	61,147.00
125 BHP	SN@16.0	Ea	64,000.00	602.00	251.00	64,853.00
150 BHP	SN@18.0	Ea	68,500.00	678.00	283.00	69,461.00
200 BHP	SN@18.0	Ea	90,100.00	678.00	283.00	91,061.00
250 BHP	SN@26.0	Ea	96,300.00	979.00	409.00	97,688.00
300 BHP	SN@26.0	Ea	109,000.00	979.00	409.00	110,388.00
350 BHP	SN@26.0	Ea	112,000.00	979.00	409.00	113,388.00
400 BHP	SN@32.0	Ea	134,000.00	1,200.00	503.00	135,703.00
500 BHP	SN@32.0	Ea	149,000.00	1,200.00	503.00	150,703.00
600 BHP	SN@32.0	Ea	153,000.00	1,200.00	503.00	154,703.00
700 BHP	SN@40.0	Ea	158,000.00	1,510.00	629.00	160,139.00
Add for LEED certified boiler	—	Ea	4%	—	—	—
Add for LEED inspection	—	Ea	—	—	—	2,000.00
Add for central monitor/ recorder	—	Ea	3,500.00	—	—	3,500.00
Add for recording Thermostat	—	Ea	350.00	—	—	350.00

Packaged hot water firebox oil or gas fired boiler.
Costs shown are based on boiler horsepower rating (BHP). Set in place only. Make additional allowances for associated appurtenances and pipe connections.

Description	Craft@Hrs	Unit	Material $	Labor $	Equipment $	Total $
13 BHP	SN@8.00	Ea	16,400.00	301.00	129.00	16,830.00
16 BHP	SN@8.00	Ea	16,700.00	301.00	129.00	17,130.00
19 BHP	SN@8.00	Ea	17,200.00	301.00	129.00	17,630.00
22 BHP	SN@8.00	Ea	17,600.00	301.00	129.00	18,030.00
28 BHP	SN@8.00	Ea	18,600.00	301.00	129.00	19,030.00
34 BHP	SN@8.00	Ea	21,200.00	301.00	129.00	21,630.00
40 BHP	SN@8.00	Ea	23,400.00	301.00	129.00	23,830.00
46 BHP	SN@8.00	Ea	26,800.00	301.00	129.00	27,230.00
52 BHP	SN@8.00	Ea	31,300.00	301.00	129.00	31,730.00
61 BHP	SN@8.00	Ea	34,400.00	301.00	129.00	34,830.00
70 BHP	SN@8.00	Ea	36,900.00	301.00	129.00	37,330.00
80 BHP	SN@8.00	Ea	38,100.00	301.00	129.00	38,530.00
100 BHP	SN@8.00	Ea	42,900.00	301.00	129.00	43,330.00
125 BHP	SN@16.0	Ea	51,100.00	602.00	256.00	51,958.00
150 BHP	SN@16.0	Ea	54,900.00	602.00	256.00	55,758.00
Add for LEED certified boiler	—	Ea	4%	—	—	—
Add for LEED inspection	—	Ea	—	—	—	2,000.00
Add for central monitor/ recorder	—	Ea	2,500.00	—	—	2,500.00
Add for recording Thermostat	—	Ea	350.00	—	—	350.00

Commercial Boilers

Description	Craft@Hrs	Unit	Material $	Labor $	Equipment $	Total $

Packaged watertube industrial high pressure boilers. Costs shown are based on boiler horsepower rating (BHP). Set in place only. Make additional allowances for associated appurtenances and pipe connections.

Description	Craft@Hrs	Unit	Material $	Labor $	Equipment $	Total $
480 BHP	SN@32.0	Ea	310,000.00	1,200.00	504.00	311,704.00
526 BHP	SN@32.0	Ea	318,000.00	1,200.00	504.00	319,704.00
572 BHP	SN@32.0	Ea	335,000.00	1,200.00	504.00	336,704.00
618 BHP	SN@32.0	Ea	340,000.00	1,200.00	504.00	341,704.00
664 BHP	SN@32.0	Ea	357,000.00	1,200.00	504.00	358,704.00
710 BHP	SN@64.0	Ea	362,000.00	2,410.00	1,010.00	365,420.00
747 BHP	SN@64.0	Ea	367,000.00	2,410.00	1,010.00	370,420.00
792 BHP	SN@64.0	Ea	379,000.00	2,410.00	1,010.00	382,420.00
838 BHP	SN@64.0	Ea	390,000.00	2,410.00	1,010.00	393,420.00
884 BHP	SN@64.0	Ea	401,000.00	2,410.00	1,010.00	404,420.00
930 BHP	SN@64.0	Ea	407,000.00	2,410.00	1,010.00	410,420.00
1,000 BHP	SN@64.0	Ea	414,000.00	2,410.00	1,010.00	417,420.00
Add for LEED certified boiler	—	Ea	4%	—	—	—
Add for LEED inspection	—	Ea	—	—	—	2,000.00
Add for central monitor/ recorder	—	Ea	4,300.00	—	—	4,300.00
Add for recording Thermostat	—	Ea	350.00	—	—	350.00

Gas fired hot water steel boilers, 12 PSIG Mid-efficiency (82%) atmospheric draft, standing pilot light, c/w primary loop circulating pump. Labor includes setting in place and connecting pump and controls. Make additional allowances for distribution piping connections, safety devices, flue, gas connections, wiring, etc.

Description	Craft@Hrs	Unit	Material $	Labor $	Equipment $	Total $
50,000 Btu	P1@4.00	Ea	1,175.00	143.00	504.00	1,822.00
70,000 Btu	P1@4.25	Ea	1,350.00	152.00	504.00	2,006.00
105,000 Btu	P1@4.50	Ea	1,490.00	161.00	504.00	2,155.00
140,000 Btu	P1@4.75	Ea	1,700.00	170.00	504.00	2,374.00
175,000 Btu	P1@5.00	Ea	1,900.00	179.00	504.00	2,583.00
210,000 Btu	P1@5.50	Ea	2,400.00	197.00	1,010.00	3,607.00
245,000 Btu	P1@7.00	Ea	2,680.00	250.00	1,010.00	3,940.00
Electronic spark ignition Add	P1@—	Ea	270.00			270.00
Add for LEED certified boiler	—	Ea	15%	—	—	—
Add for LEED inspection	—	Ea	—	—	—	2,000.00
Add for central monitor/ recorder	—	Ea	1,100.00	—	—	1,100.00
Add for recording Thermostat	—	Ea	150.00	—	—	150.00

Gas fired hot water cast iron boilers, 12 PSIG Mid-efficiency (82%) direct vent, forced draft electronic spark ignition, c/w primary loop circulating pump. Labor includes setting in place and connecting pump and controls. Make additional allowances for distribution piping connections, safety devices, flue, gas connections, wiring, etc.

50,000 Btu	P1@4.00	Ea	1,480.00	143.00	504.00	2,127.00
65,000 Btu	P1@4.25	Ea	1,650.00	152.00	504.00	2,306.00
100,000 Btu	P1@4.50	Ea	1,800.00	161.00	504.00	2,465.00
130,000 Btu	P1@4.75	Ea	2,000.00	170.00	504.00	2,674.00
170,000 Btu	P1@5.00	Ea	2,300.00	179.00	504.00	2,983.00
200,000 Btu	P1@5.50	Ea	2,550.00	197.00	1,010.00	3,757.00
235,000 Btu	P1@7.00	Ea	2,800.00	250.00	1,010.00	4,060.00
Add for LEED certified boiler	—	Ea	15%	—	—	—
Add for LEED inspection	—	Ea	—	—	—	2,000.00
Add for central monitor/ recorder	—	Ea	1,100.00	—	—	1,100.00
Add for recording Thermostat	—	Ea	150.00	—	—	150.00

Steam heating boilers for residential and light commercial applications. Parker or equal 15 PSIG Spiral watertube type, with ASTM SA-53 grade tubing, full American Society of Mechanical Engineers "M", "H" and "S" ratings and Underwriters Laboratories certifications. Specifically designed for "twinning" and multiple unit additions and retrofits. Packaged type units with ASME rated pressure relief valve, low water cutoff valve, main and check valves, pressure gauge, and pre-wired Fireye/McDonnell Miller combustion and steam regulator controls. Also includes pre-assembled FM/IRI approved natural gas combustion train w/Fisher or equal diaphragm regulator valve, check trim and vent line fittings or pre-assembled FM/IRI fuel oil combustion train with oil preheater, valved recirculation loop, filter check trim and fuel oil pump. Does not include sheet metal stack nor vents, installation permits and approvals, nor inspection costs. (bhp = boiler horse power)

37,500 Btu (1 bhp)	P1@3.00	Ea	1,800.00	107.00	1,907.00
50,000 Btu (1.5 bhp)	P1@3.00	Ea	2,200.00	107.00	2,307.00
100,000 Btu (2.0 bhp)	P1@4.00	Ea	2,700.00	143.00	2,843.00
150,000 Btu (2.5 bhp)	P1@4.25	Ea	3,400.00	152.00	3,552.00
200,000 Btu (3.0 bhp)	P1@5.00	Ea	3,600.00	179.00	3,779.00
Add for LEED certified boiler	—	Ea	15%	—	—
Add for LEED inspection	—	Ea	—	—	2,000.00
Add for central monitor/ recorder	—	Ea	1,150.00	—	1,150.00
Add for recording Thermostat	—	Ea	150.00	—	150.00

Description	Craft@Hrs	Unit	Material $	Labor $	Equipment $	Total $

Commercial hot water boiler connection. Includes pipe, fittings, water feed and pressure relief (drains), pipe insulation, valves, gauges, vents and thermometers to connect boiler to distribution system.

Description	Craft@Hrs	Unit	Material $	Labor $	Equipment $	Total $
2½" supply	SN@18.0	Ea	1,840.00	678.00	283.00	2,801.00
3" supply	SN@20.5	Ea	2,310.00	772.00	322.00	3,404.00
4" supply	SN@26.0	Ea	3,360.00	979.00	409.00	4,748.00
6" supply	SN@32.0	Ea	4,910.00	1,200.00	503.00	6,613.00
8" supply	SN@39.5	Ea	6,560.00	1,490.00	613.00	8,663.00
10" supply	SN@48.0	Ea	9,400.00	1,810.00	754.00	11,964.00
12" supply	SN@54.0	Ea	11,600.00	2,030.00	849.00	14,479.00

Commercial steam boiler connection. Includes pipe, fittings, water feed and pressure relief (drains), pipe insulation, valves, gauges, vents and thermometers to connect boiler to distribution system.

Description	Craft@Hrs	Unit	Material $	Labor $	Equipment $	Total $
2½" supply	SN@34.0	Ea	3,530.00	1,280.00	534.00	5,344.00
3" supply	SN@38.0	Ea	3,720.00	1,430.00	597.00	5,747.00
4" supply	SN@43.0	Ea	4,320.00	1,620.00	677.00	6,617.00
6" supply	SN@53.0	Ea	5,930.00	2,000.00	833.00	8,763.00
8" supply	SN@62.0	Ea	7,640.00	2,330.00	974.00	10,944.00
10" supply	SN@70.0	Ea	10,400.00	2,640.00	1,100.00	14,140.00
12" supply	SN@82.0	Ea	13,100.00	3,090.00	1,290.00	17,480.00

Additional costs for industrial, commercial and institutional packaged boilers. See more detailed installation costs and boiler trim costs in the sections that follow.

Description	Craft@Hrs	Unit	Material $	Labor $	Equipment $	Total $
Form and pour an interior slab	CF@0.15	SF	4.85	5.25	—	10.10
Place vibration pad and anchors	CF@.10	SF	5.00	3.50	—	8.50
Boltdown of boiler	P1@3.00	Ea	—	107.00	—	107.00
Fuel train and feedwater lines	P1@0.20	LF	10.00	7.15	—	17.15
Mount interior boiler drain	P1@.500	LF	7.76	17.90	—	25.66
Bore pipe hole in concrete wall	P1@.250	Ea	—	8.94	—	8.94
Bore stack vent in concrete wall	P1@.250	Ea	—	8.94	—	8.94
Mount and edge-seal stack	SN@0.10	LF	8.50	3.77	110.00	122.27
Mount circulating pump	P1@.450	Ea	10.00	16.10	—	26.10
Expansion tank or condensate return lines	P1@0.25	LF	3.25	8.94	—	12.19
Electrical wiring	BE@0.15	LF	1.85	5.94	—	7.79
Combustion controls	BE@0.20	LF	3.50	7.92	—	11.42
Install flue gas Monitoring system	BE@0.20	LF	3.50	7.92	—	11.42

Description	Craft@Hrs	Unit	Material $	Labor $	Equipment $	Total $

Additional costs for steam heating boilers for residential and light commercial applications. See more detailed installation costs and boiler trim costs in the sections that follow.

Description	Craft@Hrs	Unit	Material $	Labor $	Equipment $	Total $
Form and pour a 4' x '4 interior slab	CF@3.00	Ea	97.00	105.00	—	202.00
Place 4' x 4' x 1/2" vibration pads	CF@.750	Ea	38.80	26.30	—	65.10
Boltdown of boiler	P1@1.00	Ea	—	35.80	—	35.80
Fuel train and feedwater lines	P1@2.50	Ea	—	89.40	16.00	105.40
Mount interior boiler Drain	P1@.500	LF	7.76	17.90	—	25.66
Bore piping hole in basement wall	P1@.250	Ea	—	8.94	—	8.94
Bore stack vent in concrete wall	P1@.250	Ea	—	8.94	—	8.94
Mount and edge-seal stack	P1@.500	Ea	40.00	17.90	—	57.90
Mount circulating pump	P1@.450	Ea	—	16.10	—	16.10
Install expansion tank						
2.1 gallon	P1@.300	Ea	32.59	10.70	21.00	64.29
4.5 gallon	P1@.300	Ea	54.32	10.70	21.00	86.02

Commercial Boiler Components and Accessories

Costs for Combustion Trains for Boilers. Factory Mutual, Industrial Risk Insurers, and Underwriters' Laboratories certified, US Green Building Council rated, LFUE (Levelized Fuel Utilization Efficiency) rated at 90% or better. Includes the burner, burner head, burner mount flange, associated refractory and gasketing, main fuel valve, draft damper and fan (for a power burner), or burner face plate (for an atmospheric or induced-draft burner). Note: Note: BHP multiplied by 33.5 is the approximate required 1000 Btu per hour input (MBH input). Cost of fuel line relief vent piping and trip valve to outside of the building are not required in all jurisdictions and are not assumed in these costs.

Description	Craft@Hrs	Unit	Material $	Labor $	Equipment $	Total $

Combustion train for Scotch marine firetube boilers. Costs shown are based on boiler horsepower rating (BHP).

Description	Craft@Hrs	Unit	Material $	Labor $	Equipment $	Total $
20 BHP	SN@8.00	Ea	2,590.00	301.00	127.00	3,018.00
30 BHP	SN@8.00	Ea	3,520.00	301.00	127.00	3,948.00
40 BHP	SN@8.00	Ea	4,810.00	301.00	127.00	5,238.00
50 BHP	SN@8.00	Ea	5,960.00	301.00	127.00	6,388.00
60 BHP	SN@8.00	Ea	6,630.00	301.00	127.00	7,058.00
70 BHP	SN@8.00	Ea	7,250.00	301.00	127.00	7,678.00
80 BHP	SN@8.00	Ea	7,400.00	301.00	127.00	7,828.00
100 BHP	SN@8.00	Ea	8,070.00	301.00	127.00	8,498.00
125 BHP	SN@8.00	Ea	8,380.00	301.00	127.00	8,808.00
150 BHP	SN@8.00	Ea	9,280.00	301.00	127.00	9,708.00
200 BHP	SN@8.00	Ea	9,670.00	301.00	127.00	10,098.00
250 BHP	SN@8.00	Ea	10,600.00	301.00	127.00	11,028.00
300 BHP	SN@8.00	Ea	12,000.00	301.00	127.00	12,428.00
350 BHP	SN@8.00	Ea	12,200.00	301.00	127.00	12,628.00
400 BHP	SN@8.00	Ea	12,500.00	301.00	127.00	12,928.00
500 BHP	SN@16.0	Ea	13,600.00	602.00	251.00	14,453.00
600 BHP	SN@16.0	Ea	14,600.00	602.00	251.00	15,453.00
750 BHP	SN@32.0	Ea	15,500.00	1,200.00	347.00	17,047.00
800 BHP	SN@32.0	Ea	16,200.00	1,200.00	347.00	17,747.00

Combustion train for package firebox boilers. Costs shown are based on boiler horsepower rating (BHP).

Description	Craft@Hrs	Unit	Material $	Labor $	Equipment $	Total $
13.4 BHP	SN@8.00	Ea	1,340.00	301.00	127.00	1,768.00
16.4 BHP	SN@8.00	Ea	1,770.00	301.00	127.00	2,198.00
17.4 BHP	SN@8.00	Ea	2,100.00	301.00	127.00	2,528.00
22.4 BHP	SN@8.00	Ea	2,510.00	301.00	127.00	2,938.00
28.4 BHP	SN@8.00	Ea	2,670.00	301.00	127.00	3,098.00
34.4 BHP	SN@8.00	Ea	2,950.00	301.00	127.00	3,378.00
40.3 BHP	SN@8.00	Ea	3,200.00	301.00	127.00	3,628.00
46.3 BHP	SN@8.00	Ea	3,540.00	301.00	127.00	3,968.00
52.3 BHP	SN@8.00	Ea	3,940.00	301.00	127.00	4,368.00
61.1 BHP	SN@8.00	Ea	4,280.00	301.00	127.00	4,708.00
70.0 BHP	SN@8.00	Ea	4,370.00	301.00	127.00	4,798.00
80.0 BHP	SN@8.00	Ea	4,540.00	301.00	127.00	4,968.00
100.0 BHP	SN@8.00	Ea	4,810.00	301.00	127.00	5,238.00
125.0 BHP	SN@8.00	Ea	5,140.00	301.00	127.00	5,568.00
150.0 BHP	SN@8.00	Ea	5,580.00	301.00	127.00	6,008.00

Description	Craft@Hrs	Unit	Material $	Labor $	Equipment $	Total $

Natural gas fuel train piping for firetube and watertube boilers. Meets requirements of Underwriters' Laboratories. Shipped pre-assembled. Add the cost of venting and piping to a gas meter. Costs shown are based on boiler horsepower rating (BHP).

Description	Craft@Hrs	Unit	Material $	Labor $	Equipment $	Total $
20 BHP	SN@8.00	Ea	677.00	301.00	127.00	1,105.00
30 BHP	SN@8.00	Ea	927.00	301.00	127.00	1,355.00
40 BHP	SN@8.00	Ea	1,420.00	301.00	127.00	1,848.00
50 BHP	SN@8.00	Ea	1,690.00	301.00	127.00	2,118.00
60 BHP	SN@8.00	Ea	1,850.00	301.00	127.00	2,278.00
70 BHP	SN@8.00	Ea	1,950.00	301.00	127.00	2,378.00
80 BHP	SN@8.00	Ea	2,120.00	301.00	127.00	2,548.00
100 BHP	SN@8.00	Ea	2,460.00	301.00	127.00	2,888.00
125 BHP	SN@8.00	Ea	2,930.00	301.00	127.00	3,358.00
150 BHP	SN@8.00	Ea	3,480.00	301.00	127.00	3,908.00
200 BHP	SN@8.00	Ea	4,150.00	301.00	127.00	4,578.00
250 BHP	SN@8.00	Ea	5,140.00	301.00	127.00	5,568.00
300 BHP	SN@8.00	Ea	5,410.00	301.00	127.00	5,838.00
350 BHP	SN@8.00	Ea	5,800.00	301.00	127.00	6,228.00
400 BHP	SN@8.00	Ea	6,240.00	301.00	127.00	6,668.00
500 BHP	SN@16.0	Ea	7,000.00	602.00	251.00	7,853.00
600 BHP	SN@16.0	Ea	8,070.00	602.00	251.00	8,923.00
750 BHP	SN@16.0	Ea	8,730.00	602.00	251.00	9,583.00
800 BHP	SN@16.0	Ea	9,140.00	602.00	251.00	9,993.00

Natural gas fuel train piping for package firebox boilers. Costs shown are based on boiler horsepower rating (BHP).

Description	Craft@Hrs	Unit	Material $	Labor $	Equipment $	Total $
13 BHP	SN@8.00	Ea	905.00	301.00	127.00	1,333.00
16 BHP	SN@8.00	Ea	1,040.00	301.00	127.00	1,468.00
19 BHP	SN@8.00	Ea	1,110.00	301.00	127.00	1,538.00
22 BHP	SN@8.00	Ea	1,220.00	301.00	127.00	1,648.00
28 BHP	SN@8.00	Ea	1,290.00	301.00	127.00	1,718.00
34 BHP	SN@8.00	Ea	1,420.00	301.00	127.00	1,848.00
40 BHP	SN@8.00	Ea	1,690.00	301.00	127.00	2,118.00
45 BHP	SN@8.00	Ea	1,770.00	301.00	127.00	2,198.00
50 BHP	SN@8.00	Ea	2,120.00	301.00	127.00	2,548.00
60 BHP	SN@8.00	Ea	2,410.00	301.00	127.00	2,838.00
70 BHP	SN@8.00	Ea	2,710.00	301.00	127.00	3,138.00
80 BHP	SN@8.00	Ea	2,840.00	301.00	127.00	3,268.00
100 BHP	SN@8.00	Ea	3,400.00	301.00	127.00	3,828.00
125 BHP	SN@8.00	Ea	3,530.00	301.00	127.00	3,958.00
150 BHP	SN@8.00	Ea	3,850.00	301.00	127.00	4,278.00
Add for code approval:						
Factory Mutual	—	%	7.0	—	—	—
Industrial Risk	—	%	10.0	—	—	—

Commercial Boiler Components and Accessories

Oil Fuel Train Piping. Costs shown include recirculating pump, electric oil preheater, regulating valve, check valves, isolating pump stop cocks, oil line pressure relief valve and pressure gauges. Add the cost of concrete pump pad, electrical connections, relief line piping back to the oil tank and tank connecting piping. Add for number 4-6 fuel oil as shown.

Description	Craft@Hrs	Unit	Material $	Labor $	Equipment $	Total $

Oil fuel train piping for boilers. Costs shown are based on boiler horsepower rating (BHP) using number 2 fuel oil.

Description	Craft@Hrs	Unit	Material $	Labor $	Equipment $	Total $
20 to 30 BHP	SN@4.00	Ea	1,440.00	151.00	62.40	1,653.40
40 to 60 BHP	SN@4.00	Ea	1,620.00	151.00	62.40	1,833.40
70 to 80 BHP	SN@4.00	Ea	1,760.00	151.00	62.40	1,973.40
100 BHP	SN@8.00	Ea	2,270.00	301.00	127.00	2,698.00
125 to 150 BHP	SN@8.00	Ea	2,930.00	301.00	127.00	3,358.00
200 BHP	SN@8.00	Ea	3,860.00	301.00	127.00	4,288.00
250 BHP	SN@8.00	Ea	4,570.00	301.00	127.00	4,998.00
300 BHP	SN@16.0	Ea	4,880.00	602.00	251.00	5,733.00
350 BHP	SN@16.0	Ea	5,460.00	602.00	251.00	6,313.00
400 BHP	SN@16.0	Ea	6,560.00	602.00	251.00	7,413.00
500 BHP	SN@32.0	Ea	6,640.00	1,200.00	503.00	8,343.00
600 BHP	SN@32.0	Ea	6,960.00	1,200.00	503.00	8,663.00
750 to 800	SN@32.0	Ea	7,600.00	1,200.00	503.00	9,303.00
900 to 1000	SN@32.0	Ea	8,990.00	1,200.00	503.00	10,693.00
Add for #4-6 oil	—	%	10.0	—	—	—
Add for tie-in with:						
dual-fuel burner	—	%	5.0	—	—	—

Costs for Electrical Service for Boilers, Subcontract. Costs shown include typical main power tie-in and fusing, circuit breaker panel, and main bus. Assumes controls, fan, blower switchgear and control console are pre-wired electrical service. These costs include the electrical subcontractor's overhead and profit.

Description	Craft@Hrs	Unit	Material $	Labor $	Total $

Electrical service for firetube and watertube boilers. Costs shown are based on boiler horsepower rating (BHP).

Description	Craft@Hrs	Unit	Material $	Labor $	Total $
20 to 100 BHP	—	Ea	—	—	1,290.00
125 to 200 BHP	—	Ea	—	—	1,550.00
250 to 350 BHP	—	Ea	—	—	1,760.00
400 to 500 BHP	—	Ea	—	—	2,020.00
500 to 600 BHP	—	Ea	—	—	2,920.00
750 to 800 BHP	—	Ea	—	—	5,260.00

Electrical service for package firebox boilers. Costs shown are based on boiler horsepower rating (BHP).

Description	Craft@Hrs	Unit	Material $	Labor $	Total $
13 to 70 BHP	—	Ea	—	—	1,250.00
80 to 125 BHP	—	Ea	—	—	1,330.00
150 BHP	—	Ea	—	—	1,570.00

Costs for Refractory for Boilers, Subcontract. For package Scotch marine firetube, firebox and watertube boilers. Most new boilers are shipped from the factory with the refractory brick installed. If refractory brick is installed on the job site to repair shipping damage or to create firedoor seals, use the figures below. These costs include the refractory subcontractor's overhead and profit. Costs shown are per 100 pounds of refractory used.

Description	Craft@Hrs	Unit	Material $	Labor $	Equipment $	Total $
Repairs and seals	—	100lb	—	—	—	215.00

Costs for Boiler Feedwater Pumps. Costs shown are for the pump only. Add the cost of piping, pump actuating valves and concrete pump base.

Feedwater pumps for firetube and watertube boilers. Costs shown are based on boiler horsepower rating (BHP).

20 to 40 BHP	SN@8.00	Ea	9,600.00	301.00	127.00	10,028.00
50 to 80 BHP	SN@8.00	Ea	11,600.00	301.00	127.00	12,028.00
100 BHP	SN@8.00	Ea	12,900.00	301.00	127.00	13,328.00
125 to 200 BHP	SN@16.0	Ea	13,300.00	602.00	251.00	14,153.00
250 to 400 BHP	SN@32.0	Ea	14,600.00	1,200.00	503.00	16,303.00
500 BHP	SN@32.0	Ea	15,500.00	1,200.00	503.00	17,203.00
600 BHP	SN@32.0	Ea	16,000.00	1,200.00	503.00	17,703.00
750 BHP	SN@32.0	Ea	16,500.00	1,200.00	503.00	18,203.00
800 BHP	SN@32.0	Ea	22,400.00	1,200.00	503.00	24,103.00

Feedwater pumps for package firebox boilers. Costs shown are based on boiler horsepower rating (BHP).

13 to 35 BHP	SN@8.00	Ea	5,160.00	301.00	127.00	5,588.00
40 to 45 BHP	SN@8.00	Ea	6,060.00	301.00	127.00	6,488.00
50 to 60 BHP	SN@8.00	Ea	7,600.00	301.00	127.00	8,028.00
70 to 80 BHP	SN@8.00	Ea	9,050.00	301.00	127.00	9,478.00
100 BHP	SN@8.00	Ea	11,600.00	301.00	127.00	12,028.00
125 BHP	SN@16.0	Ea	11,600.00	602.00	251.00	12,453.00
150 BHP	SN@16.0	Ea	12,900.00	602.00	251.00	13,753.00

Costs for Natural Gas Combustion Controls for Boilers. Instrumentation Society of America, Underwriters' Laboratories, Factory Mutual, and Industrial Risk Insurers certified. Fireye or equal. Includes automatic recycling or continuous pilot, spark igniter and coil (for electric ignition), timer to control the purge cycle, pilot, main burner (hi-lo-off) or main burner (proportional control), flame safeguard and sensor and combustion control panel.

Natural gas combustion controls for firetube and watertube boilers.
Costs shown are based on boiler horsepower rating (BHP).

20 to 80 BHP	SN@4.00	Ea	3,940.00	151.00	62.40	4,153.40
100 BHP	SN@4.00	Ea	4,370.00	151.00	62.40	4,583.40
125 to 200 BHP	SN@8.00	Ea	4,370.00	301.00	127.00	4,798.00
250 BHP	SN@16.0	Ea	4,370.00	602.00	251.00	5,223.00
300 BHP	SN@16.0	Ea	5,230.00	602.00	251.00	6,083.00
350 or 400 BHP	SN@16.0	Ea	6,140.00	602.00	251.00	6,993.00
500 BHP	SN@32.0	Ea	6,140.00	1,200.00	503.00	7,843.00
600 BHP	SN@32.0	Ea	7,330.00	1,200.00	503.00	9,033.00
750 BHP	SN@64.0	Ea	8,550.00	2,410.00	1,010.00	11,970.00
800 BHP	SN@64.0	Ea	9,940.00	2,410.00	1,010.00	13,360.00

Commercial Boiler Components and Accessories

Description	Craft@Hrs	Unit	Material $	Labor $	Equipment $	Total $

Natural gas combustion controls for firebox boilers. Costs shown are based on boiler horsepower rating (BHP).

Description	Craft@Hrs	Unit	Material $	Labor $	Equipment $	Total $
13.4-80 BHP	SN@4.00	Ea	3,770.00	151.00	62.40	3,983.40
100.0 BHP	SN@4.00	Ea	4,370.00	151.00	62.40	4,583.40
126.9 BHP	SN@8.00	Ea	4,370.00	301.00	127.00	4,798.00
150.0 BHP	SN@8.00	Ea	8,140.00	301.00	127.00	8,568.00
Add if burner controls are for:						
#2 oil-fired	—	%	—	—	—	+10.0
Heavy oil-fired	—	%	—	—	—	+20.0
Dual fuel burners	—	%	—	—	—	+70.0

Costs for Water Softening Systems for Use with Boilers. Costs shown include the metering valve, tank, pump, and zeolite media. Add the cost of piping to connect the water softening system to the main feedwater tank and pump.

Water softening system for firetube and watertube boilers. Costs shown are based on boiler horsepower rating (BHP).

Description	Craft@Hrs	Unit	Material $	Labor $	Equipment $	Total $
20 to 80 BHP	SN@4.00	Ea	6,300.00	151.00	62.40	6,513.40
100 BHP	SN@4.00	Ea	7,390.00	151.00	62.40	7,603.40
125 BHP	SN@8.00	Ea	8,560.00	301.00	127.00	8,988.00
150 or 200 BHP	SN@8.00	Ea	10,900.00	301.00	127.00	11,328.00
250 BHP	SN@8.00	Ea	12,900.00	301.00	127.00	13,328.00
300 or 350 BHP	SN@16.0	Ea	16,500.00	602.00	251.00	17,353.00
400 or 500 BHP	SN@16.0	Ea	20,600.00	602.00	251.00	21,453.00
600 BHP	SN@32.0	Ea	24,900.00	1,200.00	503.00	26,603.00
750 or 800 BHP	SN@64.0	Ea	29,000.00	2,410.00	1,010.00	32,420.00

Water softening system for package firebox boilers. Costs shown are based on boiler horsepower rating (BHP).

Description	Craft@Hrs	Unit	Material $	Labor $	Equipment $	Total $
13.4 to 80 BHP	SN@4.00	Ea	7,410.00	151.00	62.40	7,623.40
100.0 BHP	SN@4.00	Ea	7,410.00	151.00	62.40	7,623.40
126.9 BHP	SN@8.00	Ea	8,710.00	301.00	127.00	9,138.00
150.0 BHP	SN@8.00	Ea	10,100.00	301.00	127.00	10,528.00

Costs for Start-up, Shakedown, and Calibration of Boilers. Costs shown include minor changes and adjustments to meet requirements of the boiler inspector, adjustment of combustion efficiency at the burner by a control technician, pump and metering calibration testing and steam or hot water metering on start-up.

Start-up, shakedown & calibration of firetube and watertube boilers. Costs shown are based on boiler horsepower rating (BHP.) If a factory technician performs the start-up, add this cost at a minimum of $800 (8 hours at $100.00 per hour).

Description	Craft@Hrs	Unit	Material $	Labor $	Equipment $	Total $
13.4 to 100 BHP	S2@4.00	Ea	—	141.00	—	141.00
125 to 200 BHP	S2@8.00	Ea	—	283.00	—	283.00
250 to 400 BHP	S2@16.0	Ea	—	566.00	—	566.00
500 to 600 BHP	S2@24.0	Ea	—	849.00	—	849.00
750 to 800 BHP	S2@30.0	Ea	—	1,060.00	—	1,060.00

Costs for Stack Economizer Units for Boilers. Costs shown include piping connection to existing boiler, economizer fin tube coils, feedwater pump, check valves, thermostatic regulating valve, stack dampers, pressure relief valve, stack mount flanges and insulating enclosure, but no architectural modifications.

Description	Craft@Hrs	Unit	Material	Labor	Equipment	Total

Stack waste heat recovery system for boilers. Costs shown are based on boiler horsepower rating (BHP).

Description	Craft@Hrs	Unit	Material	Labor	Equipment	Total
500 BHP	SN@40.0	Ea	36,700.00	1,510.00	624.00	38,834.00
750 BHP	SN@48.0	Ea	45,800.00	1,810.00	751.00	48,361.00
800 BHP	SN@48.0	Ea	49,400.00	1,810.00	751.00	51,961.00
1,000 BHP	SN@56.0	Ea	61,600.00	2,110.00	880.00	64,590.00

Pollution control modules for gas-fired industrial package watertube boilers.
N0x, S0x, VM (volatile materials) and particulates emissions reduction module. Free-standing unit. Verantis, Croll-Reynolds or equal. Designed to meet or exceed 2010 California Code, Tier IV, EPA and European ISO standards. Either vertical counter-flow or packed-bed type. For multi-dwelling, process and industrial applications. By boiler rating in millions of Btu's per hour (MMBtu/hr). See installation costs below. Oil-fired boilers will add about 15% to the installed cost, including a second stage reduction module, second stage pump, sulfur extraction, venturi particulate separator and sludge tank. To calculate the annual lease rate for a complete installation, divide the total installed cost by five. Then add 2% per year to cover vendor-supplied maintenance.

Description	Craft@Hrs	Unit	Material	Labor	Equipment	Total
30 BHP, 1.0 MMBtu/Hr	—	Ea	35,500.00	—	—	35,500.00
50 BHP, 1.7 MMBTu/Hr	—	Ea	63,000.00	—	—	63,000.00
100 BHP, 3.4 MMBtu/Hr	—	Ea	112,500.00	—	—	112,500.00
125 BHP 4.1 MMBtu/Hr	—	Ea	117,000.00	—	—	117,000.00
150 BHP 5.0 MMBtu/Hr	—	Ea	125,750.00	—	—	125,750.00
200 BHP 6.7 MMBtu/Hr	—	Ea	135,200.00	—	—	135,200.00
400 BHP 13.4 MMBtu/Hr	—	Ea	140,000.00	—	—	140,000.00
600 BHP 20.1 MMBtu/Hr	—	Ea	155,000.00	—	—	155,000.00
800 BHP 26.8 MMBtu/Hr	—	Ea	165,000.00	—	—	165,000.00
1,000 BHP 33.5 MMBtu/Hr	—	Ea	180,000.00	—	—	180,000.00

Commercial Boiler Components and Accessories

Description	Craft@Hrs	Unit	Material $	Labor $	Equipment $	Total $

Installation costs for pollution control modules for gas-fired industrial package watertube boilers.

Description	Craft@Hrs	Unit	Material $	Labor $	Equipment $	Total $
Form and pour foundation	ER@.04	SF	2.05	1.52	—	3.57
Position main housing, fans and pumps	ER@4.0	Ea	—	152.00	—	152.00
Tie-down and leveling	ER@4.0	Ea	500.00	152.00	—	652.00
Mount piping system frame supports	ER@6.0	Ea	1.50	229.00	—	230.50
Rig and mount plenum and ducts	SN@5.0	Ea	—	188.00	—	188.00
Rig and mount tank for emission reduction, fluid and housing inlet metering valve	ER@4.0	Ea	—	152.00	—	152.00
Weld systems piping	ER@8.0	Ea	2.10	305.00	—	307.10
Wire electrical system	BE@8.0	Ea	1.80	317.00	—	318.80
Wire controls system	BE@16.0	Ea	1.90	633.00	—	634.90
Level and balance rotating equipment	P1@8.0	Ea	200.00	286.00	—	486.00
Calibration and test	ER@2.0	Ea	—	76.20	—	76.20
Witnessed run trials (add the permit fee)	P1@4.0	Ea	—	143.00	—	143.00

Costs for Condensate Receiver and Pumping Units for Boilers. Costs for Condensate Receiver and Pumping Units for Boilers. Costs shown include a stainless steel factory assembled tank and pump with motor, float switch, condensate inlet, feedwater tank, connection, level gauge, thermometer, solenoid valve for direct boiler feed. No concrete mounting pad required.

Description	Craft@Hrs	Unit	Material	Labor	Equipment	Total

Condensate receiver and pumping unit for boilers. Unit sizing and costs are based on storage for one minute at the designed steam return rate. Set in place only. Make additional allowances for pipe and electrical connections.

Description	Craft@Hrs	Unit	Material	Labor	Equipment	Total
750 lbs per hour	SN@4.00	Ea	4,800.00	151.00	62.40	5,013.40
1,500 lbs per hour	SN@4.00	Ea	4,800.00	151.00	62.40	5,013.40
3,000 lbs per hour	SN@4.00	Ea	6,090.00	151.00	62.40	6,303.40
5,000 lbs per hour	SN@6.00	Ea	6,720.00	226.00	93.50	7,039.50
6,250 lbs per hour	SN@6.00	Ea	7,490.00	226.00	93.50	7,809.50

Costs for Deaerator/Condenser Units for Boilers. Pressurized jet spray type. Costs shown include an ASME storage receiver, stand, jet spray deaerator head, makeup control valve, makeup controller, pressure control valve, level controls, pressure and temperature controls, pumped condensate inlet, pumps and motor, steam inlet and system accessories. Add for a concrete mounting pad and piping to the condensate return line (inlet) and piping to the boiler feedwater system (outlet).

Description	Craft@Hrs	Unit	Material	Labor	Equipment	Total

Deaerator/condenser unit for boilers. Costs shown are based on boiler horsepower rating (BHP).

Description	Craft@Hrs	Unit	Material	Labor	Equipment	Total
100 BHP	SN@8.00	Ea	50,300.00	301.00	127.00	50,728.00
200 BHP	SN@12.0	Ea	52,000.00	452.00	189.00	52,641.00
300 BHP	SN@14.0	Ea	55,400.00	527.00	220.00	56,147.00
400 BHP	SN@16.0	Ea	62,700.00	602.00	256.00	63,558.00
600 BHP	SN@24.0	Ea	67,200.00	904.00	385.00	68,489.00
800 BHP	SN@36.0	Ea	73,500.00	1,360.00	569.00	75,429.00
1,000 BHP	SN@48.0	Ea	85,700.00	1,810.00	751.00	88,261.00

Rolairtrol type air separators. Priced by capacity in gallons per minute (GPM) and by pipe connection size. Includes hanging separator but no connection, (2" is a screwed connection, 2-1/2" and up are flanged connections) using small tools.

Description	Craft@Hrs	Unit	Material	Labor	Equipment	Total
56 GPM, 2"	S2@2.50	Ea	902.00	88.40	39.60	1,030.00
90 GPM, 2½"	S2@2.50	Ea	980.00	88.40	39.60	1,108.00
170 GPM, 3"	S2@3.00	Ea	1,290.00	106.00	47.10	1,443.10
300 GPM, 4"	S2@3.50	Ea	1,930.00	124.00	55.00	2,109.00
500 GPM, 6"	S2@4.75	Ea	2,510.00	168.00	74.70	2,752.70
700 GPM, 6"	S2@4.75	Ea	3,260.00	168.00	74.70	3,502.70
1,300 GPM, 8	SN@6.00	Ea	4,380.00	226.00	94.30	4,700.30
2,000 GPM, 10"	SN@6.00	Ea	6,130.00	226.00	94.30	6,450.30
2,750 GPM, 12"	SN@12.0	Ea	7,950.00	452.00	189.00	8,591.00

Costs for Chemical Feed Duplex Pump, Packaged Units for Boilers. Costs shown include the chemical tanks, electronic metering, valve and stand. Add the cost of piping to the feedwater tank.

Chemical feed duplex pump for boilers. Costs shown are based on boiler horsepower rating (BHP).

Description	Craft@Hrs	Unit	Material	Labor	Equipment	Total
20 to 100 BHP	SN@3.00	Ea	1,850.00	113.00	47.60	2,010.60
25 to 200 BHP	SN@3.00	Ea	2,410.00	113.00	47.60	2,570.60
200 to 400 BHP	SN@6.00	Ea	2,410.00	226.00	93.50	2,729.50
500 to 600 BHP	SN@6.00	Ea	5,480.00	226.00	93.50	5,799.50
700 to 800 BHP	SN@8.00	Ea	6,120.00	301.00	127.00	6,548.00
1,000 BHP	SN@10.0	Ea	8,270.00	377.00	157.00	8,804.00

Chemical feed simple in-line system for hot water heating system.

Costs shown include 3 gallon pot feeder, two 3/4" isolation valves, 10' of 3/4" pipe and two 3/4" threadolets. Chemicals are drawn in through suction created by hot water system circulation pumps.

Description	Craft@Hrs	Unit	Material	Labor	Equipment	Total
50 to 20 BHP	SN@3.00	Ea	428.00	113.00	47.60	588.60

Commercial Boiler Components and Accessories

Description	Craft@Hrs	Unit	Material $	Labor $	Equipment $	Total $

Costs for Packaged Boiler Feedwater Systems for Boilers. Costs shown include duplex pumps with heater assembly, thermal lining, pressure gauges, makeup feeder valve, ASME feedwater tank and stand, single phase electric motor and level gauge. Add the costs for a concrete mounting pad and piping to the boilers.

Packaged boiler feedwater systems. Costs shown are based on boiler horsepower rating (BHP).

Description	Craft@Hrs	Unit	Material $	Labor $	Equipment $	Total $
15 to 50 BHP	SN@5.00	Ea	12,900.00	188.00	78.90	13,166.90
60 to 80 BHP	SN@5.00	Ea	14,200.00	188.00	78.90	14,466.90
100 BHP	SN@6.00	Ea	14,500.00	226.00	93.50	14,819.50
125 to 150 BHP	SN@6.00	Ea	15,500.00	226.00	93.50	15,819.50
200 BHP	SN@8.00	Ea	15,800.00	301.00	127.00	16,228.00
250 BHP	SN@8.00	Ea	16,600.00	301.00	127.00	17,028.00
300 BHP	SN@8.00	Ea	17,600.00	301.00	127.00	18,028.00
400 BHP	SN@8.00	Ea	18,200.00	301.00	127.00	18,628.00
500 BHP	SN@8.00	Ea	22,200.00	301.00	127.00	22,628.00
600 BHP	SN@12.0	Ea	23,100.00	452.00	189.00	23,741.00
750 BHP	SN@12.0	Ea	23,000.00	452.00	189.00	23,641.00

Costs for Continuous Blowdown Heat Recovery Systems. Costs shown include a flash receiver (50 PSI ASME rated tank), plate and frame heat exchanger, pneumatic level controller and valve, air filter regulator, water gauge, pressure gauge, backflush piping, pressure relief valve connected to mud drum and drain. Add for concrete pad mount, piping and electrical wiring.

Description	Craft@Hrs	Unit	Material $	Labor $	Equipment $	Total $

Continuous blowdown heat recovery systems. Costs shown by pounds per hour (lbs per hr) of rated makeup capacity.

Description	Craft@Hrs	Unit	Material $	Labor $	Equipment $	Total $
5,000 lbs per hr	SN@8.00	Ea	43,300.00	301.00	127.00	43,728.00
10,000 lbs per hr	SN@8.00	Ea	85,500.00	301.00	127.00	85,928.00
25,000 lbs per hr	SN@8.00	Ea	108,000.00	301.00	127.00	108,428.00
50,000 lbs per hr	SN@8.00	Ea	173,000.00	301.00	127.00	173,428.00

Costs for Boiler Stacks. Costs shown assume a 20-foot stack height and include rigging, breeching, connection of stack sections, guy wires where required, vendor-supplied stack supports and insulation. Add the cost of a concrete base.

Boiler stack. Costs shown assume a 20-foot stack height.

Description	Craft@Hrs	Unit	Material $	Labor $	Equipment $	Total $
10" diameter	SN@12.0	Ea	3,070.00	452.00	189.00	3,711.00
12" diameter	SN@12.0	Ea	3,830.00	452.00	189.00	4,471.00
16" diameter	SN@16.0	Ea	4,580.00	602.00	251.00	5,433.00
20" diameter	SN@20.0	Ea	5,760.00	753.00	314.00	6,827.00
24" diameter	SN@24.0	Ea	7,600.00	904.00	377.00	8,881.00
30" diameter	SN@32.0	Ea	9,020.00	1,200.00	503.00	10,723.00
56" diameter	SN@40.0	Ea	17,800.00	1,510.00	629.00	19,939.00
Add for each 10' section (or portion thereof) beyond 20' high						
per 10' sect.	—	%	—	—	—	15.0

Description	Craft@Hrs	Unit	Material	Labor	Equipment	Total

Costs for Tankless Indirect Water Heater. Provides domestic hot water from boiler coil insert. Costs shown include heat exchanger, coil, circulating pump, thermostatically controlled valve, bypass piping and air bleeder valve. Add the cost of an equipment base, electrical wiring, piping and an insulated hot water storage tank.

Tankless indirect water heater. Costs by gallons per hour (GPH).

Description	Craft@Hrs	Unit	Material	Labor	Equipment	Total
240 GPH	SN@8.00	Ea	17,800.00	301.00	127.00	18,228.00
360 GPH	SN@8.00	Ea	22,400.00	301.00	127.00	22,828.00
480 GPH	SN@10.0	Ea	22,900.00	377.00	157.00	23,434.00
840 GPH	SN@12.0	Ea	51,900.00	452.00	189.00	52,541.00
920 GPH	SN@16.0	Ea	57,900.00	602.00	251.00	58,753.00
1,240 GPH	SN@20.0	Ea	64,800.00	753.00	314.00	65,867.00

Costs for Boiler Trim. Costs shown include the pressure relief valve, level gauge (for steam systems), level controller & low water cutoff, pressure and temperature gauge, feedwater valve & actuator, main valve, blowdown valve, cleanout manholes and gaskets.

Boiler trim. Costs shown are per boiler horsepower rating (BHP).

Description	Craft@Hrs	Unit	Material	Labor	Equipment	Total
20 to 100 BHP	SN@4.00	Ea	2,860.00	151.00	62.40	3,073.40
100 to 200 BHP	SN@6.00	Ea	3,220.00	226.00	93.50	3,539.50
200 to 400 BHP	SN@8.00	Ea	3,970.00	301.00	127.00	4,398.00
400 to 600 BHP	SN@8.00	Ea	6,050.00	301.00	127.00	6,478.00
000 to 800 BHP	SN@8.00	Ea	7,140.00	301.00	127.00	7,568.00
800 to 1,000 BHP	SN@10.0	Ea	7,150.00	377.00	157.00	7,684.00

Description	Craft@Hrs	Unit	Material $	Labor $	Total $

Heat reclaimer condensate return & feedwater system, stainless steel non-siphon, including flash tank with level control and vacuum breaker, overflow relief drain, simplex centrifugal pump, pre-assembled piping to feedwater tank, structural steel stand and frame, check trim and pre-wired controls. Also provides potable hot water as part of the condensed steam return cycle. Capacity rating is in GPH (gallons per hour of condensate)

Description	Craft@Hrs	Unit	Material $	Labor $	Total $
33 gal. ½ HP pump	P1@3.00	Ea	900.00	107.00	1,007.00
46 gal, ¾ HP pump	P1@4.00	Ea	1,100.00	143.00	1,243.00

Vacuum breakers, rough brass, threaded, to be mounted on finned radiation tube inlet valve fitting

Description	Craft@Hrs	Unit	Material $	Labor $	Total $
1/2" vacuum breaker	P1@.25	Ea	42.00	8.94	50.94

Steam traps, float valve Sarco type or equal w/drain fitting, brazed

Description	Craft@Hrs	Unit	Material $	Labor $	Total $
3/4" with check valve	P1@.25	Ea	40.00	8.94	48.94
1.0" with check valve	P1@.45	Ea	64.00	16.10	80.10

Baseboard fin tube radiation, per linear foot, 15 PSIG test thinwall copper tube with aluminum fins, braze fitted wall mounted

Description	Craft@Hrs	Unit	Material $	Labor $	Total $
¾" element with 8" cover	P1@.280	LF	10.20	10.00	20.20
1" element with 8" cover	P1@.350	LF	16.55	12.50	29.05
Add for corners, fillers and caps, average per foot			—	15.0%	

Commercial Boiler Components and Accessories

Description	Craft@Hrs	Unit	Material $	Labor $	Equipment $	Total $

Solenoid valves, zone thermostatically-controlled steam electric, 15 PSIG rating, braze fitting. Thermostatic Honeywell-type or equal solenoid relay valves

Description	Craft@Hrs	Unit	Material $	Labor $	Equipment $	Total $
3/4", brazed	P1@.25	Ea	12.60	8.94	—	21.54
1" brazed	P1@.25	Ea	12.60	8.94	—	21.54

Additional costs for boiler pumps, condenser units, feedwater systems, heat recovery and oil fuel train piping.

Description	Craft@Hrs	Unit	Material $	Labor $	Equipment $	Total $
Form and pour concrete pad	CF@2.50	Ea	97.00	87.60	—	184.60
Place vibration pads	CF@.750	Ea	38.80	26.30	—	65.10
Mount interior drain	P1@.500	LF	7.76	17.90	—	25.66
Bore hole through wall or pad	P1@.250	Ea	—	8.94	—	8.94
Piping to load (typical)	P1@.500	LF	7.76	17.90	—	25.66
Bore burner stack vent in exterior wall	P1@.250	Ea	—	8.94	—	8.94
Mounting and edge-sealing stack	P1@.500	Ea	40.00	17.90	—	57.90

Description	Craft@Hrs	Unit	Material $	Labor $	Equipment $	Total $

Emissions sensing and recording equipment. For stack gas combustion monitoring of boilers. Monitors and records NOx, SOx, CO, H2S, volatile materials and particulates emissions to meet 2011 EPA greenhouse gas regulations. CO2 monitoring and recording is required on boilers producing 25,000 tons or more a year of carbon emissions. Signal Group Limited or equal. Includes a stack-mounted pulse type infra-red single-beam light source with solid-state detector. Sensitivity range is variable between .1 ppm and 1,000 ppm. Includes a hot-wire radiation source, rotating shutter with reference target for continuous unmanned recalibration, 240 x 64 pixel LCD display with remote panel-mounted solid-state digital analyzer and lockout type inspector-only access recording circuitry. Cost includes calibration and testing by a factory technician. Add the cost of the permit, inspection, test and documentation.

Description	Craft@Hrs	Unit	Material $	Labor $	Equipment $	Total $
Analyzer and recorder installation to remote panel	BE@3.00	Ea	25,500.00	119.00	—	25,619.00
Infrared sensor unit installation to stack	BE@2.00	Ea	8,000.00	79.20	—	8,079.20
Wiring of sensor to analyzer and recorder	BE@8.00	Ea	500.00	317.00	—	817.00
Circuit testing and grounding	BE@2.00	Ea	—	79.20	—	79.20

Description	Craft@Hrs	Unit	Material $	Labor $	Equipment $	Total $

Base mounted, bronze fitted centrifugal pump, set in place only. Make additional allowances for pipe connection assembly.

Description	Craft@Hrs	Unit	Material $	Labor $	Equipment $	Total $
1/2 HP	P1@1.50	Ea	1,140.00	53.60	—	1,193.60
3/4 HP	P1@1.75	Ea	1,450.00	62.60	—	1,512.60
1 HP	P1@2.00	Ea	1,770.00	71.50	—	1,841.50
1½ HP	P1@2.40	Ea	2,150.00	85.80	—	2,235.80
2 HP	P1@2.70	Ea	2,650.00	96.60	—	2,746.60
3 HP	P1@3.00	Ea	3,160.00	107.00	—	3,267.00
5 HP	P1@3.50	Ea	3,600.00	125.00	—	3,725.00
7½ HP	P1@4.00	Ea	3,980.00	143.00	—	4,123.00
10 HP	P1@4.50	Ea	5,810.00	161.00	—	5,971.00
15 HP	ER@5.00	Ea	6,720.00	191.00	79.20	6,990.20
20 HP	ER@6.00	Ea	7,780.00	229.00	93.50	8,102.50
25 HP	ER@7.00	Ea	8,110.00	267.00	110.00	8,487.00
30 HP	ER@7.40	Ea	9,650.00	282.00	116.00	10,048.00
40 HP	ER@7.70	Ea	9,560.00	293.00	121.00	9,974.00
50 HP	ER@8.00	Ea	11,600.00	305.00	127.00	12,032.00

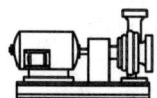

Base mounted centrifugal pump connection assembly. Includes pipe, fittings, pipe insulation, valves, strainers, suction guides and gauges.

Description	Craft@Hrs	Unit	Material $	Labor $	Equipment $	Total $
2½" supply	P1@12.0	Ea	1,960.00	429.00	—	2,389.00
3" supply	P1@13.5	Ea	2,270.00	483.00	—	2,753.00
4" supply	P1@16.0	Ea	3,110.00	572.00	—	3,682.00
6" supply	P1@22.0	Ea	5,140.00	787.00	—	5,927.00
8" supply	P1@28.0	Ea	7,470.00	1,000.00	—	8,470.00
10" supply	P1@35.0	Ea	12,300.00	1,250.00	—	13,550.00
12" supply	P1@42.0	Ea	16,400.00	1,500.00	—	17,900.00

Cast iron line-mounted centrifugal pump, including connection and supports.

Description	Craft@Hrs	Unit	Material $	Labor $	Equipment $	Total $
1/6 HP	P1@1.25	Ea	294.00	44.70	—	338.70
1/4 HP	P1@1.40	Ea	376.00	50.10	—	426.10
1/3 HP	P1@1.50	Ea	466.00	53.60	—	519.60
1/2 HP	P1@1.70	Ea	594.00	60.80	—	654.80
3/4 HP	P1@1.80	Ea	801.00	64.40	—	865.40
1 HP	P1@2.00	Ea	950.00	71.50	—	1,021.50

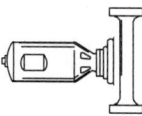

Heat Exchangers and Connections

Description	Craft@Hrs	Unit	Material $	Labor $	Equipment $	Total $

2-pass shell and tube type heat exchanger, set in place only. Make additional allowances for pipe connection assembly.

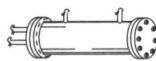

Description	Craft@Hrs	Unit	Material $	Labor $	Equipment $	Total $
4" dia. x 30"	SN@2.00	Ea	1,380.00	75.30	31.10	1,486.40
4" dia. x 60"	SN@2.50	Ea	1,610.00	94.10	38.50	1,742.60
6" dia. x 30"	SN@3.25	Ea	1,990.00	122.00	51.40	2,163.40
6" dia. x 60"	SN@3.50	Ea	2,460.00	132.00	55.00	2,647.00
8" dia. x 30"	SN@4.25	Ea	3,050.00	160.00	66.00	3,276.00
8" dia. x 60"	SN@4.50	Ea	3,940.00	169.00	71.50	4,180.50
10" dia. x 30"	SN@5.50	Ea	4,520.00	207.00	86.20	4,813.20
10" dia. x 60"	SN@5.75	Ea	5,980.00	216.00	89.90	6,285.90
10" dia. x 90"	SN@6.00	Ea	7,450.00	226.00	93.50	7,769.50
12" dia. x 60"	SN@6.50	Ea	7,830.00	245.00	103.00	8,178.00
12" dia. x 90"	SN@7.00	Ea	10,000.00	264.00	110.00	10,374.00
12" dia. x 120"	SN@7.25	Ea	12,200.00	273.00	113.00	12,586.00
14" dia. x 60"	SN@7.50	Ea	11,600.00	282.00	118.00	12,000.00
14" dia. x 90"	SN@8.00	Ea	14,400.00	301.00	127.00	14,828.00
14" dia. x 120"	SN@8.25	Ea	17,400.00	311.00	130.00	17,841.00
16" dia. x 60"	SN@8.50	Ea	14,300.00	320.00	134.00	14,754.00
16" dia. x 90"	SN@9.00	Ea	17,800.00	339.00	141.00	18,280.00
16" dia. x 120"	SN@9.25	Ea	20,200.00	348.00	145.00	20,693.00

Heat exchanger connection assembly, hot water to hot water. Includes pipe, fittings, pipe insulation, valves, thermometers gauges and vents.

Description	Craft@Hrs	Unit	Material $	Labor $	Equipment $	Total $
1½" supply	P1@32.0	Ea	3,160.00	1,140.00	1,760.00	6,060.00
2" supply	P1@38.0	Ea	3,860.00	1,360.00	2,090.00	7,310.00
2½" supply	P1@50.0	Ea	5,050.00	1,790.00	2,750.00	9,590.00
3" supply	P1@55.0	Ea	5,300.00	1,970.00	3,030.00	10,300.00
4" supply	P1@62.0	Ea	6,450.00	2,220.00	3,410.00	12,080.00
6" supply	P1@85.0	Ea	10,500.00	3,040.00	4,680.00	18,220.00
8" supply	P1@102.	Ea	14,800.00	3,650.00	5,610.00	24,060.00
10" supply	P1@130.	Ea	19,400.00	4,650.00	7,150.00	31,200.00

Heat exchanger connection assembly, steam to hot water. Includes pipe, fittings, pipe insulation, valves, thermometers gauges and vents.

Description	Craft@Hrs	Unit	Material $	Labor $	Equipment $	Total $
1½" supply	P1@22.0	Ea	3,610.00	787.00	1,210.00	5,607.00
2" supply	P1@26.0	Ea	4,210.00	930.00	1,430.00	6,570.00
2½" supply	P1@41.0	Ea	5,220.00	1,470.00	2,260.00	8,950.00
3" supply	P1@44.0	Ea	6,050.00	1,570.00	2,420.00	10,040.00
4" supply	P1@54.5	Ea	6,880.00	1,950.00	2,990.00	11,820.00
6" supply	P1@67.0	Ea	11,000.00	2,400.00	3,690.00	17,090.00
8" supply	P1@80.0	Ea	15,500.00	2,860.00	4,400.00	22,760.00

Description	Craft@Hrs	Unit	Material $	Labor $	Equipment $	Total $

Ceiling-hung horizontal, 2 pipe hydronic fan coil unit with cabinet, set in place only. Make additional allowances for pipe connection assembly.

200 CFM	P1@2.00	Ea	767.00	.71.50	—	838.50
400 CFM	P1@2.50	Ea	909.00	89.40	—	998.40
600 CFM	P1@3.00	Ea	1,070.00	107.00	—	1,177.00
800 CFM	P1@3.25	Ea	1,350.00	116.00	—	1,466.00
1,000 CFM	P1@3.50	Ea	1,690.00	125.00	—	1,815.00

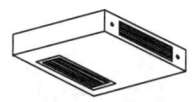

Less cabinet: Deduct 20% from material cost.
Floor-mounted (vertical): Add 10% to material cost.

Vertical, 4 pipe hydronic fan coil unit, set in place only. Make additional allowances for pipe connection assembly.

200 CFM	P1@2.15	Ea	872.00	76.90	—	948.90
400 CFM	P1@2.75	Ea	1,010.00	98.30	—	1,108.30
600 CFM	P1@3.25	Ea	1,190.00	116.00	—	1,306.00
800 CFM	P1@3.45	Ea	1,470.00	123.00	—	1,593.00
1,000 CFM	P1@3.60	Ea	1,850.00	129.00	—	1,979.00

2 pipe fan coil unit connection. Coil connection includes type L copper pipe and wrought fittings, pipe insulation, condensate drain, isolation valves and 2-way control valve.

1/2" supply	P1@2.50	Ea	129.00	89.40	—	218.40
3/4" supply	P1@2.80	Ea	155.00	100.00	—	255.00
1" supply	P1@3.40	Ea	235.00	122.00	—	357.00

4 pipe fan coil unit connection. Coil connection includes type L copper pipe and wrought fittings, pipe insulation, condensate drain, isolation valves and 3-way control valve.

1/2" supply	P1@2.75	Ea	155.00	98.30	—	253.30
3/4" supply	P1@2.95	Ea	180.00	105.00	—	285.00

Reheat Coils and Connections

Description	Craft@Hrs	Unit	Material $	Labor $	Equipment $	Total $

Duct-mounted flanged hot water reheat coil, set in place only. Make
additional allowances for pipe connection assembly.

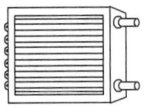

Description	Craft@Hrs	Unit	Material $	Labor $	Equipment $	Total $
12" x 6"	SN@1.00	Ea	320.00	37.70	15.70	373.40
12" x 8"	SN@1.00	Ea	336.00	37.70	15.70	389.40
12" x 10"	SN@1.00	Ea	352.00	37.70	15.70	405.40
12" x 12"	SN@1.00	Ea	382.00	37.70	15.70	435.40
18" x 6"	SN@1.25	Ea	405.00	47.10	19.60	471.70
18" x 12"	SN@1.25	Ea	416.00	47.10	19.60	482.70
18" x 18"	SN@1.25	Ea	446.00	47.10	19.60	512.70
24" x 12"	SN@1.50	Ea	483.00	56.50	23.70	563.20
24" x 18"	SN@1.50	Ea	506.00	56.50	23.70	586.20
24" x 24"	SN@1.75	Ea	568.00	65.90	27.50	661.40

Duct-mounted hot water reheat coil connection assembly. Includes
pipe, fittings, pipe insulation, valves and strainers.

Description	Craft@Hrs	Unit	Material $	Labor $	Equipment $	Total $
1/2" steel	P1@2.25	Ea	164.00	80.50	—	244.50
3/4" steel	P1@2.35	Ea	194.00	84.00	—	278.00
1" steel	P1@2.65	Ea	253.00	94.80	—	347.80
1¼" steel	P1@3.00	Ea	307.00	107.00	—	414.00
1½" steel	P1@3.25	Ea	327.00	116.00	—	443.00
2" steel	P1@3.75	Ea	365.00	134.00	—	499.00

Duct-mounted slip-in electric reheat coil, set in place only. Make
additional allowances for electrical connection.

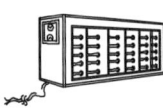

Description	Craft@Hrs	Unit	Material $	Labor $	Equipment $	Total $
12" x 6"	SN@1.10	Ea	713.00	41.40	17.30	771.70
12" x 8"	SN@1.20	Ea	812.00	45.20	18.90	876.10
12" x 10"	SN@1.30	Ea	987.00	48.90	20.40	1,056.30
12" x 12"	SN@1.40	Ea	1,070.00	52.70	22.00	1,144.70
18" x 6"	SN@1.50	Ea	1,190.00	56.50	23.70	1,270.20
18" x 12"	SN@1.60	Ea	1,290.00	60.20	25.10	1,375.30
18" x 18"	SN@1.70	Ea	1,380.00	64.00	26.70	1,470.70
24" x 12"	SN@1.90	Ea	1,460.00	71.50	29.90	1,561.40
24" x 18"	SN@2.10	Ea	1,540.00	79.10	33.00	1,652.10
24" x 24"	SN@2.40	Ea	1,900.00	90.40	37.70	2,028.10

Description	Craft@Hrs	Unit	Material $	Labor $	Equipment $	Total $

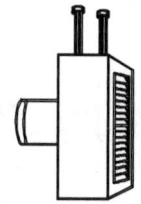

Gas-fired unit heater, hang in place only. Make additional allowances for gas connection assembly.

Description	Craft@Hrs	Unit	Material $	Labor $	Equipment $	Total $
60mbh, 200cfm	P1@2.00	Ea	1,420.00	71.50	—	1,491.50
100mbh, 400cfm	P1@2.50	Ea	1,660.00	89.40	—	1,749.40
150mbh, 600cfm	P1@3.00	Ea	2,010.00	107.00	—	2,117.00
175mbh, 800cfm	P1@3.25	Ea	2,180.00	116.00	—	2,296.00
200mbh, 1000cfm	P1@3.50	Ea	2,340.00	125.00	—	2,465.00
250mbh, 1500cfm	P1@3.75	Ea	2,740.00	134.00	—	2,874.00
300mbh, 2000cfm	P1@4.00	Ea	3,060.00	143.00	—	3,203.00

Hot water unit heater, hang in place only. Make additional allowances for pipe connection assembly.

Description	Craft@Hrs	Unit	Material $	Labor $	Equipment $	Total $
8 mbh, 400cfm	P1@2.00	Ea	377.00	71.50	110.00	558.50
12 mbh, 400cfm	P1@2.00	Ea	485.00	71.50	110.00	666.50
16 mbh, 400 cfm	P1@2.00	Ea	510.00	71.50	110.00	691.50
30mbh, 700cfm	P1@3.00	Ea	684.00	107.00	165.00	956.00
43mbh, 1040cfm	P1@3.50	Ea	841.00	125.00	193.00	1,159.00
57mbh, 1370cfm	P1@4.00	Ea	906.00	143.00	220.00	1,269.00
68mbh, 1715cfm	P1@4.00	Ea	1,420.00	143.00	220.00	1,783.00
105mbh, 2360cfm	P1@5.00	Ea	1,690.00	179.00	275.00	2,144.00
123mbh, 2900cfm	P1@6.00	Ea	2,170.00	215.00	330.00	2,715.00
140mbh, 3530cfm	P1@6.00	Ea	2,510.00	215.00	330.00	3,055.00
156mbh, 4250cfm	P1@6.00	Ea	2,640.00	215.00	330.00	3,185.00
210mbh, 4450cfm	P1@8.00	Ea	4,690.00	286.00	440.00	5,416.00
123mbh, 5200cfm	P1@6.00	Ea	3,870.00	215.00	330.00	4,415.00
257mbh 5350cfm	P1@8.00	Ea	7,930.00	286.00	440.00	8,656.00

Hot water unit heater connection. Includes schedule 40 threaded and coupled supply and return piping with malleable iron fittings, pipe insulation, valves and vents Make additional allowances for electrical connection.

Description	Craft@Hrs	Unit	Material $	Labor $	Equipment $	Total $
3/4" supply	P1@2.50	Ea	218.00	89.40	138.00	445.40
1" supply	P1@3.00	Ea	256.00	107.00	165.00	528.00
1¼" supply	P1@3.30	Ea	304.00	118.00	182.00	604.00
1½" supply	P1@3.75	Ea	352.00	134.00	207.00	693.00
2" supply	P1@4.10	Ea	456.00	147.00	226.00	829.00

Gas-fired unit heater flue connection

Description	Craft@Hrs	Unit	Material $	Labor $	Equipment $	Total $
2" b vent	SN@.090	LF	4.51	3.39	—	7.90
3" b vent	SN@.100	LF	5.61	3.77	—	9.38
4" b vent	SN@.110	LF	7.48	4.14	—	11.62
6" b vent	SN@.130	LF	8.51	4.89	—	13.40

Gas-fired unit heater fuel connection. Assembly includes 8' of black steel pipe, malleable fittings, plug valve and regulator.

Description	Craft@Hrs	Unit	Material $	Labor $	Equipment $	Total $
3/4" gas conn.	P1@1.30	Ea	79.10	46.50	—	125.60
1" gas conn.	P1@1.60	Ea	86.10	57.20	—	143.30
1¼" gas conn.	P1@2.10	Ea	154.00	75.10	—	229.10
1½" gas conn.	P1@2.50	Ea	177.00	89.40	—	266.40

Chillers and Chiller Connections

Description	Craft@Hrs	Unit	Material $	Labor $	Equipment $	Total $

Centrifugal water-cooled chiller, set in place only. Unit comes factory assembled complete. Make additional allowances for cooling tower and pipe connections.

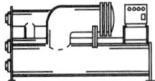

Description	Craft@Hrs	Unit	Material $	Labor $	Equipment $	Total $
100 tons	SN@20.0	Ea	77,800.00	753.00	314.00	78,867.00
150 tons	SN@20.0	Ea	117,000.00	753.00	314.00	118,067.00
200 tons	SN@20.0	Ea	126,000.00	753.00	314.00	127,067.00
250 tons	SN@20.0	Ea	149,000.00	753.00	393.00	150,146.00
300 tons	SN@25.0	Ea	160,000.00	941.00	393.00	161,334.00
350 tons	SN@25.0	Ea	164,000.00	941.00	393.00	165,334.00
400 tons	SN@25.0	Ea	194,000.00	941.00	393.00	195,334.00
500 tons	SN@25.0	Ea	244,000.00	941.00	393.00	245,334.00

Reciprocating water-cooled chiller, set in place only. Unit comes factory assembled complete. Make additional allowances for cooling tower and pipe connections.

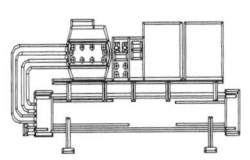

Description	Craft@Hrs	Unit	Material $	Labor $	Equipment $	Total $
20 tons	SN@12.0	Ea	23,800.00	452.00	189.00	24,441.00
25 tons	SN@12.0	Ea	27,200.00	452.00	189.00	27,841.00
30 tons	SN@12.0	Ea	29,600.00	452.00	189.00	30,241.00
40 tons	SN@12.0	Ea	35,700.00	452.00	189.00	36,341.00
50 tons	SN@16.0	Ea	50,500.00	602.00	251.00	51,353.00
60 tons	SN@16.0	Ea	64,900.00	602.00	251.00	65,753.00
75 tons	SN@16.0	Ea	69,500.00	602.00	251.00	70,353.00
100 tons	SN@20.0	Ea	74,100.00	753.00	314.00	75,167.00
125 tons	SN@20.0	Ea	88,800.00	753.00	314.00	89,867.00
150 tons	SN@20.0	Ea	103,000.00	753.00	314.00	104,067.00

Water-cooled chiller connection assembly. Includes condenser and chilled water pipe, fittings, pipe insulation, valves, thermometers and gauges.

Description	Craft@Hrs	Unit	Material $	Labor $	Equipment $	Total $
2½" chiller conn.	SN@36.0	Ea	2,490.00	1,360.00	567.00	4,417.00
3" chiller conn.	SN@38.0	Ea	2,730.00	1,430.00	603.00	4,763.00
4" chiller conn.	SN@51.0	Ea	3,640.00	1,920.00	801.00	6,361.00
6" chiller conn.	SN@65.0	Ea	6,110.00	2,450.00	1,020.00	9,580.00
8" chiller conn.	SN@77.0	Ea	9,270.00	2,900.00	1,210.00	13,380.00
10" chiller conn.	SN@100.	Ea	17,000.00	3,770.00	1,570.00	22,340.00
12" chiller conn.	SN@125.	Ea	19,700.00	4,710.00	1,960.00	26,370.00

Condensing Units and Cooling Towers

Description	Craft@Hrs	Unit	Material $	Labor $	Equipment $	Total $

Air-cooled condensing unit, set in place only. Unit comes factory assembled complete. Make additional allowances for pipe and electrical connections.

Description	Craft@Hrs	Unit	Material $	Labor $	Equipment $	Total $
3 tons	SN@2.00	Ea	2,960.00	75.30	31.40	3,066.70
5 tons	SN@3.00	Ea	5,070.00	113.00	47.10	5,230.10
7½ tons	SN@4.00	Ea	6,550.00	151.00	62.90	6,763.90
10 tons	SN@4.50	Ea	8,750.00	169.00	70.80	8,989.80
12½ tons	SN@5.00	Ea	10,300.00	188.00	78.70	10,566.70
15 tons	SN@5.50	Ea	11,800.00	207.00	86.40	12,093.40
20 tons	SN@6.00	Ea	15,600.00	226.00	94.30	15,920.30
25 tons	SN@7.00	Ea	19,700.00	264.00	110.00	20,074.00
30 tons	SN@8.00	Ea	23,700.00	301.00	125.00	24,126.00
40 tons	SN@9.00	Ea	30,800.00	339.00	142.00	31,281.00
50 tons	SN@10.0	Ea	38,700.00	377.00	157.00	39,234.00

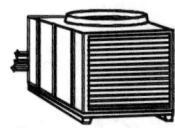

Galvanized steel cooling tower. Assemble and set in place only. Add for pipe connections.

Description	Craft@Hrs	Unit	Material $	Labor $	Equipment $	Total $
10 tons	SN@16.0	Ea	3,000.00	602.00	251.00	3,853.00
15 tons	SN@16.0	Ea	3,400.00	602.00	251.00	4,253.00
20 tons	SN@18.0	Ea	5,340.00	678.00	283.00	0,301.00
25 tons	SN@20.0	Ea	6,590.00	753.00	314.00	7,657.00
30 tons	SN@25.0	Ea	7,510.00	941.00	393.00	8,844.00
40 tons	SN@30.0	Ea	9,600.00	1,130.00	471.00	11,201.00
50 tons	SN@35.0	Ea	10,500.00	1,320.00	550.00	12,370.00
60 tons	SN@40.0	Ea	11,000.00	1,510.00	629.00	13,139.00
80 tons	SN@45.0	Ea	12,100.00	1,690.00	707.00	14,497.00
100 tons	SN@50.0	Ea	14,300.00	1,880.00	787.00	16,967.00
125 tons	SN@60.0	Ea	17,300.00	2,260.00	943.00	20,503.00
150 tons	SN@65.0	Ea	20,100.00	2,450.00	1,020.00	23,570.00
175 tons	SN@70.0	Ea	22,700.00	2,640.00	1,100.00	26,440.00
200 tons	SN@70.0	Ea	25,000.00	2,640.00	1,100.00	28,740.00
300 tons	SN@75.0	Ea	36,000.00	2,820.00	1,180.00	40,000.00
400 tons	SN@80.0	Ea	45,600.00	3,010.00	1,260.00	49,870.00
500 tons	SN@80.0	Ea	54,800.00	3,010.00	1,260.00	59,070.00

Redwood or fir induced-draft cooling tower. Assemble and set in place only. Add for pipe connections.

Description	Craft@Hrs	Unit	Material $	Labor $	Equipment $	Total $
100 tons	SN@75.0	Ea	9,300.00	2,820.00	1,180.00	13,300.00
200 tons	SN@80.0	Ea	15,600.00	3,010.00	1,260.00	19,870.00
300 tons	SN@85.0	Ea	22,500.00	3,200.00	1,340.00	27,040.00
400 tons	SN@90.0	Ea	25,500.00	3,390.00	1,410.00	30,300.00
500 tons	SN@90.0	Ea	29,800.00	3,390.00	1,410.00	34,600.00

Cooling Towers and Cooling Tower Connections

Description	Craft@Hrs	Unit	Material $	Labor $	Equipment $	Total $

Redwood or fir forced-draft cooling tower. Assemble and set in place only. Add for pipe connections.

Description	Craft@Hrs	Unit	Material $	Labor $	Equipment $	Total $
100 tons	SN@75.0	Ea	15,300.00	2,820.00	1,180.00	19,300.00
200 tons	SN@80.0	Ea	29,200.00	3,010.00	1,260.00	33,470.00
300 tons	SN@85.0	Ea	34,100.00	3,200.00	1,340.00	38,640.00
400 tons	SN@90.0	Ea	45,100.00	3,390.00	1,410.00	49,900.00
500 tons	SN@90.0	Ea	54,900.00	3,390.00	1,410.00	59,700.00

Cooling tower connection assembly. Includes condenser water pipe, fittings, valves, thermometers and gauges.

Description	Craft@Hrs	Unit	Material $	Labor $	Equipment $	Total $
2½" connection	SN@42.0	Ea	4,480.00	1,580.00	660.00	6,720.00
3" connection	SN@47.0	Ea	5,090.00	1,770.00	734.00	7,594.00
4" connection	SN@62.0	Ea	7,310.00	2,330.00	972.00	10,612.00
6" connection	SN@96.0	Ea	11,500.00	3,610.00	1,510.00	16,620.00
8" connection	SN@118	Ea	16,700.00	4,440.00	1,850.00	22,990.00
10" connection	SN@150	Ea	22,800.00	5,650.00	2,370.00	30,820.00
12" connection	SN@215	Ea	27,800.00	8,090.00	3,370.00	39,260.00
14" connection	SN@260	Ea	32,000.00	9,790.00	4,090.00	45,880.00

Carbon Steel, Schedule 40 with 150# Fittings & Butt-Welded Joints

Schedule 40 carbon steel (ASTM A-53 and A-120) pipe with 150 pound carbon steel welding fittings is commonly used for heating hot water, chilled water, glycol and low pressure steam systems where operating pressures do not exceed 250 PSIG.

Schedule 40 (ASTM A-53B SML) steel pipe with carbon steel welding fittings can also be used for refrigerant piping systems.

The cost estimates in this section are based on the conditions, limitations and wage rates described in the section "How to Use This Book" beginning on page 5.

Equipment cost, where shown, is $110 per hour for a 10-ton hydraulic truck-mounted crane.

Description	Craft@Hrs	Unit	Material $	Labor $	Equipment $	Total $

Schedule 40 carbon steel butt-welded horizontal pipe assembly.

Horizontally hung in a building. Assembly includes fittings and hanger assemblies. Based on a reducing tee and a 45 degree elbow every 16 feet for 1/2" pipe. A reducing tee and a 45 degree elbow every 50 feet for 12" pipe. Hangers spaced to meet plumbing code. *(Distance between fittings in the pipe assembly increases as pipe diameter increases.)* Use these figures for preliminary estimates.

Description	Craft@Hrs	Unit	Material $	Labor $	Equipment $	Total $
1/2"	P1@.144	LF	1.58	5.15	—	6.73
3/4"	P1@.170	LF	1.71	6.08	—	7.79
1"	P1@.200	LF	2.09	7.37	—	9.46
1¼"	P1@.208	LF	2.78	7.44	—	10.22
1½"	P1@.238	LF	2.99	8.51	—	11.50
2"	P1@.286	LF	3.95	10.20	—	14.15
2½"	P1@.344	LF	6.94	12.30	—	19.24
3"	P1@.390	LF	8.68	13.90	—	22.58
4"	P1@.507	LF	12.00	18.10	—	30.10
6"	ER@.712	LF	19.30	27.10	11.20	57.60
8"	ER@.847	LF	33.50	32.30	13.30	79.10
10"	ER@1.02	LF	48.90	38.90	16.10	103.90
12"	ER@1.25	LF	62.60	47.60	19.60	129.80

Schedule 40 carbon steel butt-welded vertical pipe assembly.
Riser assembly, including fittings and riser clamps. Based on a reducing tee, and a riser clamp on every floor. Use these figures for preliminary estimates.

Description	Craft@Hrs	Unit	Material $	Labor $	Equipment $	Total $
1/2"	P1@.104	LF	1.19	3.72	—	4.91
3/4"	P1@.128	LF	1.34	4.58	—	5.92
1"	P1@.162	LF	1.72	5.79	—	7.51
1¼"	P1@.184	LF	2.99	6.58	—	9.57
1½"	P1@.204	LF	3.34	7.30	—	10.64
2"	P1@.247	LF	4.44	8.83	—	13.27
2½"	P1@.304	LF	7.25	10.90	—	18.15
3"	P1@.351	LF	8.75	12.60	—	21.35
4"	P1@.505	LF	11.60	18.10	—	29.70
6"	ER@.737	LF	20.80	28.10	11.60	60.50
8"	ER@.878	LF	37.50	33.50	13.90	84.90
10"	ER@1.07	LF	57.10	40.80	16.80	114.70
12"	ER@1.32	LF	75.30	50.30	20.70	146.30

Carbon Steel, Schedule 40 with 150# Fittings & Butt-Welded Joints

Description	Craft@Hrs	Unit	Material $	Labor $	Equipment $	Total $

Schedule 40 carbon steel pipe, plain end, butt-welded joints, pipe only

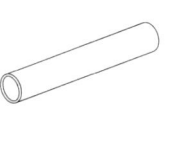

Description	Craft@Hrs	Unit	Material $	Labor $	Equipment $	Total $
1/2"	P1@.070	LF	.71	2.50	—	3.21
3/4"	P1@.080	LF	.81	2.86	—	3.67
1"	P1@.100	LF	1.19	3.58	—	4.77
1¼"	P1@.110	LF	1.55	3.93	—	5.48
1½"	P1@.120	LF	1.87	4.29	—	6.16
2"	P1@.140	LF	2.51	5.01	—	7.52
2½"	P1@.170	LF	4.69	6.08	—	10.77
3"	P1@.190	LF	6.16	6.79	—	12.95
4"	P1@.300	LF	8.77	10.70	—	19.47
6"	ER@.420	LF	15.10	16.00	6.60	37.70
8"	ER@.510	LF	19.70	19.40	8.10	47.20
10"	ER@.600	LF	28.10	22.90	9.40	60.40
12"	ER@.760	LF	33.90	29.00	11.90	74.80

Schedule 40 carbon steel 45 degree ell, butt-welded joints

Description	Craft@Hrs	Unit	Material $	Labor $	Equipment $	Total $
1/2"	P1@.220	Ea	2.43	7.87	—	10.30
3/4"	P1@.330	Ea	2.43	11.80	—	14.23
1"	P1@.440	Ea	2.43	15.70	—	18.13
1¼"	P1@.560	Ea	2.56	20.00	—	22.56
1½"	P1@.710	Ea	2.56	25.40	—	27.96
2"	P1@.890	Ea	2.64	31.80	—	34.44
2½"	P1@1.11	Ea	3.82	39.70	—	43.52
3"	P1@1.33	Ea	4.62	47.60	—	52.22
4"	P1@1.78	Ea	6.80	63.70	—	70.50
6"	ER@2.67	Ea	17.50	102.00	42.10	161.60
8"	ER@3.20	Ea	41.40	122.00	50.30	213.70
10"	ER@4.00	Ea	63.60	152.00	62.90	278.50
12"	ER@4.80	Ea	92.40	183.00	75.40	350.80

Schedule 40 carbon steel 90 degree ell, butt-welded joints

Description	Craft@Hrs	Unit	Material $	Labor $	Equipment $	Total $
1/2"	P1@.220	Ea	2.37	7.87	—	10.24
3/4"	P1@.330	Ea	2.37	11.80	—	14.17
1"	P1@.440	Ea	2.37	15.70	—	18.07
1¼"	P1@.560	Ea	2.51	20.00	—	22.51
1½"	P1@.670	Ea	2.51	24.00	—	26.51
2"	P1@.890	Ea	2.76	31.80	—	34.56
2½"	P1@1.11	Ea	4.38	39.70	—	44.08
3"	P1@1.33	Ea	5.87	47.60	—	53.47
4"	P1@1.78	Ea	9.45	63.70	—	73.15
6"	ER@2.67	Ea	26.70	102.00	42.10	170.80
8"	ER@3.20	Ea	61.00	122.00	50.30	233.30
10"	ER@4.00	Ea	109.00	152.00	62.90	323.90
12"	ER@4.80	Ea	162.00	183.00	75.40	420.40

Carbon Steel, Schedule 40 with 150# Fittings & Butt-Welded Joints

Description	Craft@Hrs	Unit	Material $	Labor $	Equipment $	Total $

Schedule 40 carbon steel tee, butt-welded joints

Description	Craft@Hrs	Unit	Material $	Labor $	Equipment $	Total $
1/2"	P1@.330	Ea	7.49	11.80	—	19.29
3/4"	P1@.500	Ea	7.49	17.90	—	25.39
1"	P1@.670	Ea	7.49	24.00	—	31.49
1¼"	P1@.830	Ea	8.33	29.70	—	38.03
1½"	P1@1.00	Ea	8.75	35.80	—	44.55
2"	P1@1.33	Ea	9.59	47.60	—	57.19
2½"	P1@1.67	Ea	13.40	59.70	—	73.10
3"	P1@2.00	Ea	14.90	71.50	—	86.40
4"	P1@2.67	Ea	20.40	95.50	—	115.90
6"	ER@4.00	Ea	37.30	152.00	62.90	252.20
8"	ER@4.80	Ea	86.80	183.00	75.40	345.20
10"	ER@6.00	Ea	149.00	229.00	94.30	472.30
12"	ER@7.20	Ea	224.00	274.00	113.00	611.00

Schedule 40 carbon steel reducing tee, butt-welded joints

Description	Craft@Hrs	Unit	Material $	Labor $	Equipment $	Total $
1½" x 1/2"	P1@.940	Ea	11.40	33.60	—	45.00
1½" x 3/4"	P1@.940	Ea	11.40	33.60	—	45.00
1½" x 1"	P1@.940	Ea	11.40	33.60	—	45.00
1½" x 1¼"	P1@.940	Ea	11.40	33.60	—	45.00
2" x 3/4"	P1@1.22	Ea	15.20	43.60	—	58.80
2" x 1"	P1@1.22	Ea	15.20	43.60	—	58.80
2" x 1¼"	P1@1.22	Ea	15.20	43.60	—	58.80
2" x 1½"	P1@1.22	Ea	15.20	43.60	—	58.80
2½" x 1"	P1@1.56	Ea	19.00	55.80	—	74.80
2½" x 1¼"	P1@1.56	Ea	19.00	55.80	—	74.80
2½" x 1½"	P1@1.56	Ea	19.00	55.80	—	74.80
2½" x 2"	P1@1.56	Ea	19.00	55.80	—	74.80
3" x 1¼"	P1@1.89	Ea	19.40	67.60	—	87.00
3" x 2"	P1@1.89	Ea	19.40	67.60	—	87.00
3" x 2½"	P1@1.89	Ea	19.40	67.60	—	87.00
4" x 1½"	P1@2.44	Ea	20.80	87.30	—	108.10
4" x 2"	P1@2.44	Ea	20.80	87.30	—	108.10
4" x 2½"	P1@2.44	Ea	20.80	87.30	—	108.10
4" x 3"	P1@2.44	Ea	20.80	87.30	—	108.10
6" x 2½"	ER@3.78	Ea	46.90	144.00	59.40	250.30
6" x 3"	ER@3.78	Ea	46.90	144.00	59.40	250.30
6" x 4"	ER@3.78	Ea	46.90	144.00	59.40	250.30
8" x 3"	ER@4.40	Ea	116.00	168.00	69.20	353.20
8" x 4"	ER@4.40	Ea	116.00	168.00	69.20	353.20
8" x 6"	ER@4.40	Ea	116.00	168.00	69.20	353.20
10" x 4"	ER@5.60	Ea	195.00	213.00	88.00	496.00
10" x 6"	ER@5.60	Ea	195.00	213.00	88.00	496.00
10" x 8"	ER@5.60	Ea	195.00	213.00	88.00	496.00
12" x 6"	ER@6.80	Ea	291.00	259.00	106.90	656.90
12" x 8"	ER@6.80	Ea	291.00	259.00	106.90	656.90
12" x 10"	ER@6.80	Ea	291.00	259.00	106.90	656.90

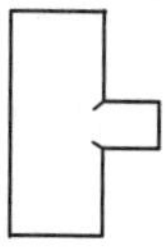

Description	Craft@Hrs	Unit	Material $	Labor $	Equipment $	Total $

Schedule 40 carbon steel concentric reducer, butt-welded joints

Description	Craft@Hrs	Unit	Material $	Labor $	Equipment $	Total $
3/4"	P1@.295	Ea	5.08	10.50	—	15.58
1"	P1@.395	Ea	6.06	14.10	—	20.16
1¼"	P1@.500	Ea	6.41	17.90	—	24.31
1½"	P1@.600	Ea	7.13	21.50	—	28.63
2"	P1@.800	Ea	6.65	28.60	—	35.25
2½"	P1@1.00	Ea	10.50	35.80	—	46.30
3"	P1@1.20	Ea	8.39	42.90	—	51.29
4"	P1@1.60	Ea	11.90	57.20	—	69.10
6"	ER@2.40	Ea	22.00	91.50	37.70	151.20
8"	ER@2.88	Ea	42.10	110.00	45.30	197.40
10"	ER@3.60	Ea	64.30	137.00	56.70	258.00
12"	ER@4.30	Ea	98.80	164.00	67.70	330.50

Schedule 40 carbon steel eccentric reducer, butt-welded joints

Description	Craft@Hrs	Unit	Material $	Labor $	Equipment $	Total $
3/4"	P1@.295	Ea	6.23	10.50	—	16.73
1"	P1@.395	Ea	8.40	14.10	—	22.50
1¼"	P1@.500	Ea	9.73	17.90	—	27.63
1½"	P1@.600	Ea	10.30	21.50	—	31.80
2"	P1@.800	Ea	7.94	28.60	—	36.54
2½"	P1@1.00	Ea	12.70	35.80	—	48.50
3"	P1@1.20	Ea	11.30	42.90	—	54.20
4"	P1@1.60	Ea	16.00	57.20	—	73.20
6"	ER@2.40	Ea	31.90	91.50	37.70	161.10
8"	ER@2.88	Ea	58.70	110.00	45.30	214.00
10"	ER@3.60	Ea	101.00	137.00	56.70	294.70
12"	ER@4.30	Ea	153.00	164.00	67.70	384.70

Schedule 40 carbon steel cap, butt-welded joints

Description	Craft@Hrs	Unit	Material $	Labor $	Equipment $	Total $
1/2"	P1@.150	Ea	5.74	5.36	—	11.10
3/4"	P1@.230	Ea	6.15	8.22	—	14.37
1"	P1@.300	Ea	6.56	10.70	—	17.26
1¼"	P1@.390	Ea	8.20	13.90	—	22.10
1½"	P1@.470	Ea	8.20	16.80	—	25.00
2"	P1@.620	Ea	8.20	22.20	—	30.40
2½"	P1@.770	Ea	10.20	27.50	—	37.70
3"	P1@.930	Ea	9.69	33.30	—	42.99
4"	P1@1.25	Ea	12.50	44.70	—	57.20
6"	ER@1.87	Ea	14.50	71.30	29.40	115.20
8"	ER@2.24	Ea	26.10	85.40	35.20	146.70
10"	ER@2.80	Ea	44.10	107.00	44.00	195.10
12"	ER@3.36	Ea	55.20	128.00	52.80	236.00

Carbon Steel, Schedule 40 with 150# Fittings & Butt-Welded Joints

Description	Craft@Hrs	Unit	Material $	Labor $	Equipment $	Total $

Schedule 40 carbon steel union, butt-welded joints

Description	Craft@Hrs	Unit	Material $	Labor $	Equipment $	Total $
1/2"	P1@.220	Ea	4.13	7.87	—	12.00
3/4"	P1@.330	Ea	5.35	11.80	—	17.15
1"	P1@.440	Ea	7.39	15.70	—	23.09
1¼"	P1@.560	Ea	15.60	20.00	—	35.60
1½"	P1@.670	Ea	17.10	24.00	—	41.10
2"	P1@.890	Ea	22.90	31.80	—	54.70
2½"	P1@1.11	Ea	61.00	39.70	—	100.70
3"	P1@1.33	Ea	72.50	47.60	—	120.10

Carbon steel weldolet, Schedule 40

Description	Craft@Hrs	Unit	Material $	Labor $	Equipment $	Total $
1/2"	P1@.330	Ea	5.76	11.80	—	17.56
3/4"	P1@.500	Ea	6.35	17.90	—	24.25
1"	P1@.670	Ea	7.09	24.00	—	31.09
1¼"	P1@.830	Ea	10.80	29.70	—	40.50
1½"	P1@1.00	Ea	10.80	35.80	—	46.60
2"	P1@1.33	Ea	11.80	47.60	—	59.40
2½"	P1@1.67	Ea	32.10	59.70	—	91.80
3"	P1@2.00	Ea	34.70	71.50	—	106.20
4"	P1@2.67	Ea	45.00	95.50	—	110.50
6"	P1@4.00	Ea	184.00	143.00	—	327.00
8"	P1@4.80	Ea	350.00	172.00	—	522.00
10"	P1@6.00	Ea	479.00	215.00	—	694.00
12"	P1@7.20	Ea	587.00	257.00	—	844.00

Carbon steel threadolet, Schedule 40

Description	Craft@Hrs	Unit	Material $	Labor $	Equipment $	Total $
3/4"	P1@.330	Ea	3.83	11.80	—	15.63
1"	P1@.440	Ea	4.54	15.70	—	20.24
1¼"	P1@.560	Ea	7.86	20.00	—	27.86
1½"	P1@.670	Ea	8.86	24.00	—	32.86
2"	P1@.890	Ea	10.30	31.80	—	42.10
2½"	P1@1.11	Ea	39.50	39.70	—	79.20
3"	P1@1.33	Ea	43.30	47.60	—	90.90
4"	P1@1.78	Ea	85.00	63.70	—	148.70

150# forged steel companion flange, weld neck

Description	Craft@Hrs	Unit	Material $	Labor $	Equipment $	Total $
2½"	P1@.610	Ea	22.80	21.80	—	44.60
3"	P1@.730	Ea	21.20	26.10	—	47.30
4"	P1@.980	Ea	29.30	35.00	—	64.30
6"	ER@1.47	Ea	47.00	56.00	23.10	126.10
8"	ER@1.77	Ea	95.00	67.50	27.80	190.30
10"	ER@2.20	Ea	157.00	83.80	34.70	275.50
12"	ER@2.64	Ea	228.00	101.00	41.50	370.50

Carbon Steel, Schedule 40 with 150# Fittings & Butt-Welded Joints

Description	Craft@Hrs	Unit	Material $	Labor $	Equipment $	Total $

Class 125 bronze body gate valve, threaded ends

1/2"	P1@.210	Ea	15.40	7.51	—	22.91
3/4"	P1@.250	Ea	19.30	8.94	—	28.24
1"	P1@.300	Ea	27.10	10.70	—	37.80
1¼"	P1@.400	Ea	34.30	14.30	—	48.60
1½"	P1@.450	Ea	46.10	16.10	—	62.20
2"	P1@.500	Ea	77.00	17.90	—	94.90
2½"	P1@.750	Ea	128.00	26.80	—	154.80
3"	P1@.950	Ea	184.00	34.00	—	218.00

Class 125 iron body gate valve, flanged ends

2"	P1@.500	Ea	295.00	17.90	—	312.90
2½"	P1@.600	Ea	398.00	21.50	—	419.50
3"	P1@.750	Ea	432.00	26.80	—	458.80
4"	P1@1.35	Ea	634.00	48.30	—	682.30
6"	ER@2.50	Ea	1,210.00	95.30	39.30	1,344.60
8"	ER@3.00	Ea	1,990.00	114.00	47.10	2,151.10
10"	ER@4.00	Ea	3,250.00	152.00	62.90	3,464.90
12"	ER@4.50	Ea	4,500.00	171.00	70.80	4,741.80

Class 125 bronze body globe valve, threaded ends

1/2"	P1@.210	Ea	28.60	7.51	—	36.11
3/4"	P1@.250	Ea	38.10	8.94	—	47.04
1"	P1@.300	Ea	54.70	10.70	—	65.40
1¼"	P1@.400	Ea	76.90	14.30	—	91.20
1½"	P1@.450	Ea	104.00	16.10	—	120.10
2"	P1@.500	Ea	168.00	17.90	—	185.90

Class 125 iron body globe valve, flanged ends

2½"	P1@.600	Ea	334.00	21.50	—	355.50
3"	P1@.750	Ea	401.00	26.80	—	427.80
4"	P1@1.35	Ea	538.00	48.30	—	586.30
6"	ER@2.50	Ea	943.00	95.30	—	1,038.30
8"	ER@3.00	Ea	1,450.00	114.00	—	1,564.00
10"	ER@4.00	Ea	2,210.00	152.00	—	2,362.00

200 PSIG iron body butterfly valve, lug type, lever operated

2"	P1@.450	Ea	128.00	16.10	—	144.10
2½"	P1@.450	Ea	132.00	16.10	—	148.10
3"	P1@.550	Ea	139.00	19.70	—	158.70
4"	P1@.550	Ea	173.00	19.70	—	192.70
6"	ER@.800	Ea	279.00	30.50	12.70	322.20
8"	ER@.800	Ea	382.00	30.50	12.70	425.20
10"	ER@.900	Ea	530.00	34.30	14.10	578.40
12"	ER@1.00	Ea	694.00	38.10	15.70	747.80

Carbon Steel, Schedule 40 with 150# Fittings & Butt-Welded Joints

Description	Craft@Hrs	Unit	Material $	Labor $	Equipment $	Total $

200 PSIG iron body butterfly valve, wafer type, lever operated

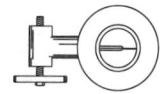

Description	Craft@Hrs	Unit	Material $	Labor $	Equipment $	Total $
2"	P1@.450	Ea	116.00	16.10	—	132.10
2½"	P1@.450	Ea	119.00	16.10	—	135.10
3"	P1@.550	Ea	128.00	19.70	—	147.70
4"	P1@.550	Ea	155.00	19.70	—	174.70
6"	ER@.800	Ea	259.00	30.50	12.70	302.20
8"	ER@.800	Ea	359.00	30.50	12.70	402.20
10"	ER@.900	Ea	502.00	34.30	14.10	550.40
12"	ER@1.00	Ea	659.00	38.10	15.70	712.80

Class 125 bronze body swing check valve, threaded

Description	Craft@Hrs	Unit	Material $	Labor $	Equipment $	Total $
1/2"	P1@.210	Ea	16.90	7.51	—	24.41
3/4"	P1@.250	Ea	24.20	8.94	—	33.14
1"	P1@.300	Ea	31.70	10.70	—	42.40
1¼"	P1@.400	Ea	45.00	14.30	—	59.30
1½"	P1@.450	Ea	63.50	16.10	—	79.60
2"	P1@.500	Ea	106.00	17.90	—	123.90

Class 125 iron body swing check valve, flanged joints

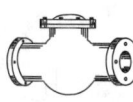

Description	Craft@Hrs	Unit	Material $	Labor $	Equipment $	Total $
2"	P1@.500	Ea	148.00	17.90	—	165.90
2½"	P1@.600	Ea	188.00	21.50	—	209.50
3"	P1@.750	Ea	233.00	26.80	—	259.80
4"	P1@1.35	Ea	343.00	48.30	—	391.30
6"	ER@2.50	Ea	666.00	95.30	39.30	800.60
8"	ER@3.00	Ea	1,190.00	114.00	47.10	1,351.10
10"	ER@4.00	Ea	1,960.00	152.00	62.90	2,174.90
12"	ER@4.50	Ea	2,660.00	171.00	70.80	2,901.80

Class 125 iron body silent check valve, flanged joints

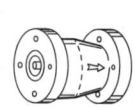

Description	Craft@Hrs	Unit	Material $	Labor $	Equipment $	Total $
2"	P1@.500	Ea	108.00	17.90	—	125.90
2½"	P1@.600	Ea	125.00	21.50	—	146.50
3"	P1@.750	Ea	140.00	26.80	—	166.80
4"	P1@1.35	Ea	184.00	48.30	—	232.30
6"	ER@2.50	Ea	330.00	95.30	39.30	464.60
8"	ER@3.00	Ea	595.00	114.00	47.10	756.10
10"	ER@4.00	Ea	924.00	152.00	62.90	1,138.90

Class 125 bronze body strainer, threaded ends

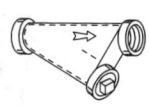

Description	Craft@Hrs	Unit	Material $	Labor $	Equipment $	Total $
1/2"	P1@.210	Ea	26.40	7.51	—	33.91
3/4"	P1@.250	Ea	34.70	8.94	—	43.64
1"	P1@.300	Ea	42.50	10.70	—	53.20
1¼"	P1@.400	Ea	59.30	14.30	—	73.60
1½"	P1@.450	Ea	89.30	16.10	—	105.40
2"	P1@.500	Ea	154.00	17.90	—	171.90

Carbon Steel, Schedule 40 with 150# Fittings & Butt-Welded Joints

Description	Craft@Hrs	Unit	Material $	Labor $	Equipment $	Total $

Class 125 iron body strainer, flanged

Description	Craft@Hrs	Unit	Material $	Labor $	Equipment $	Total $
2"	P1@.500	Ea	118.00	17.90	—	135.90
2½"	P1@.600	Ea	132.00	21.50	—	153.50
3"	P1@.750	Ea	152.00	26.80	—	178.80
4"	P1@1.35	Ea	260.00	48.30	—	308.30
6"	ER@2.50	Ea	528.00	95.30	39.30	662.60
8"	ER@3.00	Ea	892.00	114.00	47.10	1,053.10

Installation of 2-way control valve, threaded joints

Description	Craft@Hrs	Unit	Material $	Labor $	Equipment $	Total $
1/2"	P1@.210	Ea	—	7.51	—	7.51
3/4"	P1@.250	Ea	—	8.94	—	8.94
1"	P1@.300	Ea	—	10.70	—	10.70
1¼"	P1@.400	Ea	—	14.30	—	14.30
1½"	P1@.450	Ea	—	16.10	—	16.10
2"	P1@.500	Ea	—	17.90	—	17.90
2½"	P1@.830	Ea	—	29.70	—	29.70
3"	P1@.990	Ea	—	35.40	—	35.40

Installation of 2-way control valve, flanged joints

Description	Craft@Hrs	Unit	Material $	Labor $	Equipment $	Total $
2"	P1@.500	Ea	—	17.90	—	17.90
2½"	P1@.600	Ea	—	21.50	—	21.50
3"	P1@.750	Ea	—	26.80	—	26.80
4"	P1@1.35	Ea	—	48.30	—	48.30
6"	P1@2.50	Ea	—	89.40	—	89.40
8"	P1@3.00	Ea	—	107.00	—	107.00
10"	P1@4.00	Ea	—	143.00	—	143.00
12"	P1@4.50	Ea	—	161.00	—	161.00

Installation of 3-way control valve, threaded joints

Description	Craft@Hrs	Unit	Material $	Labor $	Equipment $	Total $
1/2"	P1@.260	Ea	—	9.30	—	9.30
3/4"	P1@.365	Ea	—	13.10	—	13.10
1"	P1@.475	Ea	—	17.00	—	17.00
1¼"	P1@.575	Ea	—	20.60	—	20.60
1½"	P1@.680	Ea	—	24.30	—	24.30
2"	P1@.910	Ea	—	32.50	—	32.50
2½"	P1@1.12	Ea	—	40.10	—	40.10
3"	P1@1.33	Ea	—	47.60	—	47.60

Installation of 3-way control valve, flanged joints

Description	Craft@Hrs	Unit	Material $	Labor $	Equipment $	Total $
2"	P1@.910	Ea	—	32.50	—	32.50
2½"	P1@1.12	Ea	—	40.10	—	40.10
3"	P1@1.33	Ea	—	47.60	—	47.60
4"	P1@2.00	Ea	—	71.50	—	71.50
6"	P1@3.70	Ea	—	132.00	—	132.00
8"	P1@4.40	Ea	—	157.00	—	157.00
10"	P1@5.90	Ea	—	211.00	—	211.00
12"	P1@6.50	Ea	—	232.00	—	232.00

Carbon Steel, Schedule 40 with 150# Fittings & Butt-Welded Joints

Description	Craft@Hrs	Unit	Material $	Labor $	Equipment $	Total $
Bolt and gasket sets						
2"	P1@.500	Ea	3.86	17.90	—	21.76
2½"	P1@.650	Ea	4.51	23.20	—	27.71
3"	P1@.750	Ea	7.64	26.80	—	34.44
4"	P1@1.00	Ea	12.70	35.80	—	48.50
6"	P1@1.20	Ea	21.20	42.90	—	64.10
8"	P1@1.25	Ea	23.80	44.70	—	68.50
10"	P1@1.70	Ea	40.90	60.80	—	101.70
12"	P1@2.20	Ea	46.20	78.70	—	124.90
Thermometer with well						
7"	P1@.250	Ea	164.00	8.94	—	172.94
9"	P1@.250	Ea	217.00	8.94	—	225.94
Dial type pressure gauge						
2½"	P1@.200	Ea	33.10	7.15	—	40.25
3½"	P1@.200	Ea	43.80	7.15	—	50.95
Pressure/temperature tap						
tap	P1@.150	Ea	15.70	5.36	—	21.06
Hanger with swivel assembly						
2"	P1@.300	Ea	6.05	10.70	—	16.75
2½"	P1@.350	Ea	7.82	12.50	—	20.32
3"	P1@.350	Ea	9.67	12.50	—	22.17
4"	P1@.350	Ea	14.90	12.50	—	27.40
5"	P1@.450	Ea	16.70	16.10	—	32.80
6"	P1@.450	Ea	21.80	16.10	—	37.90
8"	P1@.450	Ea	28.90	16.10	—	45.00
10"	P1@.550	Ea	39.70	19.70	—	59.40
12"	P1@.600	Ea	52.80	21.50	—	74.30
Riser clamp						
2"	P1@.115	Ea	4.84	4.11	—	8.95
2½"	P1@.120	Ea	5.09	4.29	—	9.38
3"	P1@.120	Ea	5.52	4.29	—	9.81
4"	P1@.125	Ea	7.03	4.47	—	11.50
5"	P1@.180	Ea	10.10	6.44	—	16.54
6"	P1@.200	Ea	12.20	7.15	—	19.35
8"	P1@.200	Ea	19.80	7.15	—	26.95
10"	P1@.250	Ea	29.40	8.94	—	38.34
12"	P1@.250	Ea	34.90	8.94	—	43.84

Carbon Steel, Schedule 40 with 150# M.I. Fittings & Threaded Joints

Schedule 40 carbon steel (ASTM A-53 and A-120) pipe with 150 pound malleable iron threaded fittings is commonly used for heating hot water, chilled water, glycol and low pressure steam systems where operating pressures do not exceed 125 PSIG and temperatures are below 250 degrees F., for pipe sizes 2" and smaller. For pipe sizes 2 ½" and larger, wrought steel fittings with either welded or grooved joints should be used.

Schedule 40 (ASTM A-53B SML) steel pipe with carbon steel welding fittings can also be used for refrigerant piping systems.

This section has been arranged to save the estimator's time by including all normally-used system components such as pipe, fittings, valves, hanger assemblies, riser clamps and miscellaneous items under one heading. Additional items can be found under "Plumbing and Piping Specialties." The cost estimates in this section are based on the conditions, limitations and wage rates described in the section "How to Use This Book" beginning on page 5.

Description	Craft@Hrs	Unit	Material $	Labor $	Equipment $	Total $

Schedule 40 carbon steel threaded horizontal pipe assembly.

Horizontally hung in a building. Assembly includes fittings and hanger assemblies. Based on a reducing tee and a 45 degree elbow every 16 feet for 1/2" pipe. A reducing tee and a 45 degree elbow every 30 feet for 4" pipe. Hangers spaced to meet plumbing code. *(Distance between fittings in the pipe assembly increases as pipe diameter increases.)* Use these figures for preliminary estimates.

Description	Craft@Hrs	Unit	Material $	Labor $	Equipment $	Total $
1/2"	P1@.119	LF	1.41	4.26	—	5.67
3/4"	P1@.132	LF	1.66	4.72	—	6.38
1"	P1@.144	LF	2.27	5.15	—	7.42
1¼"	P1@.160	LF	2.96	5.72	—	8.68
1½"	P1@.184	LF	3.29	6.58	—	9.87
2"	P1@.200	LF	4.99	7.15	—	12.14
2½"	P1@.236	LF	9.83	8.44	—	18.27
3"	P1@.271	LF	13.50	9.69	—	23.19
4"	P1@.398	LF	20.10	14.20	—	34.30

Schedule 40 carbon steel threaded vertical pipe assembly.
Riser assembly, including fittings and riser clamps. Based on a reducing tee, and a riser clamp on every floor. Use these figures for preliminary estimates.

Description	Craft@Hrs	Unit	Material $	Labor $	Equipment $	Total $
1/2"	P1@.082	LF	1.08	2.93	—	4.01
3/4"	P1@.096	LF	1.37	3.43	—	4.80
1"	P1@.121	LF	1.92	4.33	—	6.25
1¼"	P1@.138	LF	2.76	4.93	—	7.69
1½"	P1@.156	LF	3.13	5.58	—	8.71
2"	P1@.173	LF	4.69	6.19	—	10.88
2½"	P1@.200	LF	9.65	7.15	—	16.80
3"	P1@.236	LF	14.10	8.44	—	22.54
4"	P1@.354	LF	18.60	12.70	—	31.30

Carbon Steel, Schedule 40 with 150# M.I. Fittings & Threaded Joints

Description	Craft@Hrs	Unit	Material $	Labor $	Equipment $	Total $

Schedule 40 carbon steel pipe, threaded and coupled, pipe only

Description	Craft@Hrs	Unit	Material $	Labor $	Equipment $	Total $
1/2"	P1@.060	LF	.97	2.15	—	3.12
3/4"	P1@.070	LF	1.08	2.50	—	3.58
1"	P1@.090	LF	1.65	3.22	—	4.87
1¼"	P1@.100	LF	2.13	3.58	—	5.71
1½"	P1@.110	LF	2.53	3.93	—	6.46
2"	P1@.120	LF	3.75	4.29	—	8.04
2½"	P1@.130	LF	5.78	4.65	—	10.43
3"	P1@.150	LF	7.67	5.36	—	13.03
4"	P1@.260	LF	10.90	9.30	—	20.20

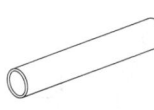

150# malleable iron 45 degree ell, threaded joints

Description	Craft@Hrs	Unit	Material $	Labor $	Equipment $	Total $
1/2"	P1@.120	Ea	1.52	4.29	—	5.81
3/4"	P1@.130	Ea	1.87	4.65	—	6.52
1"	P1@.180	Ea	2.38	6.44	—	8.82
1¼"	P1@.240	Ea	4.20	8.58	—	12.78
1½"	P1@.300	Ea	5.18	10.70	—	15.88
2"	P1@.380	Ea	7.75	13.60	—	21.35
2½"	P1@.510	Ea	22.40	18.20	—	40.60
3"	P1@.640	Ea	29.10	22.90	—	52.00
4"	P1@.850	Ea	56.70	30.40	—	87.10

150# malleable iron 90 degree ell, threaded joints

Description	Craft@Hrs	Unit	Material $	Labor $	Equipment $	Total $
1/2"	P1@.120	Ea	.93	4.29	—	5.22
3/4"	P1@.130	Ea	1.16	4.65	—	5.81
1"	P1@.180	Ea	1.98	6.44	—	8.42
1¼"	P1@.240	Ea	3.24	8.58	—	11.82
1½"	P1@.300	Ea	4.28	10.70	—	14.98
2"	P1@.380	Ea	7.30	13.60	—	20.90
2½"	P1@.510	Ea	16.00	18.20	—	34.20
3"	P1@.640	Ea	23.80	22.90	—	46.70
4"	P1@.850	Ea	55.30	30.40	—	85.70

150# malleable iron tee, threaded joints

Description	Craft@Hrs	Unit	Material $	Labor $	Equipment $	Total $
1/2"	P1@.180	Ea	1.26	6.44	—	7.70
3/4"	P1@.190	Ea	1.82	6.79	—	8.61
1"	P1@.230	Ea	3.09	8.22	—	11.31
1¼"	P1@.310	Ea	5.07	11.10	—	16.17
1½"	P1@.390	Ea	6.22	13.90	—	20.12
2"	P1@.490	Ea	10.50	17.50	—	28.00
2½"	P1@.660	Ea	22.30	23.60	—	45.90
3"	P1@.830	Ea	32.60	29.70	—	62.30
4"	P1@1.10	Ea	78.20	39.30	—	117.50

Carbon Steel, Schedule 40 with 150# M.I. Fittings & Threaded Joints

Description	Craft@Hrs	Unit	Material $	Labor $	Equipment $	Total $

150# malleable iron reducing tee, threaded joints

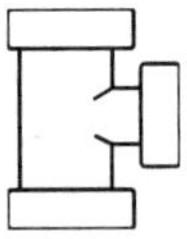

Description	Craft@Hrs	Unit	Material $	Labor $	Equipment $	Total $
3/4" x 1/2"	P1@.180	Ea	2.69	6.44	—	9.13
1" x 1/2"	P1@.220	Ea	3.42	7.87	—	11.29
1" x 3/4"	P1@.220	Ea	3.42	7.87	—	11.29
1¼" x 1/2"	P1@.290	Ea	6.22	10.40	—	16.62
1¼" x 3/4"	P1@.290	Ea	5.46	10.40	—	15.86
1¼" x 1"	P1@.290	Ea	5.91	10.40	—	16.31
1½" x 1/2"	P1@.370	Ea	7.29	13.20	—	20.49
1½" x 3/4"	P1@.370	Ea	6.88	13.20	—	20.08
1½" x 1"	P1@.370	Ea	6.88	13.20	—	20.08
1½" x 1¼"	P1@.370	Ea	8.51	13.20	—	21.71
2" x 1"	P1@.460	Ea	10.80	16.40	—	27.20
2" x 1¼"	P1@.460	Ea	11.70	16.40	—	28.10
2" x 1½"	P1@.460	Ea	11.70	16.40	—	28.10
4" x 1½"	P1@1.05	Ea	80.90	37.50	—	118.40
4" x 2"	P1@1.05	Ea	80.90	37.50	—	118.40
4" x 3"	P1@1.05	Ea	80.90	37.50	—	118.40

150# malleable iron reducer, threaded joints

Description	Craft@Hrs	Unit	Material $	Labor $	Equipment $	Total $
3/4" x 1/2"	P1@.130	Ea	1.71	4.65	—	6.36
1" x 1/2"	P1@.160	Ea	2.66	5.72	—	8.38
1" x 3/4"	P1@.160	Ea	2.66	5.72	—	8.38
1¼" x 3/4"	P1@.210	Ea	2.66	7.51	—	10.17
1¼" x 1"	P1@.210	Ea	3.31	7.51	—	10.82
1½" x 1"	P1@.270	Ea	4.85	9.66	—	14.51
1½" x 1¼"	P1@.270	Ea	4.20	9.66	—	13.86
2" x 1¼"	P1@.340	Ea	6.94	12.20	—	19.14
2" x 1½"	P1@.340	Ea	6.09	12.20	—	18.29
3" x 2"	P1@.530	Ea	18.10	19.00	—	37.10
3" x 2½"	P1@.530	Ea	21.00	19.00	—	40.00
4" x 3"	P1@.750	Ea	21.00	26.80	—	47.80

150# malleable iron cross, threaded joints

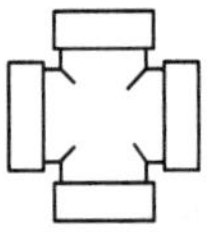

Description	Craft@Hrs	Unit	Material $	Labor $	Equipment $	Total $
1/2"	P1@.280	Ea	4.58	10.00	—	14.58
3/4"	P1@.320	Ea	5.60	11.40	—	17.00
1"	P1@.360	Ea	6.88	12.90	—	19.78
1¼"	P1@.480	Ea	11.00	17.20	—	28.20
1½"	P1@.600	Ea	13.70	21.50	—	35.20
2"	P1@.700	Ea	22.70	25.00	—	47.70
2½"	P1@1.02	Ea	41.90	36.50	—	78.40
3"	P1@1.28	Ea	58.10	45.80	—	103.90

Carbon Steel, Schedule 40 with 150# M.I. Fittings & Threaded Joints

Description	Craft@Hrs	Unit	Material $	Labor $	Equipment $	Total $

150# malleable iron cap, threaded joint

Description	Craft@Hrs	Unit	Material $	Labor $	Equipment $	Total $
1/2"	P1@.090	Ea	.95	3.22	—	4.17
3/4"	P1@.100	Ea	1.27	3.58	—	4.85
1"	P1@.140	Ea	1.58	5.01	—	6.59
1¼"	P1@.180	Ea	2.03	6.44	—	8.47
1½"	P1@.230	Ea	2.76	8.22	—	10.98
2"	P1@.290	Ea	4.07	10.40	—	14.47
2½"	P1@.380	Ea	9.10	13.60	—	22.70
3"	P1@.480	Ea	13.40	17.20	—	30.60
4"	P1@.640	Ea	22.90	22.90	—	45.80

150# malleable iron plug, threaded joint

Description	Craft@Hrs	Unit	Material $	Labor $	Equipment $	Total $
1/2"	P1@.090	Ea	.75	3.22	—	3.97
3/4"	P1@.100	Ea	.81	3.58	—	4.39
1"	P1@.140	Ea	.81	5.01	—	5.82
1¼"	P1@.180	Ea	1.28	6.44	—	7.72
1½"	P1@.230	Ea	1.85	8.22	—	10.07
2"	P1@.290	Ea	2.14	10.40	—	12.54
2½"	P1@.380	Ea	5.21	13.60	—	18.81
3"	P1@.480	Ea	6.06	17.20	—	23.26
4"	P1@.040	Ea	11.00	22.90	—	33.90

150# malleable iron union, threaded joints

Description	Craft@Hrs	Unit	Material $	Labor $	Equipment $	Total $
1/2"	P1@.140	Ea	4.16	5.01	—	9.17
3/4"	P1@.150	Ea	4.77	5.36	—	10.13
1"	P1@.210	Ea	6.23	7.51	—	13.74
1¼"	P1@.280	Ea	8.91	10.00	—	18.91
1½"	P1@.360	Ea	11.00	12.90	—	23.90
2"	P1@.450	Ea	12.80	16.10	—	28.90
2½"	P1@.610	Ea	38.20	21.80	—	60.00
3"	P1@.760	Ea	46.00	27.20	—	73.20

150# malleable iron coupling, threaded joints

Description	Craft@Hrs	Unit	Material $	Labor $	Equipment $	Total $
1/2"	P1@.120	Ea	1.27	4.29	—	5.56
3/4"	P1@.130	Ea	1.50	4.65	—	6.15
1"	P1@.180	Ea	2.28	6.44	—	8.72
1¼"	P1@.240	Ea	2.92	8.58	—	11.50
1½"	P1@.300	Ea	3.91	10.70	—	14.61
2"	P1@.380	Ea	5.69	13.60	—	19.29

Carbon Steel, Schedule 40 with 150# M.I. Fittings & Threaded Joints

Description	Craft@Hrs	Unit	Material $	Labor $	Equipment $	Total $

Steel pipe nipples — Close nipple, standard right hand thread

Description	Craft@Hrs	Unit	Material $	Labor $	Equipment $	Total $
1/2"	P1@.025	Ea	.51	.89	—	1.40
3/4"	P1@.025	Ea	.61	.89	—	1.50
1"	P1@.025	Ea	.93	.89	—	1.82
1¼"	P1@.025	Ea	1.16	.89	—	2.05
1½"	P1@.025	Ea	1.39	.89	—	2.28
2"	P1@.025	Ea	1.77	.89	—	2.66
2½"	P1@.030	Ea	6.94	1.07	—	8.01
3"	P1@.030	Ea	7.51	1.07	—	8.58
4"	P1@.040	Ea	11.90	1.43	—	13.33
6"	P1@.050	Ea	60.60	1.79	—	62.39
8"	P1@.065	Ea	106.00	2.32	—	108.32

Steel pipe nipples — 1-1/2" long, standard right hand thread

Description	Craft@Hrs	Unit	Material $	Labor $	Equipment $	Total $
1/2"	P1@.025	Ea	.55	.89	—	1.44
3/4"	P1@.025	Ea	.62	.89	—	1.51

Steel pipe nipples — 2" long, standard right hand thread

Description	Craft@Hrs	Unit	Material $	Labor $	Equipment $	Total $
1/2"	P1@.025	Ea	.55	.89	—	1.44
3/4"	P1@.025	Ea	.62	.89	—	1.51
1"	P1@.025	Ea	.93	.89	—	1.82
1¼"	P1@.025	Ea	1.16	.89	—	2.05
1½"	P1@.025	Ea	1.39	.89	—	2.28

Steel pipe nipples — 2-1/2" long, standard right hand thread

Description	Craft@Hrs	Unit	Material $	Labor $	Equipment $	Total $
1/2"	P1@.025	Ea	.57	.89	—	1.46
3/4"	P1@.025	Ea	.66	.89	—	1.55
1"	P1@.025	Ea	1.07	.89	—	1.96
1¼"	P1@.025	Ea	1.37	.89	—	2.26
1½"	P1@.025	Ea	1.71	.89	—	2.60
2"	P1@.025	Ea	2.00	.89	—	2.89

Steel pipe nipples — 3" long, standard right hand thread

Description	Craft@Hrs	Unit	Material $	Labor $	Equipment $	Total $
1/2"	P1@.025	Ea	.65	.89	—	1.54
3/4"	P1@.025	Ea	.76	.89	—	1.65
1"	P1@.025	Ea	1.11	.89	—	2.00
1¼"	P1@.025	Ea	1.45	.89	—	2.34
1½"	P1@.025	Ea	1.72	.89	—	2.61
2"	P1@.030	Ea	2.14	1.07	—	3.21
2½"	P1@.030	Ea	7.22	1.07	—	8.29
3"	P1@.035	Ea	8.12	1.25	—	9.37

Carbon Steel, Schedule 40 with 150# M.I. Fittings & Threaded Joints

Description	Craft@Hrs	Unit	Material $	Labor $	Equipment $	Total $

Steel pipe nipples — 3-1/2" long, standard right hand thread

Description	Craft@Hrs	Unit	Material $	Labor $	Equipment $	Total $
1/2"	P1@.025	Ea	.74	.89	—	1.63
3/4"	P1@.025	Ea	.89	.89	—	1.78
1"	P1@.025	Ea	1.27	.89	—	2.16
1¼"	P1@.025	Ea	1.64	.89	—	2.53
1½"	P1@.030	Ea	2.06	1.07	—	3.13
2"	P1@.030	Ea	2.52	1.07	—	3.59
2½"	P1@.035	Ea	9.51	1.25	—	10.76
3"	P1@.035	Ea	11.00	1.25	—	12.25
4"	P1@.040	Ea	15.00	1.43	—	16.43

Steel pipe nipples — 4" long, standard right hand thread

Description	Craft@Hrs	Unit	Material $	Labor $	Equipment $	Total $
1/2"	P1@.025	Ea	.79	.89	—	1.68
3/4"	P1@.025	Ea	.93	.89	—	1.82
1"	P1@.025	Ea	1.35	.89	—	2.24
1¼"	P1@.030	Ea	1.68	1.07	—	2.75
1½"	P1@.030	Ea	2.10	1.07	—	3.17
2"	P1@.030	Ea	2.66	1.07	—	3.73
2½"	P1@.035	Ea	8.30	1.25	—	9.55
3"	P1@.040	Ea	9.31	1.43	—	10.74
4"	P1@.040	Ea	13.50	1.43	—	14.93

Steel pipe nipples — 4-1/2" long, standard right hand thread

Description	Craft@Hrs	Unit	Material $	Labor $	Equipment $	Total $
1/2"	P1@.025	Ea	.93	.89	—	1.82
3/4"	P1@.025	Ea	1.06	.89	—	1.95
1"	P1@.025	Ea	1.45	.89	—	2.34
1¼"	P1@.030	Ea	2.27	1.07	—	3.34
1½"	P1@.030	Ea	2.75	1.07	—	3.82
2"	P1@.030	Ea	2.95	1.07	—	4.02
2½"	P1@.035	Ea	10.60	1.25	—	11.85
3"	P1@.040	Ea	13.30	1.43	—	14.73
4"	P1@.050	Ea	16.90	1.79	—	18.69
6"	P1@.070	Ea	72.80	2.50	—	75.30

Steel pipe nipples — 5" long, standard right hand thread

Description	Craft@Hrs	Unit	Material $	Labor $	Equipment $	Total $
1/2"	P1@.025	Ea	.95	.89	—	1.84
3/4"	P1@.025	Ea	1.06	.89	—	1.95
1"	P1@.030	Ea	1.58	1.07	—	2.65
1¼"	P1@.030	Ea	2.04	1.07	—	3.11
1½"	P1@.030	Ea	2.45	1.07	—	3.52
2"	P1@.035	Ea	3.07	1.25	—	4.32
2½"	P1@.040	Ea	11.20	1.43	—	12.63
3"	P1@.045	Ea	11.70	1.61	—	13.31
4"	P1@.050	Ea	18.30	1.79	—	20.09
6"	P1@.075	Ea	73.90	2.68	—	76.58
8"	P1@.100	Ea	123.00	3.58	—	126.58

Carbon Steel, Schedule 40 with 150# M.I. Fittings & Threaded Joints

Description	Craft@Hrs	Unit	Material $	Labor $	Equipment $	Total $

Steel pipe nipples — 6" long, standard right hand thread

Description	Craft@Hrs	Unit	Material $	Labor $	Equipment $	Total $
1/2"	P1@.025	Ea	1.06	.89	—	1.95
3/4"	P1@.025	Ea	1.29	.89	—	2.18
1"	P1@.030	Ea	1.72	1.07	—	2.79
1¼"	P1@.030	Ea	2.15	1.07	—	3.22
1½"	P1@.030	Ea	2.60	1.07	—	3.67
2"	P1@.035	Ea	3.51	1.25	—	4.76
2½"	P1@.040	Ea	10.10	1.43	—	11.53
3"	P1@.045	Ea	11.20	1.61	—	12.81
4"	P1@.060	Ea	16.90	2.15	—	19.05
6"	P1@.085	Ea	82.60	3.04	—	85.64
8"	P1@.120	Ea	131.00	4.29	—	135.29

Steel pipe nipples — 8" long, standard right hand thread

Description	Craft@Hrs	Unit	Material $	Labor $	Equipment $	Total $
2½"	P1@.050	Ea	13.10	1.79	—	14.89
3"	P1@.055	Ea	16.20	1.97	—	18.17
4"	P1@.070	Ea	25.90	2.50	—	28.40
6"	P1@.100	Ea	86.70	3.58	—	90.28
8"	P1@.150	Ea	148.00	5.36	—	153.36

Steel pipe nipples — 10" long, standard right hand thread

Description	Craft@Hrs	Unit	Material $	Labor $	Equipment $	Total $
2½"	P1@.055	Ea	15.90	1.97	—	17.87
3"	P1@.060	Ea	19.20	2.15	—	21.35
4"	P1@.080	Ea	31.30	2.86	—	34.16
6"	P1@.115	Ea	96.40	4.11	—	100.51
8"	P1@.165	Ea	167.00	5.90	—	172.90

Steel pipe nipples — 12" long, standard right hand thread

Description	Craft@Hrs	Unit	Material $	Labor $	Equipment $	Total $
2½"	P1@.060	Ea	15.40	2.15	—	17.55
3"	P1@.065	Ea	19.00	2.32	—	21.32
4"	P1@.080	Ea	30.80	2.86	—	33.66
6"	P1@.130	Ea	110.00	4.65	—	114.65
8"	P1@.190	Ea	183.00	6.79	—	189.79

Class 125 bronze body gate valve, threaded ends

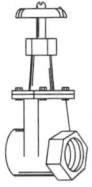

Description	Craft@Hrs	Unit	Material $	Labor $	Equipment $	Total $
1/2"	P1@.210	Ea	15.40	7.51	—	22.91
3/4"	P1@.250	Ea	19.30	8.94	—	28.24
1"	P1@.300	Ea	27.10	10.70	—	37.80
1¼"	P1@.400	Ea	34.30	14.30	—	48.60
1½"	P1@.450	Ea	46.10	16.10	—	62.20
2"	P1@.500	Ea	77.00	17.90	—	94.90
2½"	P1@.750	Ea	128.00	26.80	—	154.80
3"	P1@.950	Ea	183.00	34.00	—	217.00

Carbon Steel, Schedule 40 with 150# M.I. Fittings & Threaded Joints

Description	Craft@Hrs	Unit	Material $	Labor $	Equipment $	Total $

Class 125 iron body gate valve, flanged ends

2"	P1@.500	Ea	264.00	17.90	—	281.90
2½"	P1@.600	Ea	398.00	21.50	—	419.50
3"	P1@.750	Ea	432.00	26.80	—	458.80
4"	P1@1.35	Ea	634.00	48.30	—	682.30

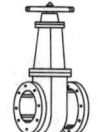

Class 125 bronze body globe valve, threaded ends

1/2"	P1@.210	Ea	28.60	7.51	—	36.11
3/4"	P1@.250	Ea	38.10	8.94	—	47.04
1"	P1@.300	Ea	54.70	10.70	—	65.40
1¼"	P1@.400	Ea	76.90	14.30	—	91.20
1½"	P1@.450	Ea	104.00	16.10	—	120.10
2"	P1@.500	Ea	168.00	17.90	—	185.90

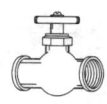

Class 125 iron body globe valve, flanged ends

2"	P1@.500	Ea	215.00	17.90	—	232.90
2½"	P1@.600	Ea	334.00	21.50	—	355.50
3"	P1@.750	Ea	401.00	26.80	—	427.80
4"	P1@1.35	Ea	538.00	48.30	—	586.30

200 PSIG iron body butterfly valve, lug type, lever operated

2"	P1@.450	Ea	128.00	16.10	—	144.10
2½"	P1@.450	Ea	132.00	16.10	—	148.10
3"	P1@.550	Ea	139.00	19.70	—	158.70
4"	P1@.550	Ea	173.00	19.70	—	192.70

200 PSIG iron body butterfly valve, wafer type, lever operated

2"	P1@.450	Ea	116.00	16.10	—	132.10
2½"	P1@.450	Ea	119.00	16.10	—	135.10
3"	P1@.550	Ea	128.00	19.70	—	147.70
4"	P1@.550	Ea	155.00	19.70	—	174.70

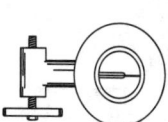

Class 125 bronze body 2-piece ball valve, threaded ends

1/2"	P1@.210	Ea	7.53	7.51	—	15.04
3/4"	P1@.250	Ea	10.00	8.94	—	18.94
1"	P1@.300	Ea	17.30	10.70	—	28.00
1¼"	P1@.400	Ea	28.90	14.30	—	43.20
1½"	P1@.450	Ea	39.00	16.10	—	55.10
2"	P1@.500	Ea	49.40	17.90	—	67.30
3"	P1@.625	Ea	338.00	22.40	—	360.40
4"	P1@.690	Ea	443.00	24.70	—	467.70

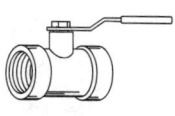

Carbon Steel, Schedule 40 with 150# M.I. Fittings & Threaded Joints

Description	Craft@Hrs	Unit	Material $	Labor $	Equipment $	Total $

Class 125 bronze body swing check valve, threaded

1/2"	P1@.210	Ea	16.90	7.51	—	24.41
3/4"	P1@.250	Ea	24.20	8.94	—	33.14
1"	P1@.300	Ea	31.70	10.70	—	42.40
1¼"	P1@.400	Ea	45.00	14.30	—	59.30
1½"	P1@.450	Ea	63.50	16.10	—	79.60
2"	P1@.500	Ea	106.00	17.90	—	123.90

Class 125 iron body swing check valve, flanged ends

2"	P1@.500	Ea	148.00	17.90	—	165.90
2½"	P1@.600	Ea	188.00	21.50	—	209.50
3"	P1@.750	Ea	233.00	26.80	—	259.80
4"	P1@1.35	Ea	343.00	48.30	—	391.30

Class 125 iron body silent check valve, wafer type

2"	P1@.500	Ea	108.00	17.90	—	125.90
2½"	P1@.600	Ea	120.00	21.50	—	141.50
3"	P1@.750	Ea	138.00	26.80	—	164.80
4"	P1@1.35	Ea	185.00	48.30	—	233.30

Class 125 bronze body strainer, threaded ends

1/2"	P1@.210	Ea	26.40	7.51	—	33.91
3/4"	P1@.250	Ea	34.70	8.94	—	43.64
1"	P1@.300	Ea	42.50	10.70	—	53.20
1¼"	P1@.400	Ea	59.30	14.30	—	73.60
1½"	P1@.450	Ea	89.30	16.10	—	105.40
2"	P1@.500	Ea	154.00	17.90	—	171.90

Class 125 iron body strainer, flanged ends

2"	P1@.500	Ea	118.00	17.90	—	135.90
2½"	P1@.600	Ea	132.00	21.50	—	153.50
3"	P1@.750	Ea	152.00	26.80	—	178.80
4"	P1@1.35	Ea	458.00	48.30	—	506.30

Installation of 2-way control valve, threaded joints

1/2"	P1@.210	Ea	—	7.51	—	7.51
3/4"	P1@.275	Ea	—	9.83	—	9.83
1"	P1@.350	Ea	—	12.50	—	12.50
1¼"	P1@.430	Ea	—	15.40	—	15.40
1½"	P1@.505	Ea	—	18.10	—	18.10
2"	P1@.675	Ea	—	24.10	—	24.10
2½"	P1@.830	Ea	—	29.70	—	29.70
3"	P1@.990	Ea	—	35.40	—	35.40

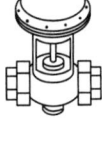

Carbon Steel, Schedule 40 with 150# M.I. Fittings & Threaded Joints

Description	Craft@Hrs	Unit	Material $	Labor $	Equipment $	Total $

Installation of 3-way control valve, threaded joints

Description	Craft@Hrs	Unit	Material $	Labor $	Equipment $	Total $
1/2"	P1@.260	Ea	—	9.30	—	9.30
3/4"	P1@.365	Ea	—	13.10	—	13.10
1"	P1@.475	Ea	—	17.00	—	17.00
1¼"	P1@.575	Ea	—	20.60	—	20.60
1½"	P1@.680	Ea	—	24.30	—	24.30
2"	P1@.910	Ea	—	32.50	—	32.50
2½"	P1@1.12	Ea	—	40.10	—	40.10
3"	P1@1.33	Ea	—	47.60	—	47.60

Companion flange, threaded

Description	Craft@Hrs	Unit	Material $	Labor $	Equipment $	Total $
2"	P1@.290	Ea	39.90	10.40	—	50.30
2½"	P1@.380	Ea	61.60	13.60	—	75.20
3"	P1@.460	Ea	66.50	16.40	—	82.90
4"	P1@.600	Ea	91.40	21.50	—	112.90

Bolt and gasket sets

Description	Craft@Hrs	Unit	Material $	Labor $	Equipment $	Total $
2"	P1@.500	Ea	3.86	17.90	—	21.76
2½"	P1@.650	Ea	4.51	23.20	—	27.71
3"	P1@.750	Ea	7.64	26.80	—	34.44
4"	P1@1.00	Ea	12.70	35.80	—	48.50

Thermometer with well

Description	Craft@Hrs	Unit	Material $	Labor $	Equipment $	Total $
7"	P1@.250	Ea	159.00	8.94	—	167.94
9"	P1@.250	Ea	164.00	8.94	—	172.94

Dial type pressure gauge

Description	Craft@Hrs	Unit	Material $	Labor $	Equipment $	Total $
2½"	P1@.200	Ea	33.10	7.15	—	40.25
3½"	P1@.200	Ea	43.80	7.15	—	50.95

Pressure/temperature tap

Description	Craft@Hrs	Unit	Material $	Labor $	Equipment $	Total $
tap	P1@.150	Ea	15.70	5.36	—	21.06

Hanger with swivel assembly

Description	Craft@Hrs	Unit	Material $	Labor $	Equipment $	Total $
1/2"	P1@.250	Ea	4.13	8.94	—	13.07
3/4"	P1@.250	Ea	4.33	8.94	—	13.27
1"	P1@.250	Ea	4.57	8.94	—	13.51
1¼"	P1@.300	Ea	4.77	10.70	—	15.47
1½"	P1@.300	Ea	5.51	10.70	—	16.21
2"	P1@.300	Ea	6.05	10.70	—	16.75
2½"	P1@.350	Ea	7.82	12.50	—	20.32
3"	P1@.350	Ea	9.67	12.50	—	22.17
4"	P1@.350	Ea	15.20	12.50	—	27.70

Carbon Steel, Schedule 40 with 150# M.I. Fittings & Threaded Joints

Description	Craft@Hrs	Unit	Material $	Labor $	Equipment $	Total $
Riser clamp						
1/2"	P1@.100	Ea	2.41	3.58	—	5.99
3/4"	P1@.100	Ea	3.55	3.58	—	7.13
1"	P1@.100	Ea	3.59	3.58	—	7.17
1¼"	P1@.105	Ea	4.33	3.75	—	8.08
1½"	P1@.110	Ea	4.58	3.93	—	8.51
2"	P1@.115	Ea	4.84	4.11	—	8.95
2½"	P1@.120	Ea	5.09	4.29	—	9.38
3"	P1@.120	Ea	5.52	4.29	—	9.81
4"	P1@.125	Ea	7.03	4.47	—	11.50

Schedule 5 carbon steel (ASTM A-53, A135 and A-795) pipe with pressfit fittings. The pressfit jointing method offers economy, speed and reliability for joining 3/4" to 2" diameter pipe for fire protection, heating and cooling water systems. *(Use roll-grooved pipe and fittings for 2 ½" and larger pipe diameters.)*

The pressfit system for carbon steel is BOCA approved. Listed by SBCCI PST and ESI and is UL/ULC listed and FM approved for 175 PSI (1200 kPa) on fire protection services and is rated to 300 PSI (2065 kPa) for heating water systems.

The pressfit system requires no special preparation of the pipe ends before assembly. Pipe should be square cut (+/-0.030") and de-burred. Pressfit system products are designed only for use on approved Schedule 5 carbon steel pipe.

O-ring selection as specified by the mechanical design engineer. Type O o-rings are rated for application temperatures of +20 degrees F to +300 degrees F. Type O o-rings are recommended for many oxidizing acids, petroleum oils, halogenated hydrocarbons, lubricants, hydraulic fluids, organic liquids and air with hydrocarbons within the specified temperature range. Not recommended for steam services.

Standard type E o-rings are rated for application temperatures of –30 degrees F to +230 degrees F. Type E o-rings are recommended for hot water service within the specified temperature range plus a variety of dilute acids, oil free air and many chemical services.

Type T o-rings are rated for application temperatures of –20 degrees F to +180 degrees F. Type T o-rings are recommended for petroleum products, air with oil vapors, vegetable or mineral oils within the specified temperature range. Not recommended for hot water services over 150 degrees F., or steam, or hot dry air over 140 degrees F.

A specialized hand held electric pressfit tool is required to make the pipe/fitting connections. Allow a rental of $300.00 per month or $2,900.00 to purchase the tool.

Description	Craft@Hrs	Unit	Material $	Labor $	Equipment $	Total $

Schedule 5 carbon steel pipe, plain end, pressfit joints, pipe only. No hangers or fittings.

Description	Craft@Hrs	Unit	Material $	Labor $	Equipment $	Total $
3/4"	P1@.030	LF	1.18	1.07	—	2.25
1"	P1@.040	LF	1.61	1.43	—	3.04
1¼"	P1@.040	LF	2.06	1.43	—	3.49
1½"	P1@.050	LF	2.45	1.79	—	4.24
2"	P1@.050	LF	2.96	1.79	—	4.75

45 degree pressfit ell, with Type O o-ring (P x P)

Description	Craft@Hrs	Unit	Material $	Labor $	Equipment $	Total $
3/4"	P1@.170	Ea	19.60	6.08	—	25.68
1"	P1@.180	Ea	23.60	6.44	—	30.04
1¼"	P1@.190	Ea	32.50	6.79	—	39.29
1½"	P1@.200	Ea	41.30	7.15	—	48.45
2"	P1@.200	Ea	55.90	7.15	—	63.05

Carbon Steel, Schedule 5 with Pressfit Fittings

Description	Craft@Hrs	Unit	Material $	Labor $	Equipment $	Total $

90 degree standard radius pressfit ell, with Type O o-ring (P x P)

Description	Craft@Hrs	Unit	Material $	Labor $	Equipment $	Total $
3/4"	P1@.170	Ea	20.60	6.08	—	26.68
1"	P1@.180	Ea	23.20	6.44	—	29.64
1¼"	P1@.190	Ea	32.50	6.79	—	39.29
1½"	P1@.200	Ea	39.70	7.15	—	46.85
2"	P1@.200	Ea	55.00	7.15	—	62.15

90 degree short radius pressfit ell, with Type O o-ring (P x P)

Description	Craft@Hrs	Unit	Material $	Labor $	Equipment $	Total $
3/4"	P1@.170	Ea	23.60	6.08	—	29.68
1"	P1@.180	Ea	28.50	6.44	—	34.94
1¼"	P1@.190	Ea	43.40	6.79	—	50.19
1½"	P1@.200	Ea	53.70	7.15	—	60.85
2"	P1@.200	Ea	70.30	7.15	—	77.45

90 degree standard radius reducing pressfit x FIP ell, with Type O o-ring (P x F)

Description	Craft@Hrs	Unit	Material $	Labor $	Equipment $	Total $
1" x 1/2"	P1@.185	Ea	26.60	6.62	—	33.22
1" x 3/4"	P1@.185	Ea	26.90	6.62	—	33.52
1¼" x 1/2"	P1@.195	Ea	35.50	6.97	—	42.47
1¼" x 1/2"	P1@.195	Ea	36.50	6.97	—	43.47
1½" x 1/2"	P1@.215	Ea	44.50	7.69	—	52.19
1½" x 3/4"	P1@.215	Ea	45.00	7.69	—	52.69

90 degree short radius reducing pressfit x FIP ell, with Type O o-ring (P x F)

Description	Craft@Hrs	Unit	Material $	Labor $	Equipment $	Total $
1" x 3/4"	P1@.185	Ea	24.30	6.62	—	30.92
3/4" x 1"	P1@.185	Ea	24.40	6.62	—	31.02
1¼" x 3/4"	P1@.195	Ea	33.90	6.97	—	40.87
1½" x 3/4"	P1@.215	Ea	43.00	7.69	—	50.69

Standard coupling, pressfit, with Type O o-ring (P x P)

Description	Craft@Hrs	Unit	Material $	Labor $	Equipment $	Total $
3/4"	P1@.170	Ea	15.30	6.08	—	21.38
1"	P1@.180	Ea	20.00	6.44	—	26.44
1¼"	P1@.190	Ea	27.70	6.79	—	34.49
1½"	P1@.200	Ea	34.00	7.15	—	41.15
2"	P1@.200	Ea	45.70	7.15	—	52.85

Description	Craft@Hrs	Unit	Material $	Labor $	Equipment $	Total $

Slip coupling, pressfit, with Type O o-ring (P x P)

Description	Craft@Hrs	Unit	Material $	Labor $	Equipment $	Total $
3/4"	P1@.170	Ea	20.80	6.08	—	26.88
1"	P1@.180	Ea	26.60	6.44	—	33.04
1¼"	P1@.190	Ea	35.50	6.79	—	42.29
1½"	P1@.200	Ea	43.90	7.15	—	51.05
2"	P1@.200	Ea	58.40	7.15	—	65.55

Tee, pressfit, with Type O o-ring (P x P)

Description	Craft@Hrs	Unit	Material $	Labor $	Equipment $	Total $
3/4"	P1@.190	Ea	23.50	6.79	—	30.29
1"	P1@.220	Ea	27.80	7.87	—	35.67
1¼"	P1@.240	Ea	36.80	8.58	—	45.38
1½"	P1@.255	Ea	44.90	9.12	—	54.02
2"	P1@.270	Ea	61.30	9.66	—	70.96

Reducing tee, pressfit, with Type O o-ring (P x P x P)

Description	Craft@Hrs	Unit	Material $	Labor $	Equipment $	Total $
1" x 3/4"	P1@.220	Ea	18.70	7.87	—	26.57
1¼" x 3/4"	P1@.240	Ea	37.70	8.58	—	46.28
1¼" x 1"	P1@.240	Ea	45.00	8.58	—	53.58
1½" x 3/4"	P1@.255	Ea	43.90	9.12	—	53.02
1½" x 1"	P1@.255	Ea	50.80	9.12	—	59.92
2" x 3/4"	P1@.270	Ea	58.00	9.66	—	67.66
2" x 1"	P1@.270	Ea	66.90	9.66	—	76.56
2" x 1½"	P1@.270	Ea	70.30	9.66	—	79.96

Male threaded pressfit adapter with Type O o-ring (P x M)

Description	Craft@Hrs	Unit	Material $	Labor $	Equipment $	Total $
3/4" x 1/2"	P1@.180	Ea	14.30	6.44	—	20.74
3/4" x 3/4"	P1@.190	Ea	15.30	6.79	—	22.09
3/4" x 1"	P1@.190	Ea	17.00	6.79	—	23.79
1" x 3/4"	P1@.190	Ea	19.90	6.79	—	26.69
1" x 1"	P1@.190	Ea	21.30	6.79	—	28.09
1¼" x 1¼"	P1@.200	Ea	29.60	7.15	—	36.75
1½" x 1½"	P1@.215	Ea	36.90	7.69	—	44.59
2" x 2"	P1@.225	Ea	47.80	8.05	—	55.85

Female threaded pressfit adapter with Type O o-ring (P x F)

Description	Craft@Hrs	Unit	Material $	Labor $	Equipment $	Total $
3/4" x 1/2"	P1@.180	Ea	14.30	6.44	—	20.74
3/4" x 3/4"	P1@.190	Ea	15.30	6.79	—	22.09
1" x 1/2"	P1@.190	Ea	15.90	6.79	—	22.69
1" x 3/4"	P1@.190	Ea	21.30	6.79	—	28.09
1" x 1"	P1@.200	Ea	21.90	7.15	—	29.05

Carbon Steel, Schedule 80 with 300# Fittings & Butt-Welded Joints

Schedule 80 carbon steel (ASTM A-53 and A-120) pipe with 300 pound carbon steel welding fittings is commonly used for steam condensate and re-circulating water systems where operating pressures do not exceed 700 PSIG.

The cost estimates in this section are based on the conditions, limitations and wage rates described in the section "How to Use This Book" beginning on page 5.

Equipment cost, where shown, is $110 per hour for a 10-ton hydraulic truck-mounted crane.

Description	Craft@Hrs	Unit	Material $	Labor $	Equipment $	Total $

Schedule 80 carbon steel butt-welded horizontal pipe assembly.

Horizontally hung in a building. Assembly includes fittings and hanger assemblies. Based on a reducing tee and a 45 degree elbow every 16 feet for 1/2" pipe. A reducing tee and a 45 degree elbow every 50 feet for 12" pipe. Hangers spaced to meet plumbing code. *(Distance between fittings in the pipe assembly increases as pipe diameter increases.)* Use these figures for preliminary estimates.

Description	Craft@Hrs	Unit	Material $	Labor $	Equipment $	Total $
1/2"	P1@.183	LF	4.07	6.54	—	10.61
3/4"	P1@.203	LF	4.53	7.26	—	11.79
1"	P1@.241	LF	5.18	8.62	—	13.80
1¼"	P1@.246	LF	5.82	8.80	—	14.62
1½"	P1@.271	LF	6.31	9.69	—	16.00
2"	P1@.350	LF	8.21	12.50	—	20.71
2½"	P1@.437	LF	12.20	15.60	—	27.80
3"	P1@.532	LF	15.90	19.00	—	34.90
4"	P1@.601	LF	23.00	21.50	—	44.50
6"	ER@.833	LF	32.20	31.70	13.00	76.90
8"	ER@.994	LF	49.70	37.90	15.60	103.20
10"	ER@1.20	LF	84.20	45.70	18.90	148.80
12"	ER@1.43	LF	98.90	54.50	22.60	176.00

Schedule 80 carbon steel butt-welded vertical pipe assembly.
Riser assembly, including fittings and riser clamps. Based on a reducing tee, and a riser clamp on every floor. Use these figures for preliminary estimates.

Description	Craft@Hrs	Unit	Material $	Labor $	Equipment $	Total $
1/2"	P1@.153	LF	4.01	5.47	—	9.48
3/4"	P1@.163	LF	4.49	5.83	—	10.32
1"	P1@.196	LF	5.14	7.01	—	12.15
1¼"	P1@.229	LF	6.41	8.19	—	14.60
1½"	P1@.240	LF	6.85	8.58	—	15.43
2"	P1@.310	LF	8.54	11.10	—	19.64
2½"	P1@.395	LF	12.50	14.10	—	26.60
3"	P1@.190	LF	16.00	6.79	—	22.79
4"	P1@.610	LF	23.00	21.80	—	44.80
6"	ER@.877	LF	34.00	33.40	13.80	81.20
8"	ER@1.04	LF	54.20	39.60	16.30	110.10
10"	ER@1.28	LF	91.20	48.80	20.10	160.10
12"	ER@1.58	LF	112.00	60.20	24.80	197.00

Carbon Steel, Schedule 80 with 300# Fittings & Butt-Welded Joints

Description	Craft@Hrs	Unit	Material $	Labor $	Equipment $	Total $

Schedule 80 carbon steel pipe, plain end, butt-welded

Description	Craft@Hrs	Unit	Material $	Labor $	Equipment $	Total $
1/2"	P1@.080	LF	1.62	2.86	—	4.48
3/4"	P1@.090	LF	2.01	3.22	—	5.23
1"	P1@.110	LF	2.75	3.93	—	6.68
1¼"	P1@.120	LF	3.67	4.29	—	7.96
1½"	P1@.130	LF	4.42	4.65	—	9.07
2"	P1@.170	LF	6.13	6.08	—	12.21
2½"	P1@.220	LF	9.41	7.87	—	17.28
3"	P1@.280	LF	12.50	10.00	—	22.50
4"	P1@.340	LF	18.40	12.20	—	30.60
6"	ER@.460	LF	27.60	17.50	7.20	52.30
8"	ER@.560	LF	41.70	21.30	8.80	71.80
10"	ER@.660	LF	72.30	25.20	10.50	108.00
12"	ER@.840	LF	83.10	32.00	13.20	128.30

Schedule 80 carbon steel 45 degree ell, butt-welded

Description	Craft@Hrs	Unit	Material $	Labor $	Equipment $	Total $
1/2"	P1@.300	Ea	5.68	10.70	—	16.38
3/4"	P1@.440	Ea	5.68	15.70	—	21.38
1"	P1@.590	Ea	5.68	21.10	—	26.78
1¼"	P1@.740	Ea	6.26	26.50	—	32.76
1½"	P1@.890	Ea	6.51	31.80	—	38.31
2"	P1@1.18	Ea	6.11	42.20	—	48.31
2½"	P1@1.49	Ea	8.34	53.30	—	61.64
3"	P1@1.77	Ea	10.50	63.30	—	73.80
4"	P1@2.37	Ea	15.90	84.80	—	100.70
6"	ER@3.55	Ea	44.70	135.00	55.90	235.60
8"	ER@4.26	Ea	72.50	162.00	67.00	301.50
10"	ER@5.32	Ea	116.00	203.00	83.60	402.60
12"	ER@6.38	Ea	174.00	243.00	100.30	517.30

Schedule 80 carbon steel 90 degree ell, butt-welded

Description	Craft@Hrs	Unit	Material $	Labor $	Equipment $	Total $
1/2"	P1@.300	Ea	5.73	10.70	—	16.43
3/4"	P1@.440	Ea	5.73	15.70	—	21.43
1"	P1@.590	Ea	5.73	21.10	—	26.83
1¼"	P1@.740	Ea	6.58	26.50	—	33.08
1½"	P1@.890	Ea	7.07	31.80	—	38.87
2"	P1@1.18	Ea	10.60	42.20	—	52.80
2½"	P1@1.49	Ea	10.60	53.30	—	63.90
3"	P1@1.77	Ea	13.80	63.30	—	77.10
4"	P1@2.37	Ea	33.00	84.80	—	117.80
6"	ER@3.55	Ea	62.80	135.00	55.90	253.70
8"	ER@4.26	Ea	115.00	162.00	67.00	344.00
10"	ER@5.32	Ea	209.00	203.00	83.60	495.60
12"	ER@6.38	Ea	294.00	243.00	100.30	637.30

Carbon Steel, Schedule 80 with 300# Fittings & Butt-Welded Joints

Description	Craft@Hrs	Unit	Material $	Labor $	Equipment $	Total $

Schedule 80 carbon steel tee, butt-welded

Description	Craft@Hrs	Unit	Material $	Labor $	Equipment $	Total $
1/2"	P1@.440	Ea	19.20	15.70	—	34.90
3/4"	P1@.670	Ea	19.20	24.00	—	43.20
1¼"	P1@1.11	Ea	19.20	39.70	—	58.90
1½"	P1@1.33	Ea	21.70	47.60	—	69.30
2"	P1@1.77	Ea	18.70	63.30	—	82.00
2½"	P1@2.22	Ea	44.90	79.40	—	124.30
3"	P1@2.66	Ea	49.30	95.10	—	144.40
4"	P1@3.55	Ea	52.00	127.00	—	179.00
6"	ER@5.32	Ea	83.40	203.00	83.60	370.00
8"	ER@6.38	Ea	168.00	243.00	100.30	511.30
10"	ER@7.98	Ea	270.00	304.00	125.40	699.40
12"	ER@9.58	Ea	391.00	365.00	150.70	906.70

Carbon Steel, Schedule 80 with 300# Fittings & Butt-Welded Joints

Description	Craft@Hrs	Unit	Material $	Labor $	Equipment $	Total $

Schedule 80 carbon steel reducing tee, butt-welded

Description	Craft@Hrs	Unit	Material $	Labor $	Equipment $	Total $
1" x 3/4"	P1@.810	Ea	23.50	29.00	—	52.50
1¼" x 1/2"	P1@1.04	Ea	23.50	37.20	—	60.70
1¼" x 3/4"	P1@1.04	Ea	23.50	37.20	—	60.70
1¼" x 1"	P1@1.04	Ea	23.50	37.20	—	60.70
1½" x 1/2"	P1@1.26	Ea	23.50	45.10	—	68.60
1½" x 3/4"	P1@1.26	Ea	23.50	45.10	—	68.60
1½" x 1"	P1@1.26	Ea	23.50	45.10	—	68.60
1½" x 1¼"	P1@1.26	Ea	23.50	45.10	—	68.60
2" x 3/4"	P1@1.63	Ea	25.40	58.30	—	83.70
2" x 1"	P1@1.63	Ea	25.40	58.30	—	83.70
2" x 1¼"	P1@1.63	Ea	25.40	58.30	—	83.70
2" x 1½"	P1@1.63	Ea	25.40	58.30	—	83.70
2½" x 1"	P1@2.07	Ea	35.50	74.00	—	109.50
2½" x 1¼"	P1@2.07	Ea	35.50	74.00	—	109.50
2½" x 1½"	P1@2.07	Ea	35.50	74.00	—	109.50
2½" x 2"	P1@2.07	Ea	35.50	74.00	—	109.50
3" x 1¼"	P1@2.51	Ea	35.50	89.80	—	125.30
3" x 2"	P1@2.51	Ea	35.50	89.80	—	125.30
3" x 2½"	P1@2.51	Ea	35.50	89.80	—	125.30
4" x 1½"	P1@3.25	Ea	55.50	116.00	—	171.50
4" x 2"	P1@3.25	Ea	55.50	116.00	—	171.50
4" x 2½"	P1@3.25	Ea	55.50	116.00	—	171.50
4" x 3"	P1@3.25	Ea	55.50	116.00	—	171.50
6" x 2½"	ER@5.03	Ea	103.00	192.00	79.20	374.20
6" x 3"	ER@5.03	Ea	103.00	192.00	79.20	374.20
6" x 4"	ER@5.03	Ea	103.00	192.00	79.20	374.20
8" x 3"	ER@5.85	Ea	183.00	223.00	92.40	498.40
8" x 4"	ER@5.85	Ea	183.00	223.00	92.40	498.40
8" x 6"	ER@5.85	Ea	183.00	223.00	92.40	498.40
10" x 4"	ER@7.45	Ea	290.00	284.00	117.70	691.70
10" x 6"	ER@7.45	Ea	290.00	284.00	117.70	691.70
10" x 8"	ER@7.45	Ea	290.00	284.00	117.70	691.70
12" x 6"	ER@9.04	Ea	531.00	345.00	141.90	1,017.90
12" x 8"	ER@9.04	Ea	531.00	345.00	141.90	1,017.90
12" x 10"	ER@9.04	Ea	531.00	345.00	141.90	1,017.90

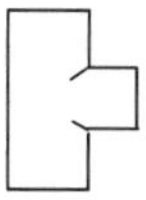

Schedule 80 carbon steel concentric reducer, butt-welded

Description	Craft@Hrs	Unit	Material $	Labor $	Equipment $	Total $
3/4"	P1@.360	Ea	11.40	12.90	—	24.30
1"	P1@.500	Ea	12.20	17.90	—	30.10
1¼"	P1@.665	Ea	10.60	23.80	—	34.40
1½"	P1@.800	Ea	11.30	28.60	—	39.90
2"	P1@1.00	Ea	9.76	35.80	—	45.56
2½"	P1@1.33	Ea	12.80	47.60	—	60.40
3"	P1@1.63	Ea	12.20	58.30	—	70.50
4"	P1@2.00	Ea	16.20	71.50	—	87.70
6"	ER@2.96	Ea	35.50	113.00	46.50	195.00
8"	ER@3.90	Ea	52.40	149.00	61.30	262.70
10"	ER@4.79	Ea	72.50	183.00	75.40	330.90
12"	ER@5.85	Ea	106.00	223.00	92.10	421.10

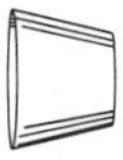

Description	Craft@Hrs	Unit	Material $	Labor $	Equipment $	Total $

Schedule 80 carbon steel eccentric reducer, butt-welded

Description	Craft@Hrs	Unit	Material $	Labor $	Equipment $	Total $
3/4"	P1@.360	Ea	14.10	12.90	—	27.00
1"	P1@.500	Ea	18.30	17.90	—	36.20
1¼"	P1@.665	Ea	14.70	23.80	—	38.50
1½"	P1@.800	Ea	17.40	28.60	—	46.00
2"	P1@1.00	Ea	13.90	35.80	—	49.70
2½"	P1@1.33	Ea	16.90	47.60	—	64.50
3"	P1@1.63	Ea	16.20	58.30	—	74.50
4"	P1@2.00	Ea	22.80	71.50	—	94.30
6"	ER@2.96	Ea	52.10	113.00	46.50	211.60
8"	ER@3.90	Ea	77.60	149.00	61.30	287.90
10"	ER@4.79	Ea	136.00	183.00	75.40	394.40
12"	ER@5.85	Ea	210.00	223.00	94.30	527.30

Schedule 80 carbon steel cap, butt-welded

Description	Craft@Hrs	Unit	Material $	Labor $	Equipment $	Total $
1/2"	P1@.180	Ea	11.30	6.44	—	17.74
3/4"	P1@.270	Ea	15.80	9.66	—	25.46
1"	P1@.360	Ea	7.31	12.90	—	20.21
1¼"	P1@.460	Ea	7.31	16.40	—	23.71
1½"	P1@.560	Ea	8.00	20.00	—	28.00
2"	P1@.740	Ea	6.92	26.50	—	33.42
2½"	P1@.920	Ea	7.76	32.90	—	40.66
3"	P1@1.11	Ea	8.27	39.70	—	47.97
4"	P1@1.50	Ea	11.00	53.60	—	64.60
6"	ER@2.24	Ea	20.60	85.40	35.20	141.20
8"	ER@2.68	Ea	31.90	102.00	42.10	176.00
10"	ER@3.36	Ea	47.20	128.00	52.80	228.00
12"	ER@4.00	Ea	62.60	152.00	62.90	277.50

Schedule 80 carbon steel union, butt-welded

Description	Craft@Hrs	Unit	Material $	Labor $	Equipment $	Total $
1/2"	P1@.300	Ea	9.52	10.70	—	20.22
3/4"	P1@.440	Ea	11.40	15.70	—	27.10
1"	P1@.590	Ea	15.00	21.10	—	36.10
1¼"	P1@.740	Ea	25.40	26.50	—	51.90
1½"	P1@.890	Ea	27.60	31.80	—	59.40
2"	P1@1.18	Ea	36.80	42.20	—	79.00
2½"	P1@1.49	Ea	80.20	53.30	—	133.50
3"	P1@1.77	Ea	100.00	63.30	—	163.30

Carbon steel weldolet

Description	Craft@Hrs	Unit	Material $	Labor $	Equipment $	Total $
1/2"	P1@.440	Ea	10.40	15.70	—	26.10
3/4"	P1@.670	Ea	10.70	24.00	—	34.70
1"	P1@.890	Ea	11.60	31.80	—	43.40
1¼"	P1@1.11	Ea	15.00	39.70	—	54.70
1½"	P1@1.33	Ea	15.00	47.60	—	62.60
2"	P1@1.77	Ea	15.50	63.30	—	78.80
2½"	P1@2.22	Ea	35.40	79.40	—	114.80
3"	P1@2.66	Ea	36.50	95.10	—	131.60
4"	P1@3.55	Ea	81.00	127.00	—	208.00

Carbon Steel, Schedule 80 with 300# Fittings & Butt-Welded Joints

Description	Craft@Hrs	Unit	Material $	Labor $	Equipment $	Total $

Carbon steel threadolet

Description	Craft@Hrs	Unit	Material $	Labor $	Equipment $	Total $
1/2"	P1@.300	Ea	7.63	10.70	—	18.33
3/4"	P1@.440	Ea	8.43	15.70	—	24.13
1"	P1@.590	Ea	10.40	21.10	—	31.50
1¼"	P1@.740	Ea	46.70	26.50	—	73.20
1½"	P1@.890	Ea	46.70	31.80	—	78.50
2"	P1@1.18	Ea	61.60	42.20	—	103.80

300# forged steel slip-on companion flange, welding type

Description	Craft@Hrs	Unit	Material $	Labor $	Equipment $	Total $
2½"	P1@.810	Ea	26.00	29.00	—	55.00
3"	P1@.980	Ea	26.00	35.00	—	61.00
4"	P1@1.30	Ea	38.30	46.50	—	84.80
6"	ER@1.95	Ea	69.50	74.30	30.60	174.40
8"	ER@2.35	Ea	98.40	89.60	37.10	225.10
10"	ER@2.90	Ea	162.00	111.00	45.70	318.70
12"	ER@3.50	Ea	259.00	133.00	55.00	447.00

Class 300 bronze body gate valve, threaded joints

Description	Craft@Hrs	Unit	Material $	Labor $	Equipment $	Total $
1/2"	P1@.210	Ea	36.10	7.51	—	43.61
3/4"	P1@.250	Ea	47.10	8.94	—	56.04
1"	P1@.300	Ea	57.20	10.70	—	67.90
1¼"	P1@.400	Ea	78.00	14.30	—	92.30
1½"	P1@.450	Ea	98.30	16.10	—	114.40
2"	P1@.500	Ea	145.00	17.90	—	162.90
2½"	P1@.750	Ea	296.00	26.80	—	322.80
3"	P1@.950	Ea	420.00	34.00	—	454.00

Class 250 iron body gate valve, flanged

Description	Craft@Hrs	Unit	Material $	Labor $	Equipment $	Total $
2½"	P1@.600	Ea	899.00	21.50	—	920.50
3"	P1@.750	Ea	649.00	26.80	—	675.80
4"	P1@1.35	Ea	944.00	48.30	—	992.30
6"	ER@2.50	Ea	1,290.00	95.30	39.30	1,424.60
8"	ER@3.00	Ea	4,370.00	114.00	47.10	4,531.10
10"	ER@4.00	Ea	7,140.00	152.00	62.90	7,354.90
12"	ER@4.50	Ea	11,700.00	171.00	70.80	11,941.80

Class 300 bronze body globe valve, threaded

Description	Craft@Hrs	Unit	Material $	Labor $	Equipment $	Total $
1/2"	P1@.210	Ea	53.80	7.51	—	61.31
3/4"	P1@.250	Ea	68.40	8.94	—	77.34
1"	P1@.300	Ea	97.20	10.70	—	107.90
1¼"	P1@.400	Ea	141.00	14.30	—	155.30
1½"	P1@.450	Ea	188.00	16.10	—	204.10
2"	P1@.500	Ea	260.00	17.90	—	277.90

Carbon Steel, Schedule 80 with 300# Fittings & Butt-Welded Joints

Description	Craft@Hrs	Unit	Material $	Labor $	Equipment $	Total $

Class 250 iron body globe valve, flanged

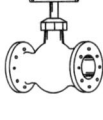

Description	Craft@Hrs	Unit	Material $	Labor $	Equipment $	Total $
2"	P1@.500	Ea	473.00	17.90	—	490.90
2½"	P1@.600	Ea	588.00	21.50	—	609.50
3"	P1@.750	Ea	700.00	26.80	—	726.80
4"	P1@1.35	Ea	965.00	48.30	—	1,013.30
6"	ER@2.50	Ea	1,900.00	95.30	39.30	2,034.60
8"	ER@3.00	Ea	4,210.00	114.00	47.10	4,371.10

200 PSIG iron body butterfly valve, lug type, lever operated

Description	Craft@Hrs	Unit	Material $	Labor $	Equipment $	Total $
2"	P1@.450	Ea	128.00	16.10	—	144.10
2½"	P1@.450	Ea	132.00	16.10	—	148.10
3"	P1@.550	Ea	139.00	19.70	—	158.70
4"	P1@.550	Ea	173.00	19.70	—	192.70
6"	ER@.800	Ea	279.00	30.50	12.70	322.20
8"	ER@.800	Ea	382.00	30.50	12.70	425.20
10"	ER@.900	Ea	530.00	34.30	14.10	578.40
12"	ER@1.00	Ea	694.00	38.10	15.70	747.80

200 PSIG iron body butterfly valve, wafer type, lever operated

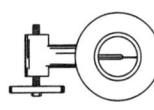

Description	Craft@Hrs	Unit	Material $	Labor $	Equipment $	Total $
2"	P1@.450	Ea	116.00	16.10	—	132.10
2½"	P1@.450	Ea	119.00	16.10	—	135.10
3"	P1@.550	Ea	128.00	19.70	—	147.70
4"	P1@.550	Ea	155.00	19.70	—	174.70
6"	ER@.800	Ea	259.00	30.50	12.70	302.20
8"	ER@.800	Ea	359.00	30.50	12.70	402.20
10"	ER@.900	Ea	502.00	34.30	14.10	550.40
12"	ER@1.00	Ea	659.00	38.10	15.70	712.80

Class 300 bronze body swing check valve, threaded

Description	Craft@Hrs	Unit	Material $	Labor $	Equipment $	Total $
1/2"	P1@.210	Ea	114.00	7.51	—	121.51
3/4"	P1@.250	Ea	135.00	8.94	—	143.94
1"	P1@.300	Ea	208.00	10.70	—	218.70
1¼"	P1@.400	Ea	238.00	14.30	—	252.30
1½"	P1@.450	Ea	341.00	16.10	—	357.10
2"	P1@.500	Ea	476.00	17.90	—	493.90

Class 250 iron body swing check valve, flanged

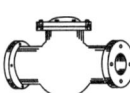

Description	Craft@Hrs	Unit	Material $	Labor $	Equipment $	Total $
2"	P1@.500	Ea	382.00	17.90	—	399.90
2½"	P1@.600	Ea	445.00	21.50	—	466.50
3"	P1@.750	Ea	553.00	26.80	—	579.80
4"	P1@1.35	Ea	678.00	48.30	—	726.30
6"	ER@2.50	Ea	1,320.00	95.30	39.30	1,454.60
8"	ER@3.00	Ea	2,470.00	114.00	47.10	2,631.10

Carbon Steel, Schedule 80 with 300# Fittings & Butt-Welded Joints

Description	Craft@Hrs	Unit	Material $	Labor $	Equipment $	Total $

Class 250 iron body silent check valve, flanged

Description	Craft@Hrs	Unit	Material $	Labor $	Equipment $	Total $
2"	P1@.500	Ea	160.00	17.90	—	177.90
2½"	P1@.600	Ea	184.00	21.50	—	205.50
3"	P1@.750	Ea	205.00	26.80	—	231.80
4"	P1@1.35	Ea	278.00	48.30	—	326.30
6"	ER@2.50	Ea	488.00	95.30	39.30	622.60
8"	ER@3.00	Ea	888.00	114.00	47.10	1,049.10
10"	ER@4.00	Ea	1,400.00	152.00	62.90	1,614.90

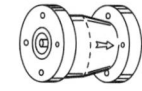

Class 250 bronze body strainer, threaded

Description	Craft@Hrs	Unit	Material $	Labor $	Equipment $	Total $
1/2"	P1@.210	Ea	35.00	7.51	—	42.51
3/4"	P1@.250	Ea	47.40	8.94	—	56.34
1"	P1@.300	Ea	60.20	10.70	—	70.90
1¼"	P1@.400	Ea	85.10	14.30	—	99.40
1½"	P1@.450	Ea	110.00	16.10	—	126.10
2"	P1@.500	Ea	189.00	17.90	—	206.90

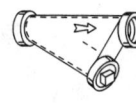

Class 250 iron body strainer, flanged

Description	Craft@Hrs	Unit	Material $	Labor $	Equipment $	Total $
2"	P1@.500	Ea	202.00	17.90	—	219.90
2½"	P1@.600	Ea	221.00	21.50	—	242.50
3"	P1@.750	Ea	258.00	26.80	—	284.80
4"	P1@1.35	Ea	458.00	48.30	—	506.30
6"	ER@2.50	Ea	733.00	95.30	39.30	867.60
8"	ER@3.00	Ea	1,340.00	114.00	47.10	1,501.10

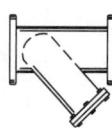

Installation of 2-way control valve, threaded joints

Description	Craft@Hrs	Unit	Material $	Labor $	Equipment $	Total $
1/2"	P1@.210	Ea	—	7.51	—	7.51
3/4"	P1@.250	Ea	—	8.94	—	8.94
1"	P1@.300	Ea	—	10.70	—	10.70
1¼"	P1@.400	Ea	—	14.30	—	14.30
1½"	P1@.450	Ea	—	16.10	—	16.10
2"	P1@.500	Ea	—	17.90	—	17.90
2½"	P1@.830	Ea	—	29.70	—	29.70
3"	P1@.990	Ea	—	35.40	—	35.40

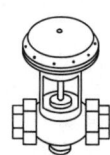

Installation of 2-way control valve, flanged joints

Description	Craft@Hrs	Unit	Material $	Labor $	Equipment $	Total $
2"	P1@.500	Ea	—	17.90	—	17.90
2½"	P1@.600	Ea	—	21.50	—	21.50
3"	P1@.750	Ea	—	26.80	—	26.80
4"	P1@1.35	Ea	—	48.30	—	48.30
6"	P1@2.50	Ea	—	89.40	—	89.40
8"	P1@3.00	Ea	—	107.00	—	107.00
10"	P1@4.00	Ea	—	143.00	—	143.00
12"	P1@4.50	Ea	—	161.00	—	161.00

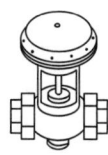

Description	Craft@Hrs	Unit	Material $	Labor $	Equipment $	Total $

Installation of 3-way control valve, threaded joints

Description	Craft@Hrs	Unit	Material $	Labor $	Equipment $	Total $
1/2"	P1@.260	Ea	—	9.30	—	9.30
3/4"	P1@.365	Ea	—	13.10	—	13.10
1"	P1@.475	Ea	—	17.00	—	17.00
1¼"	P1@.575	Ea	—	20.60	—	20.60
1½"	P1@.680	Ea	—	24.30	—	24.30
2"	P1@.910	Ea	—	32.50	—	32.50
2½"	P1@1.12	Ea	—	40.10	—	40.10
3"	P1@1.33	Ea	—	47.60	—	47.60

Installation of 3-way control valve, flanged joints

Description	Craft@Hrs	Unit	Material $	Labor $	Equipment $	Total $
2"	P1@.910	Ea	—	32.50	—	32.50
2½"	P1@1.12	Ea	—	40.10	—	40.10
3"	P1@1.33	Ea	—	47.60	—	47.60
4"	P1@2.00	Ea	—	71.50	—	71.50
6"	P1@3.70	Ea	—	132.00	—	132.00
8"	P1@4.40	Ea	—	157.00	—	157.00
10"	P1@5.90	Ea	—	211.00	—	211.00
12"	P1@6.50	Ea	—	232.00	—	232.00

Bolt and gasket set

Description	Craft@Hrs	Unit	Material $	Labor $	Equipment $	Total $
2"	P1@.500	Ea	3.86	17.90	—	21.76
2½"	P1@.650	Ea	4.51	23.20	—	27.71
3"	P1@.750	Ea	7.64	26.80	—	34.44
4"	P1@1.00	Ea	12.70	35.80	—	48.50
6"	P1@1.20	Ea	21.20	42.90	—	64.10
8"	P1@1.25	Ea	23.80	44.70	—	68.50
10"	P1@1.70	Ea	40.90	60.80	—	101.70
12"	P1@2.20	Ea	46.20	78.70	—	124.90

Thermometer with well

Description	Craft@Hrs	Unit	Material $	Labor $	Equipment $	Total $
7"	P1@.250	Ea	159.00	8.94	—	167.94
9"	P1@.250	Ea	164.00	8.94	—	172.94

Dial type pressure gauge

Description	Craft@Hrs	Unit	Material $	Labor $	Equipment $	Total $
2½"	P1@.200	Ea	33.10	7.15	—	40.25
3½"	P1@.200	Ea	43.80	7.15	—	50.95

Pressure/temperature tap

Description	Craft@Hrs	Unit	Material $	Labor $	Equipment $	Total $
tap	P1@.150	Ea	15.70	5.36	—	21.06

Carbon Steel, Schedule 80 with 300# Fittings & Butt-Welded Joints

Description	Craft@Hrs	Unit	Material $	Labor $	Equipment $	Total $

Hanger with swivel assembly

Description	Craft@Hrs	Unit	Material $	Labor $	Equipment $	Total $
1/2"	P1@.250	Ea	5.62	8.94	—	14.56
3/4"	P1@.250	Ea	6.16	8.94	—	15.10
1"	P1@.250	Ea	6.67	8.94	—	15.61
1½"	P1@.300	Ea	6.99	10.70	—	17.69
2"	P1@.300	Ea	9.09	10.70	—	19.79
2½"	P1@.350	Ea	11.70	12.50	—	24.20
3"	P1@.350	Ea	14.70	12.50	—	27.20
4"	P1@.350	Ea	25.20	12.50	—	37.70
5"	P1@.450	Ea	29.50	16.10	—	45.60
6"	P1@.450	Ea	32.80	16.10	—	48.90
8"	P1@.450	Ea	43.30	16.10	—	59.40
10"	P1@.550	Ea	59.30	19.70	—	79.00
12"	P1@.600	Ea	78.50	21.50	—	100.00

Riser clamp

Description	Craft@Hrs	Unit	Material $	Labor $	Equipment $	Total $
1/2"	P1@.100	Ea	2.41	3.58	—	5.99
3/4"	P1@.100	Ea	3.55	3.58	—	7.13
1"	P1@.100	Ea	3.59	3.58	—	7.17
1¼"	P1@.105	Ea	4.33	3.75	—	8.08
1½"	P1@.110	Ea	4.58	3.93	—	8.51
2"	P1@.115	Ea	4.84	4.11	—	8.95
2½"	P1@.120	Ea	5.09	4.29	—	9.38
3"	P1@.120	Ea	5.52	4.29	—	9.81
4"	P1@.125	Ea	7.03	4.47	—	11.50
5"	P1@.180	Ea	10.10	6.44	—	16.54
6"	P1@.200	Ea	12.20	7.15	—	19.35
8"	P1@.200	Ea	19.80	7.15	—	26.95
10"	P1@.250	Ea	29.40	8.94	—	38.34
12"	P1@.250	Ea	34.90	8.94	—	43.84

Carbon Steel, Schedule 80 with 300# M.I. Fittings & Threaded Joints

Schedule 80 carbon steel (ASTM A-53 and A-120) pipe with 300 pound malleable iron threaded fittings is commonly used for heating hot water, chilled water, steam and steam condensate systems where operating pressures do not exceed 250 PSIG for pipe sized 2 inches and smaller. For pipe sizes 2½ inches and larger wrought steel fittings either welded or grooved should be used.

The cost estimates in this section are based on the conditions, limitations and wage rates described in the section "How to Use This Book" beginning on page 5.

Equipment cost, where shown, is $110 per hour for a 10-ton hydraulic truck-mounted crane.

Description	Craft@Hrs	Unit	Material $	Labor $	Equipment $	Total $

Schedule 80 carbon steel threaded horizontal pipe assembly.
Horizontally hung in a building. Assembly includes fittings and hanger assemblies. Based on a reducing tee and a 45 degree elbow every 16 feet for 1/2" pipe and a reducing tee and a 45 degree elbow every 30 feet for 4" pipe. Hangers spaced to meet plumbing code. *(Distance between fittings in the pipe assembly increases as pipe diameter increases.)* Use these figures for preliminary estimates.

Description	Craft@Hrs	Unit	Material $	Labor $	Equipment $	Total $
1/2"	P1@.121	LF	3.36	4.33	—	7.69
3/4"	P1@.125	LF	3.83	4.47	—	8.30
1"	P1@.149	LF	4.80	5.33	—	10.13
1¼"	P1@.163	LF	5.53	5.83	—	11.36
1½"	P1@.186	LF	5.73	6.65	—	12.38
2"	P1@.210	LF	7.02	7.51	—	14.53
2½"	P1@.297	LF	9.37	10.60	—	19.97
3"	P1@.340	LF	19.90	12.20	—	32.10
4"	P1@.409	LF	30.20	14.60	—	44.80

Schedule 80 carbon steel threaded vertical pipe assembly.
Riser assembly, including fittings and riser clamps. Based on a reducing tee, and a riser clamp on every floor. Use these figures for preliminary estimates.

Description	Craft@Hrs	Unit	Material $	Labor $	Equipment $	Total $
1/2"	P1@.094	LF	2.75	3.36	—	6.11
3/4"	P1@.104	LF	3.15	3.72	—	6.87
1"	P1@.124	LF	4.14	4.43	—	8.57
1¼"	P1@.143	LF	4.84	5.11	—	9.95
1½"	P1@.161	LF	5.35	5.76	—	11.11
2"	P1@.179	LF	6.37	6.40	—	12.77
2½"	P1@.257	LF	8.91	9.19	—	18.10
3"	P1@.296	LF	26.70	10.60	—	37.30
4"	P1@.373	LF	44.50	13.30	—	57.80

Carbon Steel, Schedule 80 with 300# M.I. Fittings & Threaded Joints

Description	Craft@Hrs	Unit	Material $	Labor $	Equipment $	Total $
Schedule 80 carbon steel pipe, threaded and coupled joints						
1/2"	P1@.070	LF	1.70	2.50	—	4.20
3/4"	P1@.080	LF	2.13	2.86	—	4.99
1"	P1@.100	LF	2.90	3.58	—	6.48
1¼"	P1@.110	LF	3.80	3.93	—	7.73
1½"	P1@.120	LF	4.50	4.29	—	8.79
2"	P1@.130	LF	6.14	4.65	—	10.79
2½"	P1@.190	LF	9.66	6.79	—	16.45
3"	P1@.220	LF	12.70	7.87	—	20.57
4"	P1@.290	LF	18.40	10.40	—	28.80
300# malleable iron 45 degree ell, threaded joints						
1/2"	P1@.132	Ea	7.79	4.72	—	12.51
3/4"	P1@.143	Ea	8.65	5.11	—	13.76
1"	P1@.198	Ea	9.60	7.08	—	16.68
1¼"	P1@.264	Ea	14.80	9.44	—	24.24
1½"	P1@.330	Ea	19.30	11.80	—	31.10
2"	P1@.418	Ea	28.90	14.90	—	43.80
2½"	P1@.561	Ea	45.40	20.10	—	65.50
3"	P1@.704	Ea	60.60	25.20	—	85.80
4"	P1@.935	Ea	117.00	33.40	—	150.40
300# malleable iron 90 degree ell, threaded joints						
1/2"	P1@.132	Ea	5.99	4.72	—	10.71
3/4"	P1@.143	Ea	6.92	5.11	—	12.03
1"	P1@.198	Ea	8.83	7.08	—	15.91
1¼"	P1@.264	Ea	12.30	9.44	—	21.74
1½"	P1@.330	Ea	14.80	11.80	—	26.60
2"	P1@.418	Ea	20.70	14.90	—	35.60
2½"	P1@.561	Ea	39.30	20.10	—	59.40
3"	P1@.704	Ea	53.90	25.20	—	79.10
4"	P1@.935	Ea	108.00	33.40	—	141.40

Carbon Steel, Schedule 80 with 300# M.I. Fittings & Threaded Joints

Description	Craft@Hrs	Unit	Material $	Labor $	Equipment $	Total $

300# malleable iron tee, threaded joints

Description	Craft@Hrs	Unit	Material $	Labor $	Equipment $	Total $
1/2"	P1@.198	Ea	8.43	7.08	—	15.51
3/4"	P1@.209	Ea	9.26	7.47	—	16.73
1"	P1@.253	Ea	11.10	9.05	—	20.15
1¼"	P1@.341	Ea	14.80	12.20	—	27.00
1½"	P1@.429	Ea	18.40	15.30	—	33.70
2"	P1@.539	Ea	26.80	19.30	—	46.10
2½"	P1@.726	Ea	51.20	26.00	—	77.20
3"	P1@.913	Ea	85.80	32.60	—	118.40
4"	P1@1.21	Ea	146.00	43.30	—	189.30

300# malleable iron reducing tee, threaded joints

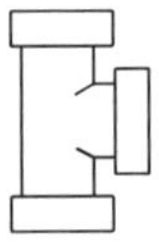

Description	Craft@Hrs	Unit	Material $	Labor $	Equipment $	Total $
3/4" x 1/2"	P1@.205	Ea	11.60	7.33	—	18.93
1" x 1/2"	P1@.205	Ea	14.60	7.33	—	21.93
1" x 3/4"	P1@.205	Ea	14.60	7.33	—	21.93
1¼" x 1/2"	P1@.310	Ea	18.90	11.10	—	30.00
1¼" x 3/4"	P1@.310	Ea	18.90	11.10	—	30.00
1¼" x 1"	P1@.310	Ea	18.90	11.10	—	30.00
1½" x 1/2"	P1@.400	Ea	23.20	14.30	—	37.50
1½" x 3/4"	P1@.400	Ea	23.20	14.30	—	37.50
1½" x 1"	P1@.400	Ea	23.20	14.30	—	37.50
1½" x 1¼"	P1@.400	Ea	23.20	14.30	—	37.50
2" x 1"	P1@.500	Ea	34.50	17.90	—	52.40
2" x 1¼"	P1@.500	Ea	34.50	17.90	—	52.40
2" x 1½"	P1@.500	Ea	34.50	17.90	—	52.40
3" x 1¼"	P1@.830	Ea	178.00	29.70	—	207.70
3" x 1½"	P1@.830	Ea	178.00	29.70	—	207.70
3" x 2"	P1@.830	Ea	178.00	29.70	—	207.70

300# malleable iron reducer, threaded joints

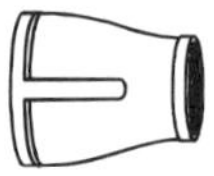

Description	Craft@Hrs	Unit	Material $	Labor $	Equipment $	Total $
3/4" x 1/2"	P1@.135	Ea	6.89	4.83	—	11.72
1" x 1/2"	P1@.170	Ea	8.23	6.08	—	14.31
1" x 3/4"	P1@.170	Ea	8.23	6.08	—	14.31
1¼" x 3/4"	P1@.220	Ea	10.40	7.87	—	18.27
1¼" x 1"	P1@.220	Ea	10.40	7.87	—	18.27
1½" x 1"	P1@.285	Ea	14.30	10.20	—	24.50
1½" x 1¼"	P1@.285	Ea	14.30	10.20	—	24.50
2" x 1¼"	P1@.360	Ea	20.60	12.90	—	33.50
2" x 1½"	P1@.360	Ea	20.60	12.90	—	33.50
3" x 2"	P1@.550	Ea	50.00	19.70	—	69.70
3" x 2½"	P1@.550	Ea	50.00	19.70	—	69.70

Carbon Steel, Schedule 80 with 300# M.I. Fittings & Threaded Joints

Description	Craft@Hrs	Unit	Material $	Labor $	Equipment $	Total $

300# malleable iron cross, threaded joints

Description	Craft@Hrs	Unit	Material $	Labor $	Equipment $	Total $
1/2"	P1@.310	Ea	20.20	11.10	—	31.30
3/4"	P1@.350	Ea	22.10	12.50	—	34.60
1"	P1@.400	Ea	25.00	14.30	—	39.30
1¼"	P1@.530	Ea	30.90	19.00	—	49.90
1½"	P1@.660	Ea	40.50	23.60	—	64.10
2"	P1@.770	Ea	58.00	27.50	—	85.50

300# malleable iron cap, threaded joint

Description	Craft@Hrs	Unit	Material $	Labor $	Equipment $	Total $
1/2"	P1@.100	Ea	4.05	3.58	—	7.63
3/4"	P1@.110	Ea	5.39	3.93	—	9.32
1"	P1@.155	Ea	6.95	5.54	—	12.49
1¼"	P1@.200	Ea	9.44	7.15	—	16.59
1½"	P1@.250	Ea	11.20	8.94	—	20.14
2"	P1@.320	Ea	15.50	11.40	—	26.90
3"	P1@.530	Ea	32.30	19.00	—	51.30
4"	P1@.705	Ea	59.30	25.20	—	84.50

300# malleable iron plug, threaded joint

Description	Craft@Hrs	Unit	Material $	Labor $	Equipment $	Total $
1/2"	P1@.000	Ea	1.00	3.22	—	4.22
3/4"	P1@.100	Ea	1.36	3.58	—	4.94
1"	P1@.140	Ea	1.62	5.01	—	6.63
1¼"	P1@.180	Ea	1.74	6.44	—	8.18
1½"	P1@.230	Ea	2.23	8.22	—	10.45
2"	P1@.290	Ea	3.45	10.40	—	13.85
2½"	P1@.380	Ea	4.86	13.60	—	18.46
3"	P1@.480	Ea	8.10	17.20	—	25.30
4"	P1@.640	Ea	23.20	22.90	—	46.10

300# malleable iron union, threaded joints

Description	Craft@Hrs	Unit	Material $	Labor $	Equipment $	Total $
1/2"	P1@.150	Ea	7.42	5.36	—	12.78
3/4"	P1@.165	Ea	8.36	5.90	—	14.26
1"	P1@.230	Ea	10.90	8.22	—	19.12
1¼"	P1@.310	Ea	16.80	11.10	—	27.90
1½"	P1@.400	Ea	18.20	14.30	—	32.50
2"	P1@.500	Ea	23.20	17.90	—	41.10
2½"	P1@.670	Ea	51.20	24.00	—	75.20
3"	P1@.840	Ea	86.10	30.00	—	116.10
4"	P1@.960	Ea	242.00	34.30	—	276.30

Carbon Steel, Schedule 80 with 300# M.I. Fittings & Threaded Joints

Description	Craft@Hrs	Unit	Material $	Labor $	Equipment $	Total $

300# malleable iron coupling, threaded joints

Description	Craft@Hrs	Unit	Material $	Labor $	Equipment $	Total $
1/2"	P1@.132	Ea	4.51	4.72	—	9.23
3/4"	P1@.143	Ea	5.27	5.11	—	10.38
1"	P1@.198	Ea	6.21	7.08	—	13.29
1¼"	P1@.264	Ea	7.42	9.44	—	16.86
1½"	P1@.330	Ea	11.10	11.80	—	22.90
2"	P1@.418	Ea	15.50	14.90	—	30.40
2½"	P1@.561	Ea	19.80	20.10	—	39.90
3"	P1@.704	Ea	32.30	25.20	—	57.50
4"	P1@.935	Ea	77.10	33.40	—	110.50

Class 300 bronze body gate valve, threaded joints

Description	Craft@Hrs	Unit	Material $	Labor $	Equipment $	Total $
1/2"	P1@.210	Ea	36.10	7.51	—	43.61
3/4"	P1@.250	Ea	46.90	8.94	—	55.84
1"	P1@.300	Ea	57.10	10.70	—	67.80
1¼"	P1@.400	Ea	77.80	14.30	—	92.10
1½"	P1@.450	Ea	98.10	16.10	—	114.20
2"	P1@.500	Ea	144.00	17.90	—	161.90

Class 250 iron body gate valve, flanged joints

Description	Craft@Hrs	Unit	Material $	Labor $	Equipment $	Total $
2"	P1@.500	Ea	899.00	17.90	—	916.90
2½"	P1@.600	Ea	649.00	21.50	—	670.50
3"	P1@.750	Ea	821.00	26.80	—	847.80
4"	P1@1.35	Ea	1,390.00	48.30	—	1,438.30

Class 300 bronze body globe valve, threaded joints

Description	Craft@Hrs	Unit	Material $	Labor $	Equipment $	Total $
1/2"	P1@.210	Ea	53.80	7.51	—	61.31
3/4"	P1@.250	Ea	68.40	8.94	—	77.34
1"	P1@.300	Ea	97.20	10.70	—	107.90
1¼"	P1@.400	Ea	141.00	14.30	—	155.30
1½"	P1@.450	Ea	188.00	16.10	—	204.10
2"	P1@.500	Ea	260.00	17.90	—	277.90

Class 250 iron body globe valve, flanged joints

Description	Craft@Hrs	Unit	Material $	Labor $	Equipment $	Total $
2"	P1@.500	Ea	473.00	17.90	—	490.90
2½"	P1@.600	Ea	588.00	21.50	—	609.50
3"	P1@.750	Ea	700.00	26.80	—	726.80
4"	P1@1.35	Ea	965.00	48.30	—	1,013.30

200 PSIG iron body butterfly valve, lug type, lever operated

Description	Craft@Hrs	Unit	Material $	Labor $	Equipment $	Total $
2"	P1@.450	Ea	128.00	16.10	—	144.10
2½"	P1@.450	Ea	132.00	16.10	—	148.10
3"	P1@.550	Ea	139.00	19.70	—	158.70
4"	P1@.550	Ea	173.00	19.70	—	192.70

Carbon Steel, Schedule 80 with 300# M.I. Fittings & Threaded Joints

Description	Craft@Hrs	Unit	Material $	Labor $	Equipment $	Total $

200 PSIG iron body butterfly valve, wafer type, lever operated

2"	P1@.450	Ea	116.00	16.10	—	132.10
2½"	P1@.450	Ea	119.00	16.10	—	135.10
3"	P1@.550	Ea	128.00	19.70	—	147.70
4"	P1@.550	Ea	155.00	19.70	—	174.70

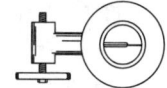

Class 150 bronze body 2-piece ball valve, threaded joints

1/2"	P1@.210	Ea	11.20	7.51	—	18.71
3/4"	P1@.250	Ea	18.20	8.94	—	27.14
1"	P1@.300	Ea	23.10	10.70	—	33.80
1¼"	P1@.400	Ea	28.90	14.30	—	43.20
1½"	P1@.450	Ea	39.00	16.10	—	55.10
2"	P1@.500	Ea	49.40	17.90	—	67.30
3"	P1@.625	Ea	338.00	22.40	—	360.40
4"	P1@.690	Ea	443.00	24.70	—	467.70

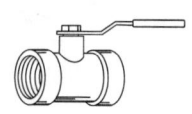

Class 300 bronze body swing check valve, threaded joints

1/2"	P1@.210	Ea	114.00	7.51	—	121.51
3/4"	P1@.250	Ea	135.00	8.94	—	143.94
1"	P1@.300	Ea	208.00	10.70	—	218.70
1¼"	P1@.400	Ea	238.00	14.30	—	252.30
1½"	P1@.450	Ea	341.00	16.10	—	357.10
2"	P1@.500	Ea	476.00	17.90	—	493.90

Class 250 iron body swing check valve, flanged joints

2"	P1@.500	Ea	382.00	17.90	—	399.90
2½"	P1@.600	Ea	445.00	21.50	—	466.50
3"	P1@.750	Ea	553.00	26.80	—	579.80
4"	P1@1.35	Ea	678.00	48.30	—	726.30

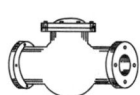

Class 250 iron body silent check valve, wafer type

2"	P1@.500	Ea	160.00	17.90	—	177.90
2½"	P1@.600	Ea	184.00	21.50	—	205.50
3"	P1@.750	Ea	205.00	26.80	—	231.80
4"	P1@1.35	Ea	278.00	48.30	—	326.30

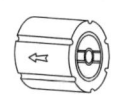

Class 250 bronze body wye strainer, threaded joints

1/2"	P1@.210	Ea	35.00	7.51	—	42.51
3/4"	P1@.250	Ea	47.40	8.94	—	56.34
1"	P1@.300	Ea	60.20	10.70	—	70.90
1¼"	P1@.400	Ea	85.10	14.30	—	99.40
1½"	P1@.450	Ea	110.00	16.10	—	126.10
2"	P1@.500	Ea	189.00	17.90	—	206.90

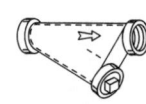

Carbon Steel, Schedule 80 with 300# M.I. Fittings & Threaded Joints

Description	Craft@Hrs	Unit	Material $	Labor $	Equipment $	Total $

Class 250 iron body strainer, flanged ends

2"	P1@.500	Ea	202.00	17.90	—	219.90
2½"	P1@.600	Ea	221.00	21.50	—	242.50
3"	P1@.750	Ea	258.00	26.80	—	284.80
4"	P1@1.35	Ea	458.00	48.30	—	506.30

Installation of 2-way control valve, threaded joints

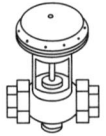

1/2"	P1@.210	Ea	—	7.51	—	7.51
3/4"	P1@.275	Ea	—	9.83	—	9.83
1"	P1@.350	Ea	—	12.50	—	12.50
1¼"	P1@.430	Ea	—	15.40	—	15.40
1½"	P1@.505	Ea	—	18.10	—	18.10
2"	P1@.675	Ea	—	24.10	—	24.10
2½"	P1@.830	Ea	—	29.70	—	29.70
3"	P1@.990	Ea	—	35.40	—	35.40

Installation of 3-way control valve, threaded joints

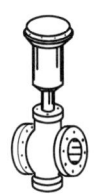

1/2"	P1@.260	Ea	—	9.30	—	9.30
3/4"	P1@.365	Ea	—	13.10	—	13.10
1"	P1@.475	Ea	—	17.00	—	17.00
1¼"	P1@.575	Ea	—	20.60	—	20.60
1½"	P1@.680	Ea	—	24.30	—	24.30
2"	P1@.910	Ea	—	32.50	—	32.50
2½"	P1@1.12	Ea	—	40.10	—	40.10
3"	P1@1.33	Ea	—	47.60	—	47.60

Threaded companion flange, 300# malleable iron

2"	P1@.290	Ea	83.30	10.40	—	93.70
2½"	P1@.380	Ea	114.00	13.60	—	127.60
3"	P1@.460	Ea	108.00	16.40	—	124.40
4"	P1@.600	Ea	201.00	21.50	—	222.50

Bolt and gasket sets

2"	P1@.500	Ea	3.86	17.90	—	21.76
2½"	P1@.650	Ea	4.51	23.20	—	27.71
3"	P1@.750	Ea	7.67	26.80	—	34.47
4"	P1@1.00	Ea	12.80	35.80	—	48.60

Carbon Steel, Schedule 80 with 300# M.I. Fittings & Threaded Joints

Description	Craft@Hrs	Unit	Material $	Labor $	Equipment $	Total $
Thermometer with well						
7"	P1@.250	Ea	159.00	8.94	—	167.94
9"	P1@.250	Ea	164.00	8.94	—	172.94
Dial type pressure gauge						
2½"	P1@.200	Ea	33.10	7.15	—	40.25
3½"	P1@.200	Ea	43.80	7.15	—	50.95
Pressure/temperature tap						
tap	P1@.150	Ea	15.70	5.36	—	21.06
Hanger with swivel assembly						
1/2"	P1@.250	Ea	4.13	8.94	—	13.07
3/4"	P1@.250	Ea	4.33	8.94	—	13.27
1"	P1@.250	Ea	4.57	8.94	—	13.51
1¼"	P1@.300	Ea	4.77	10.70	—	15.47
1½"	P1@.300	Ea	5.51	10.70	—	16.21
2"	P1@.300	Ea	6.05	10.70	—	16.75
2½"	P1@.350	Ea	7.82	12.50	—	20.32
3"	P1@.350	Ea	9.67	12.50	—	22.17
4"	P1@.350	Ea	14.90	12.50	—	27.40
Riser clamp						
1/2"	P1@.100	Ea	2.41	3.58	—	5.99
3/4"	P1@.100	Ea	3.55	3.58	—	7.13
1"	P1@.100	Ea	3.59	3.58	—	7.17
1¼"	P1@.105	Ea	4.33	3.75	—	8.08
1½"	P1@.110	Ea	4.58	3.93	—	8.51
2"	P1@.115	Ea	4.84	4.11	—	8.95
2½"	P1@.120	Ea	5.09	4.29	—	9.38
3"	P1@.120	Ea	5.52	4.29	—	9.81
4"	P1@.125	Ea	7.03	4.47	—	11.50

Carbon Steel, Schedule 160 with 3,000-6,000# Fittings

Schedule 160 carbon steel (ASTM A-106) pipe with steel fittings is commonly used for steam, steam condensate and recirculating systems.

The cost estimates in this section are based on the conditions, limitations and wage rates described in the section "How to Use This Book" beginning on page 5.

Description	Craft@Hrs	Unit	Material $	Labor $	Equipment $	Total $

Schedule 160 carbon steel threaded horizontal pipe assembly.

Horizontally hung in a building. Assembly includes fittings and hanger assemblies. Based on a reducing tee and a 45 degree elbow every 16 feet for 1/2" pipe. A reducing tee and a 45 degree elbow every 30 feet for 4" pipe. Hangers spaced to meet plumbing code. *(Distance between fittings in the pipe assembly increases as pipe diameter increases.)* Use these figures for preliminary estimates.

Description	Craft@Hrs	Unit	Material $	Labor $	Equipment $	Total $
1/2"	P1@.140	LF	4.90	5.01	—	9.91
3/4"	P1@.145	LF	5.74	5.19	—	10.93
1"	P1@.172	LF	7.88	6.15	—	14.03
1¼"	P1@.184	LF	8.63	6.58	—	15.21
1½"	P1@.207	LF	10.00	7.40	—	17.40
2"	P1@.272	LF	15.00	9.73	—	24.73
2½"	P1@.388	LF	23.10	13.90	—	37.00
3"	P1@.434	LF	34.50	15.50	—	50.00
4"	P1@.530	LF	56.50	19.00	—	75.50

Schedule 160 carbon steel socket weld horizontal pipe assembly.

Horizontally hung in a building. Assembly includes fittings and hanger assemblies. Based on a reducing tee and a 45 degree elbow every 16 feet for 1/2" pipe. A reducing tee and a 45 degree elbow every 30 feet for 4" pipe Hangers spaced to meet plumbing code. *(Distance between fittings in the pipe assembly increases as pipe diameter increases.)* Use these figures for preliminary estimates.

Description	Craft@Hrs	Unit	Material $	Labor $	Equipment $	Total $
1/2"	P1@.175	LF	4.76	6.26	—	11.02
3/4"	P1@.206	LF	5.51	7.37	—	12.88
1"	P1@.247	LF	7.59	8.83	—	16.42
1¼"	P1@.250	LF	8.42	8.94	—	17.36
1½"	P1@.284	LF	9.19	10.20	—	19.39
2"	P1@.375	LF	14.30	13.40	—	27.70
2½"	P1@.503	LF	21.50	18.00	—	39.50
3"	P1@.568	LF	33.50	20.30	—	53.80
4"	P1@.709	LF	55.30	25.40	—	80.70

Description	Craft@Hrs	Unit	Material $	Labor $	Equipment $	Total $

Schedule 160 carbon steel threaded vertical pipe assembly. Riser assembly, including fittings and riser clamps. Based on a reducing tee, and a riser clamp on every floor. Use these figures for preliminary estimates.

Description	Craft@Hrs	Unit	Material $	Labor $	Equipment $	Total $
1/2"	P1@.114	LF	4.30	4.08	—	8.38
3/4"	P1@.125	LF	5.16	4.47	—	9.63
1"	P1@.148	LF	7.13	5.29	—	12.42
1¼"	P1@.165	LF	8.74	5.90	—	14.64
1½"	P1@.183	LF	9.65	6.54	—	16.19
2"	P1@.242	LF	14.40	8.65	—	23.05
2½"	P1@.348	LF	21.50	12.40	—	33.90
3"	P1@.393	LF	33.20	14.10	—	47.30
4"	P1@.487	LF	53.00	17.40	—	70.40

Schedule 160 carbon steel socket weld vertical pipe assembly. Riser assembly, including fittings and riser clamps. Based on a reducing tee, and a riser clamp on every floor. Use these figures for preliminary estimates.

Description	Craft@Hrs	Unit	Material $	Labor $	Equipment $	Total $
1/2"	P1@.133	LF	4.20	4.76	—	8.96
3/4"	P1@.162	LF	4.96	5.79	—	10.75
1"	P1@.199	LF	6.96	7.12	—	14.08
1¼"	P1@.227	LF	8.11	8.12	·	10.23
1½"	P1@.255	LF	9.04	9.12	—	18.16
2"	P1@.341	LF	13.70	12.20	—	25.90
2½"	P1@.467	LF	20.00	16.70	—	36.70
3"	P1@.532	LF	33.50	19.00	—	52.50
4"	P1@.674	LF	52.40	24.10	—	76.50

Schedule 160 carbon steel plain end pipe

Description	Craft@Hrs	Unit	Material $	Labor $	Equipment $	Total $
1/2"	P1@.090	LF	3.99	3.22	—	7.21
3/4"	P1@.100	LF	4.62	3.58	—	8.20
1"	P1@.120	LF	6.56	4.29	—	10.85
1¼"	P1@.130	LF	7.72	4.65	—	12.37
1½"	P1@.140	LF	12.00	5.01	—	17.01
2"	P1@.190	LF	9.94	6.79	—	16.73
3"	P1@.310	LF	19.20	11.10	—	30.30
4"	P1@.380	LF	30.00	13.60	—	43.60

Description	Craft@Hrs	Unit	Material $	Labor $	Equipment $	Total $

3,000# carbon steel 45 degree ell, threaded joints

Description	Craft@Hrs	Unit	Material $	Labor $	Equipment $	Total $
1/2"	P1@.132	Ea	7.74	4.72	—	12.46
3/4"	P1@.143	Ea	8.87	5.11	—	13.98
1"	P1@.198	Ea	12.20	7.08	—	19.28
1¼"	P1@.264	Ea	16.80	9.44	—	26.24
1½"	P1@.330	Ea	25.40	11.80	—	37.20
2"	P1@.418	Ea	34.80	14.90	—	49.70
2½"	P1@.561	Ea	86.00	20.10	—	106.10
3"	P1@.704	Ea	140.00	25.20	—	165.20
4"	P1@.935	Ea	258.00	33.40	—	291.40

3,000# carbon steel 45 degree ell, socket-welded joints

Description	Craft@Hrs	Unit	Material $	Labor $	Equipment $	Total $
1/2"	P1@.300	Ea	6.28	10.70	—	16.98
3/4"	P1@.440	Ea	7.59	15.70	—	23.29
1"	P1@.590	Ea	9.89	21.10	—	30.99
1¼"	P1@.740	Ea	13.50	26.50	—	40.00
1½"	P1@.890	Ea	16.40	31.80	—	48.20
2"	P1@1.18	Ea	28.40	42.20	—	70.60
2½"	P1@1.49	Ea	73.30	53.30	—	126.60
3"	P1@1.77	Ea	123.00	63.30	—	186.30
4"	P1@2.37	Ea	244.00	84.80	—	328.80

6,000# carbon steel 45 degree ell, threaded joints

Description	Craft@Hrs	Unit	Material $	Labor $	Equipment $	Total $
1/2"	P1@.132	Ea	13.00	4.72	—	17.72
3/4"	P1@.143	Ea	17.80	5.11	—	22.91
1"	P1@.198	Ea	20.00	7.08	—	27.08
1¼"	P1@.264	Ea	53.10	9.44	—	62.54
1½"	P1@.330	Ea	53.10	11.80	—	64.90
2"	P1@.418	Ea	74.20	14.90	—	89.10
2½"	P1@.561	Ea	282.00	20.10	—	302.10
3"	P1@.704	Ea	290.00	25.20	—	315.20

6,000# carbon steel 45 degree ell, socket-welded joints

Description	Craft@Hrs	Unit	Material $	Labor $	Equipment $	Total $
1/2"	P1@.300	Ea	10.20	10.70	—	20.90
3/4"	P1@.440	Ea	11.70	15.70	—	27.40
1"	P1@.590	Ea	15.90	21.10	—	37.00
1¼"	P1@.740	Ea	57.00	26.50	—	83.50
1½"	P1@.890	Ea	38.50	31.80	—	70.30
2"	P1@1.18	Ea	47.80	42.20	—	90.00
2½"	P1@1.49	Ea	191.00	53.30	—	244.30
3"	P1@1.77	Ea	224.00	63.30	—	287.30
4"	P1@2.37	Ea	267.00	84.80	—	351.80

Description	Craft@Hrs	Unit	Material $	Labor $	Equipment $	Total $

3,000# carbon steel 90 degree ell, threaded joints

Description	Craft@Hrs	Unit	Material $	Labor $	Equipment $	Total $
1/2"	P1@.132	Ea	6.05	4.72	—	10.77
3/4"	P1@.143	Ea	7.28	5.11	—	12.39
1"	P1@.198	Ea	10.50	7.08	—	17.58
1¼"	P1@.264	Ea	17.80	9.44	—	27.24
1½"	P1@.330	Ea	26.70	11.80	—	38.50
2"	P1@.418	Ea	32.20	14.90	—	47.10
2½"	P1@.561	Ea	79.80	20.10	—	99.90
3"	P1@.704	Ea	122.00	25.20	—	147.20
4"	P1@.935	Ea	267.00	33.40	—	300.40

3,000# carbon steel 90 degree ell, socket-welded joints

Description	Craft@Hrs	Unit	Material $	Labor $	Equipment $	Total $
1/2"	P1@.300	Ea	5.99	10.70	—	16.69
3/4"	P1@.440	Ea	6.19	15.70	—	21.89
1"	P1@.590	Ea	7.95	21.10	—	29.05
1¼"	P1@.740	Ea	13.00	26.50	—	39.50
1½"	P1@.890	Ea	16.70	31.80	—	48.50
2"	P1@1.18	Ea	25.40	42.20	—	67.60
2½"	P1@1.49	Ea	61.00	53.30	—	114.30
3"	P1@1.77	Ea	101.00	63.30	—	161.30
4"	P1@2.37	Ea	296.00	84.80	—	380.80

6,000# carbon steel 90 degree ell, threaded joints

Description	Craft@Hrs	Unit	Material $	Labor $	Equipment $	Total $
1/2"	P1@.132	Ea	9.82	4.72	—	14.54
3/4"	P1@.143	Ea	12.50	5.11	—	17.61
1"	P1@.198	Ea	16.10	7.08	—	23.18
1¼"	P1@.264	Ea	27.00	9.44	—	36.44
1½"	P1@.330	Ea	41.90	11.80	—	53.70
2"	P1@.418	Ea	73.70	14.90	—	88.60
2½"	P1@.561	Ea	109.00	20.10	—	129.10
3"	P1@.704	Ea	248.00	25.20	—	273.20

6,000# carbon steel 90 degree ell, socket-welded joints

Description	Craft@Hrs	Unit	Material $	Labor $	Equipment $	Total $
1/2"	P1@.300	Ea	8.30	10.70	—	19.00
3/4"	P1@.440	Ea	9.72	15.70	—	25.42
1"	P1@.590	Ea	13.50	21.10	—	34.60
1¼"	P1@.740	Ea	19.60	26.50	—	46.10
1½"	P1@.890	Ea	29.70	31.80	—	61.50
2"	P1@1.18	Ea	43.60	42.20	—	85.80
2½"	P1@1.49	Ea	90.90	53.30	—	144.20
3"	P1@1.77	Ea	160.00	63.30	—	223.30
4"	P1@2.37	Ea	353.00	84.80	—	437.80

Carbon Steel, Schedule 160 with 3,000-6,000# Fittings

Description	Craft@Hrs	Unit	Material $	Labor $	Equipment $	Total $

3,000# carbon steel tee, threaded joints

Description	Craft@Hrs	Unit	Material $	Labor $	Equipment $	Total $
1/2"	P1@.198	Ea	7.98	7.08	—	15.06
3/4"	P1@.209	Ea	11.40	7.47	—	18.87
1"	P1@.253	Ea	14.50	9.05	—	23.55
1¼"	P1@.341	Ea	24.20	12.20	—	36.40
1½"	P1@.429	Ea	32.80	15.30	—	48.10
2"	P1@.539	Ea	43.70	19.30	—	63.00
2½"	P1@.726	Ea	105.00	26.00	—	131.00
3"	P1@.913	Ea	197.00	32.60	—	229.60
4"	P1@1.21	Ea	327.00	43.30	—	370.30

3,000# carbon steel tee, socket-welded joints

Description	Craft@Hrs	Unit	Material $	Labor $	Equipment $	Total $
1/2"	P1@.440	Ea	6.89	15.70	—	22.59
3/4"	P1@.670	Ea	8.69	24.00	—	32.69
1"	P1@.890	Ea	12.00	31.80	—	43.80
1¼"	P1@1.11	Ea	16.10	39.70	—	55.80
1½"	P1@1.33	Ea	25.10	47.60	—	72.70
2"	P1@1.77	Ea	35.20	63.30	—	98.50
2½"	P1@2.22	Ea	85.00	79.40	—	164.40
3"	P1@2.66	Ea	200.00	95.10	—	295.10
4"	P1@3.55	Ea	321.00	127.00	—	448.00

6,000# carbon steel tee, threaded joints

Description	Craft@Hrs	Unit	Material $	Labor $	Equipment $	Total $
1/2"	P1@.198	Ea	12.70	7.08	—	19.78
3/4"	P1@.209	Ea	17.00	7.47	—	24.47
1"	P1@.253	Ea	28.40	9.05	—	37.45
1¼"	P1@.341	Ea	35.40	12.20	—	47.60
1½"	P1@.429	Ea	48.20	15.30	—	63.50
2"	P1@.539	Ea	94.20	19.30	—	113.50
2½"	P1@.726	Ea	339.00	26.00	—	365.00
3"	P1@.913	Ea	342.00	32.60	—	374.60

6,000# carbon steel tee, socket-welded joints

Description	Craft@Hrs	Unit	Material $	Labor $	Equipment $	Total $
1/2"	P1@.440	Ea	11.00	15.70	—	26.70
3/4"	P1@.670	Ea	14.50	24.00	—	38.50
1"	P1@.890	Ea	18.20	31.80	—	50.00
1¼"	P1@1.11	Ea	51.80	39.70	—	91.50
1½"	P1@1.33	Ea	46.50	47.60	—	94.10
2"	P1@1.77	Ea	58.90	63.30	—	122.20
2½"	P1@2.22	Ea	125.00	79.40	—	204.40
3"	P1@2.66	Ea	209.00	95.10	—	304.10
4"	P1@3.55	Ea	397.00	127.00	—	524.00

Description	Craft@Hrs	Unit	Material $	Labor $	Equipment $	Total $

3,000# carbon steel union, threaded joints

Description	Craft@Hrs	Unit	Material $	Labor $	Equipment $	Total $
1/2"	P1@.150	Ea	8.76	5.36	—	14.12
3/4"	P1@.165	Ea	10.50	5.90	—	16.40
1"	P1@.230	Ea	13.80	8.22	—	22.02
1¼"	P1@.310	Ea	23.30	11.10	—	34.40
1½"	P1@.400	Ea	25.40	14.30	—	39.70
2"	P1@.500	Ea	33.90	17.90	—	51.80
2½"	P1@.670	Ea	73.80	24.00	—	97.80
3"	P1@.840	Ea	92.30	30.00	—	122.30

3,000# carbon steel union, socket-welded joints

Description	Craft@Hrs	Unit	Material $	Labor $	Equipment $	Total $
1/2"	P1@.300	Ea	10.50	10.70	—	21.20
3/4"	P1@.440	Ea	12.00	15.70	—	27.70
1"	P1@.590	Ea	15.40	21.10	—	36.50
1¼"	P1@.740	Ea	24.80	26.50	—	51.30
1½"	P1@.890	Ea	25.80	31.80	—	57.60
2"	P1@1.18	Ea	36.30	42.20	—	78.50
2½"	P1@1.49	Ea	84.80	53.30	—	138.10
3"	P1@1.77	Ea	108.00	63.30	—	171.30

6,000# carbon steel union, threaded joints

Description	Craft@Hrs	Unit	Material $	Labor $	Equipment $	Total $
1/2"	P1@.150	Ea	20.60	5.36	—	25.96
3/4"	P1@.165	Ea	23.30	5.90	—	29.20
1"	P1@.230	Ea	30.70	8.22	—	38.92
1¼"	P1@.310	Ea	50.10	11.10	—	61.20
1½"	P1@.400	Ea	62.00	14.30	—	76.30
2"	P1@.500	Ea	88.70	17.90	—	106.60

6,000# carbon steel union, socket-welded joints

Description	Craft@Hrs	Unit	Material $	Labor $	Equipment $	Total $
1/2"	P1@.300	Ea	21.20	10.70	—	31.90
3/4"	P1@.440	Ea	24.30	15.70	—	40.00
1"	P1@.590	Ea	35.50	21.10	—	56.60
1¼"	P1@.740	Ea	54.40	26.50	—	80.90
1½"	P1@.890	Ea	63.60	31.80	—	95.40
2"	P1@1.18	Ea	96.60	42.20	—	138.80

3,000# carbon steel reducer, socket-welded joints

Description	Craft@Hrs	Unit	Material $	Labor $	Equipment $	Total $
3/4"	P1@.360	Ea	5.72	12.90	—	18.62
1"	P1@.500	Ea	6.95	17.90	—	24.85
1¼"	P1@.665	Ea	8.05	23.80	—	31.85
1½"	P1@.800	Ea	10.10	28.60	—	38.70
2"	P1@1.00	Ea	13.80	35.80	—	49.60
2½"	P1@1.33	Ea	95.60	47.60	—	143.20
3"	P1@1.63	Ea	120.00	58.30	—	178.30

Carbon Steel, Schedule 160 with 3,000-6,000# Fittings

Description	Craft@Hrs	Unit	Material $	Labor $	Equipment $	Total $

6,000# carbon steel reducer, socket-welded joints

Description	Craft@Hrs	Unit	Material $	Labor $	Equipment $	Total $
3/4"	P1@.360	Ea	21.90	12.90	—	34.80
1"	P1@.500	Ea	22.80	17.90	—	40.70
1¼"	P1@.665	Ea	48.80	23.80	—	72.60
1½"	P1@.800	Ea	48.80	28.60	—	77.40
2"	P1@1.00	Ea	64.70	35.80	—	100.50

3,000# carbon steel cap, socket-welded joint

Description	Craft@Hrs	Unit	Material $	Labor $	Equipment $	Total $
1/2"	P1@.180	Ea	3.42	6.44	—	9.86
3/4"	P1@.270	Ea	3.99	9.66	—	13.65
1"	P1@.360	Ea	6.08	12.90	—	18.98
1¼"	P1@.460	Ea	7.13	16.40	—	23.53
1½"	P1@.560	Ea	9.71	20.00	—	29.71
2"	P1@.740	Ea	15.10	26.50	—	41.60
2½"	P1@.920	Ea	81.50	32.90	—	114.40
3"	P1@1.11	Ea	84.20	39.70	—	123.90
4"	P1@1.50	Ea	127.00	53.60	—	180.60

6,000# carbon steel cap, socket-welded joint

Description	Craft@Hrs	Unit	Material $	Labor $	Equipment $	Total $
1/2"	P1@.180	Ea	6.56	6.44	—	13.00
3/4"	P1@.270	Ea	9.55	9.66	—	19.21
1"	P1@.360	Ea	10.10	12.90	—	23.00
1¼"	P1@.460	Ea	23.70	16.40	—	40.10
1½"	P1@.560	Ea	30.70	20.00	—	50.70
2"	P1@.740	Ea	45.50	26.50	—	72.00

3,000# carbon steel weldolet

Description	Craft@Hrs	Unit	Material $	Labor $	Equipment $	Total $
1/2"	P1@.580	Ea	12.10	20.70	—	32.80
3/4"	P1@.890	Ea	12.40	31.80	—	44.20
1"	P1@1.18	Ea	13.60	42.20	—	55.80
1¼"	P1@1.48	Ea	17.00	52.90	—	69.90
1½"	P1@1.75	Ea	17.00	62.60	—	79.60
2"	P1@2.35	Ea	18.40	84.00	—	102.40
2½"	P1@2.90	Ea	41.40	104.00	—	145.40
3"	P1@3.50	Ea	42.60	125.00	—	167.60
4"	P1@4.75	Ea	52.80	170.00	—	222.80

6,000# carbon steel weldolet

Description	Craft@Hrs	Unit	Material $	Labor $	Equipment $	Total $
1/2"	P1@.690	Ea	52.80	24.70	—	77.50
3/4"	P1@1.05	Ea	55.00	37.50	—	92.50
1"	P1@1.40	Ea	57.70	50.10	—	107.80
1¼"	P1@1.75	Ea	69.50	62.60	—	132.10
1½"	P1@2.10	Ea	78.10	75.10	—	153.20
2"	P1@2.80	Ea	89.90	100.00	—	189.90
2½"	P1@3.50	Ea	131.00	125.00	—	256.00
3"	P1@4.20	Ea	147.00	150.00	—	297.00
4"	P1@5.70	Ea	250.00	204.00	—	454.00

Description	Craft@Hrs	Unit	Material $	Labor $	Equipment $	Total $

3,000# carbon steel threadolet

Description	Craft@Hrs	Unit	Material $	Labor $	Equipment $	Total $
1/2"	P1@.400	Ea	9.20	14.30	—	23.50
3/4"	P1@.580	Ea	10.20	20.70	—	30.90
1"	P1@.790	Ea	12.40	28.30	—	40.70
1¼"	P1@.980	Ea	56.30	35.00	—	91.30
1½"	P1@1.20	Ea	56.30	42.90	—	99.20
2"	P1@1.60	Ea	74.30	57.20	—	131.50

6,000# carbon steel threadolet

Description	Craft@Hrs	Unit	Material $	Labor $	Equipment $	Total $
1/2"	P1@.480	Ea	14.10	17.20	—	31.30
3/4"	P1@.690	Ea	15.20	24.70	—	39.90
1"	P1@.950	Ea	21.80	34.00	—	55.80
1¼"	P1@1.20	Ea	81.50	42.90	—	124.40
1½"	P1@1.45	Ea	81.50	51.90	—	133.40
2"	P1@1.90	Ea	133.00	67.90	—	200.90

Class 300 forged steel companion flange, weld neck

Description	Craft@Hrs	Unit	Material $	Labor $	Equipment $	Total $
1/2"	P1@.240	Ea	16.60	8.58	—	25.18
3/4"	P1@.320	Ea	16.60	11.40	—	28.00
1"	P1@.340	Ea	16.60	12.20	—	28.80
1¼"	P1@.425	Ea	21.40	15.20	—	36.60
1½"	P1@.560	Ea	21.40	20.00	—	41.40
2"	P1@.740	Ea	23.50	26.50	—	50.00
2½"	P1@.810	Ea	34.40	29.00	—	63.40
3"	P1@.980	Ea	35.30	35.00	—	70.30
4"	P1@1.30	Ea	52.00	46.50	—	98.50

Class 300 forged steel companion flange, slip on

Description	Craft@Hrs	Unit	Material $	Labor $	Equipment $	Total $
1/2"	P1@.240	Ea	13.10	8.58	—	21.68
3/4"	P1@.320	Ea	13.10	11.40	—	24.50
1"	P1@.340	Ea	13.10	12.20	—	25.30
1½"	P1@.425	Ea	19.60	15.20	—	34.80
2"	P1@.560	Ea	18.70	20.00	—	38.70
2½"	P1@.740	Ea	31.70	26.50	—	58.20
3"	P1@.810	Ea	31.70	29.00	—	60.70
4"	P1@.980	Ea	46.80	35.00	—	81.80

Class 300 cast steel gate valve, flanged

Description	Craft@Hrs	Unit	Material $	Labor $	Equipment $	Total $
2"	P1@.500	Ea	1,650.00	17.90	—	1,667.90
2½"	P1@.600	Ea	2,220.00	21.50	—	2,241.50
3"	P1@.750	Ea	2,220.00	26.80	—	2,246.80
4"	P1@1.35	Ea	3,100.00	48.30	—	3,148.30

Carbon Steel, Schedule 160 with 3,000-6,000# Fittings

Description	Craft@Hrs	Unit	Material $	Labor $	Equipment $	Total $

Class 600 cast steel gate valve, flanged

2"	P1@.500	Ea	2,730.00	17.90	—	2,747.90
2½"	P1@.600	Ea	3,920.00	21.50	—	3,941.50
3"	P1@.750	Ea	3,920.00	26.80	—	3,946.80
4"	P1@1.35	Ea	6,290.00	48.30	—	6,338.30

Class 300 cast steel body globe valve, flanged

2"	P1@.500	Ea	1,140.00	17.90	—	1,157.90
2½"	P1@.600	Ea	1,560.00	21.50	—	1,581.50
3"	P1@.750	Ea	1,560.00	26.80	—	1,586.80
4"	P1@1.35	Ea	2,140.00	48.30	—	2,188.30

Class 600 cast steel body globe valve, flanged

2"	P1@.500	Ea	1,600.00	17.90	—	1,617.90
2½"	P1@.600	Ea	2,420.00	21.50	—	2,441.50
3"	P1@.750	Ea	2,420.00	26.80	—	2,446.80
4"	P1@1.35	Ea	3,900.00	48.30	—	3,948.30

Class 300 cast steel body swing check valve, flanged

2"	P1@.500	Ea	1,110.00	17.90	—	1,127.90
2½"	P1@.600	Ea	1,680.00	21.50	—	1,701.50
3"	P1@.750	Ea	1,680.00	26.80	—	1,706.80
4"	P1@1.35	Ea	2,380.00	48.30	—	2,428.30

Class 600 cast steel body swing check valve, flanged

2"	P1@.500	Ea	1,710.00	17.90	—	1,727.90
2½"	P1@.600	Ea	2,700.00	21.50	—	2,721.50
3"	P1@.750	Ea	2,700.00	26.80	—	2,726.80
4"	P1@1.35	Ea	4,290.00	48.30	—	4,338.30

Installation of cast steel body 2-way control valve, threaded joints

1/2"	P1@.210	Ea	—	7.51	—	7.51
3/4"	P1@.250	Ea	—	8.94	—	8.94
1"	P1@.300	Ea	—	10.70	—	10.70
1¼"	P1@.400	Ea	—	14.30	—	14.30
1½"	P1@.450	Ea	—	16.10	—	16.10
2"	P1@.500	Ea	—	17.90	—	17.90
2½"	P1@.830	Ea	—	29.70	—	29.70
3"	P1@.990	Ea	—	35.40	—	35.40

Installation of cast steel body 2-way control valve, flanged joints

2"	P1@.500	Ea	—	17.90	—	17.90
2½"	P1@.600	Ea	—	21.50	—	21.50
3"	P1@.750	Ea	—	26.80	—	26.80
4"	P1@1.35	Ea	—	48.30	—	48.30

Description	Craft@Hrs	Unit	Material $	Labor $	Equipment $	Total $

Installation of cast steel body 3-way control valve, threaded joints

Description	Craft@Hrs	Unit	Material $	Labor $	Equipment $	Total $
1/2"	P1@.260	Ea	—	9.30	—	9.30
3/4"	P1@.365	Ea	—	13.10	—	13.10
1"	P1@.475	Ea	—	17.00	—	17.00
1¼"	P1@.575	Ea	—	20.60	—	20.60
1½"	P1@.680	Ea	—	24.30	—	24.30
2"	P1@.910	Ea	—	32.50	—	32.50
2½"	P1@1.12	Ea	—	40.10	—	40.10
3"	P1@1.33	Ea	—	47.60	—	47.60

Installation of cast steel body 3-way control valve, flanged joints

Description	Craft@Hrs	Unit	Material $	Labor $	Equipment $	Total $
2"	P1@.910	Ea	—	32.50	—	32.50
2½"	P1@1.12	Ea	—	40.10	—	40.10
3"	P1@1.33	Ea	—	47.60	—	47.60
4"	P1@2.00	Ea	—	71.50	—	71.50

300# bolt and gasket set, ring face

Description	Craft@Hrs	Unit	Material $	Labor $	Equipment $	Total $
2"	P1@.500	Ea	7.67	17.90	—	25.57
2½"	P1@.650	Ea	8.96	23.20	—	32.16
3"	P1@.760	Ea	15.40	26.80	—	42.20
4"	P1@1.00	Ea	25.00	35.80	—	60.80

300# bolt and gasket set, full face

Description	Craft@Hrs	Unit	Material $	Labor $	Equipment $	Total $
2"	P1@.500	Ea	15.40	17.90	—	33.30
2½"	P1@.650	Ea	17.90	23.20	—	41.10
3"	P1@.750	Ea	30.80	26.80	—	57.60
4"	P1@1.00	Ea	50.20	35.80	—	86.00

Thermometer with well

Description	Craft@Hrs	Unit	Material $	Labor $	Equipment $	Total $
7"	P1@.250	Ea	159.00	8.94	—	167.94
9"	P1@.250	Ea	164.00	8.94	—	172.94

Dial type pressure gauge

Description	Craft@Hrs	Unit	Material $	Labor $	Equipment $	Total $
2½"	P1@.200	Ea	33.10	7.15	—	40.25
3½"	P1@.200	Ea	43.80	7.15	—	50.95

Pressure/temperature tap

Description	Craft@Hrs	Unit	Material $	Labor $	Equipment $	Total $
tap	P1@.150	Ea	15.70	5.36	—	21.06

Carbon Steel, Schedule 160 with 3,000-6,000# Fittings

Description	Craft@Hrs	Unit	Material $	Labor $	Equipment $	Total $

Hanger with swivel assembly

Description	Craft@Hrs	Unit	Material $	Labor $	Equipment $	Total $
1/2"	P1@.250	Ea	5.62	8.94	—	14.56
3/4"	P1@.250	Ea	6.16	8.94	—	15.10
1"	P1@.250	Ea	6.45	8.94	—	15.39
1¼"	P1@.300	Ea	6.99	10.70	—	17.69
1½"	P1@.300	Ea	9.09	10.70	—	19.79
2"	P1@.300	Ea	11.70	10.70	—	22.40
2½"	P1@.350	Ea	14.70	12.50	—	27.20
3"	P1@.350	Ea	25.20	12.50	—	37.70
4"	P1@.350	Ea	32.80	12.50	—	45.30

Riser clamp

Description	Craft@Hrs	Unit	Material $	Labor $	Equipment $	Total $
1/2"	P1@.100	Ea	2.41	3.58	—	5.99
3/4"	P1@.100	Ea	3.55	3.58	—	7.13
1¼"	P1@.105	Ea	4.33	3.75	—	8.08
1½"	P1@.110	Ea	4.58	3.93	—	8.51
2"	P1@.115	Ea	4.84	4.11	—	8.95
2½"	P1@.120	Ea	5.09	4.29	—	9.38
3"	P1@.120	Ea	5.52	4.29	—	9.81
4"	P1@.125	Ea	7.03	4.47	—	11.50

Carbon Steel, Schedule 40 with Roll-Grooved Joints

Schedule 40 carbon steel (ASTM A-53 and A-120) roll-grooved pipe with factory grooved malleable iron, ductile iron, or steel fittings is commonly used for heating hot water, chilled water and condenser water systems.

Consult the manufacturers for maximum operating temperature/pressure ratings for the various combinations of pipe and fittings and system applications.

The cost estimates in this section are based on the conditions, limitations and wage rates described in the section "How to Use This Book" beginning on page 5.

Equipment cost, where shown, is $110 per hour for a 10-ton hydraulic truck-mounted crane.

Description	Craft@Hrs	Unit	Material $	Labor $	Equipment $	Total $

Schedule 40 carbon steel roll-grooved horizontal pipe assembly.
Horizontally hung in a building. Assembly includes fittings, couplings and hanger assemblies. Based on a reducing tee and a 45 degree elbow every 16 feet for 1/2" pipe. A reducing tee and a 45 degree elbow every 50 feet for 12" pipe. Hangers spaced to meet plumbing code. *(Distance between fittings in the pipe assembly increases as pipe diameter increases.)* Use these figures for preliminary estimates.

Description	Craft@Hrs	Unit	Material $	Labor $	Equipment $	Total $
2"	P1@.183	LF	8.33	6.54	—	14.87
2½"	P1@.192	LF	10.90	6.87	—	17.77
3"	P1@.220	LF	12.50	7.87	—	20.37
4"	P1@.262	LF	17.90	9.37	—	27.27
6"	ER@.347	LF	33.20	13.20	5.50	51.90
8"	ER@.414	LF	48.30	15.80	6.40	70.50
10"	ER@.530	LF	71.90	20.20	8.30	100.40
12"	ER@.617	LF	128.00	23.50	9.70	161.20

Schedule 40 carbon steel roll-grooved vertical pipe assembly.
Riser assembly, including fittings, couplings, and riser clamps. Based on a reducing tee, and a riser clamp on every floor. Use these figures for preliminary estimates.

Description	Craft@Hrs	Unit	Material $	Labor $	Equipment $	Total $
2"	P1@.120	LF	7.31	4.29	—	11.60
2½"	P1@.123	LF	10.40	4.40	—	14.80
3"	P1@.150	LF	10.80	5.36	—	16.16
4"	P1@.209	LF	16.30	7.47	—	23.77
6"	ER@.294	LF	31.60	11.20	4.60	47.40
8"	ER@.395	LF	51.80	15.10	6.20	73.10
10"	ER@.523	LF	71.90	19.90	8.30	100.10
12"	ER@.608	LF	117.00	23.20	9.50	149.70

Schedule 40 carbon steel pipe with roll-grooved joints

Description	Craft@Hrs	Unit	Material $	Labor $	Equipment $	Total $
2"	P1@.100	LF	3.31	3.58	—	6.89
2½"	P1@.115	LF	5.20	4.11	—	9.31
3"	P1@.135	LF	7.08	4.83	—	11.91
4"	P1@.190	LF	10.20	6.79	—	16.99
5"	ER@.240	LF	19.10	9.15	3.90	32.15
6"	ER@.315	LF	28.90	12.00	5.00	45.90
8"	ER@.370	LF	28.90	14.10	5.90	48.90
10"	ER@.460	LF	40.50	17.50	7.20	65.20
12"	ER@.550	LF	63.80	21.00	8.60	93.40

Carbon Steel, Schedule 40 with Roll-Grooved Joints

Description	Craft@Hrs	Unit	Material $	Labor $	Equipment $	Total $

Schedule 40 carbon steel 45 degree ell with roll-grooved joints

Description	Craft@Hrs	Unit	Material $	Labor $	Equipment $	Total $
2"	P1@.170	Ea	23.50	6.08	—	29.58
2½"	P1@.185	Ea	31.90	6.62	—	38.52
3"	P1@.200	Ea	42.10	7.15	—	49.25
4"	P1@.230	Ea	62.20	8.22	—	70.42
5"	ER@.250	Ea	150.00	9.53	3.90	163.43
6"	ER@.350	Ea	176.00	13.30	5.50	194.80
8"	ER@.500	Ea	368.00	19.10	7.90	395.00
10"	ER@.750	Ea	498.00	28.60	11.80	538.40
12"	ER@1.00	Ea	872.00	38.10	15.70	925.80

Schedule 40 carbon steel 90 degree ell with roll-grooved joints

Description	Craft@Hrs	Unit	Material $	Labor $	Equipment $	Total $
2"	P1@.170	Ea	23.50	6.08	—	29.58
2½"	P1@.185	Ea	31.90	6.62	—	38.52
3"	P1@.200	Ea	42.10	7.15	—	49.25
4"	P1@.230	Ea	62.20	8.22	—	70.42
5"	ER@.250	Ea	150.00	9.53	3.90	163.43
6"	ER@.350	Ea	176.00	13.30	5.50	194.80
8"	ER@.500	Ea	368.00	19.10	7.90	395.00
10"	ER@.750	Ea	571.00	28.60	11.80	611.40
12"	ER@1.00	Ea	921.00	38.10	15.70	974.80

Schedule 40 carbon steel tee with roll-grooved joints

Description	Craft@Hrs	Unit	Material $	Labor $	Equipment $	Total $
2"	P1@.200	Ea	35.90	7.15	—	43.05
2½"	P1@.210	Ea	49.20	7.51	—	56.71
3"	P1@.230	Ea	68.60	8.22	—	76.82
4"	P1@.250	Ea	105.00	8.94	—	113.94
5"	ER@.300	Ea	248.00	11.40	4.80	264.20
6"	ER@.500	Ea	288.00	19.10	7.90	315.00
8"	ER@.750	Ea	630.00	28.60	11.80	670.40
10"	ER@1.00	Ea	1,120.00	38.10	15.70	1,173.80
12"	ER@1.50	Ea	1,550.00	57.20	23.70	1,630.90

Carbon Steel, Schedule 40 with Roll-Grooved Joints

Description	Craft@Hrs	Unit	Material $	Labor $	Equipment $	Total $

Schedule 40 carbon steel reducing tee with roll-grooved joints

Description	Craft@Hrs	Unit	Material $	Labor $	Equipment $	Total $
3" x 2"	P1@.230	Ea	93.00	8.22	—	101.22
4" x 2"	P1@.240	Ea	126.00	8.58	—	134.58
4" x 2½"	P1@.245	Ea	126.00	8.76	—	134.76
4" x 3"	P1@.250	Ea	126.00	8.94	—	134.94
5" x 2"	P1@.260	Ea	230.00	9.30	—	239.30
5" x 3"	P1@.280	Ea	252.00	10.00	—	262.00
5" x 4"	P1@.300	Ea	273.00	10.70	—	283.70
6" x 3"	ER@.400	Ea	300.00	15.20	6.20	321.40
6" x 4"	ER@.450	Ea	300.00	17.10	7.20	324.30
6" x 5"	ER@.500	Ea	300.00	19.10	7.90	327.00
8" x 3"	ER@.600	Ea	432.00	22.90	9.40	464.30
8" x 4"	ER@.650	Ea	630.00	24.80	10.30	665.10
8" x 5"	ER@.700	Ea	630.00	26.70	11.00	667.70
8" x 6"	ER@.750	Ea	630.00	28.60	11.80	670.40
10" x 4"	ER@.800	Ea	673.00	30.50	12.70	716.20
10" x 5"	ER@.900	Ea	686.00	34.30	14.10	734.40
10" x 6"	ER@1.00	Ea	686.00	38.10	15.70	739.80
10" x 8"	ER@1.10	Ea	686.00	41.90	17.30	745.20
12" x 6"	ER@1.20	Ea	100.00	45.70	18.90	164.60
12" x 8"	ER@1.35	Ea	1,030.00	51.40	21.20	1,102.60
12" x 10"	ER@1.50	Ea	1,080.00	57.20	23.70	1,160.90

Schedule 40 carbon steel male adapter

Description	Craft@Hrs	Unit	Material $	Labor $	Equipment $	Total $
2"	P1@.170	Ea	15.20	6.08	—	21.28
2½"	P1@.185	Ea	18.10	6.62	—	24.72
3"	P1@.200	Ea	22.50	7.15	—	29.65
4"	P1@.230	Ea	37.80	8.22	—	46.02
5"	ER@.290	Ea	79.60	11.10	4.60	95.30
6"	ER@.400	Ea	188.00	15.20	6.20	209.40

Schedule 40 carbon steel female adapter

Description	Craft@Hrs	Unit	Material $	Labor $	Equipment $	Total $
2"	P1@.170	Ea	34.70	6.08	—	40.78
3"	P1@.200	Ea	53.60	7.15	—	60.75
4"	P1@.230	Ea	74.80	8.22	—	83.02

Schedule 40 carbon steel reducer, roll-grooved joints

Description	Craft@Hrs	Unit	Material $	Labor $	Equipment $	Total $
2"	P1@.160	Ea	32.40	5.72	—	38.12
3"	P1@.170	Ea	44.90	6.08	—	50.98
4"	P1@.210	Ea	54.50	7.51	—	62.01
5"	ER@.260	Ea	75.90	9.91	4.00	89.81
6"	ER@.310	Ea	88.20	11.80	5.00	105.00
8"	ER@.450	Ea	230.00	17.10	7.20	254.30
10"	ER@.690	Ea	402.00	26.30	10.80	439.10
12"	ER@.920	Ea	696.00	35.10	14.50	745.60

Carbon Steel, Schedule 40 with Roll-Grooved Joints

Description	Craft@Hrs	Unit	Material $	Labor $	Equipment $	Total $

Schedule 40 carbon steel cap, roll-grooved joint

Description	Craft@Hrs	Unit	Material $	Labor $	Equipment $	Total $
2"	P1@.110	Ea	16.80	3.93	—	20.73
2½"	P1@.120	Ea	26.30	4.29	—	30.59
3"	P1@.130	Ea	26.30	4.65	—	30.95
4"	P1@.140	Ea	28.40	5.01	—	33.41
5"	ER@.150	Ea	65.80	5.72	2.40	73.92
6"	ER@.170	Ea	68.60	6.48	2.80	77.88
8"	ER@.200	Ea	131.00	7.62	3.10	141.72
10"	ER@.225	Ea	239.00	8.57	3.50	251.07
12"	ER@.250	Ea	385.00	9.53	3.90	398.43

Flange with gasket, grooved joint

Description	Craft@Hrs	Unit	Material $	Labor $	Equipment $	Total $
2"	P1@.280	Ea	50.30	10.00	—	60.30
2½"	P1@.340	Ea	59.40	12.20	—	71.60
3"	P1@.390	Ea	67.30	13.90	—	81.20
4"	P1@.560	Ea	97.30	20.00	—	117.30
6"	ER@.780	Ea	121.00	29.70	12.30	163.00
8"	ER@1.10	Ea	143.00	41.90	17.30	202.20
10"	ER@1.40	Ea	181.00	53.40	22.00	256.40
12"	ER@1.70	Ea	250.00	64.80	26.70	341.50

Roll-grooved coupling with gasket

Description	Craft@Hrs	Unit	Material $	Labor $	Equipment $	Total $
2"	P1@.300	Ea	26.00	10.70	—	36.70
2½"	P1@.350	Ea	31.10	12.50	—	43.60
3"	P1@.400	Ea	34.70	14.30	—	49.00
4"	P1@.500	Ea	50.30	17.90	—	68.20
5"	ER@.600	Ea	77.10	22.90	9.40	109.40
6"	ER@.700	Ea	91.00	26.70	11.00	128.70
8"	ER@.900	Ea	150.00	34.30	14.10	198.40
10"	ER@1.10	Ea	209.00	41.90	17.30	268.20
12"	ER@1.35	Ea	239.00	51.40	21.20	311.60

Class 125 iron body gate valve, flanged ends

Description	Craft@Hrs	Unit	Material $	Labor $	Equipment $	Total $
2"	P1@.500	Ea	295.00	17.90	—	312.90
2½"	P1@.600	Ea	398.00	21.50	—	419.50
3"	P1@.750	Ea	432.00	26.80	—	458.80
4"	P1@1.35	Ea	634.00	48.30	—	682.30
6"	ER@2.50	Ea	1,210.00	95.30	39.30	1,344.60
8"	ER@3.00	Ea	1,990.00	114.00	47.10	2,151.10
10"	ER@4.00	Ea	3,250.00	152.00	62.90	3,464.90
12"	ER@4.50	Ea	4,500.00	171.00	69.90	4,740.90

Description	Craft@Hrs	Unit	Material $	Labor $	Equipment $	Total $

Class 125 iron body globe valve, flanged joints

Description	Craft@Hrs	Unit	Material $	Labor $	Equipment $	Total $
2"	P1@.500	Ea	215.00	17.90	—	232.90
2½"	P1@.600	Ea	334.00	21.50	—	355.50
3"	P1@.750	Ea	401.00	26.80	—	427.80
4"	P1@1.35	Ea	538.00	48.30	—	586.30
6"	ER@2.50	Ea	943.00	95.30	39.30	1,077.60
8"	ER@3.00	Ea	1,450.00	114.00	47.10	1,611.10
10"	ER@4.00	Ea	2,210.00	152.00	62.90	2,424.90

200 PSIG iron body butterfly valve, lug type, lever operated

Description	Craft@Hrs	Unit	Material $	Labor $	Equipment $	Total $
2"	P1@.450	Ea	128.00	16.10	—	144.10
2½"	P1@.450	Ea	132.00	16.10	—	148.10
3"	P1@.550	Ea	139.00	19.70	—	158.70
4"	P1@.550	Ea	173.00	19.70	—	192.70
6"	ER@.800	Ea	279.00	30.50	12.70	322.20
8"	ER@.800	Ea	382.00	30.50	12.70	425.20
10"	ER@.900	Ea	530.00	34.30	14.10	578.40
12"	ER@1.00	Ea	694.00	38.10	15.70	747.80

200 PSIG iron body butterfly valve, wafer type, lever operated

Description	Craft@Hrs	Unit	Material $	Labor $	Equipment $	Total $
2"	P1@.450	Ea	116.00	16.10	—	132.10
2½"	P1@.450	Ea	119.00	16.10	—	135.10
3"	P1@.550	Ea	128.00	19.70	—	147.70
4"	P1@.550	Ea	155.00	19.70	—	174.70
6"	ER@.800	Ea	259.00	30.50	12.70	302.20
8"	ER@.800	Ea	359.00	30.50	12.70	402.20
10"	ER@.900	Ea	502.00	34.30	14.10	550.40
12"	ER@1.00	Ea	659.00	38.10	15.70	712.80

Class 125 iron body swing check valve, flanged ends

Description	Craft@Hrs	Unit	Material $	Labor $	Equipment $	Total $
2"	P1@.500	Ea	148.00	17.90	—	165.90
2½"	P1@.600	Ea	188.00	21.50	—	209.50
3"	P1@.750	Ea	233.00	26.80	—	259.80
4"	P1@1.35	Ea	343.00	48.30	—	391.30
6"	ER@2.50	Ea	666.00	95.30	39.30	800.60
8"	ER@3.00	Ea	1,190.00	114.00	47.10	1,351.10
10"	ER@4.00	Ea	1,960.00	152.00	62.90	2,174.90
12"	ER@4.50	Ea	2,660.00	171.00	69.90	2,900.90

Class 125 iron body silent check valve, flanged joints

Description	Craft@Hrs	Unit	Material $	Labor $	Equipment $	Total $
2"	P1@.500	Ea	108.00	17.90	—	125.90
2½"	P1@.600	Ea	125.00	21.50	—	146.50
3"	P1@.750	Ea	140.00	26.80	—	166.80
4"	P1@1.35	Ea	184.00	48.30	—	232.30
5"	ER@2.00	Ea	266.00	76.20	31.40	373.60
6"	ER@2.50	Ea	330.00	95.30	39.30	464.60
8"	ER@3.00	Ea	595.00	114.00	47.10	756.10
10"	ER@4.00	Ea	924.00	152.00	62.90	1,138.90

Carbon Steel, Schedule 40 with Roll-Grooved Joints

Description	Craft@Hrs	Unit	Material $	Labor $	Equipment $	Total $
Class 125 iron body strainer, flanged						
2"	P1@.500	Ea	118.00	17.90	—	135.90
2½"	P1@.600	Ea	132.00	21.50	—	153.50
3"	P1@.750	Ea	152.00	26.80	—	178.80
4"	P1@1.35	Ea	260.00	48.30	—	308.30
6"	ER@2.50	Ea	528.00	95.30	39.30	662.60
8"	ER@3.00	Ea	892.00	114.00	47.10	1,053.10
Installation of 2-way control valve, flanged joints						
2"	P1@.500	Ea	—	17.90	—	17.90
2½"	P1@.600	Ea	—	21.50	—	21.50
3"	P1@.750	Ea	—	26.80	—	26.80
4"	P1@1.35	Ea	—	48.30	—	48.30
6"	P1@2.50	Ea	—	89.40	—	89.40
8"	P1@3.00	Ea	—	107.00	—	107.00
10"	P1@4.00	Ea	—	143.00	—	143.00
12"	P1@4.50	Ea	—	161.00	—	161.00
Installation of 3-way control valve, flanged joints						
2"	P1@.910	Ea	—	32.50	—	32.50
2½"	P1@1.12	Ea	—	40.10	—	40.10
3"	P1@1.33	Ea	—	47.60	—	47.60
4"	P1@2.00	Ea	—	71.50	—	71.50
6"	P1@3.70	Ea	—	132.00	—	132.00
8"	P1@4.40	Ea	—	157.00	—	157.00
10"	P1@5.90	Ea	—	211.00	—	211.00
12"	P1@6.50	Ea	—	232.00	—	232.00
Bolt and gasket sets						
2"	P1@.500	Ea	3.86	17.90	—	21.76
2½"	P1@.650	Ea	4.51	23.20	—	27.71
3"	P1@.750	Ea	7.67	26.80	—	34.47
4"	P1@1.00	Ea	12.80	35.80	—	48.60
6"	P1@1.20	Ea	21.20	42.90	—	64.10
8"	P1@1.25	Ea	23.80	44.70	—	68.50
10"	P1@1.70	Ea	41.20	60.80	—	102.00
12"	P1@2.20	Ea	46.20	78.70	—	124.90
Thermometer with well						
7"	P1@.250	Ea	159.00	8.94	—	167.94
9"	P1@.250	Ea	164.00	8.94	—	172.94
Dial type pressure gauge						
2½"	P1@.200	Ea	33.10	7.15	—	40.25
3½"	P1@.200	Ea	43.80	7.15	—	50.95

Carbon Steel, Schedule 40 with Roll-Grooved Joints

Description	Craft@Hrs	Unit	Material $	Labor $	Equipment $	Total $
Pressure/temperature tap						
tap	P1@.150	Ea	15.70	5.36	—	21.06
Hanger with swivel assembly						
2"	P1@.300	Ea	6.05	10.70	—	16.75
2½"	P1@.350	Ea	7.82	12.50	—	20.32
3"	P1@.350	Ea	9.67	12.50	—	22.17
4"	P1@.350	Ea	14.90	12.50	—	27.40
5"	P1@.450	Ea	16.70	16.10	—	32.80
6"	P1@.450	Ea	21.80	16.10	—	37.90
8"	P1@.450	Ea	28.90	16.10	—	45.00
10"	P1@.550	Ea	39.70	19.70	—	59.40
12"	P1@.600	Ea	52.80	21.50	—	74.30
Riser clamp						
2"	P1@.115	Ea	4.84	4.11	—	8.95
2½"	P1@.120	Ea	5.09	4.29	—	9.38
3"	P1@.120	Ea	5.52	4.29	—	9.81
4"	P1@.125	Ea	7.03	4.47	—	11.50
5"	P1@.180	Ea	10.10	6.44	—	16.54
6"	P1@.200	Ea	12.20	7.15	—	19.35
8"	P1@.200	Ea	19.80	7.15	—	26.95
10"	P1@.250	Ea	29.90	8.94	—	38.84
12"	P1@.250	Ea	34.90	8.94	—	43.84

Carbon Steel, Schedule 10 with Roll-Grooved Joints

Schedule 10 carbon steel (ASTM A-53) roll-grooved pipe with factory grooved malleable iron, ductile iron, or steel fittings is commonly used for sprinkler systems, heating hot water, chilled water and condenser water systems.

Consult the manufacturers for maximum operating temperature/pressure ratings for the various combinations of pipe and fittings and system applications.

The cost estimates in this section are based on the conditions, limitations and wage rates described in the section "How to Use This Book" beginning on page 5.

Equipment cost, where shown, is $110 per hour for a 10-ton hydraulic truck-mounted crane.

Description	Craft@Hrs	Unit	Material $	Labor $	Equipment $	Total $

Schedule 10 carbon steel roll-grooved horizontal pipe assembly.
Horizontally hung in a building. Assembly includes fittings, couplings and hanger assemblies. Based on a reducing tee and a 45 degree elbow every 16 feet for 1/2" pipe. A reducing tee and a 45 degree elbow every 50 feet for 12" pipe. Hangers spaced to meet plumbing code. *(Distance between fittings in the pipe assembly increases as pipe diameter increases.)* Use these figures for preliminary estimates.

Description	Craft@Hrs	Unit	Material $	Labor $	Equipment $	Total $
2"	P1@.178	LF	7.41	6.37	—	13.78
2½"	P1@.188	LF	9.27	6.72	—	15.99
3"	P1@.215	LF	10.70	7.69	—	18.39
4"	P1@.255	LF	15.20	9.12	—	24.32
6"	ER@.339	LF	28.90	12.90	5.30	47.10
8"	ER@.404	LF	44.00	15.40	6.40	65.80
10"	ER@.517	LF	65.70	19.70	8.10	93.50
12"	ER@.602	LF	118.00	22.90	9.50	150.40

Schedule 10 carbon steel roll-grooved vertical pipe assembly.
Riser assembly, including fittings, couplings, and riser clamps. Based on a reducing tee, and a riser clamp on every floor. Use these figures for preliminary estimates.

Description	Craft@Hrs	Unit	Material $	Labor $	Equipment $	Total $
2"	P1@.117	LF	7.09	4.18	—	11.27
2½"	P1@.120	LF	9.65	4.29	—	13.94
3"	P1@.146	LF	9.91	5.22	—	15.13
4"	P1@.204	LF	15.10	7.30	—	22.40
6"	ER@.286	LF	29.90	10.90	4.20	45.00
8"	ER@.385	LF	52.60	14.70	6.10	73.40
10"	ER@.510	LF	73.00	19.40	8.10	100.50
12"	ER@.593	LF	118.00	22.60	9.40	150.00

Description	Craft@Hrs	Unit	Material $	Labor $	Equipment $	Total $

Schedule 10 carbon steel pipe, roll-grooved joints

Description	Craft@Hrs	Unit	Material $	Labor $	Equipment $	Total $
2"	P1@.080	LF	2.16	2.86	—	5.02
2½"	P1@.100	LF	2.94	3.58	—	6.52
3"	P1@.115	LF	3.57	4.11	—	7.68
4"	P1@.160	LF	4.73	5.72	—	10.45
5"	ER@.200	LF	8.12	7.62	3.10	18.84
6"	ER@.260	LF	8.58	9.91	4.00	22.49
8"	ER@.300	LF	11.80	11.40	4.80	28.00
10"	ER@.380	LF	16.10	14.50	6.10	36.70
12"	ER@.450	LF	22.70	17.10	7.20	47.00

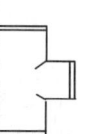

Schedule 10 carbon steel 45 degree ell, roll-grooved joints

Description	Craft@Hrs	Unit	Material $	Labor $	Equipment $	Total $
2"	P1@.170	Ea	23.50	6.08	—	29.58
2½"	P1@.185	Ea	31.90	6.62	—	38.52
3"	P1@.200	Ea	42.10	7.15	—	49.25
4"	P1@.230	Ea	62.20	8.22	—	70.42
5"	ER@.250	Ea	150.00	9.53	3.90	163.43
6"	ER@.350	Ea	176.00	13.30	5.50	194.80
8"	ER@.500	Ea	368.00	19.10	7.90	395.00
10"	ER@.750	Ea	498.00	28.60	11.80	538.40
12"	ER@1.00	Ea	872.00	38.10	15.70	925.80

Schedule 10 carbon steel 90 degree ell, roll-grooved joints

Description	Craft@Hrs	Unit	Material $	Labor $	Equipment $	Total $
2"	P1@.170	Ea	23.50	6.08	—	29.58
2½"	P1@.185	Ea	31.90	6.62	—	38.52
3"	P1@.200	Ea	42.10	7.15	—	49.25
4"	P1@.230	Ea	62.20	8.22	—	70.42
5"	ER@.250	Ea	150.00	9.53	3.90	163.43
6"	ER@.350	Ea	176.00	13.30	5.50	194.80
8"	ER@.500	Ea	368.00	19.10	7.90	395.00
10"	ER@.750	Ea	571.00	28.60	11.80	611.40
12"	ER@1.00	Ea	921.00	38.10	15.70	974.80

Schedule 10 carbon steel tee, roll-grooved joints

Description	Craft@Hrs	Unit	Material $	Labor $	Equipment $	Total $
2"	P1@.200	Ea	35.90	7.15	—	43.05
2½"	P1@.210	Ea	49.20	7.51	—	56.71
3"	P1@.230	Ea	68.60	8.22	—	76.82
4"	P1@.250	Ea	105.00	8.94	—	113.94
5"	ER@.300	Ea	248.00	11.40	4.80	264.20
6"	ER@.500	Ea	288.00	19.10	7.90	315.00
8"	ER@.750	Ea	630.00	28.60	11.80	670.40
10"	ER@1.00	Ea	1,120.00	38.10	15.70	1,173.80
12"	ER@1.50	Ea	1,550.00	57.20	23.70	1,630.90

Carbon Steel, Schedule 10 with Roll-Grooved Joints

Description	Craft@Hrs	Unit	Material $	Labor $	Equipment $	Total $

Schedule 10 carbon steel reducing tee, roll-grooved joints

Description	Craft@Hrs	Unit	Material $	Labor $	Equipment $	Total $
3" x 2"	P1@.230	Ea	93.00	8.22	—	101.22
4" x 2"	P1@.240	Ea	126.00	8.58	—	134.58
4" x 2½"	P1@.245	Ea	126.00	8.76	—	134.76
4" x 3"	P1@.250	Ea	126.00	8.94	—	134.94
5" x 2"	P1@.260	Ea	230.00	9.30	—	239.30
5" x 3"	P1@.280	Ea	252.00	10.00	—	262.00
5" x 4"	P1@.300	Ea	273.00	10.70	—	283.70
6" x 3"	ER@.400	Ea	300.00	15.20	6.20	321.40
6" x 4"	ER@.450	Ea	300.00	17.10	7.20	324.30
6" x 5"	ER@.500	Ea	300.00	19.10	7.90	327.00
8" x 3"	ER@.600	Ea	432.00	22.90	9.40	464.30
8" x 4"	ER@.650	Ea	630.00	24.80	10.30	665.10
8" x 5"	ER@.700	Ea	630.00	26.70	11.00	667.70
8" x 6"	ER@.750	Ea	630.00	28.60	11.80	670.40
10" x 4"	ER@.800	Ea	673.00	30.50	12.70	716.20
10" x 5"	ER@.900	Ea	686.00	34.30	14.10	734.40
10" x 6"	ER@1.00	Ea	686.00	38.10	15.70	739.80
10" x 8"	ER@1.10	Ea	686.00	41.90	17.30	745.20
12" x 6"	ER@1.20	Ea	100.00	45.70	18.90	164.60
12" x 8"	ER@1.35	Ea	1,030.00	51.40	19.60	1,101.00
12" x 10"	ER@1.50	Ea	1,080.00	57.20	23.70	1,160.90

Schedule 10 carbon steel male adapter

Description	Craft@Hrs	Unit	Material $	Labor $	Equipment $	Total $
2"	P1@.170	Ea	15.20	6.08	—	21.28
2½"	P1@.185	Ea	18.10	6.62	—	24.72
3"	P1@.200	Ea	22.50	7.15	—	29.65
4"	P1@.230	Ea	37.80	8.22	—	46.02
5"	ER@.290	Ea	79.60	11.10	4.60	95.30
6"	ER@.400	Ea	188.00	15.20	6.20	209.40

Schedule 10 carbon steel female adapter

Description	Craft@Hrs	Unit	Material $	Labor $	Equipment $	Total $
2"	P1@.170	Ea	34.70	6.08	—	40.78
3"	P1@.200	Ea	53.60	7.15	—	60.75
4"	P1@.230	Ea	74.80	8.22	—	83.02

Schedule 10 carbon steel reducer, roll-grooved joints

Description	Craft@Hrs	Unit	Material $	Labor $	Equipment $	Total $
2"	P1@.160	Ea	32.40	5.72	—	38.12
3"	P1@.170	Ea	44.90	6.08	—	50.98
4"	P1@.210	Ea	54.50	7.51	—	62.01
5"	ER@.260	Ea	75.90	9.91	4.00	89.81
6"	ER@.310	Ea	88.20	11.80	5.00	105.00
8"	ER@.450	Ea	230.00	17.10	7.20	254.30
10"	ER@.690	Ea	402.00	26.30	10.80	439.10
12"	ER@.920	Ea	696.00	35.10	14.50	745.60

Description	Craft@Hrs	Unit	Material $	Labor $	Equipment $	Total $

Schedule 10 carbon steel cap, roll-grooved joints

Description	Craft@Hrs	Unit	Material $	Labor $	Equipment $	Total $
2"	P1@.110	Ea	16.80	3.93	—	20.73
2½"	P1@.120	Ea	26.30	4.29	—	30.59
3"	P1@.130	Ea	26.30	4.65	—	30.95
4"	P1@.140	Ea	28.40	5.01	—	33.41
5"	ER@.150	Ea	65.80	5.72	2.40	73.92
6"	ER@.170	Ea	68.60	6.48	2.80	77.88
8"	ER@.200	Ea	131.00	7.62	3.10	141.72
10"	ER@.225	Ea	239.00	8.57	3.50	251.07
12"	ER@.250	Ea	385.00	9.53	3.90	398.43

Flange with gasket, grooved joint

Description	Craft@Hrs	Unit	Material $	Labor $	Equipment $	Total $
2"	P1@.280	Ea	50.30	10.00	—	60.30
2½"	P1@.340	Ea	59.40	12.20	—	71.60
3"	P1@.390	Ea	67.30	13.90	—	81.20
4"	P1@.560	Ea	97.30	20.00	—	117.30
6"	ER@.780	Ea	121.00	29.70	12.30	163.00
8"	ER@1.10	Ea	143.00	41.90	17.30	202.20
10"	ER@1.40	Ea	181.00	53.40	22.00	256.40
12"	ER@1.70	Ea	250.00	64.80	26.70	311.60

Roll-grooved coupling with gasket

Description	Craft@Hrs	Unit	Material $	Labor $	Equipment $	Total $
2"	P1@.300	Ea	26.00	10.70	—	36.70
2½"	P1@.350	Ea	31.10	12.50	—	43.60
3"	P1@.400	Ea	34.70	14.30	—	49.00
4"	P1@.500	Ea	50.30	17.90	—	68.20
5"	ER@.600	Ea	77.10	22.90	9.40	109.40
6"	ER@.700	Ea	91.00	26.70	11.00	128.70
8"	ER@.900	Ea	150.00	34.30	14.10	198.40
10"	ER@1.10	Ea	209.00	41.90	17.30	268.20
12"	ER@1.35	Ea	239.00	51.40	21.20	311.60

Class 125 iron body gate valve, flanged joints

Description	Craft@Hrs	Unit	Material $	Labor $	Equipment $	Total $
2"	P1@.500	Ea	392.00	17.90	—	409.90
2½"	P1@.600	Ea	532.00	21.50	—	553.50
3"	P1@.750	Ea	576.00	26.80	—	602.80
4"	P1@1.35	Ea	844.00	48.30	—	892.30
5"	ER@2.00	Ea	1,620.00	76.20	31.40	1,727.60
6"	ER@2.50	Ea	1,620.00	95.30	39.30	1,754.60
8"	ER@3.00	Ea	2,640.00	114.00	47.10	2,801.10
10"	ER@4.00	Ea	4,380.00	152.00	62.90	4,594.90
12"	ER@4.50	Ea	5,990.00	171.00	70.80	6,231.80

Description	Craft@Hrs	Unit	Material $	Labor $	Equipment $	Total $

Class 125 iron body globe valve, flanged joints

Description	Craft@Hrs	Unit	Material $	Labor $	Equipment $	Total $
2"	P1@.500	Ea	215.00	17.90	—	232.90
2½"	P1@.600	Ea	331.00	21.50	—	352.50
3"	P1@.750	Ea	401.00	26.80	—	427.80
4"	P1@1.35	Ea	535.00	48.30	—	583.30
5"	ER@2.00	Ea	943.00	76.20	31.40	1,050.60
6"	ER@2.50	Ea	943.00	95.30	39.30	1,077.60
8"	ER@3.00	Ea	1,440.00	114.00	47.10	1,601.10
10"	ER@4.00	Ea	2,210.00	152.00	62.90	2,424.90

200 PSIG iron body butterfly valve, lug type, lever operated

Description	Craft@Hrs	Unit	Material $	Labor $	Equipment $	Total $
2"	P1@.450	Ea	128.00	16.10	—	144.10
2½"	P1@.450	Ea	132.00	16.10	—	148.10
3"	P1@.550	Ea	139.00	19.70	—	158.70
4"	P1@.550	Ea	173.00	19.70	—	192.70
5"	ER@.800	Ea	229.00	30.50	12.70	272.20
6"	ER@.800	Ea	279.00	30.50	12.70	322.20
8"	ER@.800	Ea	382.00	30.50	12.70	425.20
10"	ER@.900	Ea	530.00	34.30	14.10	578.40
12"	ER@1.00	Ea	694.00	38.10	15.70	747.80

200 PSIG iron body butterfly valve, wafer type, lever operated

Description	Craft@Hrs	Unit	Material $	Labor $	Equipment $	Total $
2"	P1@.450	Ea	116.00	16.10	—	132.10
2½"	P1@.450	Ea	119.00	16.10	—	135.10
3"	P1@.550	Ea	128.00	19.70	—	147.70
4"	P1@.550	Ea	155.00	19.70	—	174.70
5"	ER@.800	Ea	207.00	30.50	12.70	250.20
6"	ER@.800	Ea	259.00	30.50	12.70	302.20
8"	ER@.800	Ea	359.00	30.50	12.70	402.20
10"	ER@.900	Ea	502.00	34.30	14.10	550.40
12"	ER@1.00	Ea	659.00	38.10	15.70	712.80

Class 125 iron body swing check valve, flanged joints

Description	Craft@Hrs	Unit	Material $	Labor $	Equipment $	Total $
2"	P1@.500	Ea	148.00	17.90	—	165.90
2½"	P1@.600	Ea	188.00	21.50	—	209.50
3"	P1@.750	Ea	233.00	26.80	—	259.80
4"	P1@1.35	Ea	343.00	48.30	—	391.30
5"	ER@2.00	Ea	660.00	76.20	31.40	767.60
6"	ER@2.50	Ea	666.00	95.30	39.30	800.60
8"	ER@3.00	Ea	1,190.00	114.00	47.10	1,351.10
10"	ER@4.00	Ea	1,960.00	152.00	62.90	2,174.90
12"	ER@4.50	Ea	2,660.00	171.00	70.80	2,901.80

Carbon Steel, Schedule 10 with Roll-Grooved Joints

Description	Craft@Hrs	Unit	Material $	Labor $	Equipment $	Total $

Class 125 iron body silent check valve, flanged joints

Description	Craft@Hrs	Unit	Material $	Labor $	Equipment $	Total $
2"	P1@.500	Ea	108.00	17.90	—	125.90
2½"	P1@.600	Ea	125.00	21.50	—	146.50
3"	P1@.750	Ea	140.00	26.80	—	166.80
4"	P1@1.35	Ea	184.00	48.30	—	232.30
5"	ER@2.00	Ea	266.00	76.20	31.40	373.60
6"	ER@2.50	Ea	330.00	95.30	39.30	464.60
8"	ER@3.00	Ea	595.00	114.00	47.10	756.10
10"	ER@4.00	Ea	924.00	152.00	62.90	1,138.90

Class 125 iron body strainer, flanged

Description	Craft@Hrs	Unit	Material $	Labor $	Equipment $	Total $
2"	P1@.500	Ea	118.00	17.90	—	135.90
2½"	P1@.600	Ea	132.00	21.50	—	153.50
3"	P1@.750	Ea	152.00	26.80	—	178.80
4"	P1@1.35	Ea	260.00	48.30	—	308.30
5"	P1@2.00	Ea	528.00	71.50	31.40	630.90
6"	P1@2.50	Ea	528.00	89.40	39.30	656.70
8"	P1@3.00	Ea	892.00	107.00	47.10	1,046.10

Installation of 2-way control valve, flanged joints

Description	Craft@Hrs	Unit	Material $	Labor $	Equipment $	Total $
2"	P1@.500	Ea	—	17.90	—	17.90
2½"	P1@.600	Ea	—	21.50	—	21.50
3"	P1@.750	Ea	—	26.80	—	26.80
4"	P1@1.35	Ea	—	48.30	—	48.30
6"	P1@2.50	Ea	—	89.40	—	89.40
8"	P1@3.00	Ea	—	107.00	—	107.00
10"	P1@4.00	Ea	—	143.00	—	143.00
12"	P1@4.50	Ea	—	161.00	—	161.00

Installation of 3-way control valve, flanged joints

Description	Craft@Hrs	Unit	Material $	Labor $	Equipment $	Total $
2"	P1@.910	Ea	—	32.50	—	32.50
2½"	P1@1.12	Ea	—	40.10	—	40.10
3"	P1@1.33	Ea	—	47.60	—	47.60
4"	P1@2.00	Ea	—	71.50	—	71.50
6"	P1@3.70	Ea	—	132.00	—	132.00
8"	P1@4.40	Ea	—	157.00	—	157.00
10"	P1@5.90	Ea	—	211.00	—	211.00
12"	P1@6.50	Ea	—	232.00	—	232.00

Carbon Steel, Schedule 10 with Roll-Grooved Joints

Description	Craft@Hrs	Unit	Material $	Labor $	Equipment $	Total $
Bolt and gasket sets						
2"	P1@.500	Ea	3.86	17.90	—	21.76
2½"	P1@.650	Ea	4.51	23.20	—	27.71
3"	P1@.750	Ea	7.64	26.80	—	34.44
4"	P1@1.00	Ea	12.70	35.80	—	48.50
5"	P1@1.10	Ea	21.20	39.30	—	60.50
6"	P1@1.20	Ea	21.20	42.90	—	64.10
8"	P1@1.25	Ea	23.80	44.70	—	68.50
10"	P1@1.70	Ea	40.90	60.80	—	101.70
12"	P1@2.20	Ea	46.20	78.70	—	124.90
Thermometer with well						
7"	P1@.250	Ea	159.00	8.94	—	167.94
9"	P1@.250	Ea	164.00	8.94	—	172.94
Dial type pressure gauge						
2½"	P1@.200	Ea	33.10	7.15	—	40.25
3½"	P1@.200	Ea	43.80	7.15	—	50.95
Pressure/temperature tap						
tap	P1@.150	Ea	15.70	5.36	—	21.06
Hanger with swivel assembly						
2"	P1@.300	Ea	6.05	10.70	—	16.75
2½"	P1@.350	Ea	7.82	12.50	—	20.32
3"	P1@.350	Ea	9.67	12.50	—	22.17
4"	P1@.350	Ea	14.90	12.50	—	27.40
5"	P1@.450	Ea	16.70	16.10	—	32.80
6"	P1@.450	Ea	21.80	16.10	—	37.90
8"	P1@.450	Ea	28.90	16.10	—	45.00
10"	P1@.550	Ea	39.70	19.70	—	59.40
12"	P1@.600	Ea	52.80	21.50	—	74.30
Riser clamp						
2"	P1@.115	Ea	4.84	4.11	—	8.95
2½"	P1@.120	Ea	5.09	4.29	—	9.38
3"	P1@.120	Ea	5.52	4.29	—	9.81
4"	P1@.125	Ea	7.03	4.47	—	11.50
5"	P1@.180	Ea	10.10	6.44	—	16.54
6"	P1@.200	Ea	12.20	7.15	—	19.35
8"	P1@.200	Ea	19.80	7.15	—	26.95
10"	P1@.250	Ea	29.40	8.94	—	38.34
12"	P1@.250	Ea	34.90	8.94	—	43.84

Carbon Steel, Schedule 40 with Cut-Grooved Joints

Schedule 40 carbon steel (ASTM A-53 and A-120) cut-grooved pipe with factory-grooved malleable iron, ductile iron or steel fittings is commonly used for heating hot water, chilled water and condenser water systems.

Consult the manufacturers for maximum operating temperature/pressure ratings for various combinations of pipe and fittings.

This section has been arranged to save the estimator's time by including all normally-used system components such as pipe, fittings, valves, hanger assemblies, riser clamps and miscellaneous items under one heading. Additional items can be found under "Plumbing and Piping Specialties." The cost estimates in this section are based on the conditions, limitations and wage rates described in the section "How to Use This Book" beginning on page 5.

Equipment cost, where shown, is $110 per hour for a 10-ton hydraulic truck-mounted crane.

Description	Craft@Hrs	Unit	Material $	Labor $	Equipment $	Total $

Schedule 40 carbon steel pipe with cut-grooved joints

Description	Craft@Hrs	Unit	Material $	Labor $	Equipment $	Total $
2"	P1@.110	LF	3.68	3.93	—	7.61
2½"	P1@.130	LF	5.79	4.65	—	10.44
3"	P1@.150	LF	7.86	5.36	—	13.22
4"	P1@.210	LF	11.30	7.51	—	18.81
5"	ER@.270	LF	21.30	10.30	4.20	35.80
6"	ER@.350	LF	32.10	13.30	5.50	50.90
8"	ER@.410	LF	32.10	15.60	6.40	54.10
10"	ER@.510	LF	45.00	19.40	8.10	72.50
12"	ER@.610	LF	70.90	23.20	9.50	103.60

Schedule 40 carbon steel 45 degree ell with cut-grooved joints

Description	Craft@Hrs	Unit	Material $	Labor $	Equipment $	Total $
2"	P1@.170	Ea	23.50	6.08	—	29.58
2½"	P1@.185	Ea	32.10	6.62	—	38.72
3"	P1@.200	Ea	42.10	7.15	—	49.25
4"	P1@.230	Ea	62.20	8.22	—	70.42
5"	ER@.250	Ea	150.00	9.53	3.90	163.43
6"	ER@.350	Ea	176.00	13.30	5.50	194.80
8"	ER@.500	Ea	368.00	19.10	7.90	395.00
10"	ER@.750	Ea	498.00	28.60	11.80	538.40
12"	ER@1.00	Ea	872.00	38.10	15.70	925.80

Schedule 40 carbon steel 90 degree ell with cut-grooved joints

Description	Craft@Hrs	Unit	Material $	Labor $	Equipment $	Total $
2"	P1@.170	Ea	23.50	6.08	—	29.58
2½"	P1@.185	Ea	32.10	6.62	—	38.72
3"	P1@.200	Ea	42.10	7.15	—	49.25
4"	P1@.230	Ea	62.20	8.22	—	70.42
5"	ER@.250	Ea	150.00	9.53	3.90	163.43
6"	ER@.350	Ea	176.00	13.30	5.50	194.80
8"	ER@.500	Ea	368.00	19.10	7.90	395.00
10"	ER@.750	Ea	571.00	28.60	11.80	611.40
12"	ER@1.00	Ea	1,550.00	38.10	15.70	1,603.80

Carbon Steel, Schedule 40 with Cut-Grooved Joints

Description	Craft@Hrs	Unit	Material $	Labor $	Equipment $	Total $

Schedule 40 carbon steel tee with cut-grooved joints

Description	Craft@Hrs	Unit	Material $	Labor $	Equipment $	Total $
2"	P1@.200	Ea	35.90	7.15	—	43.05
2½"	P1@.210	Ea	49.20	7.51	—	56.71
3"	P1@.230	Ea	68.60	8.22	—	76.82
4"	P1@.250	Ea	105.00	8.94	—	113.94
5"	ER@.300	Ea	248.00	11.40	4.80	264.20
6"	ER@.500	Ea	288.00	19.10	7.90	315.00
8"	ER@.750	Ea	630.00	28.60	11.80	670.40
10"	ER@1.00	Ea	1,120.00	38.10	15.70	1,173.80
12"	ER@1.50	Ea	1,550.00	57.20	23.70	1,630.90

Schedule 40 carbon steel reducing tee with cut-grooved joints

Description	Craft@Hrs	Unit	Material $	Labor $	Equipment $	Total $
3" x 2"	P1@.230	Ea	93.00	8.22	—	101.22
4" x 2"	P1@.240	Ea	126.00	8.58	—	134.58
4" x 2½"	P1@.245	Ea	126.00	8.76	—	134.76
4" x 3"	P1@.250	Ea	126.00	8.94	—	134.94
5" x 2"	P1@.260	Ea	230.00	9.30	—	239.30
5" x 3"	P1@.280	Ea	252.00	10.00	—	262.00
5" x 4"	P1@.300	Ea	273.00	10.70	—	283.70
6" x 3"	ER@.400	Ea	300.00	15.20	6.20	321.40
6" x 4"	ER@.450	Ea	300.00	17.10	7.20	324.30
6" x 5"	ER@.500	Ea	300.00	19.10	7.90	327.00
8" x 3"	ER@.600	Ea	432.00	22.90	9.40	464.30
8" x 4"	ER@.650	Ea	630.00	24.80	10.30	665.10
8" x 5"	ER@.700	Ea	630.00	26.70	11.00	667.70
8" x 6"	ER@.750	Ea	630.00	28.60	11.80	670.40
10" x 4"	ER@.800	Ea	673.00	30.50	12.70	716.20
10" x 5"	ER@.900	Ea	686.00	34.30	14.10	734.40
10" x 6"	ER@1.00	Ea	686.00	38.10	15.70	739.80
10" x 8"	ER@1.10	Ea	686.00	41.90	17.30	745.20
12" x 6"	ER@1.20	Ea	100.00	45.70	18.90	164.60
12" x 8"	ER@1.35	Ea	1,030.00	51.40	21.20	1,102.60
12" x 10"	ER@1.50	Ea	1,080.00	57.20	23.70	1,160.90

Schedule 40 carbon steel male adapter

Description	Craft@Hrs	Unit	Material $	Labor $	Equipment $	Total $
2"	P1@.170	Ea	15.20	6.08	—	21.28
2½"	P1@.185	Ea	18.10	6.62	—	24.72
3"	P1@.200	Ea	22.50	7.15	—	29.65
4"	P1@.230	Ea	37.50	8.22	—	45.72
5"	ER@.290	Ea	79.60	11.10	4.60	95.30
6"	ER@.400	Ea	188.00	15.20	6.20	209.40

Schedule 40 carbon steel female adapter

Description	Craft@Hrs	Unit	Material $	Labor $	Equipment $	Total $
2"	P1@.170	Ea	34.70	6.08	—	40.78
3"	P1@.200	Ea	53.60	7.15	—	60.75
4"	P1@.230	Ea	74.80	8.22	—	83.02

Description	Craft@Hrs	Unit	Material $	Labor $	Equipment $	Total $

Schedule 40 carbon steel reducer

Description	Craft@Hrs	Unit	Material $	Labor $	Equipment $	Total $
2"	P1@.160	Ea	32.40	5.72	—	38.12
3"	P1@.170	Ea	44.90	6.08	—	50.98
4"	P1@.210	Ea	54.50	7.51	—	62.01
5"	ER@.260	Ea	75.90	9.91	4.00	89.81
6"	ER@.310	Ea	88.20	11.80	5.00	105.00
8"	ER@.450	Ea	230.00	17.10	7.20	254.30
10"	ER@.690	Ea	402.00	26.30	10.80	439.10
12"	ER@.920	Ea	696.00	35.10	14.50	745.60

Schedule 40 carbon steel cap

Description	Craft@Hrs	Unit	Material $	Labor $	Equipment $	Total $
2"	P1@.110	Ea	16.80	3.93	—	20.73
2½"	P1@.120	Ea	26.30	4.29	—	30.59
3"	P1@.130	Ea	26.30	4.65	—	30.95
4"	P1@.140	Ea	28.40	5.01	—	33.41
5"	ER@.150	Ea	65.80	5.72	2.40	73.92
6"	ER@.170	Ea	68.60	6.48	2.80	77.88
8"	ER@.200	Ea	131.00	7.62	3.10	141.72
10"	ER@.225	Ea	239.00	8.57	3.50	251.07
12"	ER@.250	Ea	385.00	9.53	3.90	398.43

Flange with gasket, grooved joint

Description	Craft@Hrs	Unit	Material $	Labor $	Equipment $	Total $
2"	P1@.280	Ea	50.30	10.00	—	60.30
2½"	P1@.340	Ea	59.40	12.20	—	71.60
3"	P1@.390	Ea	67.30	13.90	—	81.20
4"	P1@.560	Ea	97.30	20.00	—	117.30
6"	ER@.780	Ea	121.00	29.70	12.30	163.00
8"	ER@1.10	Ea	143.00	41.90	17.20	202.10
10"	ER@1.40	Ea	181.00	53.40	22.00	256.40
12"	ER@1.70	Ea	250.00	64.80	26.70	341.50

Cut-grooved joint coupling with gasket

Description	Craft@Hrs	Unit	Material $	Labor $	Equipment $	Total $
2"	P1@.300	Ea	26.00	10.70	—	36.70
2½"	P1@.350	Ea	31.10	12.50	—	43.60
3"	P1@.400	Ea	34.70	14.30	—	49.00
4"	P1@.500	Ea	50.30	17.90	—	68.20
5"	ER@.600	Ea	77.10	22.90	9.40	109.40
6"	ER@.700	Ea	91.00	26.70	11.00	128.70
8"	ER@.900	Ea	150.00	34.30	14.10	198.40
10"	ER@1.10	Ea	209.00	41.90	17.30	268.20
12"	ER@1.35	Ea	239.00	51.40	21.20	311.60

Carbon Steel, Schedule 40 with Cut-Grooved Joints

Description	Craft@Hrs	Unit	Material $	Labor $	Equipment $	Total $

Class 125 iron body gate valve, flanged ends

Description	Craft@Hrs	Unit	Material $	Labor $	Equipment $	Total $
2"	P1@.500	Ea	421.00	17.90	—	438.90
2½"	P1@.600	Ea	570.00	21.50	—	591.50
3"	P1@.750	Ea	618.00	26.80	—	644.80
4"	P1@1.35	Ea	905.00	48.30	—	953.30
6"	ER@2.50	Ea	1,740.00	95.30	39.30	1,874.60
8"	ER@3.00	Ea	2,830.00	114.00	47.10	2,991.10
10"	ER@4.00	Ea	4,670.00	152.00	62.90	4,884.90
12"	ER@4.50	Ea	6,410.00	171.00	70.80	6,651.80

Class 125 iron body globe valve, flanged joints

Description	Craft@Hrs	Unit	Material $	Labor $	Equipment $	Total $
2"	P1@.500	Ea	215.00	17.90	—	232.90
2½"	P1@.600	Ea	334.00	21.50	—	355.50
3"	P1@.750	Ea	401.00	26.80	—	427.80
4"	ER@1.35	Ea	538.00	51.40	21.30	610.70
6"	ER@2.50	Ea	943.00	95.30	39.30	1,077.60
8"	ER@3.00	Ea	1,450.00	114.00	47.10	1,611.10
10"	ER@4.00	Ea	2,210.00	152.00	62.90	2,424.90

200 PSIG iron body butterfly valve, lug type, lever operated

Description	Craft@Hrs	Unit	Material $	Labor $	Equipment $	Total $
2"	P1@.450	Ea	128.00	16.10	—	144.10
2½"	P1@.450	Ea	132.00	16.10	—	148.10
3"	P1@.550	Ea	139.00	19.70	—	158.70
4"	P1@.550	Ea	173.00	19.70	—	192.70
6"	ER@.800	Ea	279.00	30.50	12.70	322.20
8"	ER@.800	Ea	382.00	30.50	12.70	425.20
10"	ER@.900	Ea	530.00	34.30	14.10	578.40
12"	ER@1.00	Ea	694.00	38.10	15.70	747.80

200 PSIG iron body butterfly valve, wafer type, lever operated

Description	Craft@Hrs	Unit	Material $	Labor $	Equipment $	Total $
2"	P1@.450	Ea	116.00	16.10	—	132.10
2½"	P1@.450	Ea	119.00	16.10	—	135.10
3"	P1@.550	Ea	128.00	19.70	—	147.70
4"	P1@.550	Ea	155.00	19.70	—	174.70
6"	ER@.800	Ea	259.00	30.50	12.70	302.20
8"	ER@.800	Ea	359.00	30.50	12.70	402.20
10"	ER@.900	Ea	502.00	34.30	14.10	550.40
12"	ER@1.00	Ea	659.00	38.10	15.70	712.80

Class 125 iron body swing check valve, flanged ends

Description	Craft@Hrs	Unit	Material $	Labor $	Equipment $	Total $
2"	P1@.500	Ea	148.00	17.90	—	165.90
2½"	P1@.600	Ea	188.00	21.50	—	209.50
3"	P1@.750	Ea	233.00	26.80	—	259.80
4"	P1@1.35	Ea	343.00	48.30	—	391.30
6"	ER@2.50	Ea	666.00	95.30	39.30	800.60
8"	ER@3.00	Ea	1,190.00	114.00	47.10	1,351.10
10"	ER@4.00	Ea	1,960.00	152.00	62.90	2,174.90
12"	ER@4.50	Ea	2,660.00	171.00	70.80	2,901.80

Carbon Steel, Schedule 40 with Cut-Grooved Joints

Description	Craft@Hrs	Unit	Material $	Labor $	Equipment $	Total $

Class 125 iron body silent check valve, flanged joints

2"	P1@.500	Ea	108.00	17.90	—	125.90
2½"	P1@.600	Ea	125.00	21.50	—	146.50
3"	P1@.750	Ea	140.00	26.80	—	166.80
4"	P1@1.35	Ea	184.00	48.30	—	232.30
5"	ER@2.00	Ea	266.00	76.20	31.40	373.60
6"	ER@2.50	Ea	330.00	95.30	39.30	464.60
8"	ER@3.00	Ea	595.00	114.00	47.10	756.10
10"	ER@4.00	Ea	924.00	152.00	62.90	1,138.90

Class 125 iron body strainer, flanged

2"	P1@.500	Ea	118.00	17.90	—	135.90
2½"	P1@.600	Ea	132.00	21.50	—	153.50
3"	P1@.750	Ea	152.00	26.80	—	178.80
4"	P1@1.35	Ea	260.00	48.30	—	308.30
6"	ER@2.50	Ea	528.00	95.30	39.30	662.60
8"	ER@3.00	Ea	892.00	114.00	47.10	1,053.10

Installation of 2-way control valve, flanged joints

2"	P1@.500	Ea	—	17.90	—	17.90
2½"	P1@.600	Ea	—	21.50	—	21.50
3"	P1@.750	Ea	—	26.80	—	26.80
4"	P1@1.35	Ea	—	48.30	—	48.30
6"	P1@2.50	Ea	—	89.40	—	89.40
8"	P1@3.00	Ea	—	107.00	—	107.00
10"	P1@4.00	Ea	—	143.00	—	143.00
12"	P1@4.50	Ea	—	161.00	—	161.00

Installation of 3-way control valve, flanged joints

2"	P1@.910	Ea	—	32.50	—	32.50
2½"	P1@1.12	Ea	—	40.10	—	40.10
3"	P1@1.33	Ea	—	47.60	—	47.60
4"	P1@2.00	Ea	—	71.50	—	71.50
6"	P1@3.70	Ea	—	132.00	—	132.00
8"	P1@4.40	Ea	—	157.00	—	157.00
10"	P1@5.90	Ea	—	211.00	—	211.00
12"	P1@6.50	Ea	—	232.00	—	232.00

Bolt and gasket sets

2"	P1@.500	Ea	3.86	17.90	—	21.76
2½"	P1@.650	Ea	4.51	23.20	—	27.71
3"	P1@.750	Ea	7.64	26.80	—	34.44
4"	P1@1.00	Ea	12.70	35.80	—	48.50
6"	P1@1.20	Ea	21.20	42.90	—	64.10
8"	P1@1.25	Ea	23.80	44.70	—	68.50
10"	P1@1.70	Ea	40.90	60.80	—	101.70
12"	P1@2.20	Ea	46.20	78.70	—	124.90

Carbon Steel, Schedule 40 with Cut-Grooved Joints

Description	Craft@Hrs	Unit	Material $	Labor $	Equipment $	Total $
Thermometer with well						
7"	P1@.250	Ea	159.00	8.94	—	167.94
9"	P1@.250	Ea	164.00	8.94	—	172.94
Dial type pressure gauge						
2½"	P1@.200	Ea	33.10	7.15	—	40.25
3½"	P1@.200	Ea	43.80	7.15	—	50.95
Pressure/temperature tap						
tap	P1@.150	Ea	15.70	5.36	—	21.06
Hanger with swivel assembly						
2"	P1@.300	Ea	6.05	10.70	—	16.75
2½"	P1@.350	Ea	7.82	12.50	—	20.32
3"	P1@.350	Ea	9.67	12.50	—	22.17
4"	P1@.350	Ea	14.90	12.50	—	27.40
5"	P1@.450	Ea	16.70	16.10	—	32.80
6"	P1@.450	Ea	21.80	16.10	—	37.90
8"	P1@.450	Ea	28.90	16.10	—	45.00
10"	P1@.550	Ea	39.70	19.70	—	59.40
12"	P1@.600	Ea	52.80	21.50	—	74.30
Riser clamp						
2"	P1@.115	Ea	4.84	4.11	—	8.95
2½"	P1@.120	Ea	5.09	4.29	—	9.38
3"	P1@.120	Ea	5.52	4.29	—	9.81
4"	P1@.125	Ea	7.03	4.47	—	11.50
5"	P1@.180	Ea	10.10	6.44	—	16.54
6"	P1@.200	Ea	12.20	7.15	—	19.35
8"	P1@.200	Ea	19.80	7.15	—	26.95
10"	P1@.250	Ea	29.40	8.94	—	38.34
12"	P1@.250	Ea	34.90	8.94	—	43.84
Galvanized steel pipe sleeves						
2"	P1@.130	Ea	4.94	4.65	—	9.59
2½"	P1@.150	Ea	5.01	5.36	—	10.37
3"	P1@.180	Ea	5.13	6.44	—	11.57
4"	P1@.220	Ea	5.85	7.87	—	13.72
5"	P1@.250	Ea	7.25	8.94	—	16.19
6"	P1@.270	Ea	7.82	9.66	—	17.48
8"	P1@.270	Ea	9.00	9.66	—	18.66
10"	P1@.290	Ea	10.50	10.40	—	20.90
12"	P1@.310	Ea	14.20	11.10	—	25.30
14"	P1@.330	Ea	14.20	11.80	—	26.00

Description	Craft@Hrs	Unit	Material $	Labor $	Equipment $	Total $

Leadership in Energy & Environmental Design (LEED)

LEED offers an incentive to building owners to install energy-efficient and environmentally sensitive HVAC equipment. The LEED program awards points for application of the best available technology. The HVAC system can earn a maximum of 17 points toward LEED certification. A LEED-certified air conditioning system must comply with ASHRAE 90.1-2004 (efficiency) and ASHRAE 62.1-2004 (indoor air quality) standards. LEED certification also requires recording controls (Carrier ComfortView 3 or equal) to measure system performance. Zone sensor thermostats (Carrier Debonair or equal) are required to provide performance and efficiency data to the digital recorder. An indoor air quality CO_2 sensor, duct-mounted aspirator modules and a refrigerant such as Puron R-410A are also required.

Description	Craft@Hrs	Unit	Material $	Labor $	Equipment $	Total $
Add for a LEED certified system with USGBC rating	—	Ea	20%	—	—	—
Add for LEED registration and inspection	—	Ea	2,000.00	—	—	2,000.00
Add for LEED central digital performance monitor/recorder, ComfortView 3 or equal	—	Ea	2,350.00	—	—	2,350.00
Add for LEED certified zone recording thermostats, Carrier DebonAir, USB remote-ready	—	Ea	350.00	—	—	350.00
Add for IAQ (Indoor Air Quality) CO_2 sensor and duct-mounted aspirator box	—	Ea	425.00	—	—	425.00

Residential A/C Cooling System

With remote exterior condenser/compressor, field installed DX coil in existing supply air plenum. Includes electrical & control wiring, R-22 refrigerant piping & connections, programmable digital thermostat, start up and testing. Costs based on a maximum distance between outdoor condenser and coil of 60' and include a contractor mark up of 25%. Use these figures for preliminary budget purposes. Costs are based on using 650 SF per ton as the cooling requirement with a thermal envelope of R20 or greater, i.e., a 1.5-ton A/C unit will serve a 975 SF home. Typical subcontract prices.

Description	Craft@Hrs	Unit	Material $	Labor $	Equipment $	Total $
1.5 tons	—	Ea	—	—	—	2,700.00
2 tons	—	Ea	—	—	—	2,900.00
2.5 tons	—	Ea	—	—	—	3,600.00
3 tons	—	Ea	—	—	—	4,900.00
5 tons	—	Ea	—	—	—	7,400.00
Add for R410A Refrigerant	—	%	—	—	—	18.0
Add for high heat loads	—	%	—	—	—	12.0

Description	Craft@Hrs	Unit	Material $	Labor $	Equipment $	Total $

Residential forced-air heating, cooling & ventilation system

Including a remote exterior condenser/compressor, field installed R-22 refrigerant DX "A" coil in supply air plenum, a gas-fired high efficiency furnace, programmable digital thermostat, ductwork, grilles & registers, HRV (heat recovery ventilator), kitchen and bathroom exhaust fans, ventilation and exhaust ductwork and wall hoods, electrical & control wiring, gas, refrigerant & flue piping & associated connections, start up, testing, and a contractor mark up of 25%. Use these figures for preliminary budget purposes. Heating & cooling load calculations based on a thermal envelope of R20 or greater. Typical subcontract prices.

Description	Craft@Hrs	Unit	Material $	Labor $	Equipment $	Total $
900 – 1,000 SF		Ea	—	—	—	14,000.00
1,100 – 1,400 SF		Ea	—	—	—	16,000.00
1,500 – 1,750 SF		Ea	—	—	—	19,600.00
1,800 – 2000 SF		Ea	—	—	—	21,000.00
2,200 – 2,500 SF		Ea	—	—	—	23,750.00
2,600 – 3,000 SF		Ea	—	—	—	28,500.00
Deduct for mid-efficient unit		Ea	—	—	—	250.00
Deduct for electric furnace		Ea	—	—	—	250.00
Add for R410A refrigerant		%	—	—	—	3.5

Description	Craft@Hrs	Unit	Material $	Labor $	Equipment $	Total $

Packaged self-contained roof-top DX air conditioning units. EnergySmart certified, ASHRAE, US Green Building Council and Underwriters Laboratories approved. Includes cooling coils, compressor, heat rejection coils, regulator valves, refrigerant tank and remote digital single-zone control package. Set in place with a 13,000 lb. truck crane. Add installation costs from the section that follows.

Description	Craft@Hrs	Unit	Material $	Labor $	Equipment $	Total $
2-T, 800 CFM	SN@5.00	Ea	3,300.00	188.00	78.70	3,566.70
2½-T, 1,000 CFM	SN@6.00	Ea	3,950.00	226.00	94.20	4,270.20
3-T, 1,200 CFM	SN@7.00	Ea	4,730.00	264.00	110.00	5,104.00
4-T, 1,600 CFM	SN@8.00	Ea	5,780.00	301.00	157.00	6,238.00
5-T, 2,000 CFM	SN@9.00	Ea	7,250.00	339.00	157.00	7,746.00
7½-T, 3,000 CFM	SN@10.0	Ea	9,880.00	377.00	314.00	10,571.00
10-T, 4,000 CFM	SN@12.0	Ea	12,400.00	452.00	314.00	13,166.00
12-T, 5,000 CFM	SN@14.0	Ea	15,000.00	527.00	377.00	15,904.00
15-T, 6,000 CFM	SN@15.0	Ea	18,400.00	565.00	377.00	19,342.00
20-T, 8,000 CFM	SN@16.0	Ea	24,300.00	602.00	550.00	25,452.00
25-T, 10,000 CFM	SN@17.0	Ea	29,500.00	640.00	550.00	30,690.00
30-T, 12,000 CFM	SN@19.0	Ea	33,600.00	715.00	707.00	35,022.00
40-T, 16,000 CFM	SN@22.0	Ea	42,100.00	828.00	707.00	43,635.00

Roof-top DX air conditioning unit, hot water coil. EnergySmart certified, ASHRAE, US Green Building Council, and Underwriters Laboratories approved. Add for the hot water coil. Single zone controls. Add installation costs from the section that follows.

Description	Craft@Hrs	Unit	Material $	Labor $	Equipment $	Total $
2-T, 800 CFM	SN@5.00	Ea	3,210.00	188.00	79.00	3,477.00
2½-T, 1,000 CFM	SN@6.00	Ea	3,870.00	226.00	94.00	4,190.00
3-T, 1,200 CFM	SN@7.00	Ea	4,640.00	264.00	110.00	5,014.00
4-T, 1,600 CFM	SN@8.00	Ea	5,650.00	301.00	157.00	6,108.00
5-T, 2,000 CFM	SN@9.00	Ea	7,060.00	339.00	157.00	7,556.00
7½-T, 3,000 CFM	SN@10.0	Ea	10,600.00	377.00	314.00	11,291.00
10-T, 4,000 CFM	SN@12.0	Ea	12,100.00	452.00	314.00	12,866.00
12-T, 5,000 CFM	SN@14.0	Ea	14,600.00	527.00	377.00	15,504.00
15-T, 6,000 CFM	SN@15.0	Ea	17,900.00	565.00	377.00	18,842.00
20-T, 8,000 CFM	SN@16.0	Ea	23,700.00	602.00	550.00	24,852.00
25-T, 10,000 CFM	SN@17.0	Ea	28,700.00	640.00	550.00	29,890.00
30-T, 12,000 CFM	SN@19.0	Ea	33,600.00	715.00	707.00	35,022.00
40-T, 16,000 CFM	SN@22.0	Ea	42,100.00	828.00	707.00	43,635.00

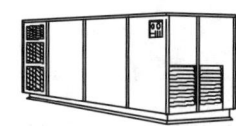

Description	Craft@Hrs	Unit	Material $	Labor $	Equipment $	Total $

Installation costs for packaged rooftop cooling

Description	Craft@Hrs	Unit	Material $	Labor $	Equipment $	Total $
Cut, frame and gasket downcomer hole in roof	S2@1.00	Ea	35.00	35.36	—	70.36
Mount duct hangers	S2@.250	Ea	3.50	8.84	—	12.34
Cut and mount sheet metal duct	S2@.350	LF	5.50	12.40	—	17.90
Apply and coat duct insulation	S2@.150	LF	2.40	5.30	—	7.70
Install piping for gas line	P1@.100	LF	3.75	3.58	—	7.33
Install piping for chilled or hot water/steam line	P1@.100	LF	2.85	3.58	—	6.43
Install electrical wiring	BE@.150	LF	1.75	5.94	—	7.69
Install HVAC controls	BE@.500	Ea	—	19.80	—	19.80
Commission and test	P1@4.00	Ea	—	143.00	—	143.00
Air balance and fine-tune	P1@4.00	Ea	—	143.00	—	143.00

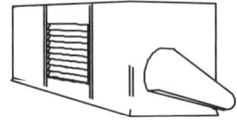

Packaged air handler with chilled water and hot water/steam coil.

Costs shown based on 400 CFM per ton cooling. Unit includes insulated single wall casing, fan section, cooling coil section, heating coil section, mixing plenum section, bag filter section, fan motor, variable pitch drive, vibration isolators and drain pan. Set in place only. Make additional allowances for coil connections, controls, motor starters and power wiring. (12,000 BTUs equals 1 ton cooling.) Use these costs for preliminary estimates.

Description	Craft@Hrs	Unit	Material $	Labor $	Equipment $	Total $
3-T, 1,200 CFM	SN@4.00	Ea	2,730.00	151.00	63.00	2,944.00
4-T, 1,600 CFM	SN@5.50	Ea	3,590.00	207.00	157.00	3,954.00
5-T, 2,000 CFM	SN@7.00	Ea	4,290.00	264.00	157.00	4,711.00
7½-T, 3,000 CFM	SN@9.00	Ea	6,030.00	339.00	314.00	6,683.00
10-T, 4,000 CFM	SN@11.0	Ea	7,660.00	414.00	314.00	8,388.00
12-T, 5,000 CFM	SN@13.0	Ea	9,020.00	489.00	377.00	9,886.00
15-T, 6,000 CFM	SN@14.0	Ea	10,100.00	527.00	377.00	11,004.00
20-T, 8,000 CFM	SN@15.0	Ea	12,000.00	565.00	550.00	13,115.00
25-T, 10,000 CFM	SN@16.0	Ea	11,900.00	602.00	550.00	13,052.00
30-T, 12,000 CFM	SN@18.0	Ea	15,200.00	678.00	707.00	16,585.00
40-T, 16,000 CFM	SN@21.0	Ea	18,700.00	791.00	707.00	20,198.00

Description	Craft@Hrs	Unit	Material $	Labor $	Equipment $	Total $

Air handling unit accessories and options. Use these costs for preliminary estimates.

Description	Craft@Hrs	Unit	Material $	Labor $	Equipment $	Total $
DDC controls per zone complete	SN@2.75	Ea	915.00	104.00	43.30	1,062.30
Electric controls per zone complete	SN@2.75	Ea	487.00	104.00	43.30	634.30
Pneumatic controls per zone complete	SN@2.75	Ea	458.00	104.00	43.30	605.30
Variable speed drive 5 HP	SN@4.00	Ea	3,660.00	151.00	62.90	3,873.90
Variable speed drive 7.5HP	SN@4.00	Ea	4,270.00	151.00	62.90	4,483.90
Variable speed drive 10 HP	SN@6.00	Ea	4,870.00	226.00	94.20	5,190.20
Variable speed drive 15 HP	SN@8.00	Ea	5,490.00	301.00	94.20	5,885.20
Variable speed drive 20 HP	SN@12.0	Ea	6,100.00	452.00	189.00	6,741.00
Variable speed drive 25 HP	SN@16.0	Ea	6,710.00	602.00	251.00	7,563.00
Variable speed drive 30 HP	SN@18.0	Ea	7,930.00	678.00	282.00	8,890.00
Variable speed drive 40 HP	SN@22.0	Ea	9,740.00	828.00	347.00	10,915.00
Variable speed drive 50 HP	SN@26.0	Ea	11,600.00	979.00	409.00	12,988.00
Variable Inlet vanes 1,000-1,500 CFM	SN@4.00	Ea	184.00	151.00	62.90	397.90
Variable Inlet vanes 1,600-2,500 CFM	SN@4.00	Ea	396.00	151.00	62.90	609.90
Variable Inlet vanes 2,600-5,000 CFM	SN@6.00	Ea	1,100.00	226.00	94.20	1,420.20
Variable Inlet vanes 6,000-10,000 CFM	SN@8.00	Ea	1,840.00	301.00	126.00	2,267.00
Variable Inlet vanes 11,000-20,000 CFM	SN@10.0	Ea	3,050.00	377.00	157.00	3,584.00

Heat Recovery Ventilators — Commercial

Heat Recovery Ventilators provide a fresh air supply to tightly sealed building envelopes. HRV's extract heat from the stale indoor air being exhausted and transfer the heat to the fresh air being drawn in to the building through the HRV. Heat recovery ventilators are also excellent dehumidifiers.

HRV unit features include energy efficient defrost cycle, cross-flow polypropylene heat exchanger, acoustically lined cabinet, outdoor air filter.

Allow 15-30 cfm per person or .04 to.05 cfm per square foot when sizing HRV unit. (ASHRAE62-19890 (cfm = cubic feet per minute)

Description	Craft@Hrs	Unit	Material $	Labor $	Equipment $	Total $

Commercial heat recovery ventilator. Set/hang in place only. Make additional allowances for duct, diffusers, controls, air balancing, electrical connections and condensate drain.

Description	Craft@Hrs	Unit	Material $	Labor $	Equipment $	Total $
HRV 700 cfm	SN@4.75	Ea	5,200.00	179.00	74.60	5,453.60
HRV 1,200 cfm	SN@5.25	Ea	5,600.00	198.00	82.50	5,880.50
HRV 2,500 cfm	SN@6.75	Ea	23,000.00	254.00	106.20	23,360.20

Swimming pool heat recovery ventilator. Set/hang in place only. Make additional allowances for duct, diffusers, controls, air balancing, electrical connections and condensate drain.

Description	Craft@Hrs	Unit	Material $	Labor $	Equipment $	Total $
HRV 700 cfm	SN@5.00	Ea	6,140.00	188.00	78.70	6,406.70
HRV 1,200 cfm	SN@5.75	Ea	6,960.00	216.00	90.40	7,266.40

Description	Craft@Hrs	Unit	Material $	Labor $	Equipment $	Total $

Conventional heat recovery ventilator. Hang in place only. Make additional allowances for duct, diffusers, controls, air balancing, electrical connections and condensate drain.

Description	Craft@Hrs	Unit	Material $	Labor $	Equipment $	Total $
HRV 65–150 cfm	S2@2.45	Ea	975.00	86.60	—	1,061.60
HRV 115–200cfm	S2@2.65	Ea	1,170.00	93.70	—	1,263.70

Compact heat recovery ventilator. Hang in place only. Make additional allowances for duct, diffusers, controls, air balancing, electrical connections and condensate drain.

Description	Craft@Hrs	Unit	Material $	Labor $	Equipment $	Total $
HRV 65–127 cfm	S2@2.45	Ea	1,030.00	86.60	—	1,116.60
HRV 115–195cfm	S2@2.65	Ea	1,170.00	93.70	—	1,263.70

High efficiency heat recovery ventilator. Hang in place only. Make additional allowances for duct, diffusers, controls, air balancing, electrical connections and condensate drain.

Description	Craft@Hrs	Unit	Material $	Labor $	Equipment $	Total $
HRV 65–127 cfm	S2@2.65	Ea	1,120.00	93.70	—	1,213.70
HRV 115–180cfm	S2@2.90	Ea	1,760.00	103.00	—	1,863.00
HRV 180–265cfm	S2@3.25	Ea	1,830.00	115.00	—	1,945.00

Heat recovery ventilator controls

Description	Craft@Hrs	Unit	Material $	Labor $	Equipment $	Total $
HRV basic control	S2@1.00	Ea	89.40	35.40	—	124.80
HRV std. control	S2@1.00	Ea	142.00	35.40	—	177.40
HRV auto control	S2@1.00	Ea	192.00	35.40	—	227.40
60 Minute timer	S2@.600	Ea	59.40	21.20	—	80.60
Interlock relay	S2@.600	Ea	70.60	21.20	—	91.80

Heat recovery ventilator accessories

Description	Craft@Hrs	Unit	Material $	Labor $	Equipment $	Total $
6" diffusers	S2@.500	Ea	15.00	17.70	—	32.70
6" wall hoods	S2@1.50	Ea	26.10	53.00	—	79.10
6" tee fittings	S2@.350	Ea	3.49	12.40	—	15.89
6" flex duct	S2@.025	LF	.82	.88	—	1.70
6" flex insul duct	S2@.025	LF	1.60	.88	—	2.48
HRV filters	S2@.450	Ea	21.40	15.90	—	37.30

Description	Craft@Hrs	Unit	Material $	Labor $	Equipment $	Total $

Installation costs for heat recovery ventilators

Description	Craft@Hrs	Unit	Material $	Labor $	Equipment $	Total $
Cut, frame and gasket, mount duct hangers	S2@.250	Ea	3.50	8.84	—	12.34
Cut and mount sheet metal duct	S2@.350	LF	5.50	12.40	—	17.90
Apply and coat duct insulation	S2@.150	LF	2.40	5.30	—	7.70
Install piping for gas line	P1@.100	LF	3.75	3.58	—	7.33
Install piping for chilled or hot water/steam	P1@.100	LF	2.85	3.58	—	6.43
Install electrical wiring	BE@.150	LF	1.75	5.94	—	7.69
Install HVAC controls	BE@.500	Ea	—	19.80	—	19.80
Commission and test	P1@4.00	Ea	—	143.00	—	143.00
Run air balance and fine tune system	P1@4.00	Ea	—	143.00	—	143.00

Engineering drawings of coil piping details have a bad reputation among HVAC contractors and estimators. They're notorious for what they leave out. They'll rarely show more than one coil bank, no matter how big the system is. Furthermore, the drawings hardly ever call out sizes for either the piping or the control valves. Don't be taken in by the apparent simplicity of the system as shown in these drawings. It's likely to be only the tip of the iceberg. For example, unless the air handling capacity of the system is less than 16,000 CFM, the single coil bank shown won't be adequate. Add one or two more coil banks and you're looking at a lot more piping — and a more complex system that takes longer to install.

You probably haven't even decided which equipment supplier to use. This is hardly the time for you to start researching heating and cooling coils. Nevertheless, you need better, more complete and realistic data to come up with a competitive estimate.

We'll see how and where to track down the hard data that you have to take the time to find. I'll also pass along a few tips on estimating water coil piping. They'll help you avoid leaving something out of your estimates — a real pitfall for any beginner. It's those little (but essential) items, so easily overlooked, that are so deadly to a profit margin. Finally, check the next two pages of diagrams with tables. The data given there, combined with the data you collected earlier, forms the basis for informed guesswork.

The Hard Data

There are two things you absolutely must know to estimate water coil piping. First, the size of the branch piping to the coils; second, the CFM rating of the air handling units for the system.

To find the pipe sizes, look at either the floor plans or the details for the equipment room. They'll list the sizes of the branch run-out pipes. Once you know them, you can make a good guess at the right size for the control valves. (See the diagrams and tables on pages 292 and 293.) The only information you need is the CFM ratings of the air handling units and the branch piping sizes to the coils. If the system's capacity is over 16,000 CFM, you need two or three coil banks.

A Few Tips on Estimating Water Coil Piping

1) Coil connection sizes seldom match branch pipe sizes. Be sure to include the reducing fittings you'll need in any estimate.

2) Two-way and three-way control valves are both usually one pipe size smaller than the pipe where they're installed. That means you'll need either two or three reducing fittings per control valve. Be sure you include their cost in your estimates.

Using the Diagrams

In the following diagrams, for clarity, some items are not included. These items are: balance valves, shut-off valves, reducers, strainers, gauges and gauge taps. Any details you need about these items for your estimate are in the engineer's coil piping details.

Water Coil Piping

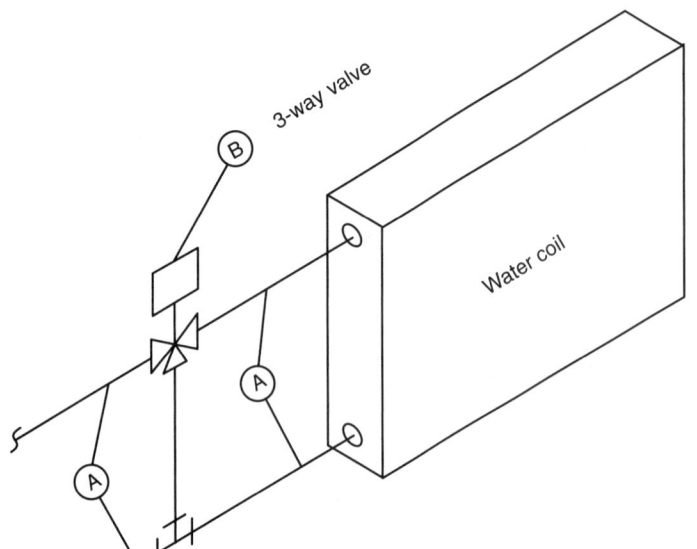

If Ⓐ is:	Then Ⓑ is:
1"	¾"
1¼"	1"
1½"	1¼"
2"	1½"
2½"	2"
3"	2½"

Typical water coil piping for A.H. units up to 16,000 CFM

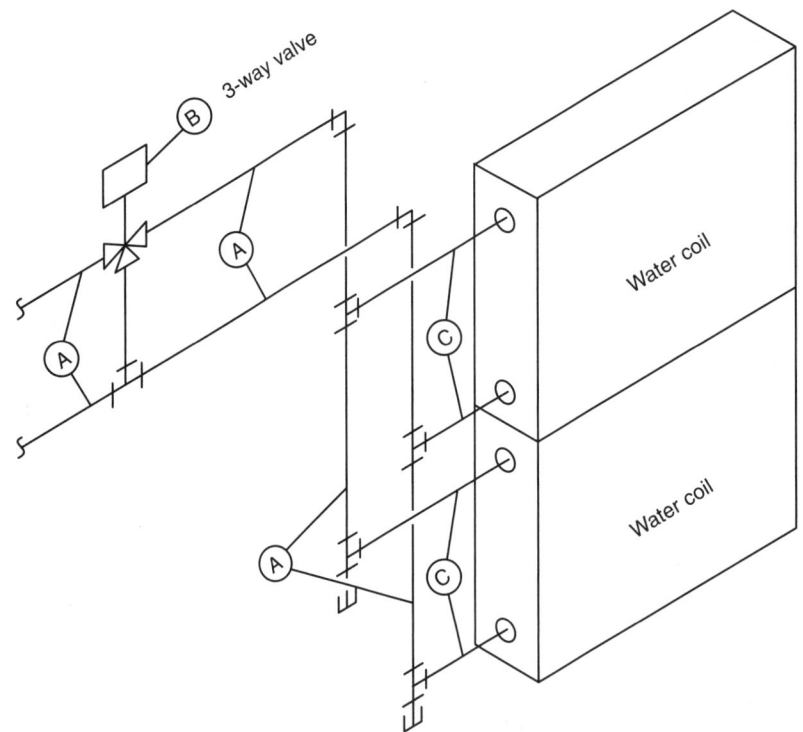

If Ⓐ is:	Then Ⓑ is:	And Ⓒ is:
2½"	2"	2"
3"	2½"	2½"
4"	3"	3"

Typical water coil piping for A.H. units from 16,000 to 26,000 CFM

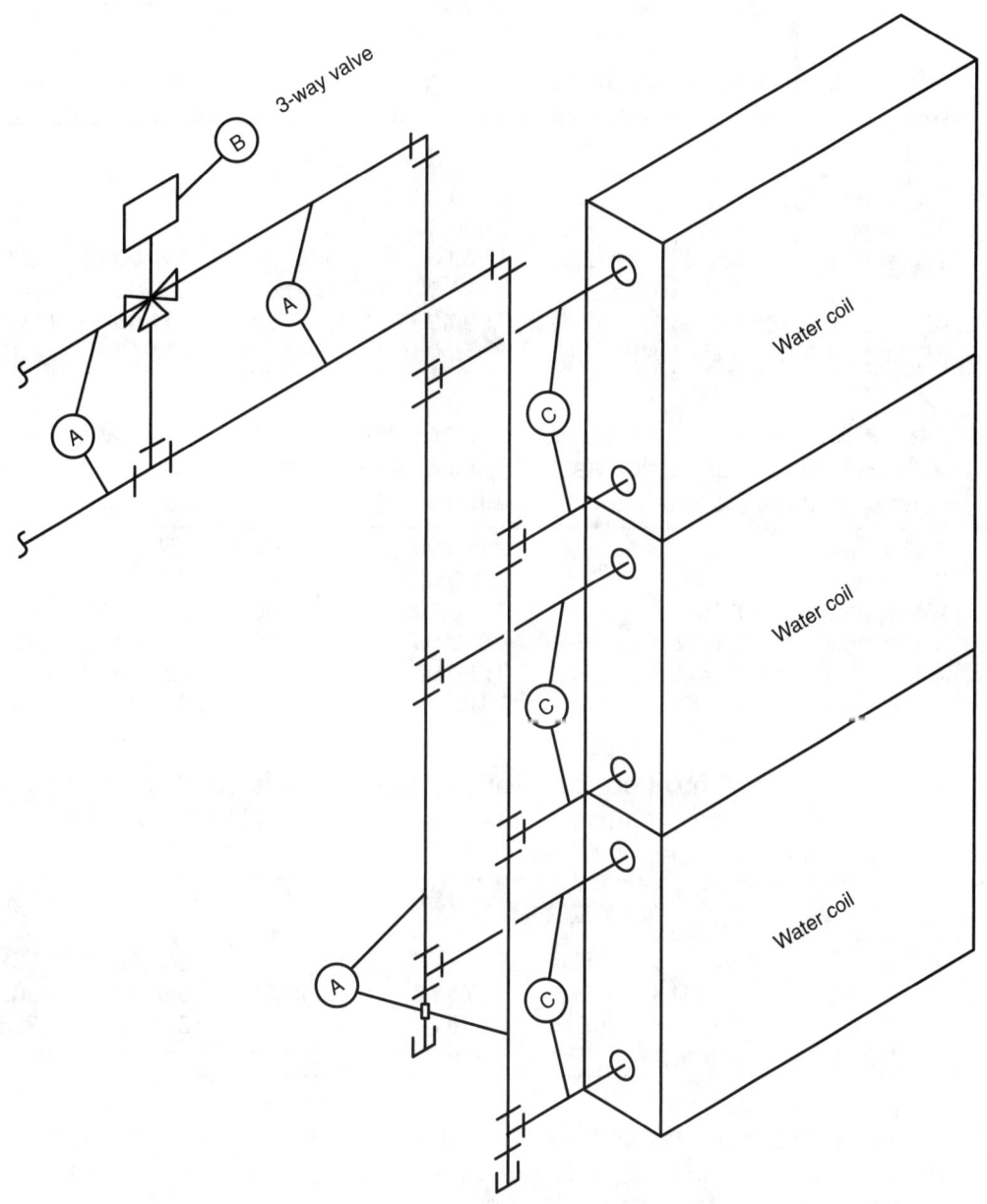

If (A) is:	Then (B) is:	And (C) is:
4"	3"	2½"
6"	4"	3"
8"	6"	4"
10"	8"	6"

Typical water coil piping for A.H. units over 26,000 CFM

Air Handling Unit Coil Connections

Description	Craft@Hrs	Unit	Material $	Labor $	Equipment $	Total $

Air handling unit coil connection, one row coil bank, non-regulated flow. Connection assembly includes pipe, fittings, pipe insulation, valves, gauges, thermometers and vents.

Description	Craft@Hrs	Unit	Material $	Labor $	Equipment $	Total $
1½" supply	SN@14.0	Ea	1,550.00	527.00	220.00	2,297.00
2" supply	SN@18.0	Ea	1,900.00	678.00	282.00	2,860.00
2½" supply	SN@28.0	Ea	2,540.00	1,050.00	440.00	4,030.00
3" supply	SN@31.5	Ea	2,790.00	1,190.00	495.00	4,475.00
4" supply	SN@34.0	Ea	2,990.00	1,280.00	534.00	4,804.00
6" supply	SN@41.0	Ea	4,310.00	1,540.00	644.00	6,494.00

Air handling unit coil connection, one row coil bank, 2-way control valve design. Connection assembly includes pipe and fittings, pipe insulation, valves, gauges, thermometers and vents.

Description	Craft@Hrs	Unit	Material $	Labor $	Equipment $	Total $
1½" supply	SN@16.5	Ea	1,750.00	621.00	259.00	2,630.00
2" supply	SN@21.0	Ea	2,260.00	791.00	330.00	3,381.00
2½" supply	SN@33.6	Ea	2,730.00	1,270.00	528.00	4,528.00
3" supply	SN@36.0	Ea	2,940.00	1,360.00	567.00	4,867.00
4" supply	SN@43.8	Ea	3,560.00	1,650.00	688.00	5,898.00
6" supply	SN@54.0	Ea	5,720.00	2,030.00	849.00	8,599.00

Air handling unit coil connection, one row coil bank, 3-way control valve design. Connection assembly includes pipe and fittings, pipe insulation, valves, gauges, thermometers and vents.

Description	Craft@Hrs	Unit	Material $	Labor $	Equipment $	Total $
1½" supply	SN@18.0	Ea	2,020.00	678.00	282.00	2,980.00
2" supply	SN@26.0	Ea	2,510.00	979.00	409.00	3,898.00
2½" supply	SN@41.5	Ea	3,140.00	1,560.00	653.00	5,353.00
3" supply	SN@43.8	Ea	3,470.00	1,650.00	688.00	5,808.00
4" supply	SN@54.6	Ea	4,480.00	2,060.00	858.00	7,398.00
6" supply	SN@61.4	Ea	6,370.00	2,310.00	965.00	9,645.00

Air handling unit coil connection, two row coil bank, non-regulated flow. Connection assembly includes pipe and fittings, pipe insulation, valves, gauges, thermometers and vents.

Description	Craft@Hrs	Unit	Material $	Labor $	Equipment $	Total $
2½" supply	SN@48.0	Ea	3,800.00	1,810.00	754.00	6,364.00
3" supply	SN@51.5	Ea	4,650.00	1,940.00	809.00	7,399.00
4" supply	SN@63.6	Ea	6,130.00	2,390.00	999.00	9,519.00
6" supply	SN@79.2	Ea	8,970.00	2,980.00	1,250.00	13,200.00

Air handling unit coil connection, two row coil bank, 2-way control valve design. Connection assembly includes pipe and fittings, pipe insulation, valves, gauges, thermometers and vents.

Description	Craft@Hrs	Unit	Material $	Labor $	Equipment $	Total $
2½" supply	SN@54.0	Ea	4,270.00	2,030.00	849.00	7,149.00
3" supply	SN@58.5	Ea	5,370.00	2,200.00	919.00	8,489.00
4" supply	SN@67.0	Ea	6,600.00	2,520.00	1,050.00	10,170.00
6" supply	SN@83.5	Ea	9,110.00	3,140.00	1,310.00	13,560.00

Description	Craft@Hrs	Unit	Material $	Labor $	Equipment $	Total $

Air handling unit coil connection, two row coil bank, 3-way control valve design. Connection assembly includes pipe and fittings, pipe insulation, valves, gauges, thermometers and vents.

Description	Craft@Hrs	Unit	Material $	Labor $	Equipment $	Total $
2½" supply	SN@58.8	Ea	4,480.00	2,210.00	924.00	7,614.00
3" supply	SN@64.0	Ea	5,640.00	2,410.00	1,007.00	9,057.00
4" supply	SN@72.0	Ea	7,130.00	2,710.00	1,130.00	10,970.00
6" supply	SN@87.0	Ea	9,500.00	3,280.00	1,370.00	14,150.00

Air handling unit coil connection, three row coil bank, non-regulated flow. Connection assembly includes pipe and fittings, pipe insulation, valves, gauges, thermometers and vents.

Description	Craft@Hrs	Unit	Material $	Labor $	Equipment $	Total $
4" supply	SN@72.0	Ea	6,470.00	2,710.00	1,130.00	10,310.00
6" supply	SN@94.0	Ea	11,200.00	3,540.00	1,480.00	16,220.00
8" supply	SN@120.	Ea	15,200.00	4,520.00	1,890.00	21,610.00

Air handling unit coil connection, three row coil bank, 2-way control valve design. Connection assembly includes pipe and fittings, pipe insulation, valves, gauges, thermometers and vents.

Description	Craft@Hrs	Unit	Material $	Labor $	Equipment $	Total $
4" supply	SN@78.0	Ea	6,860.00	2,910.00	1,230.00	11,000.00
6" supply	SN@105.	Ea	12,300.00	3,950.00	1,650.00	17,900.00
8" supply	SN@132.	Ea	17,000.00	4,970.00	2,070.00	24,040.00

Air handling unit coil connection, three row coil bank, 3-way control valve design. Connection assembly includes pipe and fittings, pipe insulation, valves, gauges, thermometers and vents.

Description	Craft@Hrs	Unit	Material $	Labor $	Equipment $	Total $
4" supply	SN@82.0	Ea	7,960.00	3,090.00	1,290.00	12,340.00
6" supply	SN@112.	Ea	12,700.00	4,220.00	1,760.00	18,680.00
8" supply	SN@140.	Ea	19,000.00	5,270.00	2,200.00	26,470.00

Description	Craft@Hrs	Unit	Material $	Labor $	Equipment $	Total $

Residential gas-fired upflow furnace. Set in place only. Make additional allowances for gas and electrical connections. Costs shown per thousand BTU per hour (MBH) Mid-efficiency, 80% AFUE.

Description	Craft@Hrs	Unit	Material $	Labor $	Equipment $	Total $
50 MBH input	SN@2.00	Ea	809.00	75.30	31.40	915.70
75 MBH input	SN@2.00	Ea	1,060.00	75.30	31.40	1,166.70
100 MBH input	SN@2.00	Ea	1,230.00	75.30	31.40	1,336.70
125 MBH input	SN@2.00	Ea	1,400.00	75.30	31.40	1,506.70

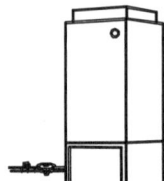

Residential combination gas fired furnace and A/C unit. With remote exterior condenser/compressor. Set in place only. Make additional allowances for gas, electrical and DX connections. Costs shown per thousand BTU per hour (MBH) Mid-efficiency, 80% AFUE.

Description	Craft@Hrs	Unit	Material $	Labor $	Equipment $	Total $
30 MBH, 1.5 tons	SN@2.50	Ea	2,410.00	94.10	39.20	2,543.30
36 MBH, 2 tons	SN@2.75	Ea	2,860.00	104.00	43.30	3,007.30
42 MBH, 2 tons	SN@3.00	Ea	3,210.00	113.00	47.10	3,370.10
48 MBH, 2.5 tons	SN@3.50	Ea	3,750.00	132.00	55.00	3,937.00
60 MBH, 3 tons	SN@4.00	Ea	4,480.00	151.00	62.90	4,693.90

High efficiency, condensing gas furnace. Direct drive, atmospheric draft, electronic ignition. Set in place only. Make additional allowances for flue, gas and electrical connections. Costs shown per thousand BTU per hour (MBH) High efficiency 94% AFUE.

Description	Craft@Hrs	Unit	Material $	Labor $	Equipment $	Total $
40 MBH input	S2@2.00	Ea	1,670.00	70.70	—	1,740.70
60 MBH input	S2@2.00	Ea	1,670.00	70.70	—	1,740.70
70 MBH input	S2@2.00	Ea	1,610.00	70.70	—	1,680.70
90 MBH input	S2@2.00	Ea	1,610.00	70.70	—	1,680.70
100 MBH input	S2@2.00	Ea	2,290.00	70.70	—	2,360.70
120 MBH input	S2@2.00	Ea	2,370.00	70.70	—	2,440.70
2 speed fan & limit control kit	S2@1.25	Ea	32.80	44.20	—	77.00
Propane conversion kit	S2@1.75	Ea	127.00	61.90	—	188.90

Description	Craft@Hrs	Unit	Material $	Labor $	Equipment $	Total $

Central dehumidification system Upflow or counterflow, 400 CFM, 30 gallon per day baseload capacity at 0.5 water column pressure differential. With inlet filter, drain pan, drain line, heater plate, cooling draft exchanger, humidistat control sensor, control relay, actuator, cabinet, two 24" by 24" by 4' long duct transitions, gaskets, hangers and ground fault protection. Whirlpool UGD160UH or equal. Installation assumes rental of an appliance dolly ($14 per day), 3,000 lb. come-a-long ($16 per day) and a 1/2-ton chain hoist ($21 per day). Add the cost of control wiring and electrical wiring.

Description	Craft@Hrs	Unit	Material $	Labor $	Equipment $	Total $
Central dehumidification system	SW@0.50	Ea	1,500.00	20.50	51.00	1,571.50
Cut, frame and gasket hole in wall, ceiling or roof	SW@1.00	Ea	35.00	41.00	—	76.00
Mount duct hangers	SW@0.25	Ea	3.50	10.30	—	13.80
Cut and mount sheet metal duct	SW@0.35	LF	5.50	14.40	—	19.90
Apply and coat duct insulation	SW@0.15	LF	2.40	6.16	—	8.56
Install electrical wiring	BE@0.15	LF	1.75	5.94	—	7.69
Install dehumidifier controls	BE@0.50	Ea	—	19.80	—	19.80
Test and balance	SW@1.50	Ea	—	61.60		61.60

Wall furnace, mid efficiency, pilot light ignition. Gas fired, atmospheric draft, (radiant heat, no fan). Set in place only. Make additional allowances for gas and flue connections. Costs shown per thousand BTU per hour (MBH). Mid efficiency 80% AFUE.

Description	Craft@Hrs	Unit	Material $	Labor $	Equipment $	Total $
35 MBH input	S2@2.75	Ea	1,130.00	97.20	—	1,227.20
50 MBH input	S2@2.75	Ea	1,280.00	97.20	—	1,377.20
65 MBH input	S2@2.75	Ea	1,290.00	97.20	—	1,387.20
Fan blower kit	S2@1.75	Ea	183.00	61.90	—	244.90

Wall furnace, mid efficiency, electronic ignition. Gas fired, atmospheric draft, (radiant heat, no fan). Set in place only. Make additional allowances for gas and flue connections. Costs shown per thousand BTU per hour (MBH). Mid efficiency 80% AFUE.

Description	Craft@Hrs	Unit	Material $	Labor $	Equipment $	Total $
35 MBH input	S2@2.75	Ea	1,410.00	97.20	—	1,507.20
65 MBH input	S2@2.75	Ea	1,590.00	97.20	—	1,687.20
Fan blower kit	S2@1.75	Ea	183.00	61.90	—	244.90

Wall furnace, direct vent (included). Gas fired, pilot light ignition, mid efficiency. (radiant heat, no fan). Set in place only. Make additional allowances for vent, gas and electrical connections. Costs shown per thousand BTU per hour (MBH). Mid efficiency 80% AFUE.

Description	Craft@Hrs	Unit	Material $	Labor $	Equipment $	Total $
15 MBH input	S2@2.75	Ea	760.00	97.20	—	857.20
25 MBH input	S2@2.75	Ea	920.00	97.20	—	1,017.20
33 MBH input	S2@2.75	Ea	980.00	97.20	—	1,077.20
40 MBH input	S2@2.75	Ea	1,340.00	97.20	—	1,437.20
65 MBH input	S2@2.75	Ea	1,480.00	97.20	—	1,577.20
Direct vent kit	S2@2.50	Ea	129.00	88.40	—	217.40
Fan blower kit	S2@1.75	Ea	183.00	61.90	—	244.90

Unit Heaters

Description	Craft@Hrs	Unit	Material $	Labor $	Equipment $	Total $

Unit heater, gas, direct fired, mid efficiency, propeller fan, atmospheric draft, single stage, standing pilot, aluminized burner and heat exchanger. Hang in place only. Make additional allowances for flue, gas and electrical connections. Costs shown per thousand BTU per hour (MBH). Mid efficiency 78% AFUE.

Description	Craft@Hrs	Unit	Material $	Labor $	Equipment $	Total $
30 MBH input	S2@4.00	Ea	933.00	141.00	—	1,074.00
50 MBH input	S2@4.00	Ea	1,030.00	141.00	—	1,171.00
75 MBH input	S2@4.00	Ea	1,150.00	141.00	—	1,291.00
100 MBH input	S2@4.00	Ea	1,280.00	141.00	—	1,421.00
125 MBH input	S2@4.00	Ea	1,520.00	141.00	—	1,661.00
145 MBH input	S2@4.00	Ea	1,580.00	141.00	—	1,721.00
175 MBH input	S2@4.00	Ea	1,730.00	141.00	—	1,871.00
200 MBH input	S2@4.00	Ea	1,910.00	141.00	—	2,051.00
250 MBH input	S2@4.00	Ea	2,010.00	141.00	—	2,151.00
300 MBH input	S2@4.00	Ea	2,610.00	141.00	—	2,751.00
350 MBH input	S2@4.00	Ea	3,080.00	141.00	—	3,221.00
400 MBH input	S2@4.00	Ea	3,500.00	141.00	—	3,641.00

Unit heater, gas fired, high efficiency, propeller fan, power vented (included), direct fired, single stage, intermittent pilot ignition, aluminized burner and heat exchanger. Hang in place and install power vent. Make additional allowances for gas and electrical connections. Costs shown per thousand BTU per hour (MBH). High efficiency 85% AFUE.

Description	Craft@Hrs	Unit	Material $	Labor $	Equipment $	Total $
30 MBH input	S2@5.25	Ea	1,380.00	186.00	—	1,566.00
50 MBH input	S2@5.25	Ea	1,470.00	186.00	—	1,656.00
75 MBH input	S2@5.25	Ea	2,350.00	186.00	—	2,536.00
100 MBH input	S2@5.25	Ea	1,780.00	186.00	—	1,966.00
125 MBH input	S2@5.25	Ea	1,980.00	186.00	—	2,166.00
145 MBH input	S2@5.25	Ea	2,100.00	186.00	—	2,286.00
175 MBH input	S2@5.25	Ea	2,220.00	186.00	—	2,406.00
200 MBH input	S2@5.25	Ea	2,370.00	186.00	—	2,556.00
250 MBH input	S2@5.25	Ea	2,740.00	186.00	—	2,926.00
300 MBH input	S2@5.25	Ea	3,230.00	186.00	—	3,416.00
350 MBH input	S2@5.25	Ea	3,610.00	186.00	—	3,796.00
400 MBH input	S2@5.25	Ea	3,960.00	186.00	—	4,146.00

Unit heater, gas, indirect fired, mid efficiency, propeller fan, power vented (included), single stage, intermittent pilot ignition, aluminized burner and heat exchanger, Hang in place and install power vent. Make additional allowances for gas and electrical connections. Costs shown per thousand BTU per hour (MBH). Mid efficiency 78% AFUE.

Description	Craft@Hrs	Unit	Material $	Labor $	Equipment $	Total $
130 MBH input	S2@5.25	Ea	2,520.00	186.00	—	2,706.00
150 MBH input	S2@5.25	Ea	2,740.00	186.00	—	2,926.00
170 MBH input	S2@5.25	Ea	2,840.00	186.00	—	3,026.00
225 MBH input	S2@5.25	Ea	3,170.00	186.00	—	3,356.00
280 MBH input	S2@5.25	Ea	3,780.00	186.00	—	3,966.00
340 MBH input	S2@5.25	Ea	4,270.00	186.00	—	4,456.00

Description	Craft@Hrs	Unit	Material $	Labor $	Equipment $	Total $

High intensity infrared heater, gas fired. Ceramic grid. Automatic spark ignition. Including reflector. Approved for unvented commercial indoor installations. Not for use in residential applications. Labor includes hang in place only. Make additional allowances for gas and 110 volt electrical connections. Refer to local gas code for proper installation requirements.

Description	Craft@Hrs	Unit	Material $	Labor $	Equipment $	Total $
30,000 BTU	P1@1.20	Ea	461.00	42.90	66.00	569.90
60,000 BTU	P1@1.80	Ea	588.00	64.40	99.00	751.40
100,000 BTU	P1@2.50	Ea	1,390.00	89.40	138.00	1,617.40
130,000 BTU	P1@2.75	Ea	1,720.00	98.30	151.00	1,969.30
160,000 BTU	P1@2.75	Ea	1,960.00	98.30	151.00	2,209.30
Add for propane —		Ea	106.00	—	—	106.00

High intensity infrared heater, gas fired. Ceramic grid. Millivolt standing pilot ignition. Approved for unvented commercial indoor installations. Not for use in residential applications. Labor includes hang in place only. Make additional allowances for gas connections. Millivolt standing pilot does not require electrical source.

Description	Craft@Hrs	Unit	Material $	Labor $	Equipment $	Total $
30,000 BTU	P1@1.20	Ea	1,070.00	42.90	66.00	1,178.90
60,000 BTU	P1@1.80	Ea	1,390.00	64.40	99.00	1,553.40
100,000 BTU	P1@2.50	Ea	1,740.00	89.40	138.00	1,967.40
130,000 BTU	P1@2.75	Ea	2,080.00	98.30	151.00	2,329.30
160,000 BTU	P1@2.75	Ea	2,410.00	98.30	151.00	2,659.30
Add for propane —		Ea	120.00	—	—	120.00

Infrared tube heater, burner unit only, gas fired, negative pressure type. Labor includes hang in place only. Not for use in residential applications. Make additional allowances for burner tube, burner tube reflector, venting, gas piping and electrical requirements.

Description	Craft@Hrs	Unit	Material $	Labor $	Equipment $	Total $
30,000 BTU	P1@1.20	Ea	1,320.00	42.90	66.00	1,428.90
60,000 BTU	P1@1.80	Ea	1,250.00	64.40	99.00	1,413.40
100,000 BTU	P1@2.50	Ea	1,440.00	89.40	138.00	1,667.40
150,000 BTU	P1@2.75	Ea	1,530.00	98.30	151.00	1,779.30
175,000 BTU	P1@2.75	Ea	1,780.00	98.30	151.00	2,029.30
200,000 BTU	P1@2.75	Ea	2,050.00	98.30	151.00	2,299.30
Add for propane —		Ea	120.00	—	—	120.00

Infrared tube heater, 4" diameter, 10' lengths. 16 gauge burner tube and .024" mill finish aluminum reflector. Labor includes hang in place only. Not for use in residential applications. Make additional allowances for burner unit, venting, gas piping and electrical requirements.

Description	Craft@Hrs	Unit	Material $	Labor $	Equipment $	Total $
Burner tubing	P1@.200	LF	5.45	7.15	—	12.60
Tubing U-bend	P1@.850	Ea	194.00	30.40	—	224.40
Reflector shield	P1@.200	LF	5.45	7.15	—	12.60
Vent terminal	P1@.600	Ea	85.70	21.50	—	107.20

Heat Pump Systems

Heat Pumps

In general, heat pumps are more effective in milder winter climates. In colder winter climates, secondary/supplemental electric heating elements (or equivalent heating source) are recommended.

A geothermal heat pump system's efficient operation is largely dependent upon proper design, layout, sizing and installation of the ground loop piping system for closed loop systems. In the case of well-to-well, the efficient operation of open loop systems is largely dependent upon the ability of the supply and return, or dump wells to meet the water volume demand of the heat pumps.

Rules of thumb when sizing loop & well systems:

Heating capacity, when considered or measured in tons cooling or refrigeration, is generally 3 to 4 times that of required cooling capacity for climates within 100 miles of the 49th parallel, i.e., if 2 tons of cooling is required, allow 6-8 tons of heating capacity (1 ton = 12,000 Btu).

Use 250 square feet per ton, e.g., a 2,000-square-foot home requires a 6-ton heat pump (heating capacity).

(Calculation based on R-20 wall and R-40 attic insulation, minimum dual pane argon charged windows.)

Closed Vertical Loop:

Vertical Bore Hole Loop — 3/4" or 1" diameter, socket fusioned loop = 200'-250' total bore hole depth per ton. For example, a 6-ton unit requires between 1200'-1500' (twelve to fifteen 5" bore holes, each 100' deep) with the associated 12-15 loop drops, spaced a minimum of 20' apart, connected to an adequately sized header, immersed in a bentonite slurry and applied with a tremie tube.

Closed Horizontal Loop:

Straight run — 3/4" or 1" diameter manifolded loop = 400 horizontal feet per ton. For example, a 6-ton unit requires 2,400 horizontal feet (eight 300' runs, 150' out, 150' back, minimum 2' radius return bend, placed a minimum 10' apart, minimum 4' buried depth, or buried below potential winter frost line).

Slinky — 3/4" or 1" diameter manifolded loop = 400 horizontal feet per ton. For example, a 6-ton unit requires 2,400 horizontal feet (eight 300' runs, 50' out, 50' back, coiled, placed a minimum 20' apart, minimum 4' buried depth, or buried below potential winter frost line. (Below 7' at 100 miles, or farther, north of the 49th parallel.)

Open Loop:

Well-to-Well — 2.5 gallons per minute per ton, each well (two required: supply & return). For example, a 6-ton unit requires 12-15 gallons per minute capacity per well.

Leadership in Energy & Environmental Design (LEED)

LEED offers an incentive to building owners to install energy-efficient and environmentally sensitive HVAC equipment. The LEED program awards points for application of the best available technology. The HVAC system can earn a maximum of 17 points toward LEED certification. A LEED-certified air conditioning system must comply with ASHRAE 90.1-2004 (efficiency) and ASHRAE 62.1-2004 (indoor air quality) standards. LEED certification also requires recording controls (Carrier ComfortView 3 or equal) to measure system performance. Zone sensor thermostats (Carrier Debonair or equal) are required to provide performance and efficiency data to the digital recorder. An indoor air quality CO_2 sensor, duct-mounted aspirator modules and a refrigerant such as Puron R-410A are also required.

Description	Craft@Hrs	Unit	Material $	Labor $	Equipment $	Total $

Geothermal ground or water source heat pump (open or closed loop), w/active cooling. Glycol/water mixture or water to air thermal transfer. The refrigeration circuit operates in the heating and cooling mode as required (sometimes referred to as active heating & cooling or mechanical heating & cooling). This unit is suited for both, open or closed loop systems. These units provide a higher degree of cooling capacity than the open loop, passive "free" cooling units. See table below. Optional desuperheater system provides domestic hot water temperature pre-heat or boost. Make additional allowances for the thermal source, i.e., horizontal ground loop, vertical bore hole ground loop, supply and return water wells, pond or lake source loop, etc. Make additional allowances for connections, ancillary components, start up and testing. Based on R-22 Refrigerant.

Description	Craft@Hrs	Unit	Material $	Labor $	Equipment $	Total $
Heat Pump, Open or Closed loop, 1/2 Ton	P1@2.25	Ea	1,500.00	80.50	—	1,580.50
Heat Pump, Open or Closed loop, 3/4 Ton	P1@2.25	Ea	1,550.00	80.50	—	1,630.50
Heat Pump, Open or Closed loop, 1 Ton	P1@2.25	Ea	1,630.00	80.50	—	1,710.50
Heat Pump, Open or Closed loop, 1.5 Ton	P1@2.25	Ea	1,890.00	80.50	—	1,970.50
Heat Pump, Open or Closed loop, 2 Ton	P1@2.50	Ea	1,950.00	89.40	—	2,039.40
Heat Pump, Open or Closed loop, 2.5 Ton	P1@2.50	Ea	2,420.00	89.40	—	2,509.40
Heat Pump, Open or Closed loop, 3 Ton	P1@3.00	Ea	2,510.00	107.00	—	2,617.00
Heat Pump, Open or Closed loop, 4 Ton	P1@3.50	Ea	2,880.00	125.00	—	3,005.00
Heat Pump, Open or Closed loop, 5 Ton	P1@4.00	Ea	5,650.00	143.00	—	5,793.00
Heat Pump, Open or Closed loop, 7.5Ton	P1@4.75	Ea	6,080.00	170.00	—	6,250.00
Optional desuperheater coil	P1@.000	Ea	717.00	—	—	717.00
Optional supplementary electric coil 5 KW	P1@.000	Ea	154.00	—	—	154.00
Optional supplementary electric coil 10 KW	P1@.000	Ea	182.00	—	—	182.00
Optional supplementary electric coil 15 KW	P1@.000	Ea	397.00	—	—	397.00
Optional supplementary electric coil 20 KW	P1@.000	Ea	441.00	—	—	441.00
Optional supplementary electric coil 25 KW	P1@.000	Ea	541.00	—	—	541.00
Water to water thermal transfer additional	P1@.000	Ea	353.00	—	—	353.00
R 410A Refrigerant – Add	P1@.000	LB	24.30	—	—	24.30

Description	Craft@Hrs	Unit	Material $	Labor $	Equipment $	Total $

Geothermal water source heat pump (open loop), w/free cooling.

Water to air thermal transfer. The refrigeration circuit is only operated in the heating mode while passive "free" cooling is provided by pumping cold well water directly to a unit mounted coil that by-passes the refrigeration (compressor) unit. This unit is suited to areas where the temperature of ground water supply is below 50°F. Optional desuperheater system provides domestic hot water temperature pre-heat or boost. Make additional allowances for the water source, i.e., supply and return water wells, pond or lake source, etc. Make additional allowances for connections, ancillary components, start up and testing.

Description	Craft@Hrs	Unit	Material $	Labor $	Equipment $	Total $
Heat Pump, Open Loop, Free Cooling, 2 Ton	P1@2.50	Ea	2,680.00	89.40	—	2,769.40
Heat Pump, Open Loop, Free Cooling, 3 Ton	P1@3.00	Ea	3,270.00	107.00	—	3,377.00
Heat Pump, Open Loop, Free Cooling, 4 Ton	P1@3.50	Ea	3,660.00	125.00	—	3,785.00
Heat Pump, Open Loop, Free Cooling, 5 Ton	P1@4.00	Ea	6,460.00	143.00	—	6,603.00
Heat Pump, Open Loop, Free Cooling, 7.5 Ton	P1@4.75	Ea	7,010.00	170.00	—	7,180.00
Optional desuperheater coil	P1@.000	Ea	717.00	—	—	717.00
Optional supplementary electric coil 5 KW	P1@.000	Ea	154.00	—	—	154.00
Optional supplementary electric coil 10 KW	P1@.000	Ea	182.00	—	—	182.00
Optional supplementary electric coil 15 KW	P1@.000	Ea	397.00	—	—	397.00
Optional supplementary electric coil 20 KW	P1@.000	Ea	441.00	—	—	441.00
Optional supplementary electric coil 25 KW	P1@.000	Ea	541.00	—	—	541.00
Water to water thermal transfer additional	P1@.000	Ea	353.00	—	—	353.00

Description	Craft@Hrs	Unit	Material $	Labor $	Equipment $	Total $

Geothermal water source heat pump (closed refrigerant loop). Water to air thermal transfer. Optional desuperheater system provides domestic hot water temperature pre-heat or boost. Make additional allowances for the heat transfer loop, i.e., refrigerant loop. Make additional allowances for connections, ancillary components, start up and testing.

Description	Craft@Hrs	Unit	Material $	Labor $	Equipment $	Total $
Heat Pump, 2 Ton	P1@2.50	Ea	1,600.00	89.40	—	1,689.40
Heat Pump, 3 Ton	P1@3.00	Ea	2,100.00	107.00	—	2,207.00
Heat Pump, 4 Ton	P1@3.50	Ea	2,430.00	125.00	—	2,555.00
Heat Pump, 5 Ton	P1@4.00	Ea	5,190.00	143.00	—	5,333.00
Heat Pump, 7.5 Ton	P1@4.75	Ea	5,570.00	170.00	—	5,740.00
Optional desuperheater coil	P1@.000	Ea	717.00	—	—	717.00
Optional supplementary electric coil 5 KW	P1@.000	Ea	154.00	—	—	154.00
Optional supplementary electric coil 10 KW	P1@.000	Ea	182.00	—	—	182.00
Optional supplementary electric coil 15 KW	P1@.000	Ea	397.00	—	—	397.00
Optional supplementary electric coil 20 KW	P1@.000	Ea	441.00	—	—	441.00
Optional supplementary electric coil 25 KW	P1@.000	Ea	541.00	—	—	541.00

LEED certification for geothermal heat pumps

Description	Craft@Hrs	Unit	Material $	Labor $	Equipment $	Total $
Add for a LEED certified heat pump with USGBC rating	—	Ea	20%	—	—	—
Add for LEED registration and inspection	—	Ea	2,000.00	—	—	2,000.00
Add for LEED central digital performance monitor/recorder, ComfortView 3 or equal	—	Ea	2,350.00	—	—	2,350.00
Add for LEED certified zone recording thermostats, Carrier DebonAir, USB remote-ready	—	Ea	350.00	—	—	350.00
Add for IAQ (Indoor Air Quality) CO_2 sensor and duct-mounted aspirator box	—	Ea	425.00	—	—	425.00

Heat Pump Systems

Description	Craft@Hrs	Unit	Material $	Labor $	Equipment $	Total $

Heat pump, residential air to air forced air aspirated split system.

Rated in nominal tonnage. Agency rating is ARI and CSA. Installed in with an air handling unit (mandatory) and filtration system (optional). Trane or equal. Rated at 13 SEER (Seasonal Energy Efficiency Rating). Includes an automatic reversing valve for switching between heating and cooling modes, semiconductor thermocouple sensor for monitoring indoor/outdoor temperature differential, interior thermostat, controller relay, digital controller module, centrifugal compressor and 208/230 V, 60 Hz, single phase motor drive. Must be installed with two fan-aspirated pre-charged R22 refrigerant evaporator coils, one interior in the air handling unit and one in the weatherproof enclosure mounted outside on a concrete slab. Assumes rental of an appliance dolly ($14 per day), a 3,000 lb. come-a-long ($16 per day) and a 1-ton chain hoist ($21 a day).

Description	Craft@Hrs	Unit	Material $	Labor $	Equipment $	Total $
Mounting and bolting interior heat pump unit in air handler enclosure						
2.0-ton unit	P1@.750	Ea	2,037.00	26.80	50.49	2,114.29
3.0-ton unit	P1@.750	Ea	2,231.00	26.80	50.49	2,308.29
3.5-ton unit	P1@.750	Ea	2,716.00	26.80	50.49	2,793.29
4.0-ton unit	P1@.750	Ea	2,910.00	26.80	50.49	2,987.29
5.0-ton unit	P1@.750	Ea	3,395.00	26.80	50.49	3,472.29
Mounting of interior evaporator coil in air handler						
duct	P1@1.20	Ea	—	42.90	20.79	63.69
Tie-in of connecting interior ductwork to house air ductwork manifold and						
exhaust	P1@4.50	Ea	145.50	161.00	—	306.50
Cutting and brazing of copper evaporator coil						
piping and drain	P1@4.00	EA	7.76	143.00	—	150.76
Wiring of controls	BE@3.00	EA	1.46	119.00	—	120.46
Calibration and test	BE@2.00	Ea	—	79.20	—	79.20
Form and pour 4' x 4' x 2" slab,						
exterior	CF@2.50	Ea	97.00	87.60	—	184.60
Form and pour 4' x 4' x 2" slab,						
interior	CF@3.00	Ea	97.00	105.00	—	202.00
Place 4' x 4' x 1/2" vibration						
pads	CF@.750	Ea	38.80	26.30	—	65.10
Mount interior evaporator						
drain	P1@.500	Ea	7.76	17.90	—	25.66
Bore piping hole through external						
coil	P1@.250	Ea	—	8.94	—	8.94
Mount and bolt external						
coil	P1@.450	Ea	29.10	16.10	—	45.20

Description	Craft@Hrs	Unit	Material $	Labor $	Equipment $	Total $

Heat pump, split system, single speed blower, single phase scroll compressor - 208 volt, 13 SEER (Seasonal Energy Efficiency Rating)

Air to air thermal transfer. Material costs include outdoor stat / low ambient, short cycle timer, high-pressure cut out and digital heat/cool thermostat with manual change over. Labor includes setting outdoor condenser, indoor blower and coil sections in place only. Make additional allowances for refrigerant piping / connections, electrical / temperature control wiring / connections and evaporator coil condensate drain.

Description	Craft@Hrs	Unit	Material $	Labor $	Equipment $	Total $
1.5 tons cooling	S2@4.00	Ea	2,690.00	141.00	—	2,831.00
2 tons cooling	S2@4.25	Ea	2,760.00	150.00	—	2,910.00
2.5 tons cooling	S2@4.50	Ea	3,200.00	159.00	—	3,359.00
3 tons cooling	S2@5.00	Ea	3,590.00	177.00	—	3,767.00
3.5 tons cooling	S2@5.75	Ea	3,930.00	203.00	—	4,133.00
4 tons cooling	S2@6.25	Ea	4,360.00	221.00	—	4,581.00
5 tons cooling	S2@7.50	Ea	5,050.00	265.00	—	5,315.00

Heat pump, split system, single speed blower, three phase scroll compressor - 230 volt, 13 SEER (Seasonal Energy Efficiency Rating)

Air to air thermal transfer. Material costs include outdoor stat / low ambient, short cycle timer, high-pressure cut out and digital heat/cool thermostat with manual change over. Labor includes setting outdoor condenser, indoor blower and coil sections in place only. Make additional allowances for refrigerant piping / connections, electrical / temperature control wiring / connections and evaporator coil condensate drain.

Description	Craft@Hrs	Unit	Material $	Labor $	Equipment $	Total $
3.5 tons cooling	S2@5.75	Ea	4,390.00	203.00	—	4,593.00
4 tons cooling	S2@6.25	Ea	4,790.00	221.00	—	5,011.00
5 tons cooling	S2@7.50	Ea	5,490.00	265.00	—	5,755.00

Heat pump, split system, variable speed blower, single phase scroll compressor - 208 volt, 13 SEER (Seasonal Energy Efficiency Rating)

Air to air thermal transfer. Material costs include outdoor stat / low ambient, short cycle timer, high-pressure cut out and digital heat/cool thermostat with manual change over. Labor includes setting outdoor condenser, indoor blower and coil sections in place only. Make additional allowances for refrigerant piping / connections, electrical / temperature control wiring / connections and evaporator coil condensate drain.

Description	Craft@Hrs	Unit	Material $	Labor $	Equipment $	Total $
1.5 tons cooling	S2@4.00	Ea	3,210.00	141.00	—	3,351.00
2 tons cooling	S2@4.25	Ea	3,260.00	150.00	—	3,410.00
2.5 tons cooling	S2@4.50	Ea	3,760.00	159.00	—	3,919.00
3 tons cooling	S2@5.00	Ea	4,150.00	177.00	—	4,327.00
3.5 tons cooling	S2@5.75	Ea	4,480.00	203.00	—	4,683.00
4 tons cooling	S2@6.25	Ea	5,050.00	221.00	—	5,271.00
5 tons cooling	S2@7.50	Ea	5,730.00	265.00	—	5,995.00

Heat Pump Systems

Description	Craft@Hrs	Unit	Material $	Labor $	Equipment $	Total $

Heat pump, split system, variable speed blower, three phase scroll compressor - 230 volt, 13 SEER (Seasonal Energy Efficiency Rating)

Air to air thermal transfer. Material costs include outdoor stat / low ambient, short cycle timer, high-pressure cut out and digital heat/cool thermostat with manual change over. Labor includes setting outdoor condenser, indoor blower and coil sections in place only. Make additional allowances for refrigerant piping / connections, electrical / temperature control wiring / connections and evaporator coil condensate drain.

Description	Craft@Hrs	Unit	Material $	Labor $	Equipment $	Total $
3.5 tons cooling	S2@5.75	Ea	4,970.00	203.00	—	5,173.00
4 tons cooling	S2@6.25	Ea	5,480.00	221.00	—	5,701.00
5 tons cooling	S2@7.50	Ea	6,200.00	265.00	—	6,465.00

Heat pump, split system, single speed blower, single phase scroll compressor - 208 volt, 14 SEER (Seasonal Energy Efficiency Rating)

Air to air thermal transfer. Material costs include outdoor stat / low ambient, short cycle timer, high-pressure cut out and digital heat/cool thermostat with manual change over. Labor includes setting outdoor condenser, indoor blower and coil sections in place only. Make additional allowances for refrigerant piping / connections, electrical / temperature control wiring / connections and evaporator coil condensate drain.

Description	Craft@Hrs	Unit	Material $	Labor $	Equipment $	Total $
2 tons cooling	S2@4.25	Ea	3,720.00	150.00	—	3,870.00
2.5 tons cooling	S2@4.50	Ea	4,200.00	159.00	—	4,359.00
3 tons cooling	S2@5.00	Ea	4,480.00	177.00	—	4,657.00
3.5 tons cooling	S2@5.75	Ea	5,120.00	203.00	—	5,323.00
4 tons cooling	S2@6.25	Ea	5,700.00	221.00	—	5,921.00
5 tons cooling	S2@7.50	Ea	6,220.00	265.00	—	6,485.00

Heat pump, split system, variable speed blower, single phase scroll compressor - 208 volt, 14 SEER (Seasonal Energy Efficiency Rating)

Air to air thermal transfer. Material costs include outdoor stat / low ambient, short cycle timer, high-pressure cut out and digital heat/cool thermostat with manual change over. Labor includes setting outdoor condenser, indoor blower and coil sections in place only. Make additional allowances for refrigerant piping / connections, electrical / temperature control wiring / connections and evaporator coil condensate drain.

Description	Craft@Hrs	Unit	Material $	Labor $	Equipment $	Total $
2 tons cooling	S2@4.25	Ea	4,230.00	150.00	—	4,380.00
2.5 tons cooling	S2@4.50	Ea	4,620.00	159.00	—	4,779.00
3 tons cooling	S2@5.00	Ea	4,980.00	177.00	—	5,157.00
3.5 tons cooling	S2@5.75	Ea	5,590.00	203.00	—	5,793.00
4 tons cooling	S2@6.25	Ea	6,050.00	221.00	—	6,271.00
5 tons cooling	S2@7.50	Ea	8,910.00	265.00	—	9,175.00

Description	Craft@Hrs	Unit	Material $	Labor $	Equipment $	Total $

Heat pump system accessories

Description	Craft@Hrs	Unit	Material $	Labor $	Equipment $	Total $
Fossil fuel kit	S2@2.50	Ea	183.00	88.40	—	271.40
High pressure cut out kit	S2@.500	Ea	63.90	17.70	—	81.60
Outdoor stat/Low ambient sensors	S2@.500	Ea	59.90	17.70	—	77.60
Short cycle protection	S2@.500	Ea	47.50	17.70	—	65.20
Digital heat/cool thermostat w/ manual change over	S2@.750	Ea	134.00	26.50	—	160.50
Digital, programmable heat/cool thermostat w/ auto change over	S2@.750	Ea	423.00	26.50	—	449.50

Heat pump, supplemental electric heating coil

Set in place only. Make additional allowances for electrical connections.

Description	Craft@Hrs	Unit	Material $	Labor $	Equipment $	Total $
5 KW, 240 Volt	S2@1.50	Ea	153.00	53.00	—	206.00
7.5 KW 240 Volt	S2@1.75	Ea	183.00	61.90	—	244.90
10 KW 240 Volt	S2@2.00	Ea	214.00	70.70	—	284.70
12.5 KW, 240 Volt	S2@2.25	Ea	358.00	79.60	—	437.60
15 KW, 240 Volt	S2@2.75	Ea	449.00	97.20	—	546.20
20 KW, 240 Volt	S2@3.50	Ea	502.00	124.00	—	626.00
25 KW 240 Volt	S2@4.25	Ea	603.00	150.00	—	753.00

LEED certification for geothermal heat pumps

Description	Craft@Hrs	Unit	Material $	Labor $	Equipment $	Total $
Add for a LEED certified heat pump with USGBC rating	—	Ea	20%	—	—	—
Add for LEED registration and inspection	—	Ea	2,000.00	—	—	2,000.00
Add for LEED central digital performance monitor/recorder, ComfortView 3 or equal	—	Ea	2,350.00	—	—	2,350.00
Add for LEED certified zone recording thermostats, Carrier DebonAir, USB remote-ready	—	Ea	350.00	—	—	350.00
Add for IAQ (Indoor Air Quality) CO_2 sensor and duct-mounted aspirator box	—	Ea	425.00	—	—	425.00

Water Pump Systems

Description	Craft@Hrs	Unit	Material $	Labor $	Equipment $	Total $

Water pump, submersible. Domestic water, geothermal open loop, well, lake or pond applications. Stainless steel jacket with multiple stage impellers and integral foot valve.

Description	Craft@Hrs	Unit	Material $	Labor $	Equipment $	Total $
Submersible Pump, 120/240 Volt, 1/2 HP	P1@2.00	Ea	473.00	71.50	—	544.50
Submersible Pump, 240 Volt, 3/4 HP	P1@2.00	Ea	588.00	71.50	—	659.50
Submersible Pump, 240 Volt, 1 HP	P1@2.00	Ea	788.00	71.50	—	859.50
Submersible Pump, 240 Volt, 1-1/2 HP	P1@2.00	Ea	1,040.00	71.50	—	1,111.50
Submersible Pump, 240 Volt, 2 HP	P1@2.00	Ea	1,210.00	71.50	—	1,281.50
Submersible Pump, 240 Volt, 3 HP	P1@2.00	Ea	1,340.00	71.50	—	1,411.50
Submersible Pump, 240 Volt, 5 HP	P1@2.00	Ea	1,580.00	71.50	—	1,651.50
Pump Controller - Constant pressure, VSD	P1@1.50	Ea	798.00	53.60	—	851.60
Pump Control Box, 120 Volt	P1@.600	Ea	47.30	21.50	—	68.80
Pump Control Box, 240 Volt	P1@.600	Ea	57.80	21.50	—	79.30
Pump Stand, PVC	P1@1.00	Ea	126.00	35.80	—	161.80
Pump Stand, Galvanized Steel	P1@1.35	Ea	173.00	48.30	—	221.30

Water pump, jet. Domestic water well, lake and pond applications. 110 volt, cast iron body, single stage impeller, including pressure switch & pressure gauge. Use for shallow depths to 25' maximum lift. Deep well applications to 120' maximum lift.

Description	Craft@Hrs	Unit	Material $	Labor $	Equipment $	Total $
Submersible Pump, shallow depth 1/2 HP	P1@2.00	Ea	200.00	71.50	—	271.50
Submersible Pump, shallow depth 3/4 HP	P1@2.00	Ea	231.00	71.50	—	302.50
Submersible Pump, deep well 1/2 HP	P1@2.00	Ea	273.00	71.50	—	344.50
Submersible Pump, deep well 3/4 HP	P1@2.00	Ea	315.00	71.50	—	386.50
Optional pump mounted pressure tank 5 gallon	P1@.500	Ea	84.00	17.90	—	101.90

Description	Craft@Hrs	Unit	Material $	Labor $	Equipment $	Total $

Water pump system accessories. Pressure tanks are factory pre-charged. VSD = variable speed drive. Typical subcontract costs.

Description	Craft@Hrs	Unit	Material $	Labor $	Equipment $	Total $
Pressure tank, bladder type, 5 gallon	P1@1.00	Ea	73.50	35.80	—	109.30
Pressure tank, bladder type, 10 gallon	P1@2.00	Ea	84.00	71.50	—	155.50
Pressure tank, bladder type, 20 gallon	P1@2.00	Ea	99.80	71.50	—	171.30
Pressure tank, bladder type, 30 gallon	P1@2.00	Ea	142.00	71.50	—	213.50
Pressure tank, bladder type, 40 gallon	P1@2.00	Ea	189.00	71.50	—	260.50
Brass tank tee, 3/4" or 1" w/ union	P1@.300	Ea	44.10	10.70	—	54.80
Brass tank tee, 3/4" or 1" w/o union	P1@.300	Ea	35.70	10.70	—	46.40
Constant pressure VSD pump controller	P1@1.50	Ea	798.00	53.60	—	851.60
Pressure Switch, 110/220 Volt	P1@.350	Ea	33.60	12.50	—	46.10
Pressure gauge, 0-100 psi, 1/4"	P1@.200	Ea	8.40	7.15	—	15.55
Sediment faucet (hose bibb), 1/2"	P1@1.50	Ea	6.83	53.60	—	60.43
Double injector deep well venturi, 1", Brass	P1@.600	Ea	179.00	21.50	—	200.50
Double injector deep well venturi, 1¼", Brass	P1@.650	Ea	205.00	23.20	—	228.20
Double injector deep well venturi, 1½", Brass	P1@.700	Ea	231.00	25.00	—	256.00
Double injector deep well venturi, 1", Plastic	P1@.550	Ea	57.80	19.70	—	77.50
Double injector deep well venturi, 1¼", Plastic	P1@.600	Ea	71.40	21.50	—	92.90
Double injector deep well venturi, 1½", Plastic	P1@.650	Ea	77.70	23.20	—	100.90
Foot valve, 1", Brass	P1@.400	Ea	60.90	14.30	—	75.20
Foot valve, 1¼", Brass	P1@.400	Ea	63.00	14.30	—	77.30
Foot valve, 1½", Brass	P1@.400	Ea	67.20	14.30	—	81.50
Foot valve, 1", Plastic	P1@.400	Ea	18.90	14.30	—	33.20
Foot valve, 1¼", Plastic	P1@.400	Ea	23.10	14.30	—	37.40
Foot valve, 1½", Plastic	P1@.400	Ea	25.70	14.30	—	40.00
Barbed adapter, FIP/MIP, 1" Brass	P1@.230	Ea	9.98	8.22	—	18.20
Barbed adapter, FIP/MIP, 1¼" Brass	P1@.250	Ea	11.60	8.94	—	20.54

Water Pump Systems

Description	Craft@Hrs	Unit	Material $	Labor $	Equipment $	Total $
Barbed adapter, FIP/MIP, 1½" Brass	P1@.265	Ea	13.90	9.48	—	23.38
Barbed adapter, FIP/MIP, 1" Plastic	P1@.215	Ea	9.24	7.69	—	16.93
Barbed adapter, FIP/MIP, 1¼" Plastic	P1@.220	Ea	10.10	7.87	—	17.97
Barbed adapter, FIP/MIP, 1½" Plastic	P1@.235	Ea	13.40	8.40	—	21.80
Pitless adapter, 1" x 5" Brass	P1@1.25	Ea	147.00	44.70	—	191.70
Pitless adapter, 1¼" x 5" Brass	P1@1.25	Ea	160.00	44.70	—	204.70
Pitless adapter, 1½" x 5" Brass	P1@1.25	Ea	179.00	44.70	—	223.70
Pitless adapter, 2" x 6" Brass	P1@1.80	Ea	273.00	64.40	—	337.40
Poly pipe, 75 psi, 1"	P1@.020	Ea	.23	.72	—	.95
Poly pipe, 75 psi, 1¼"	P1@.021	Ea	.25	.75	—	1.00
Poly pipe, 75 psi, 1½"	P1@.022	Ea	.27	.79	—	1.06
Poly pipe, 100 psi, 1"	P1@.021	Ea	.23	.75	—	.98
Poly pipe, 100 psi, 1¼"	P1@.022	Ea	.25	.79	—	1.04
Poly pipe, 100 psi, 1½"	P1@.023	Ea	.27	.82	—	1.09
Electrical wire, (110 volt) #12/4	P1@.015	Ea	.27	.54	—	.81
Electrical wire, (220 volt) #10/4	P1@.015	Ea	.29	.54	—	.83
Heat shrink kit	P1@.000	Ea	4.20	.00	—	4.20

Description	Craft@Hrs	Unit	Material $	Labor $	Equipment $	Total $

Geothermal water wells

A geothermal water well is constructed and operates exactly as a water well used for drinking water would. In many residential applications, the supply well of the geothermal system also functions as the domestic water supply well.

The cost of a drilled well depends on many variables, such as costs to get the drilling rig and crew to the site, the type of geologic formation the well drilling rig must drill through, the depth the driller must achieve to find water in that particular location and the distance from an existing available water source to provide the water required to drill (to refill the water truck if necessary).

Well drillers charge by the foot and usually have a 100' minimum charge out fee. They can only speculate to what depth they must drill to find water. Water being found at a particular depth nearby does not always translate to finding water at another given location at the same depth. It is important to remember that when drilling a relatively small hole in the ground, and hoping to hit what is described as a vein of water as part of an underground water table, is always a gamble. As such, it is, without proper research, difficult to determine a realistic budget for well drilling. It is always prudent to call the water authority or a well driller in the area for the most accurate information on the water and geological characteristics of a given location.

Use the following to begin to formulate a budget exclusive of costs associated with factors listed above. Because of the inherent uncertainties of the process, these prices are given only as typical subcontract costs.

Description	Craft@Hrs	Unit	Material $	Labor $	Equipment $	Total $

Drilled Well, 5" Diameter, PVC Casing, set into bedrock. Based on a 2-man crew, 1 rotary drill rig and a 1,500 gallon water truck. Minimum 100' charge in most cases. Typical subcontract costs.

Description	Craft@Hrs	Unit	Material $	Labor $	Equipment $	Total $
Drilled through till, clay, sand		LF	—	—	—	31.00
Drilled through till and boulder		LF	—	—	—	40.00
Drilled through limestone soft bedrock		LF	—	—	—	45.00
Drilled through granite hard bedrock		LF	—	—	—	90.00
Add for Galvanized Steel Casing		LF	—	—	—	11.30

Drilled Well, 6" Diameter, PVC Casing, set into bedrock. Based on a 2-man crew, 1 rotary drill rig and a 1,500 gallon water truck. Minimum 100' charge in most cases. Typical subcontract costs.

Description	Craft@Hrs	Unit	Material $	Labor $	Equipment $	Total $
Drilled through till, clay, sand		LF	—	—	—	34.40
Drilled through till and boulder		LF	—	—	—	43.80
Drilled through limestone soft bedrock		LF	—	—	—	47.50
Drilled through granite hard bedrock		LF	—	—	—	95.00
Add for Galvanized Steel Casing		LF	—	—	—	13.80

Geothermal Water Wells/Geothermal Closed Loops

Description	Craft@Hrs	Unit	Material $	Labor $	Equipment $	Total $

Drilled Well, 8" Diameter, PVC Casing, set into bedrock. Based on a 2-man crew, 1-rotary drill rig and a 1,500 gallon water truck. Minimum 100' charge in most cases. Typical subcontract costs.

Description	Craft@Hrs	Unit	Material $	Labor $	Equipment $	Total $
Drilled through till, clay, sand		LF	—	—	—	43.80
Drilled through till and boulder		LF	—	—	—	50.00
Drilled through limestone soft bedrock		LF	—	—	—	56.30
Drilled through granite hard bedrock		LF	—	—	—	113.00
Add for Galvanized Steel Casing		LF	—	—	—	27.50

Drilled Well, 10" Diameter, PVC Casing, set into bedrock. Based on a 2-man crew, 1 rotary drill rig and a 1,500 gallon water truck. Minimum 100' charge in most cases. Typical subcontract costs.

Description	Craft@Hrs	Unit	Material $	Labor $	Equipment $	Total $
Drilled through till, clay, sand		LF	—	—	—	52.50
Drilled through till and boulder		LF	—	—	—	58.80
Drilled through limestone soft bedrock		LF	—	—	—	65.00
Drilled through granite hard bedrock		LF	—	—	—	150.00
Add for Galvanized Steel Casing		LF	—	—	—	35.00

Geothermal bore hole (vertical loop), 3" diameter, drilled to a maximum of 30'. Includes installation of a polypropylene fluid loop using a 180-degree return bend with socket fusion joints. Bore hole filled with bentonite slurry through a tremie tube and mud pump. Includes socket fusioned header and supply & return mains to heat pump. Based on a minimum of 1,000' vertical feet (2,000' of pipe) Based on a 2-man crew, 1 skid-steer mounted rotary drill rig and a 50 gallon water tank. Typical subcontract costs.

Description	Craft@Hrs	Unit	Material $	Labor $	Equipment $	Total $
3/4" vertical loop		LF	—	—	—	10.60
1" vertical loop		LF	—	—	—	11.60
Add for thermally enhanced bentonite		LF	—	—	—	2.94

Geothermal horizontal loop, directionally bored. Includes installation of a polypropylene fluid loop. Based on a minimum of 2,000 horizontal feet (2,000' of pipe). Based on a 3-man crew, 1 directional boring/horizontal drill machine. Typical subcontract costs.

Description	Craft@Hrs	Unit	Material $	Labor $	Equipment $	Total $
3/4" straight horizontal loop		LF	—	—	—	6.44
1" straight horizontal loop		LF	—	—	—	7.00

Description	Craft@Hrs	Unit	Material $	Labor $	Equipment $	Total $

Geothermal horizontal loop, slinky type. Includes installation of a polypropylene fluid loop. Based on a minimum of 2,000 horizontal feet (2,000' of pipe). Based on a 3-man crew, 1 excavator backhoe. Typical subcontract costs.

Description	Craft@Hrs	Unit	Material $	Labor $	Equipment $	Total $
3/4" slinky loop		LF	—	—	—	4.88
1" slinky loop		LF	—	—	—	5.50
3/4" slinky pond loop		LF	—	—	—	3.81
1" slinky pond loop		LF	—	—	—	4.44

Biomass-Fired Blowers

Description	Craft@Hrs	Unit	Material $	Labor $	Equipment $	Total $

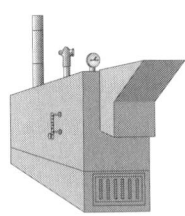

Biomass-fired hot water boilers for central heating systems. For residential, commercial and light industrial applications. 15 PSI, EPA, ASME and UL rated, with wood-pellet and/or corn fuel gravity feed hopper, ash hood cleanout, secondary natural gas or #2 oil burner, platinum stack transition emissions catalyst, forced draft fan, factory installed boiler trim, feedwater pump and valves, and tempering valve. Add the cost of the following if required: a concrete pad, optional stainless steel 2 GPM potable hot water coil insert, expansion tank and circulating pump, stack and stack liner, CO_2 fire extinguisher system and electrical connection. Add the cost of post-installation inspection by both the fire marshal and the mechanical inspector to validate federal, state and local energy tax credits or energy tax credit offsets. See pollution control stack costs below. Equipment cost is for a 2-ton capacity hydraulic crane truck. Listed by boiler horsepower (BHP) rating and input capacity in thousands of BTU per hour (MBH).

Description	Craft@Hrs	Unit	Material $	Labor $	Equipment $	Total $
3 BHP, 102 MBH	SN@24.0	Ea	6,810.00	904.00	377.00	8,091.00
3 BHP, 104 MBH	SN@24.0	Ea	8,510.00	904.00	377.00	9,791.00
6 BHP, 200 MBH	SN@28.0	Ea	11,000.00	1,050.00	440.00	12,490.00
14 BHP, 500 MBH	SN@30.0	Ea	13,000.00	1,130.00	472.00	14,602.00
27 BHP, 900 MBH	SN@44.0	Ea	18,800.00	1,660.00	692.00	21,152.00

LEED certification requires 88% boiler efficiency, recording controls, compliance with SCAQMD 1146.2 emission standards and zone thermostats.

Description	Craft@Hrs	Unit	Material $	Labor $	Equipment $	Total $
Add for LEED certified boiler with USGBC rating	—	Ea	15%	—	—	—
Add for LEED registration and inspection	—	Ea	—	—	—	2,000.00
Add for central performance monitor/ recorder	—	Ea	2,350.00	—	—	2,350.00
Add for zone recording thermostat	—	Ea	350.00	—	—	350.00

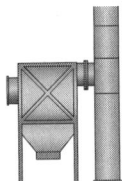

Retrofit pollution control stack for biomass hot water boilers. Stack-mounted platinum catalytic screen for biomass-fired hot water boilers for central radiant heating systems. E.P.A. and California Code standards compliant. UL rated. Includes stack mounting transition and extensions, baghouse structural steel frame, stack transition cleanout and inspection stack access bolted ports and an integral CO_2 fire extinguisher system. Add the cost of a concrete pad, electrical connection, post-installation calibration, inspection by both the fire marshal and the mechanical inspector to validate federal, state and local emission reduction tax credits and associated energy tax credit offsets. Equipment cost is for a 2-ton capacity hydraulic truck crane. Listed by boiler horsepower (BHP) rating and input capacity in thousands of BTU per hour (MBH). Use these figures to estimate the cost of adding a pollution control stack to an existing bio-mass boiler to meet environmental regulations.

Description	Craft@Hrs	Unit	Material $	Labor $	Equipment $	Total $
3 BHP, 102 MBH	SN@20.0	Ea	1,200.00	753.00	315.00	2,268.00
3 BHP, 104 MBH	SN@20.0	Ea	1,200.00	753.00	315.00	2,268.00
6 BHP, 200 MBH	SN@20.0	Ea	1,800.00	753.00	315.00	2,868.00
14 BHP, 500 MBH	SN@40.0	Ea	2,200.00	1,510.00	628.00	4,338.00
27 BHP, 900 MBH	SN@40.0	Ea	3,000.00	1,510.00	628.00	5,138.00

Description	Craft@Hrs	Unit	Material $	Labor $	Equipment $	Total $

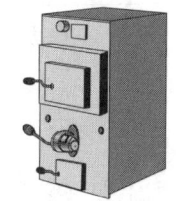

Biomass-fired water or steam boilers for multi-dwelling, process and industrial applications. 15 PSI, EPA, ASME and UL rated. Includes wood pellet and/or corn fuel auger feed hopper, ash auger cleanout, secondary natural gas or #2 oil burner, platinum stack transition emissions catalyst, forced draft fan, factory-installed boiler trim, feedwater pump and valves, and tempering valve. Add the following costs if required: a concrete pad, optional stainless steel 10 GPM potable hot water coil insert, circulating pump, expansion tank, shell and tube heat exchanger, stack and stack liner, CO_2 fire extinguisher system, pollution control baghouse and filter, electrical connection. Add the cost of post-installation inspection by both the fire marshal and the mechanical inspector to validate federal, state and local energy tax credits or energy tax credit offsets. See pollution control stack costs below. Equipment cost is for a 5-ton capacity forklift. Listed by boiler horsepower (BHP) rating and input capacity in thousands of BTU per hour (MBH).

Description	Craft@Hrs	Unit	Material $	Labor $	Equipment $	Total $
30 BHP, 1,000 MBH	SN@168	Ea	67,500.00	6,330.00	2,640.00	76,470.00
50 BHP, 1,700 MBH	SN@200	Ea	125,000.00	7,530.00	3,146.00	135,676.00
100 BHP, 3,400 MBH	SN@400	Ea	192,000.00	15,100.00	6,281.00	213,381.00
125 BHP, 4,100 MBH	SN@400	Ea	243,000.00	15,100.00	6,281.00	264,381.00
150 BHP, 5,000 MBH	SN@500	Ea	237,000.00	18,800.00	7,854.00	263,654.00
200 BHP, 6,700 MBH	SN@600	Ea	325,000.00	22,600.00	9,427.00	357,027.00

LEED certification requires 88% boiler efficiency, recording controls, compliance with SCAQMD 1146.2 emission standards and zone thermostats.

Description	Craft@Hrs	Unit	Material $	Labor $	Equipment $	Total $
Add for LEED certified boiler with USGBC rating	—	Ea	6%	—	—	—
Add for LEED registration and inspection	—	Ea	—	—	—	2,000.00
Add for central performance monitor/ recorder	—	Ea	2,350.00	—	—	2,350.00
Add for zone recording thermostat	—	Ea	350.00	—	—	350.00

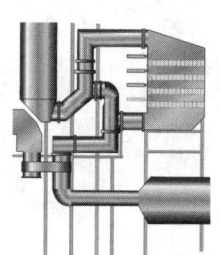

Biomass boiler stack tie-in to baghouse filter. Ionically bonded silicon-lined carbon steel. Costs assume a 20-foot vertical stack height, connection via T-shaped connector with cleanout to the baghouse inlet plenum. Includes rigging, breeching, connecting stack sections from back "turnaround" plenum of boiler to baghouse inlet plenum, vendor-supplied structural steel stack support brackets and aluminized external coating. Equipment cost assumes rental of a 2-ton hydraulic truck crane. Biomass boilers, stoves and heaters smaller than 30 BHP can use a conventional boiler stack (without a baghouse) and require only a platinum catalytic screen. Costs shown assume a 20-foot stack height and 10' horizontal run from T-connector. Cost per 10' length, including a bolt and gasket set on each end. Add 15% for each 10' section (or portion thereof) beyond 20' vertically or horizontally.

Description	Craft@Hrs	Unit	Material $	Labor $	Equipment $	Total $
10" diameter	SN@12.0	Ea	2,500.00	452.00	188.00	3,140.00
12" diameter	SN@12.0	Ea	2,800.00	452.00	188.00	3,440.00
16" diameter	SN@16.0	Ea	3,000.00	602.00	252.00	3,854.00
20" diameter	SN@20.0	Ea	3,800.00	753.00	315.00	4,868.00
24" diameter	SN@24.0	Ea	4,200.00	904.00	377.00	5,481.00
30" diameter	SN@32.0	Ea	6,200.00	1,200.00	503.00	7,903.00
56" diameter	SN@40.0	Ea	10,500.00	1,510.00	628.00	12,638.00

Biomass-Fired Blowers

Description	Craft@Hrs	Unit	Material $	Labor $	Equipment $	Total $

Retrofit pollution control stack for biomass water or steam boiler.

Stack-mounted platinum screen catalytic reactor with air filtration fabric baghouse. For use with biomass-fired hot water or low pressure steam boilers for central radiant heating or process systems. Meets EPA, NFPA, California Code and UL standards for multi-dwelling, process and industrial applications. Cost includes equipment stack transitions and fittings, transition access bolted inspection and cleanout ports, structural steel frame equipment support and ladders, bottom ash cleanout for baghouse, alarmed stack gas composition monitor and recorder, urea tank and urea pump & regulating valve and an integral CO_2 fire extinguisher system. Costs do not include the following: concrete pad, electrical tie-in, calibration and commissioning, post-installation inspection by the fire marshal and the mechanical inspector to validate federal, state and local energy tax credits or energy tax credit offsets. Equipment cost is for a 2-ton capacity hydraulic truck crane. Listed by boiler horsepower (BHP) rating and input capacity in thousands of BTU per hour (MBH). Use these figures to estimate the cost of adding a pollution control stack to an existing biomass boiler to meet environmental regulations.

Description	Craft@Hrs	Unit	Material $	Labor $	Equipment $	Total $
30 BHP, 1,000 MBH	SN@40.0	Ea	35,500.00	1,510.00	628.00	37,638.00
50 BHP, 1,700 MBH	SN@80.0	Ea	63,000.00	3,010.00	1,254.00	67,264.00
100 BHP, 3,400 MBH	SN@80.0	Ea	112,000.00	3,010.00	1,254.00	116,264.00
125 BHP, 4,100 MBH	SN@120	Ea	117,000.00	4,520.00	1,881.00	123,401.00
150 BHP, 5,000 MBH	SN@120	Ea	126,000.00	4,520.00	1,881.00	132,401.00
200 BHP, 6,700 MBH	SN@160	Ea	135,000.00	6,020.00	1,881.00	142,901.00

Biomass-fired central air space heaters for residential, multi-dwelling, and light commercial applications.

15 PSI, EPA, ASHRAE and UL rated. Includes a wood pellet and/or corn fuel gravity feed hopper, ash hood cleanout, secondary natural gas or #2 oil burner, platinum stack transition emissions catalyst, forced draft fan, ducted hot-air circulation plenum and integral three-speed circulation duct blower. Add the cost of a concrete pad, optional stainless steel 2 GPM potable hot water coil insert, expansion tank and circulating pump, stack and stack liner, CO_2 fire extinguisher system and electrical connection. These costs do not include post-installation inspection by both the fire marshal and the mechanical inspector to validate federal, state and local energy tax credits or energy tax credit offsets. Equipment cost is for a 2-ton capacity forklift. Listed by input capacity in thousands of BTU per hour (MBH).

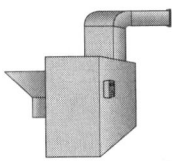

Description	Craft@Hrs	Unit	Material $	Labor $	Equipment $	Total $
90 MBH	SN@16.0	Ea	2,750.00	602.00	252.00	3,604.00
120 MBH	SN@16.0	Ea	3,550.00	602.00	252.00	4,404.00
170 MBH	SN@20.0	Ea	6,230.00	753.00	315.00	7,298.00
500 MBH	SN@30.0	Ea	9,710.00	1,130.00	472.00	11,312.00
950 MBH	SN@44.0	Ea	11,500.00	1,660.00	692.00	13,852.00

Description	Craft@Hrs	Unit	Material $	Labor $	Equipment $	Total $

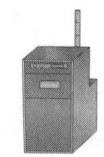

Water mist fire extinguishing systems for biomass boilers. 20 minute full saturation cycle, with automatic reset to enable instantaneous reactivation for reignition. NFPA rated. Includes high pressure spray heads, piping, pumps, valves, alarm horns, emergency halogen lighting, calibration, training, insurance inspection and safety procedure documentation. Based on remote sensor relays for the fuel hopper, fuel conveyor, combustion chamber and baghouse filter.

Description	Craft@Hrs	Unit	Material $	Labor $	Equipment $	Total $
30 BHP boiler	ST@60.0	Ea	12,000.00	2,480.00	—	14,480.00
50 BHP boiler	ST@60.0	Ea	15,000.00	2,480.00	—	17,480.00
100 BHP boiler	ST@80.0	Ea	22,000.00	3,310.00	—	25,310.00
125 BHP boiler	ST@80.0	Ea	24,000.00	3,310.00	—	27,310.00
150 BHP boiler	ST@120	Ea	27,500.00	4,960.00	—	32,460.00
200 BHP boiler	ST@120	Ea	32,000.00	4,960.00	—	36,960.00
Add for CO2 based extinguishing system		%	—	—	—	20.0
Add for Halon-based extinguishing system		%	—	—	—	30.0

Pulse type boilers for central radiant heating. 15 PSI, EPA, ASME and UL rated. Modular design for multiple installations. Natural gas only. With forced draft fan, factory installed boiler trim, feedwater pump and valves, and tempering valve. For residential, commercial and light industrial radiant heating and potable water applications. Add the cost of a concrete pad, circulating pump, stack and stack liner, CO_2 fire extinguisher system, electrical connection, and post-installation inspection by both the fire marshal and the mechanical inspector to validate federal, state and local energy tax credits or energy tax credit offsets. Equipment cost is for a 1-ton capacity hydraulic truck crane. Listed by boiler horsepower (BHP) rating and input capacity in thousands of BTU per hour (MBH).

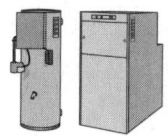

Description	Craft@Hrs	Unit	Material $	Labor $	Equipment $	Total $
3 BHP, 90 MBH	P1@24.0	Ea	3,810.00	858.00	1,320.00	5,988.00
3 BHP, 150 MBH	P1@24.0	Ea	5,510.00	858.00	1,320.00	7,688.00
6 BHP, 299 MBH	P1@28.0	Ea	7,590.00	1,000.00	1,540.00	10,130.00

LEED certification requires 88% boiler efficiency, recording controls, compliance with SCAQMD 1146.2 emission standards and zone thermostats.

Description	Craft@Hrs	Unit	Material $	Labor $	Equipment $	Total $
Add for LEED certified boiler with USGBC rating	—	Ea	15%	—	—	—
Add for LEED registration and inspection	—	Ea	—	—	—	2,000.00
Add for central performance monitor/ recorder	—	Ea	1,100.00	—	—	1,100.00
Add for zone recording thermostat	—	Ea	150.00	—	—	150.00

Biomass-Fired Blowers

Description	Craft@Hrs	Unit	Material $	Labor $	Equipment $	Total $

Passive flat-panel solar water heater. Ambient pressure, modular design for multiple installation, EPA, DOE and UL rated, with piping, tempering valve, circulating pump, controls, storage tank and heating coil. Cost does not include electrical connection, roofing modification (bracing) to accommodate the additional weight of water-filled panels, insulation nor padding between roof and panels. Add the cost of post-installation inspection by the mechanical inspector to validate federal, state and local energy tax credits or energy tax credit offsets. For larger arrays (laundries, institutional facilities, food processing plants), develop an estimate based on required daily capacity and multiply these costs by the number of panels required. Equipment cost is for a 1-ton capacity hydraulic crane truck.

Description	Craft@Hrs	Unit	Material $	Labor $	Equipment $	Total $
One 32.3 SF solar panel, 80 gallons per day	P1@16.0	Ea	3,250.00	572.00	880.00	4,702.00
Two 32.3 SF solar panels, 120 gallons per day	P1@20.0	Ea	4,530.00	715.00	1,100.00	6,345.00

Fans and Blowers

Description	Craft@Hrs	Unit	Material $	Labor $	Equipment $	Total $

Centrifugal air-foil wheel blower, with motor, drive and isolation, set in place only.

Description	Craft@Hrs	Unit	Material $	Labor $	Equipment $	Total $
1/3 HP, 12" dia.	SN@2.50	Ea	1,250.00	94.10	39.00	1,383.10
1/2 HP, 15" dia.	SN@4.50	Ea	2,190.00	169.00	71.00	2,430.00
1 HP, 18" dia.	SN@7.00	Ea	2,530.00	264.00	110.00	2,904.00
2 HP, 18" dia.	SN@7.50	Ea	3,000.00	282.00	118.00	3,400.00
3 HP, 27" dia.	SN@9.00	Ea	4,740.00	339.00	141.00	5,220.00
5 HP, 30" dia.	SN@11.0	Ea	5,590.00	414.00	173.00	6,177.00
7½ HP, 36" dia.	SN@12.0	Ea	6,770.00	452.00	189.00	7,411.00
10 HP, 40" dia.	SN@14.0	Ea	9,440.00	527.00	220.00	10,187.00
15 HP, 44" dia.	SN@17.0	Ea	12,500.00	640.00	267.00	13,407.00
20 HP, 60" dia.	SN@20.0	Ea	14,200.00	753.00	314.00	15,267.00

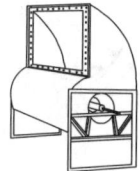

Utility centrifugal blower, with motor, drive and isolation, set in place only.

Description	Craft@Hrs	Unit	Material $	Labor $	Equipment $	Total $
1/3 HP, 10" dia.	SN@2.50	Ea	818.00	94.10	39.00	951.10
1/2 HP, 12" dia.	SN@4.50	Ea	1,070.00	169.00	71.00	1,310.00
1 HP, 15" dia.	SN@6.50	Ea	1,740.00	245.00	102.00	2,087.00
2 HP, 18" dia.	SN@7.50	Ea	1,980.00	282.00	118.00	2,380.00
3 HP, 24" dia.	SN@9.00	Ea	2,840.00	339.00	141.00	3,320.00
5 HP, 27" dia.	SN@11.0	Ea	3,270.00	414.00	173.00	3,857.00
7½ HP, 36" dia.	SN@12.0	Ea	5,260.00	452.00	189.00	5,901.00
10 HP, 40" dia.	SN@14.0	Ea	6,360.00	527.00	220.00	7,107.00
15 HP, 44" dia.	SN@17.0	Ea	6,870.00	640.00	267.00	7,777.00

Vane-axial fan, set in place only.

Description	Craft@Hrs	Unit	Material $	Labor $	Equipment $	Total $
30 HP, 36" dia.	SN@16.0	Ea	16,700.00	602.00	251.00	17,553.00
50 HP, 42" dia.	SN@24.0	Ea	18,600.00	904.00	377.00	19,881.00
100 HP, 48" dia.	SN@32.0	Ea	23,100.00	1,200.00	503.00	24,803.00
200 HP, 54" dia.	SN@40.0	Ea	32,700.00	1,510.00	629.00	34,839.00

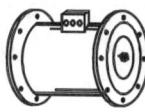

Fans and Blowers

Description	Craft@Hrs	Unit	Material $	Labor $	Equipment $	Total $

Tube-axial fan, set in place only.

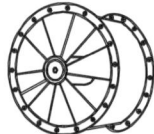

Description	Craft@Hrs	Unit	Material $	Labor $	Equipment $	Total $
15 HP, 24" dia.	SN@10.0	Ea	4,920.00	377.00	157.00	5,454.00
25 HP, 30" dia.	SN@13.0	Ea	7,640.00	489.00	204.00	8,333.00
30 HP, 36" dia.	SN@16.0	Ea	10,200.00	602.00	251.00	11,053.00
75 HP, 40" dia.	SN@22.0	Ea	15,300.00	828.00	346.00	16,474.00
100 HP, 48" dia.	SN@32.0	Ea	23,200.00	1,200.00	503.00	24,903.00

Roof exhaust fan, ASHRAE/American National Standards Institute specification 62.2-2007 and Air Conditioning Contractors of America standard ACCA 5QI-2007, "HVAC Indoor Air Quality Installation Specification 2007." US Green Building Council, Underwriters laboratories and CSA approved. Cost includes fan and bearing, shaft mount and power transmission assembly or direct drive, enclosure, mount plates, electric motor, 110 or 408, 3 phase, 60 Hz, switch or relay assembly, motor fusing and ground fault protection, fan blade guard. Set in place only. Add installation costs from the figures below.

Description	Craft@Hrs	Unit	Material $	Labor $	Equipment $	Total $
1/4 HP, 14" dia.	SN@2.50	Ea	1,300.00	94.10	39.00	1,433.10
1/3 HP, 22" dia.	SN@2.90	Ea	1,600.00	109.00	46.00	1,755.00
1/2 HP, 24" dia.	SN@3.50	Ea	1,820.00	132.00	55.00	2,007.00
1 HP, 30" dia.	SN@4.50	Ea	2,530.00	169.00	71.00	2,770.00
2 HP, 36" dia.	SN@5.00	Ea	3,620.00	188.00	79.00	3,887.00
3 HP, 40" dia.	SN@6.00	Ea	5,420.00	226.00	94.00	5,740.00
5 HP, 48" dia.	SN@7.00	Ea	7,520.00	264.00	110.00	7,894.00
Cut, frame and gasket downcomer hole in roof	S2@1.00	Ea	35.00	35.40	—	70.40
Mount duct hangers	S2@.250	Ea	3.50	8.84	—	12.34
Cut and mount sheet metal duct	S2@.350	LF	5.50	12.40	—	17.90
Apply and coat duct insulation	S2@.150	LF	2.40	5.30	—	7.70
Install electrical wiring	BE@.150	LF	1.75	5.94	—	7.69
Install HVAC controls	BE@.500	Ea	—	19.80	—	19.80
Commission and test	P1@2.00	Ea	—	71.50	—	71.50
Run air balance and fine-tune system	P1@1.00	Ea	—	35.80	—	35.80

Description	Craft@Hrs	Unit	Material $	Labor $	Equipment $	Total $

Ceiling mounted commercial ventilator. Low profile metal grille. Galvanized steel housing. Designed for continuous operation. Acoustically insulated. Can be installed horizontally, vertically or in-line. Labor includes lay-out and fasten in place only. Make additional allowances for vent duct and electrical requirements.

Description	Craft@Hrs	Unit	Material $	Labor $	Equipment $	Total $
100 CFM	P1@1.05	Ea	129.00	37.50	—	166.50
150 CFM	P1@1.15	Ea	163.00	41.10	—	204.10
200 CFM	P1@1.35	Ea	206.00	48.30	—	254.30
250 CFM	P1@1.45	Ea	259.00	51.90	—	310.90
300 CFM	P1@1.65	Ea	308.00	59.00	—	367.00
400 CFM	P1@2.15	Ea	397.00	76.90	—	473.90
500 CFM	P1@2.50	Ea	482.00	89.40	—	571.40
700 CFM	P1@2.75	Ea	672.00	98.30	—	770.30
900 CFM	P1@2.85	Ea	741.00	102.00	157.00	1,000.00
1,500 CFM	P1@3.50	Ea	1,030.00	125.00	193.00	1,348.00
2,000 CFM	P1@3.85	Ea	1,140.00	138.00	213.00	1,491.00
3,500 CFM	P1@4.25	Ea	1,290.00	152.00	235.00	1,677.00

Powered attic ventilator. Thermostatically controlled. 1/10 Hp, 120V/6-Hz motor. 8" x 24" metal housing. Labor includes lay-out and fasten in place only. Make additional allowances for electrical and roofing requirements.

Description	Craft@Hrs	Unit	Material $	Labor $	Equipment $	Total $
1,200 CFM	P1@2.00	Ea	163.00	71.50	—	234.50

Exterior wall/roof exhaust fan. Fully enclosed. Thermally protected. Lifetime lubricated 120V/6-Hz motor. Weather-resistant aluminum housing. Balance blower wheel. Neoprene vibration isolation. Includes backdraft damper, flashing plate and critter screen. 10" discharge outlet. Labor includes lay-out and fasten in place only. Make additional allowances for vent duct, electrical and roofing requirements.

Description	Craft@Hrs	Unit	Material $	Labor $	Equipment $	Total $
600 CFM	P1@2.50	Ea	482.00	89.40	—	571.40
900 CFM	P1@2.85	Ea	747.00	102.00	—	849.00

Room ventilator fan. Interior thru-wall room-to-room air transfer application. High efficiency. Manually switched operation with built-in variable speed control. 8" and 10" discharge outlets. Built in damper. Includes intake and discharge grilles. Labor includes lay-out and fasten in place only. Make additional allowances for electrical requirements.

Description	Craft@Hrs	Unit	Material $	Labor $	Equipment $	Total $
180 CFM	P1@1.25	Ea	100.00	44.70	—	144.70
270 CFM	P1@1.50	Ea	151.00	53.60	—	204.60
380 CFM	P1@1.75	Ea	206.00	62.60	—	268.60

Ceiling exhaust fan, with automatic motion response control. Adjustable (5-60 minutes) time interval operation. Manual override feature. 4" discharge outlet. Pre-wired outlet box. Balanced blower wheel. Permanently lubricated motor. Neoprene vibration isolators. Quiet operation. Suitable for washroom exhaust applications. Labor includes lay-out and fasten in place only. Make additional allowances for vent duct and electrical requirements.

Description	Craft@Hrs	Unit	Material $	Labor $	Equipment $	Total $
90 CFM	P1@1.00	Ea	155.00	35.80	—	190.80
130 CFM	P1@1.15	Ea	240.00	41.10	—	281.10

Ventilators and Residential Exhaust Fans

Description	Craft@Hrs	Unit	Material $	Labor $	Equipment $	Total $

Ceiling exhaust fan, with automatic humidity response control.
Adjustable time interval operation (5-60 minutes). Manual override feature.
4" discharge outlet. Pre-wired outlet box. Balanced blower wheel. Permanently lubricated motor. Neoprene vibration isolators. Quiet operation. Suitable for washroom exhaust applications. Labor includes lay-out and fasten in place only. Make additional allowances for vent duct and electrical requirements.

Description	Craft@Hrs	Unit	Material $	Labor $	Equipment $	Total $
90 CFM	P1@1.00	Ea	181.00	35.80	—	216.80
130 CFM	P1@1.15	Ea	266.00	41.10	—	307.10

Ceiling exhaust fan, manually switched operation. 4" discharge outlet.
Pre-wired outlet box. 4 pole motor. Suitable for washroom exhaust applications. Labor includes lay-out and fasten in place only. Make additional allowances for vent duct and electrical requirements.

Description	Craft@Hrs	Unit	Material $	Labor $	Equipment $	Total $
50 CFM	P1@1.00	Ea	34.40	35.80	—	70.20
70 CFM	P1@1.00	Ea	51.60	35.80	—	87.40
90 CFM	P1@1.00	Ea	60.30	35.80	—	96.10
110 CFM	P1@1.05	Ea	72.20	37.50	—	109.70

Ceiling exhaust fan, high efficiency. Manually switched operation. 8" and
10" discharge outlets. Labor includes lay-out and fasten in place only. Suitable for kitchens, laundry rooms, rec rooms and workshops. Make additional allowances for vent duct and electrical requirements.

Description	Craft@Hrs	Unit	Material $	Labor $	Equipment $	Total $
160 CFM	P1@1.25	Ea	65.30	44.70	—	110.00
180 CFM	P1@1.25	Ea	94.50	44.70	—	139.20
270 CFM	P1@1.50	Ea	138.00	53.60	—	191.60
350 CFM	P1@1.75	Ea	173.00	62.60	—	235.60

Discharge hood ducting kit. Including discharge hood, 4" x 10' of flexible
duct, duct clamps, 3" x 4" transition coupling.

Description	Craft@Hrs	Unit	Material $	Labor $	Equipment $	Total $
Wall hood, 4"	P1@2.00	Ea	17.30	71.50	—	88.80
Roof hood, 4"	P1@2.25	Ea	34.40	80.50	—	114.90
Wall hood, 6"	P1@2.55	Ea	77.40	91.20	—	168.60

Fan controls. Switch only.

Description	Craft@Hrs	Unit	Material $	Labor $	Equipment $	Total $
1 hour timer	BE@.155	Ea	55.00	6.14	—	61.14
Humidistat	BE@.155	Ea	37.80	6.14	—	43.94
Speed controller	BE@.155	Ea	51.60	6.14	—	57.74

Description	Craft@Hrs	Unit	Material $	Labor $	Equipment $	Total $

Air admittance valve. For ventless sanitary plumbing. Provides a water seal in traps without roof penetration and vent piping. Complies with ASHRAE/American National Standards Institute spec. 62.2-2007 and American Society for Sanitary Engineering specs 1050 and 1051.

Description	Craft@Hrs	Unit	Material $	Labor $	Equipment $	Total $
Cut, frame and gasket for wall inset and vent valve access						
door	SW@1.00	Ea	35.00	41.00	—	76.00
Mount valve	P1@.500	Ea	—	17.90	—	17.90
Connect valve to sanitary S- or						
P-trap	P1@1.00	Ea	—	35.80	—	35.80

Bathroom Fans and Heaters. Comply with ASHRAE/American National Standards Institute spec 62.2-2007 and meet new indoor air quality code requirements for dehumidification, mold and mildew prevention and radon mitigation. Continental Fan, Tamarack, Broan or equal. Includes enclosure, mount plates, 110 or 408 volt electric motor, switch or relay assembly, motor fusing, ground fault protection. Add the following installation costs, if required.

Description	Craft@Hrs	Unit	Material $	Labor $	Equipment $	Total $
Cut, frame and gasket exhaust hole in						
sidewall	SW@1.00	Ea	35.00	41.00	—	76.00
Cut and mount sheet metal						
duct	SW@.150	LF	5.50	6.16	—	11.66
Apply and coat duct						
insulation	S2@.050	LF	2.40	1.77	—	4.17
Install electrical wiring						
(typical)	BE@.150	LF	1.75	5.94	—	7.69
Install controls	BE@.500	Ea	—	19.80	—	19.80
Test and adjust	P1@.250	Ea	—	8.94	—	8.94

Bathroom exhaust fan with heater, commercial grade, Continental Fan, Broan or equal.

Description	Craft@Hrs	Unit	Material $	Labor $	Equipment $	Total $
106 CFM fan,						
4" Duct	BE@.500	Ea	185.00	19.80	—	204.80
152 CFM fan,						
4" Duct	BE@.500	Ea	210.00	19.80	—	229.80
247 CFM fan,						
6" Duct	BE@.500	Ea	350.00	19.80	—	369.80
418 CFM fan,						
8" Duct	BE@.500	Ea	450.00	19.80	—	469.80

Infrared bulb heater and fan. 4-point adjustable mounting brackets span up to 24-inches. Uses 250 watt, 120 volt R-40 infrared bulbs (not included). Heater and fan units include 70 CFM, 3.5 sones ventilator fan. Damper and duct connector included. Plastic matte-white molded grille and compact housings. CFM (cubic feet of air per minute).

Description	Craft@Hrs	Unit	Material $	Labor $	Equipment $	Total $
1-bulb, 70 CFM,						
3.5 sones	BE@1.00	Ea	52.38	39.60	—	91.98
2-bulb heater,						
7" duct	BE@1.10	Ea	65.96	43.50	—	109.46

Ventilators and Residential Exhaust Fans

Description	Craft@Hrs	Unit	Material $	Labor $	Equipment $	Total $

Heat-A-Lamp® bulb heater and fan. NuTone. Swivel-mount. Torsion spring holds plates firmly to ceiling. 250-watt radiant heat from one R-40 infrared heat lamp. Uses 4-inch duct. Adjustable socket for ceilings up to 1-inch thick. Adjustable hanger bars. Automatic reset for thermal protection. White polymeric finish. UL listed.

Description	Craft@Hrs	Unit	Material $	Labor $	Equipment $	Total $
2.6 amp, 1 lamp, 12-1/2" x 10"	BE@1.50	Ea	54.22	59.40	—	113.62
5.0 amp, 2 lamp, 15-3/8" x 11"	BE@1.50	Ea	69.84	59.40	—	129.24

Surface-mount ceiling bath resistance heater and fan. 1250 watt. 4,266 BTU. 120 volt. Chrome alloy wire element for instant heat. Built-in fan. Automatic overheat protection. Low-profile housing. Permanently lubricated motor. Mounts to standard 3-1/4-inch round or 4-inch octagonal ceiling electrical box. Satin-finish aluminum grille extends 2-3/4-inch from ceiling. 11" diameter x 2-3/4" deep.

Description	Craft@Hrs	Unit	Material $	Labor $	Equipment $	Total $
10.7 amp	BE@.750	Ea	55.19	29.70	—	84.89

Designer Series bath heater and exhaust fan. 1500-watt fan-forced heater. 120-watt light capacity, 7-watt night-light (bulbs not included). Permanently lubricated motor. Polymeric damper. Torsion-spring grille mount. Non-glare light diffusing glass lens. Fits single gang opening. Suitable for use with insulation. 100 CFM, 1.2 sones.

Description	Craft@Hrs	Unit	Material $	Labor $	Equipment $	Total $
4" duct, 100 watt lamp	BE@1.00	Ea	223.10	39.60	—	262.70

Ventilation exhausters. Includes all-weather enclosure, air balance control, software module and NFPA certified fusible plug-type vertical duct riser gravity dampers, and efficiency optimization arrays. Add the cost of roof curb, control wiring, electrical wiring, permits and final inspection. For wall or roof mount in single and multi-unit residences, commercial (i.e. hotels), institutional (i.e. hospitals and schools) buildings. Set in place only. Add installation costs from the section below. Costs assume rental of an appliance dolly ($14.00 per day), a 3,000 lb. come-a-long ($16.00 per day) and a 1/2-ton chain hoist ($21.00 a day).

Downblast ventilation exhausters. Rooftop mounted, direct drive, centrifugal type. Suitable for either ducted or nonducted applications.

Description	Craft@Hrs	Unit	Material $	Labor $	Equipment $	Total $
1/4 HP, 14" dia.	SN@2.50	Ea	1,300.00	94.10	39.00	1,433.10
1/3 HP, 22" dia.	SN@2.90	Ea	1,600.00	109.00	46.00	1,755.00
1/2 HP, 24" dia.	SN@3.50	Ea	1,820.00	132.00	55.00	2,007.00
1 HP, 30" dia.	SN@4.50	Ea	2,530.00	169.00	71.00	2,770.00
2 HP, 36" dia.	SN@5.00	Ea	3,620.00	188.00	79.00	3,887.00
3 HP, 40" dia.	SN@6.00	Ea	5,420.00	226.00	94.00	5,740.00
5 HP, 48" dia.	SN@7.00	Ea	7,520.00	264.00	110.00	7,894.00

Description	Craft@Hrs	Unit	Material $	Labor $	Equipment $	Total $

Upblast ventilation exhausters. Direct drive, centrifugal, for roof or wall mount.

Description	Craft@Hrs	Unit	Material $	Labor $	Equipment $	Total $
1/4 HP, 14" dia.	SN@2.50	Ea	1,400.00	94.10	39.00	1,533.10
1/3 HP, 22" dia.	SN@2.90	Ea	1,700.00	109.00	46.00	1,855.00
1/2 HP, 24" dia.	SN@3.50	Ea	1,950.00	132.00	55.00	2,137.00
1 HP, 30" dia.	SN@4.50	Ea	2,700.00	169.00	71.00	2,940.00
2 HP, 36" dia.	SN@5.00	Ea	3,800.00	188.00	79.00	4,067.00
3 HP, 40" dia.	SN@6.00	Ea	5,600.00	226.00	94.00	5,920.00
5 HP, 48" dia.	SN@7.00	Ea	7,700.00	264.00	110.00	8,074.00

Energy-efficient exhauster arrays. Upblast belt-drive centrifugal UL762 roof exhausters. Low profile metal grille. Galvanized steel housing. Designed for continuous operation. Acoustically insulated. Can be installed horizontally, vertically or in-line. Labor includes lay-out and fasten in place only. Make additional allowances for vent duct and electrical requirements.

Description	Craft@Hrs	Unit	Material $	Labor $	Equipment $	Total $
100 CFM	P1@1.05	Ea	200.00	37.50	—	237.50
150 CFM	P1@1.15	Ea	220.00	41.10	—	261.10
200 CFM	P1@1.35	Ea	240.00	48.30	—	288.30
250 CFM	P1@1.45	Ea	310.00	51.90	—	361.90
300 CFM	P1@1.65	Ea	350.00	59.00	—	409.00
400 CFM	P1@2.15	Ea	440.00	76.90	—	516.90
500 CFM	P1@2.50	Ea	510.00	89.40	—	599.40
700 CFM	P1@2.75	Ea	730.00	98.30	—	828.30
900 CFM	P1@2.85	Ea	790.00	102.00	157.00	1,049.00
1,500 CFM	P1@3.50	Ea	1,200.00	125.00	193.00	1,518.00
2,000 CFM	P1@3.85	Ea	1,350.00	138.00	213.00	1,701.00
3,500 CFM	P1@4.25	Ea	1,500.00	152.00	235.00	1,887.00

Installation costs for ventilation exhausters.

Description	Craft@Hrs	Unit	Material $	Labor $	Equipment $	Total $
Cut, frame and gasket downcomer hole in sidewall or roof	S2@1.00	Ea	35.00	35.40	—	70.40
Mount duct hangers	S2@.250	Ea	3.50	8.84	—	12.34
Cut & mount sheet metal duct	S2@.350	LF	5.50	12.40	—	17.90
Apply and coat duct insulation	S2@.150	LF	2.40	5.30	—	7.70
Install electrical wiring	BE@.150	LF	1.75	5.94	—	7.69
Install exhauster controls	BE@.500	Ea	—	19.80	—	19.80
Commission and test	P1@4.00	Ea	—	143.00	—	143.00
Run air balance and fine-tune system	P1@4.00	Ea	—	143.00	—	143.00

Apparatus Housing

Unitary or factory-packaged air conditioning units are often too small to meet the heating and cooling needs of a large single-zone space. In such cases, engineers design what's called a "built-up" system. They're custom fitted to the client's needs and the space. A typical "built-up" system includes the following equipment:

- centrifugal blowers or axial-flow fans
- damper assemblies
- coil banks
- filter banks

Once the system is assembled, it is enclosed in a shop-fabricated housing or casing. The illustration on page 327 shows a standard apparatus housing. The walls and roof are shop-built panels. These panels consist of a layer of thermal insulation covered on one or both sides with sheets of 18 gauge galvanized steel. An interior framework of steel angles or channels supports the panels. Seal all mating surfaces to minimize air leakage, using gaskets, caulking or sealant.

Apparatus Housing

There are two types of panels used in building apparatus housings: "double-skin" panels and "single-skin" panels. Double-skin panels have a layer of thermal insulation material sandwiched between two sheets of 18 gauge galvanized steel. Single-skin panels also have a layer of insulation. However, only the outside surface is clad in 18 gauge steel sheeting. On the inside the panel insulation is exposed.

The costs of constructing and installing apparatus housings vary greatly. This is due to the wide range of possible conditions and needs that you face on each job. Some of the many factors that influence costs here include the following:

- internal operating pressures
- external wind loads
- seismic requirements
- physical locations

That's why I recommend that you always prepare a detailed cost estimate. Then, you can take all the particulars into account. However, for budget estimating purposes only, use the following cost data:

Single-skin panels	$11.30 per SF*
Double-skin panels	$19.20 per SF*
Access doors (20" x 60")	$390.00 each

*Total area of walls and roof, less doors.

When you're estimating material costs for apparatus housings, remember to add in an extra 25 percent to cover the following costs: waste, seams, gaskets, sealants and miscellaneous assembly hardware.

Labor for Shop-Fabrication and Field Assembly of Apparatus Housings. The costs listed are manhours per square foot (SF) of panel. To find the total SF of panel, add up areas for all the panels.

Description	Shop Labor (MH per SF)	Field Labor (MH per SF)
Single-skin panels	.040	.250
Double-skin panels	.085	.300
Add for access doors	.400	.240

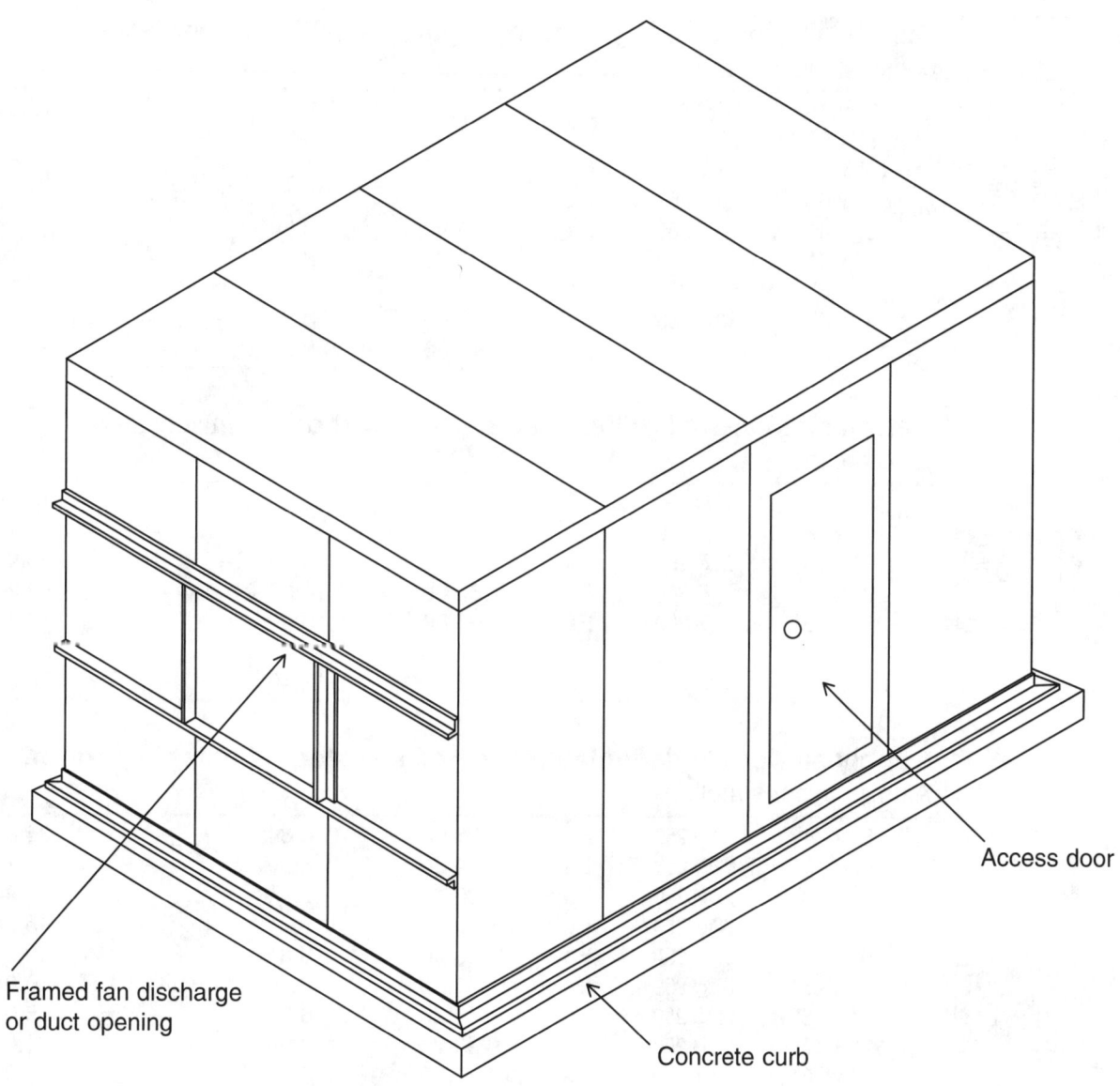

Access door

Framed fan discharge
or duct opening

Concrete curb

Figure 1 - Typical apparatus housing

Air Devices, Registers and Grilles

Description	Craft@Hrs	Unit	Material $	Labor $	Equipment $	Total $

Supply air side-wall register, fixed pattern, with damper. Steel construction.

Description	Craft@Hrs	Unit	Material $	Labor $	Equipment $	Total $
8" x 4"	S2@.250	Ea	8.70	8.84	—	17.54
8" x 6"	S2@.250	Ea	9.05	8.84	—	17.89
10" x 4"	S2@.250	Ea	9.83	8.84	—	18.67
10" x 6"	S2@.250	Ea	16.60	8.84	—	25.44
10" x 8"	S2@.250	Ea	15.90	8.84	—	24.74
12" x 6"	S2@.250	Ea	16.90	8.84	—	25.74
12" x 8"	S2@.250	Ea	13.80	8.84	—	22.64
14" x 6"	S2@.250	Ea	13.80	8.84	—	22.64
14" x 8"	S2@.250	Ea	15.80	8.84	—	24.64

Return air side-wall grille, fixed pattern, without damper. Steel construction.

Description	Craft@Hrs	Unit	Material $	Labor $	Equipment $	Total $
14" x 6"	S2@.250	Ea	7.51	8.84	—	16.35
14" x 8"	S2@.250	Ea	7.27	8.84	—	16.11
16" x 16"	S2@.250	Ea	22.10	8.84	—	30.94
20" x 10"	S2@.300	Ea	32.60	10.60	—	43.20
20" x 16"	S2@.300	Ea	35.80	10.60	—	46.40
20" x 20"	S2@.300	Ea	47.40	10.60	—	58.00
24" x 24"	S2@.350	Ea	61.60	12.40	—	74.00

Supply air double deflection side-wall register, adjustable pattern. Aluminum construction.

Description	Craft@Hrs	Unit	Material $	Labor $	Equipment $	Total $
8" x 4"	S2@.250	Ea	28.00	8.84	—	36.84
8" x 6"	S2@.250	Ea	31.30	8.84	—	40.14
10" x 4"	S2@.250	Ea	36.10	8.84	—	44.94
10" x 6"	S2@.250	Ea	37.10	8.84	—	45.94
10" x 8"	S2@.250	Ea	39.70	8.84	—	48.54
12" x 6"	S2@.250	Ea	41.20	8.84	—	50.04
12" x 8"	S2@.250	Ea	46.20	8.84	—	55.04
12" x 10"	S2@.250	Ea	52.20	8.84	—	61.04
14" x 6"	S2@.250	Ea	52.20	8.84	—	61.04
14" x 8"	S2@.250	Ea	52.80	8.84	—	61.64
14" x 10"	S2@.250	Ea	56.30	8.84	—	65.14
16" x 8"	S2@.250	Ea	56.30	8.84	—	65.14
16" x 10"	S2@.250	Ea	61.60	8.84	—	70.44
18" x 8"	S2@.300	Ea	64.20	10.60	—	74.80
18" x 10"	S2@.300	Ea	68.80	10.60	—	79.40
18" x 12"	S2@.300	Ea	72.60	10.60	—	83.20

Description	Craft@Hrs	Unit	Material $	Labor $	Equipment $	Total $

Return air single deflection side-wall grille, adjustable pattern.
Aluminum construction.

8" x 4"	S2@.250	Ea	25.50	8.84	—	34.34
8" x 6"	S2@.250	Ea	28.00	8.84	—	36.84
10" x 4"	S2@.250	Ea	30.90	8.84	—	39.74
10" x 6"	S2@.250	Ea	33.50	8.84	—	42.34
10" x 8"	S2@.250	Ea	35.70	8.84	—	44.54
12" x 6"	S2@.250	Ea	37.00	8.84	—	45.84
12" x 8"	S2@.250	Ea	41.50	8.84	—	50.34
12" x 10"	S2@.250	Ea	46.90	8.84	—	55.74
14" x 6"	S2@.250	Ea	46.90	8.84	—	55.74
14" x 8"	S2@.250	Ea	47.70	8.84	—	56.54
14" x 10"	S2@.250	Ea	50.30	8.84	—	59.14
16" x 8"	S2@.250	Ea	50.30	8.84	—	59.14
16" x 10"	S2@.300	Ea	55.50	10.60	—	66.10
18" x 8"	S2@.300	Ea	57.90	10.60	—	68.50
18" x 10"	S2@.300	Ea	62.20	10.60	—	72.80
18" x 12"	S2@.300	Ea	65.60	10.60	—	76.20

4-way adjustable pattern ceiling diffuser 24" x 24" with collar and volume damper. Steel construction.

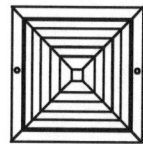

6" neck	S2@.350	Ea	72.30	12.40	—	84.70
8" neck	S2@.350	Ea	75.90	12.40	—	88.30
10" neck	S2@.350	Ea	81.70	12.40	—	94.10
12" neck	S2@.350	Ea	85.10	12.40	—	97.50
14" neck	S2@.400	Ea	102.00	14.10	—	116.10

Fixed pattern ceiling diffuser 24" x 24" with collar, without volume damper. Steel construction.

6" neck	S2@.350	Ea	43.50	12.40	—	55.90
8" neck	S2@.350	Ea	47.10	12.40	—	59.50
10" neck	S2@.350	Ea	52.60	12.40	—	65.00
12" neck	S2@.350	Ea	55.90	12.40	—	68.30
14" neck	S2@.350	Ea	72.30	12.40	—	84.70

Perforated ceiling diffuser 24" x 24" fixed pattern with collar, without volume damper. Steel construction.

6" neck	S2@.350	Ea	58.00	12.40	—	70.40
8" neck	S2@.350	Ea	60.10	12.40	—	72.50
10" neck	S2@.350	Ea	64.40	12.40	—	76.80
12" neck	S2@.350	Ea	68.80	12.40	—	81.20
14" neck	S2@.400	Ea	77.00	14.10	—	91.10
16" neck	S2@.720	Ea	130.00	25.50	—	155.50

Air Devices, Diffusers and Grilles

Description	Craft@Hrs	Unit	Material $	Labor $	Equipment $	Total $

Round 3-cone, 2-position ceiling diffuser with collar, without volume damper. Steel construction.

Description	Craft@Hrs	Unit	Material $	Labor $	Equipment $	Total $
6"	S2@.350	Ea	59.90	12.40	—	72.30
8"	S2@.350	Ea	68.80	12.40	—	81.20
10"	S2@.350	Ea	83.40	12.40	—	95.80
12"	S2@.350	Ea	97.90	12.40	—	110.30
14"	S2@.350	Ea	131.00	12.40	—	143.40

Round adjustable ceiling diffuser with collar and volume damper.
Steel construction.

Description	Craft@Hrs	Unit	Material $	Labor $	Equipment $	Total $
6"	S2@.350	Ea	90.50	12.40	—	102.90
8"	S2@.350	Ea	107.00	12.40	—	119.40
10"	S2@.350	Ea	83.40	12.40	—	95.80
12"	S2@.350	Ea	131.00	12.40	—	143.40
14"	S2@.350	Ea	244.00	12.40	—	256.40
16"	S2@.450	Ea	284.00	15.90	—	299.90
18"	S2@.450	Ea	328.00	15.90	—	343.90
20"	S2@.550	Ea	387.00	19.40	—	406.40
24"	S2@.650	Ea	533.00	23.00	—	556.00

Description	Craft@Hrs	Unit	Material $	Labor $	Equipment $	Total $

Makeup air unit. Makeup air unit Uses 100% direct-heated outdoor air for warehouse, factory and institutional applications to provide higher thermal efficiency and balanced distribution of system air while pre-heating air for use in boiler or other HVAC applications. ANSI standard Z83.4/CSA 3.7 certified. With weather-proof metal enclosure, burner, analog electric controls for burner and fan, thermostat, actuator, natural gas burner array, and either roof mount supports or sidewall external mounts and transition and gaskets. Set in place only. Add installation costs.

Description	Craft@Hrs	Unit	Material $	Labor $	Equipment $	Total $
200 to 399 CFM	SN@2.25	Ea	543.00	84.70	35.40	663.10
400 to 599 CFM	SN@2.25	Ea	598.00	84.70	35.40	718.10
600 to 799 CFM	SN@2.75	Ea	689.00	104.00	43.30	836.30
800 to 999 CFM	SN@3.00	Ea	828.00	113.00	47.10	988.10
1,000 to 1,499 CFM	SN@3.50	Ea	1,100.00	132.00	55.00	1,287.00
1,500 to 1,999 CFM	SN@4.00	Ea	1,390.00	151.00	62.90	1,603.90
4,000 to 5,999 CFM	SN@6.00	Ea	3,500.00	226.00	106.20	3,832.20
7,500 to 10,000 CFM	SN@8.00	Ea	6,000.00	301.00	106.20	6,407.20
15,000 to 25,000 CFM	SN@10.0	Ea	8,500.00	377.00	141.60	9,018.60
30,000 to 40,000 CFM	SN@16.0	Ea	12,500.00	602.00	283.20	13,385.20
60,000 to 80,000 CFM	SN@24.0	Ea	22,000.00	904.00	283.20	23,187.20

Installation costs for makeup air units.

Description	Craft@Hrs	Unit	Material $	Labor $	Equipment $	Total $
Cut, frame and gasket downcomer/ sidewall hole	S2@1.00	Ea	35.00	35.40	—	70.40
Mount duct hangers	S2@.250	Ea	3.50	8.84	—	12.34
Cut and mount sheet metal duct	S2@.350	LF	5.50	12.40	—	17.90
Apply and coat duct insulation	S2@.150	LF	2.40	5.30	—	7.70
Install piping for gas line	P1@.100	LF	3.75	3.58	—	7.33
Install electrical wiring	BE@.150	LF	1.75	5.94	—	7.69
Install makeup air unit controls	BE@.500	Ea	—	19.80	—	19.80
Commission and test	P1@4.00	Ea	—	143.00	—	143.00
Run air balance and fine-tune system	P1@4.00	Ea	—	143.00	—	143.00

Terminal Units (VAV)

Variable-air volume units regulate supply air to ducted air conditioning systems. These systems are certified to ANSI standard Z83.4/CSA 3.7 and can be either roof-mounted or upper sidewall externally mounted. Costs include a weather-proof metal enclosure, heater coil or electrical heater fixture, fan and fan controls, downcomer or sidewall entry transition and gaskets, and either roof mount supports or sidewall external mounts. Includes set in place only. Add installation costs from the table that follows.

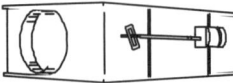

Description	Craft@Hrs	Unit	Material $	Labor $	Equipment $	Total $

Variable-air volume terminal unit with analog electric controls, thermostat, actuator and controller

Description	Craft@Hrs	Unit	Material $	Labor $	Equipment $	Total $
200- 400 CFM	SN@2.00	Ea	494.00	75.30	31.40	600.70
400- 600 CFM	SN@2.00	Ea	551.00	75.30	31.40	657.70
600- 800 CFM	SN@2.50	Ea	580.00	94.10	39.20	713.30
800-1,000 CFM	SN@2.75	Ea	628.00	104.00	43.30	775.30
1,000-1,500 CFM	SN@3.00	Ea	915.00	113.00	47.10	1,075.10
1,500-2,000 CFM	SN@3.50	Ea	1,160.00	132.00	55.00	1,347.00

Variable-air volume terminal reheat unit with analog electric controls, thermostat, actuator, controller and a two row hot water coil

Description	Craft@Hrs	Unit	Material $	Labor $	Equipment $	Total $
200- 400 CFM	SN@2.25	Ea	570.00	84.70	35.40	690.10
400- 600 CFM	SN@2.25	Ea	628.00	84.70	35.40	748.10
600- 800 CFM	SN@2.75	Ea	723.00	104.00	43.30	870.30
800-1,000 CFM	SN@3.00	Ea	869.00	113.00	47.10	1,029.10
1,000-1,500 CFM	SN@3.50	Ea	1,160.00	132.00	55.00	1,347.00
1,500-2,000 CFM	SN@4.00	Ea	1,460.00	151.00	62.90	1,673.90

Installation costs for variable-air volume units.

Description	Craft@Hrs	Unit	Material $	Labor $	Equipment $	Total $
Cut, frame and gasket downcomer hole in roof	S2@1.00	Ea	35.00	35.40	—	70.40
Mount duct hangers	S2@.250	Ea	3.50	8.84	—	12.34
Cut and mount sheet metal duct	S2@.350	LF	5.50	12.40	—	17.90
Apply and coat duct insulation	S2@.150	LF	2.40	5.30	—	7.70
Install piping for chilled or hot water/steam	P1@.100	LF	2.85	3.58	—	6.43
Install electrical wiring	BE@.150	LF	1.75	5.94	—	7.69
Install VAV controls	BE@.500	Ea	—	19.80	—	19.80
Commission and test	P1@4.00	Ea	—	143.00	—	143.00
Run air balance and fine-tune system	P1@4.00	Ea	—	143.00	—	143.00

Air balance software and control modules for variable-air volume ducted zone control For commercial, institutional and industrial applications. Johnson Controls or equal. The zone and duct-mounted sensor and response unit comprises a digital controller, actuator and differential pressure transducer. These components are wired together and securely housed inside a plenum-rated enclosure which mounts directly to the VAV "serviced zone" box. Complies with Underwriters Laboratories file E107041, CCN PAZX, UL916 Energy Management Equipment. Interfaces with Windows-based commercial desktop computers per the digital control manufacturer's specifications.

Description	Craft@Hrs	Unit	Material $	Labor $	Equipment $	Total $

Fully programmable 28 point controller, 16 universal inputs,12 universal outputs, 12 HOA switches, a battery backed real time clock, plus preconfigured function blocks for a real time clock. Stores schedules and trending information.

Description	Craft@Hrs	Unit	Material $	Labor $	Equipment $	Total $
Programmable controller	BE@.500	Ea	900.00	19.80	—	919.80

Configurable variable air volume pressure independent terminal box controller with wireless transceiver (868.3 MHz)

Description	Craft@Hrs	Unit	Material $	Labor $	Equipment $	Total $
Terminal box controller	BE@.500	Ea	2,700.00	19.80	—	2,719.80

Configurable controller, preconfigured LNS plug-in, and all input devices required for typical rooftop unit application

Description	Craft@Hrs	Unit	Material $	Labor $	Equipment $	Total $
Preconfigured controller	BE@.500	Ea	1,800.00	19.80	—	1,819.80

Variable air volume controller system software for a Windows computer. Includes a software license and support for two concurrent client browser connections with no limit on the number of concurrent users and no data tag limit.

Description	Craft@Hrs	Unit	Material $	Labor $	Equipment $	Total $
System software	BE@18.0	Ea	6,800.00	713.00	—	7,513.00

Ductwork Specialties

Description	Craft@Hrs	Unit	Material $	Labor $	Equipment $	Total $

Opposed-blade damper for rectangular ductwork

Description	Craft@Hrs	Unit	Material $	Labor $	Equipment $	Total $
12" x 6"	S2@.600	Ea	58.00	21.20	—	79.20
12" x 10"	S2@.650	Ea	68.60	23.00	—	91.60
12" x 12"	S2@.700	Ea	84.10	24.80	—	108.90
18" x 12"	S2@.800	Ea	96.90	28.30	—	125.20
18" x 18"	S2@.850	Ea	108.00	30.10	—	138.10
24" x 12"	S2@.850	Ea	113.00	30.10	—	143.10
24" x 18"	S2@.900	Ea	124.00	31.80	—	155.80
24" x 24"	S2@1.00	Ea	139.00	35.40	—	174.40
30" x 12"	S2@.900	Ea	129.00	31.80	—	160.80
30" x 18"	S2@1.00	Ea	145.00	35.40	—	180.40
30" x 24"	S2@1.10	Ea	163.00	38.90	—	201.90
30" x 30"	S2@1.20	Ea	176.00	42.40	—	218.40
36" x 18"	S2@1.10	Ea	169.00	38.90	—	207.90
36" x 36"	S2@1.40	Ea	232.00	49.50	—	281.50
42" x 24"	S2@1.30	Ea	236.00	46.00	—	282.00
42" x 42"	S2@1.50	Ea	323.00	53.00	—	376.00
48" x 24"	S2@1.40	Ea	244.00	49.50	—	293.50
48" x 48"	S2@1.60	Ea	339.00	56.60	—	395.60
54" x 24"	S2@1.45	Ea	310.00	51.30	—	361.30
54" x 54"	S2@1.70	Ea	387.00	60.10	—	447.10

Single-blade damper for rectangular ductwork

Description	Craft@Hrs	Unit	Material $	Labor $	Equipment $	Total $
12" x 6"	S2@.600	Ea	36.40	21.20	—	57.60
12" x 8"	S2@.625	Ea	39.80	22.10	—	61.90
12" x 10"	S2@.650	Ea	44.70	23.00	—	67.70
12" x 12"	S2@.700	Ea	52.50	24.80	—	77.30
18" x 12"	S2@.800	Ea	58.40	28.30	—	86.70
18" x 18"	S2@.850	Ea	66.30	30.10	—	96.40
24" x 12"	S2@.850	Ea	70.10	30.10	—	100.20
24" x 18"	S2@.900	Ea	72.90	31.80	—	104.70
24" x 24"	S2@1.00	Ea	80.10	35.40	—	115.50

Single-blade butterfly damper for round ductwork

Description	Craft@Hrs	Unit	Material $	Labor $	Equipment $	Total $
6"	S2@.600	Ea	36.40	21.20	—	57.60
8"	S2@.550	Ea	34.20	19.40	—	53.60
10"	S2@.600	Ea	39.70	21.20	—	60.90
12"	S2@.650	Ea	44.90	23.00	—	67.90
14"	S2@.700	Ea	51.10	24.80	—	75.90
16"	S2@.765	Ea	56.60	27.10	—	83.70
18"	S2@.835	Ea	62.80	29.50	—	92.30
20"	S2@.900	Ea	70.60	31.80	—	102.40

Description	Craft@Hrs	Unit	Material $	Labor $	Equipment $	Total $

Curtain-type duct fire damper, 1-1/2 hour rating, U.L. label

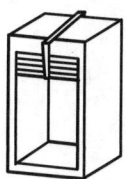

Description	Craft@Hrs	Unit	Material $	Labor $	Equipment $	Total $
12" x 6"	S2@1.00	Ea	66.70	35.40	—	102.10
12" x 8"	S2@1.00	Ea	72.20	35.40	—	107.60
12" x 10"	S2@1.05	Ea	79.00	37.10	—	116.10
12" x 12"	S2@1.10	Ea	85.30	38.90	—	124.20
18" x 12"	S2@1.20	Ea	119.00	42.40	—	161.40
18" x 18"	S2@1.30	Ea	129.00	46.00	—	175.00
24" x 12"	S2@1.40	Ea	144.00	49.50	—	193.50
24" x 18"	S2@1.50	Ea	163.00	53.00	—	216.00
24" x 24"	S2@1.60	Ea	194.00	56.60	—	250.60
30" x 12"	S2@1.50	Ea	166.00	53.00	—	219.00
30" x 18"	S2@1.80	Ea	185.00	63.60	—	248.60
30" x 24"	S2@2.00	Ea	244.00	70.70	—	314.70
30" x 30"	S2@2.30	Ea	264.00	81.30	—	345.30
36" x 18"	S2@2.00	Ea	210.00	70.70	—	280.70
36" x 36"	S2@2.40	Ea	314.00	84.90	—	398.90
42" x 24"	S2@2.40	Ea	272.00	84.90	—	356.90
42" x 42"	S2@2.90	Ea	394.00	103.00	—	497.00
48" x 24"	S2@2.90	Ea	296.00	103.00	—	399.00
48" x 48"	S2@3.50	Ea	491.00	124.00	—	615.00
54" x 24"	S2@3.00	Ea	320.00	106.00	—	426.00
54" x 54"	S2@3.80	Ea	570.00	134.00	—	704.00

Correction factors:

(1) For 22 gauge break-away connection, multiply material cost by 1.30

(2) For blades out of air stream cap, multiply material cost by 1.25

Description	Craft@Hrs	Unit	Material $	Labor $	Equipment $	Total $

Fusible plug dampers for air duct risers. Vertical riser and header air duct fusible (melting) and heat-sensing automatic counterweight closure actuators and closure dampers. For large A/C systems in multi-story institutions such as schools, hospitals and office buildings. 1-1/2 hour rating. The liner and insulation must meet the requirements of U.L. 181 and NFPA No. 90 A-96, as modified, "Standard for the Installation of Air Conditioning and Ventilation Systems."

Description	Craft@Hrs	Unit	Material $	Labor $	Equipment $	Total $
12" x 6"	S2@1.00	Ea	88.00	35.40	—	123.40
12" x 8"	S2@1.00	Ea	90.00	35.40	—	125.40
12" x 10"	S2@1.05	Ea	93.00	37.10	—	130.10
12" x 12"	S2@1.10	Ea	95.00	38.90	—	133.90
18" x 12"	S2@1.20	Ea	120.00	42.40	—	162.40
18" x 18"	S2@1.30	Ea	155.00	46.00	—	201.00
24" x 12"	S2@1.40	Ea	165.00	49.50	—	214.50
24" x 18"	S2@1.50	Ea	185.00	53.00	—	238.00
24" x 24"	S2@1.60	Ea	210.00	56.60	—	266.60
30" x 12"	S2@1.50	Ea	200.00	53.00	—	253.00
30" x 18"	S2@1.80	Ea	250.00	63.60	—	313.60
30" x 24"	S2@2.00	Ea	290.00	70.70	—	360.70
30" x 30"	S2@2.30	Ea	350.00	81.30	—	431.30
36" x 18"	S2@2.00	Ea	280.00	70.70	—	350.70
36" x 36"	S2@2.40	Ea	375.00	84.90	—	459.90
42" x 24"	S2@2.40	Ea	350.00	84.90	—	434.90
42" x 42"	S2@2.90	Ea	425.00	103.00	—	528.00
48" x 24"	S2@2.90	Ea	380.00	103.00	—	483.00
48" x 48"	S2@3.50	Ea	850.00	124.00	—	974.00
54" x 24"	S2@3.00	Ea	1,100.00	106.00	—	1,206.00
54" x 54"	S2@3.80	Ea	1,780.00	134.00	—	1,914.00
Cut, frame and gasket access hole in sidewall	S2@1.00	Ea	65.00	35.40	—	100.40
Mount duct hangers	S2@.250	Ea	3.50	8.84	—	12.34
Cut and mount sheet metal duct	S2@.350	LF	5.50	12.40	—	17.90
Apply and coat duct insulation	S2@.150	LF	2.40	5.30	—	7.70

Description	Craft@Hrs	Unit	Material $	Labor $	Equipment $	Total $

Single skin 2" fixed position turning vanes for duct

Description	Craft@Hrs	Unit	Material $	Labor $	Equipment $	Total $
12" x 6"	S2@.280	Ea	7.79	9.90	—	17.69
12" x 8"	S2@.285	Ea	9.64	10.10	—	19.74
12" x 10"	S2@.290	Ea	12.40	10.30	—	22.70
12" x 12"	S2@.300	Ea	15.40	10.60	—	26.00
18" x 12"	S2@.375	Ea	23.30	13.30	—	36.60
18" x 18"	S2@.450	Ea	28.20	15.90	—	44.10
24" x 12"	S2@.450	Ea	30.90	15.90	—	46.80
24" x 18"	S2@.500	Ea	33.70	17.70	—	51.40
24" x 24"	S2@.525	Ea	38.40	18.60	—	57.00
30" x 12"	S2@.550	Ea	38.70	19.40	—	58.10
30" x 18"	S2@.750	Ea	55.80	26.50	—	82.30
30" x 24"	S2@.800	Ea	74.20	28.30	—	102.50
30" x 30"	S2@.900	Ea	94.20	31.80	—	126.00
36" x 18"	S2@.750	Ea	66.30	26.50	—	92.80
36" x 36"	S2@.800	Ea	135.00	28.30	—	163.30
42" x 24"	S2@.900	Ea	105.00	31.80	—	136.80
42" x 42"	S2@1.00	Ea	185.00	35.40	—	220.40
48" x 24"	S2@1.10	Ea	119.00	38.90	—	157.90
48" x 48"	S2@1.50	Ea	238.00	53.00	—	291.00
54" x 24"	S2@1.15	Ea	133.00	40.70	—	173.70
54" x 54"	S2@1.25	Ea	303.00	44.20	—	347.20
60" x 24"	S2@1.30	Ea	150.00	46.00	—	196.00
60" x 36"	S2@1.40	Ea	225.00	49.50	—	274.50
60" x 60"	S2@1.50	Ea	373.00	53.00	—	426.00
72" x 36"	S2@1.65	Ea	272.00	58.30	—	330.30
72" x 54"	S2@2.40	Ea	402.00	84.90	—	486.90

Note: Labor includes cost to cut, assemble and install vanes and rails. For 4-inch air-foil vanes, multiply costs by 1.80 and labor costs by .90

Ductwork Specialties

Description	Craft@Hrs	Unit	Material $	Labor $	Equipment $	Total $

Spin-in galvanized steel collars without volume damper

Description	Craft@Hrs	Unit	Material $	Labor $	Equipment $	Total $
4"	S2@.200	Ea	5.79	7.07	—	12.86
5"	S2@.200	Ea	6.50	7.07	—	13.57
6"	S2@.200	Ea	7.17	7.07	—	14.24
7"	S2@.200	Ea	7.88	7.07	—	14.95
8"	S2@.250	Ea	9.15	8.84	—	17.99
9"	S2@.250	Ea	10.10	8.84	—	18.94
10"	S2@.300	Ea	10.90	10.60	—	21.50
12"	S2@.350	Ea	12.90	12.40	—	25.30
14"	S2@.400	Ea	15.50	14.10	—	29.60
16"	S2@.450	Ea	18.50	15.90	—	34.40
18"	S2@.500	Ea	23.00	17.70	—	40.70

Spin-in galvanized steel collar with volume damper

Description	Craft@Hrs	Unit	Material $	Labor $	Equipment $	Total $
4"	S2@.250	Ea	19.40	8.84	—	28.24
5"	S2@.250	Ea	19.80	8.84	—	28.64
6"	S2@.250	Ea	20.00	8.84	—	28.84
7"	S2@.250	Ea	20.40	8.84	—	29.24
8"	S2@.300	Ea	20.80	10.60	—	31.40
9"	S2@.300	Ea	21.00	10.60	—	31.60
10"	S2@.350	Ea	21.50	12.40	—	33.90
12"	S2@.400	Ea	23.20	14.10	—	37.30
14"	S2@.450	Ea	24.70	15.90	—	40.60
16"	S2@.500	Ea	26.30	17.70	—	44.00
18"	S2@.550	Ea	28.20	19.40	—	47.60

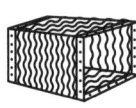

Duct-to-equipment flexible connection

Description	Craft@Hrs	Unit	Material $	Labor $	Equipment $	Total $
18" x 12"	S2@.650	Ea	17.90	23.00	—	40.90
18" x 18"	S2@.850	Ea	21.50	30.10	—	51.60
24" x 12"	S2@.850	Ea	21.50	30.10	—	51.60
24" x 18"	S2@1.00	Ea	25.00	35.40	—	60.40
24" x 24"	S2@1.30	Ea	28.80	46.00	—	74.80
30" x 12"	S2@1.00	Ea	25.00	35.40	—	60.40
30" x 18"	S2@1.30	Ea	28.80	46.00	—	74.80
30" x 30"	S2@1.60	Ea	36.00	56.60	—	92.60
36" x 12"	S2@1.30	Ea	28.80	46.00	—	74.80
36" x 24"	S2@1.60	Ea	36.00	56.60	—	92.60
36" x 36"	S2@1.80	Ea	42.60	63.60	—	106.20
48" x 24"	S2@1.80	Ea	42.60	63.60	—	106.20
48" x 48"	S2@2.00	Ea	56.50	70.70	—	127.20

The costs for fabricating and installing galvanized steel ductwork are usually based on the total weights of the duct and fittings. Because of cost differences, fitting weights should be kept separate from straight duct weights.

Round spiral duct: Normally-used duct gauge/size relationships for low-pressure systems, up to 2 inches static pressure, are as follows:

Duct Diameter (inches)	U.S. Standard Gauge
Up to 12	26
13 through 24	24
26 through 36	22

Rectangular duct: Normally-used duct gauge/size relationships for low-pressure systems, up to 2 inches static pressure, are as follows:

Duct Size (inches)	U.S. Standard Gauge
Up to 12	26
13 through 30	24
31 through 54	22
55 through 72	20

The fitting weights in the following tables are in accordance with the above gauges.

Other material costs: All weights in the tables are net weights; they do not include bracing, cleats, scrap, hangers, end closures, sealants and miscellaneous hardware. Add 15% to calculated duct/fitting weights to cover these items if ductwork is purchased. Add 20% if ducting is manufactured in contractor's shop. The additional 5% covers the costs for scrap which is already included in an outside vendor's price.

Ductwork costs: Typical costs for purchased ductwork are as follows:

Straight duct, less than 1,000 pounds per order	$2.63/lb.
Straight duct, over 1,000 pounds per order	$2.36/lb.
Fittings, less than 1,000 pounds per order	$5.31/lb.
Fittings, over 1,000 pounds per order	$4.76/lb.
Lined ducts and fittings, add	$ 1.07/lb.*
Delivery costs	$ .15/lb.

Duct sizes shown on drawings are always inside, or net, dimensions and must be increased in size when estimating lined ductwork.

If ductwork is manufactured in the contractor's fabrication shop, the above costs can be reduced by about 20%, depending on the shop's rate of productivity.

Galvanized Steel Ductwork

Installation costs: An average crew can install approximately 25 pounds of duct and fittings per manhour under normal conditions. See "Applying Correction Factors" on page 6 for situations that do not conform to the definition of a standard labor unit.

EXAMPLE:

What is the cost to furnish and install 2,585 pounds of unlined duct and 560 pounds of unlined fittings? The ductwork will be purchased from an outside vendor. A 15% allowance for miscellaneous material is included in the weights.

Material:		
2,585 pounds of straight duct x $2.36/lb.	=	$6,100.60
560 pounds of fittings x $5.31/lb.	=	2,973.60
Delivery cost: (2,585 + 560) x $.15/lb.	=	471.75
Total material cost	=	$9,545.95*
Labor:		
$\dfrac{2,585+560}{25} = 125.8\ MH; 125.8\ MH \cdot \$35.36$	=	$4,448.29
Total installed cost: $9,545.95 + $4,448.29	=	$13,994.24*

Sales tax not included.

Description	Craft@Hrs	Unit	Material $	Labor $	Equipment $	Total $

Installed ductwork per lb. (under 1,000 lbs.). Duct purchased from an independent fabrication shop, installed by this contractors' shop.
Material price includes scrap, cleats, hangers, sealant, miscellaneous hardware, delivery and fabrication shop markups.

Description	Craft@Hrs	Unit	Material $	Labor $	Equipment $	Total $
Straight duct	S2@.042	Lb	2.58	1.49	—	4.07
Duct fittings	S2@.042	Lb	5.21	1.49	—	6.70
Duct & fittings	S2@.042	Lb	3.45	1.49	—	4.94

Installed ductwork per lb. (over 1,000 lbs.). Duct purchased from an independent fabrication shop, installed by this contractors' shop.
Material price includes scrap, cleats, hangers, sealant, miscellaneous hardware, delivery and subcontractor markups.

Description	Craft@Hrs	Unit	Material $	Labor $	Equipment $	Total $
Straight duct	S2@.039	Lb	2.32	1.38	—	3.70
Duct fittings	S2@.039	Lb	4.67	1.38	—	6.05
Duct & fittings	S2@.039	Lb	3.12	1.38	—	4.50

Installed lined ductwork per lb. (under 1,000 lbs.). Duct purchased from an independent fabrication shop, installed by this contractors' shop. Material price includes scrap, cleats, hangers, sealant, miscellaneous hardware, delivery and subcontractor markups.

Description	Craft@Hrs	Unit	Material $	Labor $	Equipment $	Total $
Straight duct	S2@.042	Lb	3.96	1.49	—	5.45
Duct fittings	S2@.042	Lb	6.57	1.49	—	8.06
Duct & fittings	S2@.042	Lb	4.83	1.49	—	6.32

Installed lined ductwork per lb. (over 1,000 lbs.). Duct purchased from an independent fabrication shop, installed by this contractors' shop. Material price includes scrap, cleats, hangers, sealant, miscellaneous hardware, delivery and subcontractor markups.

Description	Craft@Hrs	Unit	Material $	Labor $	Equipment $	Total $
Straight duct	S2@.039	Lb	3.73	1.38	—	5.11
Duct fittings	S2@.039	Lb	6.09	1.38	—	7.47
Duct & fittings	S2@.039	Lb	4.49	1.38	—	5.87

Installed ductwork per lb. (under 1,000 lbs.). Duct fabricated and installed by this contractors' shop. Material price includes scrap, cleats, hangers, sealant, miscellaneous hardware and delivery.

Description	Craft@Hrs	Unit	Material $	Labor $	Equipment $	Total $
Straight duct	S2@.051	Lb	1.59	1.80	—	3.39
Duct fittings	S2@.074	Lb	1.82	2.62	—	4.44
Duct & fittings	S2@.059	Lb	1.67	2.09	—	3.76

Installed ductwork per lb. (over 1,000 lbs.). Duct fabricated and installed by this contractors' shop. Material price includes scrap, cleats, hangers, sealant, miscellaneous hardware and delivery.

Description	Craft@Hrs	Unit	Material $	Labor $	Equipment $	Total $
Straight duct	S2@.046	Lb	1.45	1.63	—	3.08
Duct fittings	S2@.067	Lb	1.68	2.37	—	4.05
Duct & fittings	S2@.053	Lb	1.52	1.87	—	3.39

Installed Ductwork Per Pound

Description	Craft@Hrs	Unit	Material $	Labor $	Equipment $	Total $

Installed lined ductwork per lb. (under 1,000 lbs.). Duct fabricated and installed by this contractors' shop. Material price includes scrap, cleats, hangers, sealant, miscellaneous hardware and delivery.

Description	Craft@Hrs	Unit	Material $	Labor $	Equipment $	Total $
Straight duct	S2@.062	Lb	2.34	2.19	—	4.53
Duct fittings	S2@.085	Lb	2.58	3.01	—	5.59
Duct & fittings	S2@.070	Lb	2.44	2.48	—	4.92

Installed lined ductwork per lb. (over 1,000 lbs.). Duct fabricated and installed by this contractors' shop. Material price includes scrap, cleats, hangers, sealant, miscellaneous hardware and delivery. Use for preliminary estimates.

Description	Craft@Hrs	Unit	Material $	Labor $	Equipment $	Total $
Straight duct	S2@.057	Lb	2.21	2.02	—	4.23
Duct fittings	S2@.078	Lb	2.47	2.76	—	5.23
Duct & fittings	S2@.064	Lb	4.01	2.26	—	6.27

Weights of Galvanized Steel Spiral Duct (pounds per LF)					
Diameter (inches)	U.S. Standard Gauge				
	26	24	22	20	18
3	.76	1.01	1.23	1.46	—
4	1.02	1.35	1.64	1.94	—
5	1.28	1.69	2.06	2.42	—
6	1.54	2.03	2.47	2.91	3.88
7	1.79	2.37	2.88	3.40	4.52
8	2.05	2.71	3.29	3.86	5.17
9	2.31	3.05	3.71	4.37	5.82
10	2.57	3.39	4.12	4.86	6.47
12	3.08	4.07	4.95	5.83	7.75
14	3.59	4.74	5.77	6.81	9.05
16	4.11	5.42	6.60	7.78	10.34
18	4.63	6.10	7.43	8.76	11.64
20	5.15	6.78	8.25	9.73	12.93
22	5.65	7.46	9.08	10.71	14.93
24	6.16	8.14	9.91	11.68	15.52
26	6.67	8.82	10.73	12.66	16.82
28	7.18	9.50	11.56	13.63	18.11
30	7.71	10.18	12.38	14.60	19.41
32	8.22	10.84	13.21	15.58	20.71
34	—	11.52	14.05	16.55	22.00
36	—	12.20	14.90	17.53	23.29

Galvanized Steel Round Spiral Fittings

Weights of Galvanized Steel Round Spiral Fittings (pounds per piece)				
Diameter (inches)	90° elbow	45° elbow	Coupling	Reducer
3	1.3	1.0	0.5	1.0
4	2.2	1.3	0.6	1.2
5	3.3	1.9	0.7	1.4
6	4.3	2.5	0.9	1.8
7	5.8	3.3	1.0	2.0
8	7.3	4.3	1.2	2.4
9	8.8	5.3	2.6	3.3
10	11.8	7.5	2.9	4.4
12	16.3	10.0	3.5	5.3
14	22.0	13.0	4.1	6.2
16	28.3	15.8	4.6	6.9
18	34.5	19.0	5.2	7.8
20	41.5	23.5	5.8	8.7
22	48.3	27.5	6.4	9.6
24	57.5	32.0	6.9	10.3
26	68.8	37.5	7.5	11.2
28	76.8	42.8	8.1	12.1
30	87.0	48.0	8.7	13.0
32	99.5	55.0	9.2	13.8
34	112.0	61.0	9.8	14.7
36	161.0	89.5	10.4	15.6

Largest run diam. (inches)	Weights of Galvanized Steel Round Spiral Fittings (pounds per piece) Tee with reducing run and branch Branch diameter (inches)													
	3	4	5	6	7	8	9	10	12	14	16	18	20	22
4	2.3													
5	2.8	3.0												
6	3.2	3.5	3.7											
7	3.7	3.9	4.1	4.5										
8	4.1	4.4	4.7	5.0	5.3									
9	4.5	4.8	5.1	5.5	5.8	6.2								
10	6.4	6.7	7.1	7.5	7.8	8.2	8.6							
12	7.7	8.1	8.5	8.9	9.4	9.9	10.3	10.7						
14	8.9	9.4	10.0	10.5	11.0	11.5	12.1	12.6	13.6					
16	10.2	10.8	11.4	12.0	12.5	13.1	13.7	14.3	15.5	16.7				
18	11.5	12.1	12.7	13.5	14.1	14.8	15.5	16.1	17.5	18.8	19.2			
20	12.7	13.4	14.2	14.9	17.7	16.4	17.2	17.9	19.4	20.9	22.4	23.9		
22	14.0	14.8	15.6	16.4	17.2	18.1	18.9	19.7	21.3	23.0	24.6	26.2	27.9	
24	15.3	16.1	17.0	17.8	18.7	19.6	20.5	21.4	23.2	25.0	26.8	28.5	30.3	32.1
26	16.5	17.5	18.5	19.5	20.4	21.3	22.3	23.3	25.2	26.2	29.1	31.1	33.0	34.9
28	17.7	18.7	19.8	20.8	21.9	22.9	23.9	25.0	27.0	29.2	31.2	33.2	35.4	37.4
30	18.9	20.0	21.1	22.2	23.3	24.5	25.5	26.7	28.9	31.1	33.3	35.5	37.7	40.0
32	20.4	21.6	22.6	24.0	25.2	26.4	27.6	28.8	31.2	33.6	36.0	38.4	40.8	42.2
34	21.6	22.9	24.2	25.4	26.7	28.0	29.3	30.5	33.1	35.6	38.1	40.7	42.2	45.6
36	29.6	31.4	33.1	34.8	36.6	38.6	40.1	41.8	45.3	48.8	52.2	55.7	59.2	62.7

Run diam. (inches)	Weights of Galvanized Steel Round Spiral Fittings (pounds per piece) Cross with reducing run and branch Branch diameter (inches)													
	3	4	5	6	7	8	9	10	12	14	16	18	20	22
4	2.6													
5	3.2	3.6												
6	3.8	4.2	4.6											
7	4.4	4.9	5.4	5.8										
8	4.9	5.4	6.0	6.5	7.0									
9	5.4	6.0	6.6	7.2	7.7	8.3								
10	7.6	8.2	8.9	9.5	10.2	10.8	11.4							
12	8.8	9.6	10.3	11.0	11.7	12.5	13.2	13.8						
14	10.1	10.9	11.7	12.6	13.4	14.2	15.1	15.9	17.5					
16	11.3	12.2	13.1	14.0	15.0	15.9	16.8	17.7	19.5	21.4				
18	12.5	13.5	14.5	15.5	16.5	17.5	18.5	19.5	21.5	23.5	25.5			
20	13.8	15.0	16.1	17.2	18.2	19.4	20.6	21.7	23.9	26.2	28.4	30.6		
22	15.2	16.4	17.7	18.9	20.1	21.4	22.6	23.8	26.2	28.7	31.2	33.6	36.1	
24	16.6	17.9	19.3	20.6	22.0	23.3	24.6	26.0	28.7	31.5	34.1	36.7	39.4	42.1
26	18.0	19.4	20.9	22.3	23.8	25.2	26.7	28.2	31.1	33.9	36.9	39.8	42.7	45.6
28	19.3	20.8	22.4	24.0	25.5	27.1	28.6	30.2	32.3	36.4	39.6	42.7	45.8	48.9
30	20.6	22.3	23.9	25.6	27.3	28.9	30.6	32.3	35.6	38.9	42.3	45.6	49.9	52.2
32	22.2	24.0	25.8	27.6	29.4	31.2	33.0	34.8	38.4	42.0	45.6	49.1	52.8	56.5
34	23.5	25.4	27.3	29.2	31.1	33.1	34.9	36.9	40.7	44.5	48.3	52.1	55.9	59.8
36	32.2	34.8	37.4	40.0	42.8	45.3	48.0	50.5	55.7	61.0	66.2	71.5	76.7	82.0

Weights of 26 Gauge Galvanized Steel Rectangular Duct (pounds per LF)									
Size (inches)	4	5	6	7	8	9	10	11	12
4	1.21								
5	1.36	1.51							
6	1.51	1.66	1.81						
7	1.66	1.81	1.96	2.11					
8	1.81	1.96	2.11	2.27	2.42				
9	1.96	2.11	2.27	2.42	2.57	2.72			
10	2.11	2.27	2.42	2.57	2.72	2.87	3.02		
11	2.27	2.42	2.57	2.72	2.87	3.02	3.17	3.32	
12	2.42	2.57	2.72	2.87	3.02	3.17	3.32	3.47	3.62

Galvanized Steel Rectangular Ductwork

Weights of 24 Gauge Galvanized Steel Rectangular Duct (pounds per LF)									
Size (inches)	14	16	18	20	22	24	26	28	30
6	3.85	4.24	4.62	5.01	5.40	5.78	6.17	6.55	6.94
8	4.24	4.62	5.01	5.40	5.78	6.17	6.55	6.94	7.32
10	4.62	5.01	5.40	5.78	6.17	6.55	6.94	7.32	7.71
12	5.01	5.40	5.78	6.17	6.55	6.94	7.32	7.71	8.09
14	5.40	5.78	6.17	6.55	6.94	7.32	7.71	8.09	8.48
16	5.78	6.17	6.55	6.94	7.32	7.71	8.09	8.48	8.86
18	6.17	6.55	6.94	7.32	7.71	8.09	8.48	8.86	9.25
20	6.55	6.94	7.32	7.71	8.09	8.48	8.86	9.25	9.64
22	6.94	7.32	7.71	8.09	8.48	8.86	9.25	9.64	10.00
24	7.32	7.71	8.09	8.48	8.86	9.25	9.64	10.00	10.40
26	7.71	8.09	8.48	8.86	9.25	9.64	10.00	10.40	10.80
28	8.09	8.48	8.86	9.25	9.64	10.00	10.40	10.80	11.20
30	8.48	8.86	9.25	9.64	10.00	10.40	10.80	11.20	11.60

Weights of 22 Gauge Galvanized Steel Rectangular Duct (pounds per LF)											
Size (inches)	32	34	36	38	40	42	44	46	48	50	52	54
8	9.37	9.84	10.3	10.8	11.3	11.7	12.2	12.7	13.1	13.6	14.1	14.5
10	9.84	10.3	10.8	11.3	11.7	12.2	12.7	13.1	13.6	14.1	14.5	15.0
12	10.3	10.8	11.3	11.7	12.2	12.7	13.1	13.6	14.1	14.5	15.0	15.5
14	10.8	11.3	11.7	12.2	12.7	13.1	13.6	14.1	14.5	15.0	15.5	15.9
16	11.3	11.7	12.2	12.7	13.1	13.6	14.1	14.5	15.0	15.5	15.9	16.4
18	11.7	12.2	12.7	13.1	13.6	14.1	14.5	15.0	15.5	15.9	16.4	16.9
20	12.2	12.7	13.1	13.6	14.1	14.5	15.0	15.5	15.9	16.4	16.9	17.3
22	12.7	13.1	13.6	14.1	14.5	15.0	15.5	15.9	16.4	16.9	17.3	17.8
24	13.1	13.6	14.1	14.5	15.0	15.5	15.9	16.4	16.9	17.3	17.8	18.3
26	13.6	14.1	14.5	15.0	15.5	15.9	16.4	16.9	17.3	17.8	18.3	18.7
28	14.1	14.5	15.0	15.5	15.9	16.4	16.9	17.3	17.8	18.3	18.7	19.2
30	14.5	15.0	15.5	15.9	16.4	16.9	17.3	17.8	18.3	18.7	19.2	19.7
32	15.0	15.5	15.9	16.4	16.9	17.3	17.8	18.3	18.7	19.2	19.7	20.2
34	15.5	15.9	16.4	16.9	17.3	17.8	18.3	18.7	19.2	19.7	20.2	20.6
36	15.9	16.4	16.9	17.3	17.8	18.3	18.7	19.2	19.7	20.2	20.6	21.1
38	16.4	16.9	17.3	17.8	18.3	18.7	19.2	19.7	20.2	20.6	21.1	21.6
40	16.9	17.3	17.8	18.3	18.7	19.2	19.7	20.2	20.6	21.1	21.6	22.0
42	17.3	17.8	18.3	18.7	19.2	19.7	20.2	20.6	21.1	21.6	22.0	22.5
44	17.8	18.3	18.7	19.2	19.7	20.2	20.6	21.1	21.6	22.0	22.5	23.0
46	18.3	18.7	19.2	19.7	20.2	20.6	21.1	21.6	22.0	22.5	23.0	23.4
48	18.7	19.2	19.7	20.2	20.6	21.1	21.6	22.0	22.5	23.0	23.4	23.9
50	19.2	19.7	20.2	20.6	21.1	21.6	22.0	22.5	23.0	23.4	23.9	24.4
52	19.7	20.2	20.6	21.1	21.6	22.0	22.5	23.0	23.4	23.9	24.4	24.8
54	20.2	20.6	21.1	21.6	22.0	22.5	23.0	23.4	23.9	24.4	24.8	25.3

Weights of 20 Gauge Galvanized Steel Rectangular Duct (pounds per LF)								
Size (inches)	56	58	60	62	64	66	68	70	72
18	20.4	21.0	21.5	22.1	22.6	23.2	23.7	24.3	24.8
20	21.0	21.5	22.1	22.6	23.2	23.7	24.3	24.8	25.4
22	21.5	22.1	22.6	23.2	23.7	24.3	24.8	25.4	25.9
24	22.1	22.6	23.2	23.7	24.3	24.8	25.4	25.9	26.5
26	22.6	23.2	23.7	24.3	24.8	25.4	25.9	26.5	27.0
28	23.2	23.7	24.3	24.8	25.4	25.9	26.5	27.0	27.6
30	23.7	24.3	24.8	25.4	25.9	26.5	27.0	27.6	28.2
32	24.3	24.8	25.4	25.9	26.5	27.0	27.6	28.2	28.7
34	24.8	25.4	25.9	26.5	27.0	27.6	28.2	28.7	29.3
36	25.4	25.9	26.5	27.0	27.6	28.2	28.7	29.3	29.8
38	25.9	26.5	27.0	27.6	28.2	28.7	29.3	29.8	30.4
40	26.5	27.0	27.6	28.2	28.7	29.3	29.8	30.4	30.9
42	27.0	27.6	28.2	28.7	29.3	29.8	30.4	30.9	31.5
44	27.6	28.2	28.7	29.3	29.0	30.4	30.9	31.5	32.0
46	28.2	28.7	29.3	29.8	30.4	30.9	31.5	32.0	32.6
48	28.7	29.3	29.8	30.4	30.9	31.5	32.0	32.6	33.1
50	29.3	29.8	30.4	30.9	31.5	32.0	32.6	33.1	33.7
52	29.8	30.4	30.9	31.5	32.0	32.6	33.1	33.7	34.2
54	30.4	30.9	31.5	32.0	32.6	33.1	33.7	34.2	34.8
56	30.9	31.5	32.0	32.6	33.1	33.7	34.2	34.8	35.3
58	31.5	32.0	32.6	33.1	33.7	34.2	34.8	35.3	35.9
60	32.0	32.6	33.1	33.7	34.2	34.8	35.3	35.9	36.4
62	32.6	33.1	33.7	34.2	34.8	35.3	35.9	36.4	37.0
64	33.1	33.7	34.2	34.8	35.3	35.9	36.4	37.0	37.5
66	33.7	34.2	34.8	35.3	35.9	36.4	37.0	37.5	38.1
68	34.2	34.8	35.3	35.9	36.4	37.0	37.5	38.1	38.6
70	34.8	35.3	35.9	36.4	37.0	37.5	38.1	38.6	39.2
72	35.3	35.9	36.4	37.0	37.5	38.1	38.6	39.2	39.7

Weights of Galvanized Steel Rectangular 90 Degree Elbow (pounds per piece)									
Size (inches)	4	6	8	10	12	14	16	18	20
4	1.05	1.45	1.95	2.60	3.30	5.30	6.50	7.80	9.25
6	1.45	1.95	2.60	3.30	3.70	5.80	7.75	8.75	9.75
8	1.95	2.60	3.30	3.70	4.25	6.50	8.50	9.75	10.4
10	2.60	3.30	3.70	4.25	4.85	7.30	9.50	10.7	11.2
12	3.30	3.70	4.25	4.85	5.60	8.20	10.5	11.7	12.2
14	5.30	5.80	6.50	7.30	8.20	9.25	11.5	12.6	13.2
16	6.50	7.50	8.50	9.50	10.5	11.5	12.5	13.4	14.4
18	7.80	8.75	9.75	10.7	11.7	12.6	13.4	14.5	15.7
20	9.25	9.75	10.4	11.2	12.2	13.8	14.4	15.7	17.2
22	10.8	11.8	12.8	13.8	14.8	25.8	16.8	17.8	18.8
24	12.6	13.6	14.6	15.6	16.6	17.6	18.6	19.6	20.5
26	14.4	15.4	16.4	17.4	18.4	19.4	20.4	21.4	22.3
28	16.4	17.4	18.4	19.4	20.4	21.4	22.4	23.4	24.3
30	18.5	19.5	20.5	21.5	22.5	23.5	24.5	25.5	26.4
32	24.3	25.7	27.0	28.4	29.7	31.1	32.4	33.7	35.0
34	28.2	29.4	30.6	31.8	33.0	34.3	35.5	36.7	37.9
36	31.3	32.5	33.7	34.9	36.1	37.3	38.5	39.7	40.9
38	34.5	35.7	36.9	38.1	39.3	40.5	41.7	42.9	44.1
40	37.9	39.1	40.3	41.5	42.7	43.9	45.1	46.3	47.5
42	41.4	42.6	43.8	45.0	46.2	47.4	48.6	49.8	51.0
44	45.1	46.4	47.6	48.8	50.0	51.2	52.4	53.6	54.8
46	49.0	50.2	51.4	52.6	53.8	55.0	56.2	57.4	58.6
48	53.0	54.2	55.4	56.6	57.8	59.0	60.2	61.4	62.6
50	57.2	58.5	59.7	60.9	62.1	63.3	64.5	65.7	66.9

Note: Elbow weights do not include turning vanes.

Weights of Galvanized Steel Rectangular 90 Degree Elbows (pounds per piece)								
Size (inches)	22	24	26	28	30	32	34	36	38
22	19.6	21.9	24.1	26.3	28.5	37.0	40.8	44.5	48.3
24	21.9	24.2	26.4	28.6	30.9	39.5	43.6	47.2	50.8
26	24.1	26.4	28.7	30.9	33.2	42.0	46.4	49.8	53.3
28	26.3	28.6	30.9	33.3	35.6	44.5	49.2	52.5	55.8
30	28.5	30.9	33.2	35.6	37.9	47.0	52.0	55.1	58.3
32	37.0	39.5	42.0	44.5	47.0	49.5	52.0	57.8	60.8
34	40.8	43.6	46.4	49.2	52.0	54.8	57.6	60.5	63.3
36	44.5	47.2	49.8	52.5	55.1	57.8	60.5	63.1	65.8
38	48.3	50.8	53.3	55.8	58.3	60.8	63.3	65.8	68.3
40	52.1	54.6	57.1	59.6	62.1	64.6	67.1	69.6	72.1
42	55.9	58.4	60.9	63.4	65.9	68.4	70.9	73.4	75.9
44	59.6	62.1	64.6	67.1	69.6	72.1	74.6	77.1	79.6
46	63.4	65.9	68.4	70.9	73.4	75.9	78.4	80.9	83.4
48	67.2	69.7	72.2	74.7	77.2	79.7	82.2	84.7	87.2
50	70.9	73.4	75.9	78.4	80.9	83.4	85.9	88.4	90.9
52	74.8	77.3	79.8	82.3	84.8	87.3	89.8	92.3	94.7
54	78.5	81.0	83.5	86.0	88.5	91.0	93.5	96.0	98.5
56	97.2	100	103	106	109	112	115	118	121
58	103	106	109	112	115	118	121	124	127
60	110	113	116	119	122	125	128	131	134
62	116	119	122	125	128	131	134	137	140
64	122	125	129	132	135	138	141	144	147
66	128	131	135	138	141	144	148	151	154
68	135	138	142	145	148	151	154	157	160
70	141	144	148	151	154	157	160	163	166
72	147	150	154	157	160	163	166	169	172

Note: Elbow weights do not include turning vanes.

Weights of Galvanized Steel Rectangular 90 Degree Elbows (pounds per piece)								
Size (inches)	40	42	44	46	48	50	52	54	56
40	74.6	78.4	82.1	85.9	89.7	93.4	97.2	102	125
42	78.4	82.3	86.2	90.1	94.0	97.9	102	106	130
44	82.1	86.2	90.3	94.4	98.5	102	106	110	135
46	85.9	90.1	94.4	98.7	102	106	110	113	140
48	89.7	94.0	98.5	102	106	110	113	117	144
50	93.4	97.9	102	106	110	113	117	122	149
52	97.2	102	106	110	113	117	122	126	154
54	102	106	110	113	117	122	126	131	159
56	126	130	134	139	145	150	155	160	165
58	132	137	142	147	151	156	161	166	170
60	137	142	147	152	157	162	167	172	176
62	143	148	153	158	163	168	173	178	182
64	150	155	160	165	169	174	179	184	188
66	156	161	166	171	176	181	186	191	195
68	163	168	173	178	182	187	192	197	201
70	170	175	180	185	189	194	199	204	208
72	176	181	185	190	194	199	204	210	215

Note: Elbow weights do not include turning vanes.

Weights of Galvanized Steel Rectangular 90 Degree Elbows (pounds per piece)								
Size (inches)	58	60	62	64	66	68	70	72
58	176	182	187	194	200	207	214	221
60	182	187	194	200	207	214	221	227
62	187	194	200	207	214	221	227	233
64	194	200	207	214	221	227	233	239
66	200	207	214	221	227	233	239	246
68	207	214	221	227	233	239	246	252
70	214	221	227	233	239	246	252	259
72	221	227	233	239	246	252	259	266

Note: Elbow weights do not include turning vanes.

Weights of Galvanized Steel Rectangular Drops and Tap-In Tees (pounds per piece)								
Size (inches)	6	8	10	12	16	20	24	28	32
6	1.81	2.11	2.42	2.72	4.24	5.01	5.78	6.55	8.90
8	2.11	2.42	2.72	3.02	4.62	5.40	6.17	6.94	9.37
10	2.42	2.72	3.02	3.32	5.01	5.78	6.55	7.32	9.84
12	2.72	3.02	3.32	3.62	5.40	6.17	6.94	7.71	10.3
16	4.24	4.62	5.01	5.40	6.17	6.94	7.71	8.48	11.3
20	5.01	5.40	5.78	6.17	6.94	7.71	8.48	9.25	12.2
24	5.78	6.17	6.55	6.94	7.71	8.48	9.25	10.0	13.1
28	6.55	6.94	7.32	7.71	8.48	9.25	10.0	10.8	14.1
32	8.90	9.37	9.84	10.3	11.3	12.2	13.1	14.1	15.0
36	9.84	10.3	10.8	11.3	12.2	13.1	14.1	15.0	15.9
40	10.8	11.3	11.7	12.2	13.1	14.1	15.0	15.9	16.9
44	11.7	12.2	12.7	13.1	14.1	15.0	15.9	16.9	17.8
48	12.7	13.1	13.6	14.1	15.0	15.9	16.9	17.8	18.7
52	13.6	14.1	14.5	15.0	15.9	16.9	17.8	18.7	19.7
56	16.0	17.6	18.2	18.8	19.6	21.0	22.1	23.2	24.3
60	18.7	19.0	19.3	19.9	21.0	22.1	23.2	24.3	25.4

Note: Weights of drops and tap-in tees do not include splitter dampers.

Description	Craft@Hrs	Unit	Material $	Labor $	Equipment $	Total $

Costs for Steel Spiral Duct. Costs per linear foot of duct, based on quantities less than 1,000 pounds. Add 50% to the material cost for lined duct. These costs include delivery, typical bracing, cleats, scrap, hangers, end closures, sealants and miscellaneous hardware.

Description	Craft@Hrs	Unit	Material $	Labor $	Equipment $	Total $

Galvanized steel spiral duct 26 gauge

Description	Craft@Hrs	Unit	Material $	Labor $	Equipment $	Total $
3"	S2@.030	LF	2.14	1.06	—	3.20
4"	S2@.041	LF	2.90	1.45	—	4.35
5"	S2@.051	LF	3.64	1.80	—	5.44
6"	S2@.061	LF	4.33	2.16	—	6.49
7"	S2@.072	LF	5.03	2.55	—	7.58
8"	S2@.082	LF	5.77	2.90	—	8.67
9"	S2@.092	LF	6.51	3.25	—	9.76
10"	S2@.103	LF	7.22	3.64	—	10.86
12"	S2@.123	LF	8.63	4.35	—	12.98
14" - 16"	S2@.174	LF	10.80	6.15	—	16.95
18" - 20"	S2@.195	LF	13.80	6.90	—	20.70
22" - 24"	S2@.236	LF	16.60	8.34	—	24.94
26" - 28"	S2@.277	LF	19.60	9.79	—	29.39
30" - 32"	S2@.318	LF	22.50	11.20	—	33.70

Galvanized steel spiral duct 24 gauge

Description	Craft@Hrs	Unit	Material $	Labor $	Equipment $	Total $
3"	S2@.041	LF	2.89	1.45	—	4.34
4"	S2@.054	LF	3.78	1.91	—	5.69
5"	S2@.067	LF	4.74	2.37	—	7.11
6"	S2@.081	LF	5.71	2.86	—	8.57
7"	S2@.095	LF	6.63	3.36	—	9.99
8"	S2@.108	LF	7.63	3.82	—	11.45
9"	S2@.122	LF	8.59	4.31	—	12.90
10"	S2@.135	LF	9.52	4.77	—	14.29
12"	S2@.163	LF	11.40	5.76	—	17.16
14" - 16"	S2@.203	LF	14.20	7.18	—	21.38
18" - 20"	S2@.257	LF	18.10	9.09	—	27.19
22" - 24"	S2@.312	LF	21.90	11.00	—	32.90
26" - 28"	S2@.366	LF	25.80	12.90	—	38.70
30" - 32"	S2@.419	LF	29.50	14.80	—	44.30
34" - 36"	S2@.474	LF	33.70	16.80	—	50.50

Galvanized Steel Spiral Duct

Description	Craft@Hrs	Unit	Material $	Labor $	Equipment $	Total $

Galvanized steel spiral duct 22 gauge

Description	Craft@Hrs	Unit	Material $	Labor $	Equipment $	Total $
3"	S2@.049	LF	3.49	1.73	—	5.22
4"	S2@.065	LF	4.63	2.30	—	6.93
5"	S2@.082	LF	5.83	2.90	—	8.73
6"	S2@.099	LF	6.93	3.50	—	10.43
7"	S2@.115	LF	8.09	4.07	—	12.16
8"	S2@.132	LF	9.21	4.67	—	13.88
9"	S2@.148	LF	10.40	5.23	—	15.63
10"	S2@.165	LF	11.60	5.83	—	17.43
12"	S2@.198	LF	14.00	7.00	—	21.00
14" - 16"	S2@.247	LF	17.30	8.73	—	26.03
18" - 20"	S2@.313	LF	22.10	11.10	—	33.20
22" - 24"	S2@.380	LF	26.60	13.40	—	40.00
26" - 28"	S2@.444	LF	31.50	15.70	—	47.20
30" - 32"	S2@.491	LF	36.30	17.40	—	53.70
34" - 36"	S2@.576	LF	41.10	20.40	—	61.50

Galvanized steel spiral duct 20 gauge

Description	Craft@Hrs	Unit	Material $	Labor $	Equipment $	Total $
3"	S2@.058	LF	4.14	2.05	—	6.19
4"	S2@.078	LF	5.49	2.76	—	8.25
5"	S2@.097	LF	6.82	3.43	—	10.25
6"	S2@.116	LF	8.19	4.10	—	12.29
7"	S2@.136	LF	9.57	4.81	—	14.38
8"	S2@.154	LF	10.90	5.45	—	16.35
9"	S2@.175	LF	12.20	6.19	—	18.39
10"	S2@.194	LF	13.70	6.86	—	20.56
12"	S2@.233	LF	16.50	8.24	—	24.74
14" - 16"	S2@.291	LF	20.30	10.30	—	30.60
18" - 20"	S2@.369	LF	26.00	13.00	—	39.00
22" - 24"	S2@.446	LF	31.50	15.80	—	47.30
26" - 28"	S2@.525	LF	37.10	18.60	—	55.70
30" - 32"	S2@.604	LF	42.30	21.40	—	63.70
34" - 36"	S2@.681	LF	47.80	24.10	—	71.90

Galvanized steel spiral duct 18 gauge

Description	Craft@Hrs	Unit	Material $	Labor $	Equipment $	Total $
6"	S2@.155	LF	10.90	5.48	—	16.38
7"	S2@.181	LF	12.70	6.40	—	19.10
8"	S2@.207	LF	14.40	7.32	—	21.72
9"	S2@.233	LF	16.40	8.24	—	24.64
10"	S2@.259	LF	18.10	9.16	—	27.26
12"	S2@.310	LF	21.90	11.00	—	32.90
14" - 16"	S2@.387	LF	27.30	13.70	—	41.00
18" - 20"	S2@.491	LF	34.80	17.40	—	52.20
22" - 24"	S2@.608	LF	42.60	21.50	—	64.10
26" - 28"	S2@.696	LF	49.00	24.60	—	73.60
30" - 32"	S2@.803	LF	56.40	28.40	—	84.80
34" - 36"	S2@.905	LF	63.70	32.00	—	95.70

Description	Craft@Hrs	Unit	Material $	Labor $	Equipment $	Total $

Costs for Steel Spiral Duct Fittings. Costs per fitting, based on quantities less than 1,000 pounds. For quantities over 1,000 pounds, deduct 15% from material costs. For lined duct add 50%. These costs include delivery, typical bracing, cleats, scrap, hangers, end closures, sealants and miscellaneous hardware.

Description	Craft@Hrs	Unit	Material $	Labor $	Equipment $	Total $

Galvanized steel spiral duct 90 degree elbows

Description	Craft@Hrs	Unit	Material $	Labor $	Equipment $	Total $
3"	S2@.052	Ea	2.64	1.84	—	4.48
4"	S2@.088	Ea	4.45	3.11	—	7.56
5"	S2@.132	Ea	6.68	4.67	—	11.35
6"	S2@.172	Ea	8.77	6.08	—	14.85
7"	S2@.232	Ea	11.80	8.20	—	20.00
8"	S2@.292	Ea	15.00	10.30	—	25.30
9"	S2@.352	Ea	18.00	12.40	—	30.40
10"	S2@.470	Ea	24.00	16.60	—	40.60
12"	S2@.653	Ea	33.20	23.10	—	56.30
14" - 16"	S2@1.01	Ea	50.90	35.70	—	86.60
18" - 20"	S2@1.52	Ea	77.00	53.70	—	130.70
22" - 24"	S2@2.11	Ea	107.00	74.60	—	181.60
26" - 28"	S2@2.91	Ea	150.00	103.00	—	253.00
30" - 32"	S2@3.73	Ea	190.00	132.00	—	322.00
34" - 36"	S2@5.47	Ea	277.00	193.00	—	470.00

Galvanized steel spiral duct 45 degree elbows

Description	Craft@Hrs	Unit	Material $	Labor $	Equipment $	Total $
3"	S2@.040	Ea	2.03	1.41	—	3.44
4"	S2@.052	Ea	2.64	1.84	—	4.48
5"	S2@.076	Ea	3.85	2.69	—	6.54
6"	S2@.100	Ea	5.09	3.54	—	8.63
7"	S2@.132	Ea	6.68	4.67	—	11.35
8"	S2@.172	Ea	8.77	6.08	—	14.85
9"	S2@.212	Ea	10.70	7.50	—	18.20
10"	S2@.300	Ea	15.20	10.60	—	25.80
12"	S2@.400	Ea	20.30	14.10	—	34.40
14" - 16"	S2@.576	Ea	29.10	20.40	—	49.50
18" - 20"	S2@.850	Ea	43.20	30.10	—	73.30
22" - 24"	S2@1.19	Ea	60.30	42.10	—	102.40
26" - 28"	S2@1.61	Ea	81.70	56.90	—	138.60
30" - 32"	S2@2.06	Ea	105.00	72.80	—	177.80
34" - 36"	S2@3.01	Ea	153.00	106.00	—	259.00

Galvanized Steel Spiral Duct Fittings

Description	Craft@Hrs	Unit	Material $	Labor $	Equipment $	Total $

Galvanized steel spiral duct coupling

Description	Craft@Hrs	Unit	Material $	Labor $	Equipment $	Total $
3"	S2@.020	Ea	2.02	.71	—	2.73
4"	S2@.024	Ea	2.47	.85	—	3.32
5"	S2@.028	Ea	2.82	.99	—	3.81
6"	S2@.036	Ea	3.64	1.27	—	4.91
7"	S2@.040	Ea	4.07	1.41	—	5.48
8"	S2@.048	Ea	4.87	1.70	—	6.57
9"	S2@.104	Ea	10.50	3.68	—	14.18
10"	S2@.116	Ea	11.80	4.10	—	15.90
12"	S2@.140	Ea	14.20	4.95	—	19.15
14" - 16"	S2@.174	Ea	17.80	6.15	—	23.95
18" - 20"	S2@.220	Ea	22.00	7.78	—	29.78
22" - 24"	S2@.266	Ea	27.00	9.41	—	36.41
26" - 28"	S2@.312	Ea	31.60	11.00	—	42.60
30" - 32"	S2@.357	Ea	36.40	12.60	—	49.00
34" - 36"	S2@.403	Ea	41.00	14.30	—	55.30

Galvanized steel spiral duct with reducer

Description	Craft@Hrs	Unit	Material $	Labor $	Equipment $	Total $
3"	S2@.040	Ea	4.07	1.41	—	5.48
4"	S2@.048	Ea	4.87	1.70	—	6.57
5"	S2@.056	Ea	5.69	1.98	—	7.67
6"	S2@.072	Ea	7.34	2.55	—	9.89
7"	S2@.080	Ea	8.09	2.83	—	10.92
8"	S2@.096	Ea	9.76	3.39	—	13.15
9"	S2@.132	Ea	13.40	4.67	—	18.07
10"	S2@.176	Ea	17.80	6.22	—	24.02
12"	S2@.212	Ea	21.50	7.50	—	29.00
14" - 16"	S2@.262	Ea	26.60	9.26	—	35.86
18" - 20"	S2@.330	Ea	33.60	11.70	—	45.30
22" - 24"	S2@.398	Ea	40.40	14.10	—	54.50
26" - 28"	S2@.465	Ea	47.30	16.40	—	63.70
30" - 32"	S2@.536	Ea	54.10	19.00	—	73.10
34" - 36"	S2@.606	Ea	61.40	21.40	—	82.80

Description	Craft@Hrs	Unit	Material $	Labor $	Equipment $	Total $

Costs for Steel Spiral Tees with Reducing Branch. Costs per tee, based on quantities less than 1,000 pounds. Deduct 15% from the material cost for quantities over 1,000 pounds. Add 50% to the material cost for lined duct. These costs include delivery, typical bracing, cleats, scrap, hangers, end closures, sealants and miscellaneous hardware.

Description	Craft@Hrs	Unit	Material $	Labor $	Equipment $	Total $

Galvanized steel spiral duct tees with 3" reducing branch

Description	Craft@Hrs	Unit	Material $	Labor $	Equipment $	Total $
4"	S2@.109	Ea	9.05	3.85	—	12.90
5"	S2@.112	Ea	10.60	3.96	—	14.56
6"	S2@.128	Ea	12.20	4.53	—	16.73
7"	S2@.148	Ea	13.80	5.23	—	19.03
8"	S2@.164	Ea	15.50	5.80	—	21.30
9"	S2@.180	Ea	16.50	6.36	—	22.86
10"	S2@.256	Ea	26.10	9.05	—	35.15
12"	S2@.308	Ea	31.00	10.90	—	41.90
14" - 16"	S2@.382	Ea	39.20	13.50	—	52.70
18" - 20"	S2@.485	Ea	49.20	17.10	—	66.30
22" - 24"	S2@.585	Ea	59.40	20.70	—	80.10
26" - 28"	S2@.683	Ea	69.30	24.20	—	93.50
30" - 32"	S2@.786	Ea	79.90	27.80	—	107.70
34" - 36"	S2@1.02	Ea	104.00	36.10	—	140.10

Galvanized steel spiral duct tees with 4" reducing branch

Description	Craft@Hrs	Unit	Material $	Labor $	Equipment $	Total $
5"	S2@.120	Ea	11.70	4.24	—	15.94
6"	S2@.140	Ea	13.80	4.95	—	18.75
7"	S2@.156	Ea	15.00	5.52	—	20.52
8"	S2@.176	Ea	17.00	6.22	—	23.22
9"	S2@.192	Ea	18.10	6.79	—	24.89
10"	S2@.268	Ea	27.60	9.48	—	37.08
12"	S2@.324	Ea	33.60	11.50	—	45.10
14" - 16"	S2@.403	Ea	41.70	14.30	—	56.00
18" - 20"	S2@.508	Ea	52.60	18.00	—	70.60
22" - 24"	S2@.619	Ea	63.90	21.90	—	85.80
26" - 28"	S2@.724	Ea	74.90	25.60	—	100.50
30" - 32"	S2@.830	Ea	85.80	29.30	—	115.10
34" - 36"	S2@1.08	Ea	112.00	38.20	—	150.20

Galvanized steel spiral duct tees with 5" reducing branch

Description	Craft@Hrs	Unit	Material $	Labor $	Equipment $	Total $
6"	S2@.148	Ea	15.00	5.23	—	20.23
7"	S2@.164	Ea	16.50	5.80	—	22.30
8"	S2@.188	Ea	18.60	6.65	—	25.25
9"	S2@.204	Ea	20.20	7.21	—	27.41
10"	S2@.284	Ea	30.00	10.00	—	40.00
12"	S2@.340	Ea	35.70	12.00	—	47.70
14" - 16"	S2@.428	Ea	44.90	15.10	—	60.00
18" - 20"	S2@.538	Ea	57.10	19.00	—	76.10
22" - 24"	S2@.651	Ea	68.20	23.00	—	91.20
26" - 28"	S2@.764	Ea	80.00	27.00	—	107.00
30" - 32"	S2@.873	Ea	92.30	30.90	—	123.20
34" - 36"	S2@1.15	Ea	120.00	40.70	—	160.70

Galvanized Steel Spiral Tees

Description	Craft@Hrs	Unit	Material $	Labor $	Equipment $	Total $

Galvanized steel spiral duct tees with 6" reducing branch

Description	Craft@Hrs	Unit	Material $	Labor $	Equipment $	Total $
7"	S2@.180	Ea	18.60	6.36	—	24.96
8"	S2@.200	Ea	20.20	7.07	—	27.27
9"	S2@.220	Ea	21.80	7.78	—	29.58
10"	S2@.300	Ea	32.10	10.60	—	42.70
12"	S2@.356	Ea	38.40	12.60	—	51.00
14" - 16"	S2@.449	Ea	47.90	15.90	—	63.80
18" - 20"	S2@.566	Ea	60.60	20.00	—	80.60
22" - 24"	S2@.683	Ea	72.80	24.20	—	97.00
26" - 28"	S2@.807	Ea	85.50	28.50	—	114.00
30" - 32"	S2@.924	Ea	98.70	32.70	—	131.40
34" - 36"	S2@1.20	Ea	128.00	42.40	—	170.40

Galvanized steel spiral duct tees with 7" reducing branch

Description	Craft@Hrs	Unit	Material $	Labor $	Equipment $	Total $
8"	S2@.164	Ea	21.80	5.80	—	27.60
9"	S2@.180	Ea	23.80	6.36	—	30.16
10"	S2@.252	Ea	33.60	8.91	—	42.51
12"	S2@.304	Ea	40.70	10.70	—	51.40
14" - 16"	S2@.381	Ea	51.00	13.50	—	64.50
18" - 20"	S2@.482	Ea	63.90	17.00	—	80.90
22" - 24"	S2@.579	Ea	77.00	20.50	—	97.50
26" - 28"	S2@.683	Ea	91.40	24.20	—	115.60
30" - 32"	S2@.784	Ea	104.00	27.70	—	131.70
34" - 36"	S2@1.02	Ea	136.00	36.10	—	172.10

Galvanized steel spiral duct tees with 8" reducing branch

Description	Craft@Hrs	Unit	Material $	Labor $	Equipment $	Total $
9"	S2@.228	Ea	30.50	8.06	—	38.56
10"	S2@.268	Ea	35.70	9.48	—	45.18
12"	S2@.324	Ea	43.00	11.50	—	54.50
14" - 16"	S2@.401	Ea	53.50	14.20	—	67.70
18" - 20"	S2@.508	Ea	67.80	18.00	—	85.80
22" - 24"	S2@.617	Ea	82.10	21.80	—	103.90
26" - 28"	S2@.722	Ea	96.50	25.50	—	122.00
30" - 32"	S2@.830	Ea	110.00	29.30	—	139.30
34" - 36"	S2@1.08	Ea	143.00	38.20	—	181.20

Galvanized steel spiral duct tees with 9" reducing branch

Description	Craft@Hrs	Unit	Material $	Labor $	Equipment $	Total $
10"	S2@.284	Ea	38.10	10.00	—	48.10
12"	S2@.340	Ea	45.20	12.00	—	57.20
14" - 16"	S2@.426	Ea	57.10	15.10	—	72.20
18" - 20"	S2@.540	Ea	71.50	19.10	—	90.60
22" - 24"	S2@.649	Ea	86.30	22.90	—	109.20
26" - 28"	S2@.760	Ea	101.00	26.90	—	127.90
30" - 32"	S2@.875	Ea	117.00	30.90	—	147.90
34" - 36"	S2@1.15	Ea	152.00	40.70	—	192.70

Description	Craft@Hrs	Unit	Material $	Labor $	Equipment $	Total $

Galvanized steel spiral duct tees with 10" reducing branch

12"	S2@.360	Ea	47.90	12.70	—	60.60
14" - 16"	S2@.447	Ea	59.40	15.80	—	75.20
18" - 20"	S2@.564	Ea	75.40	19.90	—	95.30
22" - 24"	S2@.683	Ea	91.40	24.20	—	115.60
26" - 28"	S2@.805	Ea	107.00	28.50	—	135.50
30" - 32"	S2@.924	Ea	122.00	32.70	—	154.70
34" - 36"	S2@1.20	Ea	161.00	42.40	—	203.40

Galvanized steel spiral duct tees with 12" reducing branch

14" - 16"	S2@.572	Ea	65.50	20.20	—	85.70
18" - 20"	S2@.681	Ea	82.80	24.10	—	106.90
22" - 24"	S2@.820	Ea	100.00	29.00	—	129.00
26" - 28"	S2@.965	Ea	117.00	34.10	—	151.10
30" - 32"	S2@1.11	Ea	135.00	39.20	—	174.20
34" - 36"	S2@1.44	Ea	178.00	50.90	—	228.90

Galvanized steel spiral duct tees with 14" reducing branch

14" - 16"	S2@.572	Ea	76.00	20.20	—	96.20
18" - 20"	S2@.681	Ea	90.50	24.10	—	114.60
22" - 24"	S2@.820	Ea	109.00	29.00	—	138.00
26" - 28"	S2@.965	Ea	128.00	34.10	—	162.10
30" - 32"	S2@1.11	Ea	148.00	39.20	—	187.20
34" - 36"	S2@1.44	Ea	194.00	50.90	—	244.90

Galvanized steel spiral duct tees with 16" reducing branch

18" - 20"	S2@.739	Ea	98.70	26.10	—	124.80
22" - 24"	S2@.888	Ea	119.00	31.40	—	150.40
26" - 28"	S2@1.04	Ea	139.00	36.80	—	175.80
30" - 32"	S2@1.20	Ea	159.00	42.40	—	201.40
34" - 36"	S2@1.56	Ea	209.00	55.20	—	264.20

Galvanized steel spiral duct tees with 18" reducing branch

18" - 20"	S2@.837	Ea	111.00	29.60	—	140.60
22" - 24"	S2@.954	Ea	127.00	33.70	—	160.70
26" - 28"	S2@1.12	Ea	150.00	39.60	—	189.60
30" - 32"	S2@1.29	Ea	172.00	45.60	—	217.60
34" - 36"	S2@1.68	Ea	226.00	59.40	—	285.40

Galvanized steel spiral duct tees with 20" reducing branch

22" - 24"	S2@1.02	Ea	137.00	36.10	—	173.10
26" - 28"	S2@1.20	Ea	161.00	42.40	—	203.40
30" - 32"	S2@1.38	Ea	184.00	48.80	—	232.80
34" - 36"	S2@1.81	Ea	241.00	64.00	—	305.00

Galvanized steel spiral duct tees with 22" reducing branch

22" - 24"	S2@1.57	Ea	152.00	55.50	—	207.50
26" - 28"	S2@1.32	Ea	178.00	46.70	—	224.70
30" - 32"	S2@1.48	Ea	197.00	52.30	—	249.30
34" - 36"	S2@1.92	Ea	259.00	67.90	—	326.90

Galvanized Steel Spiral Tees

Costs for Steel Spiral Tees with Reducing Run and Branch. Costs per tee, based on quantities less than 1,000 pounds. Deduct 15% from the material cost for quantities over 1,000 pounds. Add 50% to the material cost for lined duct. These costs include delivery, typical bracing, cleats, scrap, hangers, end closures, sealants and miscellaneous hardware.

Description	Craft@Hrs	Unit	Material $	Labor $	Equipment $	Total $

Galvanized steel spiral duct tees with 3" run and branch

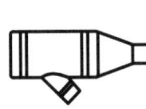

Description	Craft@Hrs	Unit	Material $	Labor $	Equipment $	Total $
4"	S2@.109	Ea	14.70	3.85	—	18.55
5"	S2@.112	Ea	15.00	3.96	—	18.96
6"	S2@.128	Ea	17.00	4.53	—	21.53
7"	S2@.148	Ea	19.70	5.23	—	24.93
8"	S2@.164	Ea	21.80	5.80	—	27.60
9"	S2@.180	Ea	23.80	6.36	—	30.16
10"	S2@.256	Ea	34.30	9.05	—	43.35
12"	S2@.308	Ea	40.90	10.90	—	51.80
14" - 16"	S2@.382	Ea	51.00	13.50	—	64.50
18" - 20"	S2@.485	Ea	64.40	17.10	—	81.50
22" - 24"	S2@.585	Ea	78.10	20.70	—	98.80
26" - 28"	S2@.683	Ea	91.40	24.20	—	115.60
30" - 32"	S2@.786	Ea	105.00	27.80	—	132.80
34" - 36"	S2@1.02	Ea	136.00	36.10	—	172.10

Galvanized steel spiral duct tees with 4" run and branch

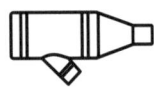

Description	Craft@Hrs	Unit	Material $	Labor $	Equipment $	Total $
5"	S2@.120	Ea	15.90	4.24	—	20.14
6"	S2@.140	Ea	18.60	4.95	—	23.55
7"	S2@.156	Ea	20.90	5.52	—	26.42
8"	S2@.176	Ea	23.30	6.22	—	29.52
9"	S2@.192	Ea	25.40	6.79	—	32.19
10"	S2@.268	Ea	35.70	9.48	—	45.18
12"	S2@.324	Ea	43.00	11.50	—	54.50
14" - 16"	S2@.403	Ea	53.80	14.30	—	68.10
18" - 20"	S2@.508	Ea	67.80	18.00	—	85.80
22" - 24"	S2@.619	Ea	82.20	21.90	—	104.10
26" - 28"	S2@.724	Ea	96.80	25.60	—	122.40
30" - 32"	S2@.830	Ea	110.00	29.30	—	139.30
34" - 36"	S2@1.08	Ea	144.00	38.20	—	182.20

Galvanized steel spiral duct tees with 5" run and branch

Description	Craft@Hrs	Unit	Material $	Labor $	Equipment $	Total $
6"	S2@.148	Ea	19.70	5.23	—	24.93
7"	S2@.164	Ea	21.80	5.80	—	27.60
8"	S2@.188	Ea	25.00	6.65	—	31.65
9"	S2@.204	Ea	27.20	7.21	—	34.41
10"	S2@.284	Ea	38.10	10.00	—	48.10
12"	S2@.340	Ea	45.20	12.00	—	57.20
14" - 16"	S2@.428	Ea	57.10	15.10	—	72.20
18" - 20"	S2@.538	Ea	71.40	19.00	—	90.40
22" - 24"	S2@.651	Ea	87.00	23.00	—	110.00
26" - 28"	S2@.764	Ea	102.00	27.00	—	129.00
30" - 32"	S2@.873	Ea	117.00	30.90	—	147.90
34" - 36"	S2@1.15	Ea	152.00	40.70	—	192.70

Description	Craft@Hrs	Unit	Material $	Labor $	Equipment $	Total $

Galvanized steel spiral duct tees with 6" run and branch

Description	Craft@Hrs	Unit	Material $	Labor $	Equipment $	Total $
7"	S2@.180	Ea	23.80	6.36	—	30.16
8"	S2@.200	Ea	26.50	7.07	—	33.57
9"	S2@.220	Ea	29.00	7.78	—	36.78
10"	S2@.300	Ea	40.00	10.60	—	50.60
12"	S2@.356	Ea	47.60	12.60	—	60.20
14" - 16"	S2@.449	Ea	59.70	15.90	—	75.60
18" - 20"	S2@.566	Ea	75.40	20.00	—	95.40
22" - 24"	S2@.683	Ea	91.40	24.20	—	115.60
26" - 28"	S2@.807	Ea	107.00	28.50	—	135.50
30" - 32"	S2@.924	Ea	122.00	32.70	—	154.70
34" - 36"	S2@1.20	Ea	161.00	42.40	—	203.40

Galvanized steel spiral duct tees with 7" run and branch

Description	Craft@Hrs	Unit	Material $	Labor $	Equipment $	Total $
8"	S2@.212	Ea	28.20	7.50	—	35.70
9"	S2@.232	Ea	31.00	8.20	—	39.20
10"	S2@.312	Ea	41.40	11.00	—	52.40
12"	S2@.376	Ea	50.10	13.30	—	63.40
14" - 16"	S2@.470	Ea	62.70	16.60	—	79.30
18" - 20"	S2@.636	Ea	84.60	22.50	—	107.10
22" - 24"	S2@.717	Ea	96.20	25.40	—	121.60
26" - 28"	S2@.845	Ea	112.00	29.90	—	141.90
30" - 32"	S2@.969	Ea	131.00	34.30	—	165.30
34" - 36"	S2@1.27	Ea	169.00	44.90	—	213.90

Galvanized steel spiral duct tees with 8" run and branch

Description	Craft@Hrs	Unit	Material $	Labor $	Equipment $	Total $
9"	S2@.248	Ea	33.20	8.77	—	41.97
10"	S2@.328	Ea	44.00	11.60	—	55.60
12"	S2@.396	Ea	52.50	14.00	—	66.50
14" - 16"	S2@.493	Ea	65.20	17.40	—	82.60
18" - 20"	S2@.623	Ea	82.80	22.00	—	104.80
22" - 24"	S2@.751	Ea	100.00	26.60	—	126.60
26" - 28"	S2@.882	Ea	118.00	31.20	—	149.20
30" - 32"	S2@1.02	Ea	135.00	36.10	—	171.10
34" - 36"	S2@1.33	Ea	178.00	47.00	—	225.00

Galvanized steel spiral duct tees with 9" run and branch

Description	Craft@Hrs	Unit	Material $	Labor $	Equipment $	Total $
10"	S2@.344	Ea	45.60	12.20	—	57.80
12"	S2@.412	Ea	54.70	14.60	—	69.30
14" - 16"	S2@.515	Ea	68.70	18.20	—	86.90
18" - 20"	S2@.653	Ea	87.00	23.10	—	110.10
22" - 24"	S2@.779	Ea	105.00	27.50	—	132.50
26" - 28"	S2@.924	Ea	122.00	32.70	—	154.70
30" - 32"	S2@1.06	Ea	141.00	37.50	—	178.50
34" - 36"	S2@1.39	Ea	184.00	49.20	—	233.20

Galvanized Steel Spiral Tees

Description	Craft@Hrs	Unit	Material $	Labor $	Equipment $	Total $

Galvanized steel spiral duct tees with 10" run and branch

Description	Craft@Hrs	Unit	Material $	Labor $	Equipment $	Total $
12"	S2@.427	Ea	28.60	15.10	—	43.70
14" - 16"	S2@.538	Ea	36.00	19.00	—	55.00
18" - 20"	S2@.681	Ea	90.40	24.10	—	114.50
22" - 24"	S2@.820	Ea	109.00	29.00	—	138.00
26" - 28"	S2@.965	Ea	128.00	34.10	—	162.10
30" - 32"	S2@1.11	Ea	148.00	39.20	—	187.20
34" - 36"	S2@1.44	Ea	192.00	50.90	—	242.90

Galvanized steel spiral duct tees with 12" run and branch

Description	Craft@Hrs	Unit	Material $	Labor $	Equipment $	Total $
14" - 16"	S2@.581	Ea	77.10	20.50	—	97.60
18" - 20"	S2@.739	Ea	98.70	26.10	—	124.80
22" - 24"	S2@.888	Ea	119.00	31.40	—	150.40
26" - 28"	S2@1.04	Ea	139.00	36.80	—	175.80
30" - 32"	S2@1.20	Ea	159.00	42.40	—	201.40
34" - 36"	S2@1.57	Ea	209.00	55.50	—	264.50

Galvanized steel spiral duct tees with 14" run and branch

Description	Craft@Hrs	Unit	Material $	Labor $	Equipment $	Total $
14" - 16"	S2@.666	Ea	88.90	23.50	—	112.40
18" - 20"	S2@.794	Ea	106.00	28.10	—	134.10
22" - 24"	S2@.959	Ea	127.00	33.90	—	160.90
26" - 28"	S2@1.11	Ea	148.00	39.20	—	187.20
30" - 32"	S2@1.29	Ea	172.00	45.60	—	217.60
34" - 36"	S2@1.69	Ea	226.00	59.80	—	285.80

Galvanized steel spiral duct tees with 16" run and branch

Description	Craft@Hrs	Unit	Material $	Labor $	Equipment $	Total $
18" - 20"	S2@.833	Ea	110.00	29.50	—	139.50
22" - 24"	S2@1.03	Ea	137.00	36.40	—	173.40
26" - 28"	S2@1.20	Ea	161.00	42.40	—	203.40
30" - 32"	S2@1.39	Ea	184.00	49.20	—	233.20
34" - 36"	S2@1.81	Ea	241.00	64.00	—	305.00

Galvanized steel spiral duct tees with 18" run and branch

Description	Craft@Hrs	Unit	Material $	Labor $	Equipment $	Total $
18" - 20"	S2@.956	Ea	127.00	33.80	—	160.80
22" - 24"	S2@1.09	Ea	144.00	38.50	—	182.50
26" - 28"	S2@1.29	Ea	171.00	45.60	—	216.60
30" - 32"	S2@1.48	Ea	197.00	52.30	—	249.30
34" - 36"	S2@1.93	Ea	259.00	68.20	—	327.20

Galvanized steel spiral duct tees with 20" run and branch

Description	Craft@Hrs	Unit	Material $	Labor $	Equipment $	Total $
22" - 24"	S2@1.16	Ea	155.00	41.00	—	196.00
26" - 28"	S2@1.37	Ea	183.00	48.40	—	231.40
30" - 32"	S2@1.57	Ea	209.00	55.50	—	264.50
34" - 36"	S2@2.03	Ea	272.00	71.80	—	343.80

Galvanized steel spiral duct tees with 22" run and branch

Description	Craft@Hrs	Unit	Material $	Labor $	Equipment $	Total $
22" - 24"	S2@1.28	Ea	171.00	45.30	—	216.30
26" - 28"	S2@1.45	Ea	194.00	51.30	—	245.30
30" - 32"	S2@1.64	Ea	219.00	58.00	—	277.00
34" - 36"	S2@2.16	Ea	287.00	76.40	—	363.40

Costs for Steel Spiral Crosses with Reducing Branches. Costs per cross, based on quantities less than 1,000 pounds. Deduct 15% from the material cost for quantities over 1,000 pounds. Add 50% to the material cost for lined duct. These costs include delivery, typical bracing, cleats, scrap, hangers, end closures, sealants and miscellaneous hardware.

Description	Craft@Hrs	Unit	Material $	Labor $	Equipment $	Total $

Galvanized steel spiral duct crosses with 3" branch

Description	Craft@Hrs	Unit	Material $	Labor $	Equipment $	Total $
4"	S2@.104	Ea	13.80	3.68	—	17.48
5"	S2@.128	Ea	17.00	4.53	—	21.53
6"	S2@.152	Ea	20.20	5.37	—	25.57
7"	S2@.176	Ea	23.30	6.22	—	29.52
8"	S2@.196	Ea	26.10	6.93	—	33.03
9"	S2@.216	Ea	28.60	7.64	—	36.24
10"	S2@.304	Ea	40.70	10.70	—	51.40
12"	S2@.352	Ea	47.30	12.40	—	59.70
14" - 16"	S2@.428	Ea	57.40	15.10	—	72.50
18" - 20"	S2@.525	Ea	70.00	18.60	—	88.60
22" - 24"	S2@.634	Ea	84.60	22.40	—	107.00
26" - 28"	S2@.745	Ea	99.40	26.30	—	125.70
30" - 32"	S2@.856	Ea	113.00	30.30	—	143.30
34" - 36"	S2@1.11	Ea	149.00	39.20	—	188.20

Galvanized steel spiral duct crosses with 4" branch

Description	Craft@Hrs	Unit	Material $	Labor $	Equipment $	Total $
5"	S2@.144	Ea	19.20	5.09	—	24.29
6"	S2@.168	Ea	22.60	5.94	—	28.54
7"	S2@.196	Ea	26.10	6.93	—	33.03
8"	S2@.216	Ea	28.60	7.64	—	36.24
9"	S2@.240	Ea	32.10	8.49	—	40.59
10"	S2@.328	Ea	44.00	11.60	—	55.60
12"	S2@.384	Ea	51.60	13.60	—	65.20
14" - 16"	S2@.461	Ea	61.50	16.30	—	77.80
18" - 20"	S2@.568	Ea	75.80	20.10	—	95.90
22" - 24"	S2@.685	Ea	91.60	24.20	—	115.80
26" - 28"	S2@.805	Ea	107.00	28.50	—	135.50
30" - 32"	S2@.927	Ea	124.00	32.80	—	156.80
34" - 36"	S2@1.20	Ea	161.00	42.40	—	203.40

Galvanized steel spiral duct crosses with 5" branch

Description	Craft@Hrs	Unit	Material $	Labor $	Equipment $	Total $
6"	S2@.184	Ea	24.60	6.51	—	31.11
7"	S2@.216	Ea	28.60	7.64	—	36.24
8"	S2@.240	Ea	32.10	8.49	—	40.59
9"	S2@.264	Ea	35.40	9.34	—	44.74
10"	S2@.356	Ea	47.60	12.60	—	60.20
12"	S2@.412	Ea	54.70	14.60	—	69.30
14" - 16"	S2@.497	Ea	65.70	17.60	—	83.30
18" - 20"	S2@.613	Ea	81.20	21.70	—	102.90
22" - 24"	S2@.741	Ea	98.80	26.20	—	125.00
26" - 28"	S2@.867	Ea	116.00	30.70	—	146.70
30" - 32"	S2@.993	Ea	133.00	35.10	—	168.10
34" - 36"	S2@1.29	Ea	172.00	45.60	—	217.60

Galvanized Steel Spiral Crosses

Description	Craft@Hrs	Unit	Material $	Labor $	Equipment $	Total $

Galvanized steel spiral duct crosses with 6" branch

Description	Craft@Hrs	Unit	Material $	Labor $	Equipment $	Total $
7"	S2@.232	Ea	31.00	8.20	—	39.20
8"	S2@.260	Ea	34.70	9.19	—	43.89
9"	S2@.288	Ea	38.40	10.20	—	48.60
10"	S2@.380	Ea	50.90	13.40	—	64.30
12"	S2@.440	Ea	58.70	15.60	—	74.30
14" - 16"	S2@.532	Ea	70.60	18.80	—	89.40
18" - 20"	S2@.653	Ea	87.00	23.10	—	110.10
22" - 24"	S2@.790	Ea	106.00	27.90	—	133.90
26" - 28"	S2@.927	Ea	124.00	32.80	—	156.80
30" - 32"	S2@1.06	Ea	141.00	37.50	—	178.50
34" - 36"	S2@1.38	Ea	184.00	48.80	—	232.80

Galvanized steel spiral duct crosses with 7" branch

Description	Craft@Hrs	Unit	Material $	Labor $	Equipment $	Total $
8"	S2@.280	Ea	37.10	9.90	—	47.00
9"	S2@.308	Ea	40.90	10.90	—	51.80
10"	S2@.408	Ea	54.40	14.40	—	68.80
12"	S2@.470	Ea	62.30	16.60	—	78.90
14" - 16"	S2@.566	Ea	75.40	20.00	—	95.40
18" - 20"	S2@.692	Ea	92.60	24.50	—	117.10
22" - 24"	S2@.841	Ea	111.00	29.70	—	140.70
26" - 28"	S2@.986	Ea	132.00	34.90	—	166.90
30" - 32"	S2@1.13	Ea	151.00	40.00	—	191.00
34" - 36"	S2@1.48	Ea	197.00	52.30	—	249.30

Galvanized steel spiral duct crosses with 8" branch

Description	Craft@Hrs	Unit	Material $	Labor $	Equipment $	Total $
9"	S2@.310	Ea	44.30	11.00	—	55.30
10"	S2@.403	Ea	58.00	14.30	—	72.30
12"	S2@.470	Ea	66.10	16.60	—	82.70
14" - 16"	S2@.561	Ea	80.00	19.80	—	99.80
18" - 20"	S2@.690	Ea	98.70	24.40	—	123.10
22" - 24"	S2@.835	Ea	119.00	29.50	—	148.50
26" - 28"	S2@.980	Ea	139.00	34.70	—	173.70
30" - 32"	S2@1.12	Ea	161.00	39.60	—	200.60
34" - 36"	S2@1.47	Ea	209.00	52.00	—	261.00

Galvanized steel spiral duct crosses with 9" branch

Description	Craft@Hrs	Unit	Material $	Labor $	Equipment $	Total $
10"	S2@.427	Ea	60.60	15.10	—	75.70
12"	S2@.495	Ea	70.10	17.50	—	87.60
14" - 16"	S2@.596	Ea	84.90	21.10	—	106.00
18" - 20"	S2@.732	Ea	104.00	25.90	—	129.90
22" - 24"	S2@.882	Ea	126.00	31.20	—	157.20
26" - 28"	S2@1.04	Ea	148.00	36.80	—	184.80
30" - 32"	S2@1.19	Ea	169.00	42.10	—	211.10
34" - 36"	S2@1.55	Ea	219.00	54.80	—	273.80

Description	Craft@Hrs	Unit	Material $	Labor $	Equipment $	Total $

Galvanized steel spiral duct crosses with 12" branch

14" - 16"	S2@.692	Ea	98.80	24.50	—	123.30
18" - 20"	S2@.850	Ea	121.00	30.10	—	151.10
22" - 24"	S2@1.03	Ea	147.00	36.40	—	183.40
26" - 28"	S2@1.18	Ea	169.00	41.70	—	210.70
30" - 32"	S2@1.38	Ea	197.00	48.80	—	245.80
34" - 36"	S2@1.80	Ea	259.00	63.60	—	322.60

Galvanized steel spiral duct crosses with 14" branch

14" - 16"	S2@1.81	Ea	50.10	64.00	—	114.10
18" - 20"	S2@2.11	Ea	58.30	74.60	—	132.90
22" - 24"	S2@2.55	Ea	70.60	90.20	—	160.80
26" - 28"	S2@2.98	Ea	82.50	105.00	—	187.50
30" - 32"	S2@3.43	Ea	94.80	121.00	—	215.80
34" - 36"	S2@4.47	Ea	124.00	158.00	—	282.00

Galvanized steel spiral duct crosses with 16" branch

18" - 20"	S2@1.08	Ea	143.00	38.20	—	181.20
22" - 24"	S2@1.30	Ea	173.00	46.00	—	210.00
26" - 28"	S2@1.53	Ea	204.00	54.10	—	258.10
30" - 32"	S2@1.76	Ea	233.00	62.20	—	295.20
34" - 36"	S2@2.31	Ea	308.00	81.70	—	389.70

Galvanized steel spiral duct crosses with 18" branch

18" - 20"	S2@1.22	Ea	163.00	43.10	—	206.10
22" - 24"	S2@1.40	Ea	187.00	49.50	—	236.50
26" - 28"	S2@1.65	Ea	219.00	58.30	—	277.30
30" - 32"	S2@1.89	Ea	251.00	66.80	—	317.80
34" - 36"	S2@2.47	Ea	332.00	87.30	—	419.30

Galvanized steel spiral duct crosses with 20" branch

22" - 24"	S2@1.51	Ea	201.00	53.40	—	254.40
26" - 28"	S2@1.77	Ea	236.00	62.60	—	298.60
30" - 32"	S2@2.05	Ea	275.00	72.50	—	347.50
34" - 36"	S2@2.65	Ea	354.00	93.70	—	447.70

Galvanized steel spiral duct crosses with 22" branch

22" - 24"	S2@1.68	Ea	226.00	59.40	—	285.40
26" - 28"	S2@1.89	Ea	251.00	66.80	—	317.80
30" - 32"	S2@2.17	Ea	287.00	76.70	—	363.70
34" - 36"	S2@2.84	Ea	381.00	100.00	—	481.00

Galvanized Steel Rectangular Ductwork

Description	Craft@Hrs	Unit	Material $	Labor $	Equipment $	Total $

Costs for Steel 26 Gauge Rectangular Duct. Costs per linear foot of duct, based on quantities less than 1,000 pounds of prefabricated duct. Deduct 15% from the material cost for quantities over 1,000 pounds. Add 50% to the material cost for lined duct. These costs include delivery, typical bracing, cleats, scrap, hangers, end closures, sealants and miscellaneous hardware.

Galvanized steel 4" rectangular duct 26 gauge

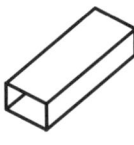

Description	Craft@Hrs	Unit	Material $	Labor $	Equipment $	Total $
4"	S2@.048	LF	3.30	1.70	—	5.00
5"	S2@.054	LF	3.64	1.91	—	5.55
6"	S2@.060	LF	4.07	2.12	—	6.19
7"	S2@.066	LF	4.48	2.33	—	6.81
8"	S2@.072	LF	4.89	2.55	—	7.44
9"	S2@.078	LF	5.25	2.76	—	8.01
10"	S2@.084	LF	5.68	2.97	—	8.65
11"	S2@.091	LF	6.11	3.22	—	9.33
12"	S2@.097	LF	6.51	3.43	—	9.94

Galvanized steel 5" rectangular duct 26 gauge

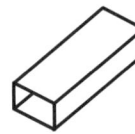

Description	Craft@Hrs	Unit	Material $	Labor $	Equipment $	Total $
5"	S2@.060	LF	4.07	2.12	—	6.19
6"	S2@.066	LF	4.48	2.33	—	6.81
7"	S2@.072	LF	4.89	2.55	—	7.44
8"	S2@.078	LF	5.25	2.76	—	8.01
9"	S2@.084	LF	5.68	2.97	—	8.65
10"	S2@.091	LF	6.11	3.22	—	9.33
11"	S2@.097	LF	6.51	3.43	—	9.94
12"	S2@.103	LF	6.90	3.64	—	10.54

Galvanized steel 6" rectangular duct 26 gauge

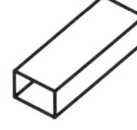

Description	Craft@Hrs	Unit	Material $	Labor $	Equipment $	Total $
6"	S2@.072	LF	4.89	2.55	—	7.44
7"	S2@.078	LF	5.25	2.76	—	8.01
8"	S2@.084	LF	5.68	2.97	—	8.65
9"	S2@.091	LF	6.11	3.22	—	9.33
10"	S2@.097	LF	6.51	3.43	—	9.94
11"	S2@.103	LF	6.90	3.64	—	10.54
12"	S2@.109	LF	7.29	3.85	—	11.14

Galvanized steel 7" rectangular duct 26 gauge

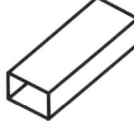

Description	Craft@Hrs	Unit	Material $	Labor $	Equipment $	Total $
7"	S2@.084	LF	5.68	2.97	—	8.65
8"	S2@.091	LF	6.11	3.22	—	9.33
9"	S2@.097	LF	6.51	3.43	—	9.94
10"	S2@.103	LF	6.90	3.64	—	10.54
11"	S2@.109	LF	7.29	3.85	—	11.14
12"	S2@.115	LF	7.70	4.07	—	11.77

Description	Craft@Hrs	Unit	Material $	Labor $	Equipment $	Total $

Galvanized steel 8" rectangular duct 26 gauge

Description	Craft@Hrs	Unit	Material $	Labor $	Equipment $	Total $
8"	S2@.097	LF	6.51	3.43	—	9.94
9"	S2@.103	LF	6.90	3.64	—	10.54
10"	S2@.109	LF	7.29	3.85	—	11.14
11"	S2@.115	LF	7.70	4.07	—	11.77
12"	S2@.121	LF	8.10	4.28	—	12.38

Galvanized steel 9" rectangular duct 26 gauge

Description	Craft@Hrs	Unit	Material $	Labor $	Equipment $	Total $
9"	S2@.109	LF	7.29	3.85	—	11.14
10"	S2@.115	LF	7.70	4.07	—	11.77
11"	S2@.121	LF	8.10	4.28	—	12.38
12"	S2@.127	LF	8.49	4.49	—	12.98

Galvanized steel 10" rectangular duct 26 gauge

Description	Craft@Hrs	Unit	Material $	Labor $	Equipment $	Total $
10"	S2@.129	LF	8.10	4.56	—	12.66
11"	S2@.127	LF	8.49	4.49	—	12.98
12"	S2@.133	LF	8.91	4.70	—	13.61

Galvanized steel 11" rectangular duct 26 gauge

Description	Craft@Hrs	Unit	Material $	Labor $	Equipment $	Total $
11"	S2@.133	LF	8.91	4.70	—	13.61
12"	S2@.139	LF	9.30	4.92	—	14.22

Galvanized steel 12" rectangular duct 26 gauge

Description	Craft@Hrs	Unit	Material $	Labor $	Equipment $	Total $
12"	S2@.145	LF	9.75	5.13	—	14.88

Galvanized Steel Rectangular Ductwork

Costs for Galvanized 24 Gauge Rectangular Duct. Costs per linear foot of duct, based on quantit[?] less than 1,000 pounds of prefabricated duct. Deduct 15% from the material cost for over 1,000 poun[?] Add 50% to the material cost for lined duct. These costs include delivery, typical bracing, cleats, scr[?] hangers, end closures.

Description	Craft@Hrs	Unit	Material $	Labor $	Equipment $	Total $

Galvanized steel 14" rectangular duct 24 gauge

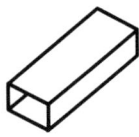

Description	Craft@Hrs	Unit	Material $	Labor $	Equipment $	Total $
6"	S2@.154	LF	10.40	5.45	—	15.85
8"	S2@.170	LF	11.40	6.01	—	17.41
10"	S2@.184	LF	12.50	6.51	—	19.01
12"	S2@.200	LF	13.60	7.07	—	20.67
14" - 16"	S2@.223	LF	15.10	7.89	—	22.99
18" - 20"	S2@.254	LF	17.20	8.98	—	26.18
22" - 24"	S2@.285	LF	19.40	10.10	—	29.50
26" - 28"	S2@.316	LF	21.50	11.20	—	32.70
30"	S2@.339	LF	23.00	12.00	—	35.00

Galvanized steel 16" rectangular duct 24 gauge

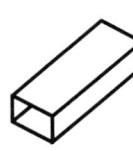

Description	Craft@Hrs	Unit	Material $	Labor $	Equipment $	Total $
6"	S2@.170	LF	11.40	6.01	—	17.41
8"	S2@.184	LF	12.50	6.51	—	19.01
10"	S2@.200	LF	13.60	7.07	—	20.67
12"	S2@.216	LF	14.70	7.64	—	22.34
14" - 16"	S2@.239	LF	16.20	8.45	—	24.65
18" - 20"	S2@.269	LF	18.30	9.51	—	27.81
22" - 24"	S2@.300	LF	20.20	10.60	—	30.80
26" - 28"	S2@.331	LF	22.60	11.70	—	34.30
30"	S2@.354	LF	24.30	12.50	—	36.80

Galvanized steel 18" rectangular duct 24 gauge

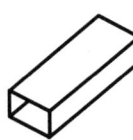

Description	Craft@Hrs	Unit	Material $	Labor $	Equipment $	Total $
6"	S2@.184	LF	12.50	6.51	—	19.01
8"	S2@.200	LF	13.60	7.07	—	20.67
10"	S2@.216	LF	14.70	7.64	—	22.34
12"	S2@.231	LF	15.70	8.17	—	23.87
14" - 16"	S2@.254	LF	17.20	8.98	—	26.18
18" - 20"	S2@.285	LF	19.40	10.10	—	29.50
22" - 24"	S2@.316	LF	21.50	11.20	—	32.70
26" - 28"	S2@.346	LF	23.30	12.20	—	35.50
30"	S2@.370	LF	25.10	13.10	—	38.20

Description	Craft@Hrs	Unit	Material $	Labor $	Equipment $	Total $

Galvanized steel 20" rectangular duct 24 gauge

Description	Craft@Hrs	Unit	Material $	Labor $	Equipment $	Total $
6"	S2@.200	LF	13.60	7.07	—	20.67
8"	S2@.216	LF	14.70	7.64	—	22.34
10"	S2@.231	LF	15.70	8.17	—	23.87
12"	S2@.247	LF	16.70	8.73	—	25.43
14" - 16"	S2@.269	LF	18.30	9.51	—	27.81
18" - 20"	S2@.300	LF	20.20	10.60	—	30.80
22" - 24"	S2@.331	LF	22.60	11.70	—	34.30
26" - 28"	S2@.362	LF	24.60	12.80	—	37.40
30"	S2@.385	LF	26.10	13.60	—	39.70

Galvanized steel 22" rectangular duct 24 gauge

Description	Craft@Hrs	Unit	Material $	Labor $	Equipment $	Total $
6"	S2@.216	LF	14.70	7.64	—	22.34
8"	S2@.231	LF	15.70	8.17	—	23.87
10"	S2@.247	LF	16.70	8.73	—	25.43
12"	S2@.262	LF	17.80	9.26	—	27.06
14" - 16"	S2@.285	LF	19.40	10.10	—	29.50
18" - 20"	S2@.316	LF	21.50	11.20	—	32.70
22" - 24"	S2@.346	LF	23.30	12.20	—	35.50
26" - 28"	S2@.377	LF	25.90	13.30	—	39.20
30"	S2@.400	LF	27.20	14.10	—	41.30

Galvanized steel 24" rectangular duct 24 gauge

Description	Craft@Hrs	Unit	Material $	Labor $	Equipment $	Total $
6"	S2@.231	LF	15.70	8.17	—	23.87
8"	S2@.247	LF	16.70	8.73	—	25.43
10"	S2@.262	LF	17.80	9.26	—	27.06
12"	S2@.278	LF	18.70	9.83	—	28.53
14" - 16"	S2@.300	LF	20.20	10.60	—	30.80
18" - 20"	S2@.331	LF	22.60	11.70	—	34.30
22" - 24"	S2@.362	LF	24.60	12.80	—	37.40
26" - 28"	S2@.392	LF	26.50	13.90	—	40.40
30"	S2@.415	LF	28.20	14.70	—	42.90

Galvanized steel 26" rectangular duct 24 gauge

Description	Craft@Hrs	Unit	Material $	Labor $	Equipment $	Total $
6"	S2@.247	LF	16.70	8.73	—	25.43
8"	S2@.262	LF	17.80	9.26	—	27.06
10"	S2@.278	LF	18.70	9.83	—	28.53
12"	S2@.292	LF	19.80	10.30	—	30.10
14" - 16"	S2@.316	LF	21.50	11.20	—	32.70
18" - 20"	S2@.346	LF	23.30	12.20	—	35.50
22" - 24"	S2@.377	LF	25.90	13.30	—	39.20
26" - 28"	S2@.407	LF	27.60	14.40	—	42.00
30"	S2@.431	LF	29.00	15.20	—	44.20

Galvanized Steel Rectangular Ductwork

Description	Craft@Hrs	Unit	Material $	Labor $	Equipment $	Total $

Galvanized steel 28" rectangular duct 24 gauge

Description	Craft@Hrs	Unit	Material $	Labor $	Equipment $	Total $
6"	S2@.262	LF	17.80	9.26	—	27.06
8"	S2@.278	LF	18.70	9.83	—	28.53
10"	S2@.292	LF	19.80	10.30	—	30.10
12"	S2@.308	LF	20.90	10.90	—	31.80
14" - 16"	S2@.331	LF	22.60	11.70	—	34.30
18" - 20"	S2@.362	LF	24.60	12.80	—	37.40
22" - 24"	S2@.392	LF	26.50	13.90	—	40.40
26" - 28"	S2@.423	LF	28.70	15.00	—	43.70
30"	S2@.448	LF	30.50	15.80	—	46.30

Galvanized steel 30" rectangular duct 24 gauge

Description	Craft@Hrs	Unit	Material $	Labor $	Equipment $	Total $
6"	S2@.278	LF	18.70	9.83	—	28.53
8"	S2@.292	LF	19.80	10.30	—	30.10
10"	S2@.308	LF	20.90	10.90	—	31.80
12"	S2@.323	LF	21.90	11.40	—	33.30
14" - 16"	S2@.346	LF	23.30	12.20	—	35.50
18" - 20"	S2@.377	LF	25.90	13.30	—	39.20
22" - 24"	S2@.407	LF	27.60	14.40	—	42.00
26" - 28"	S2@.440	LF	29.70	15.60	—	45.30
30"	S2@.465	LF	31.40	16.40	—	47.80

Costs for Galvanized 22 Gauge Rectangular Duct. Costs per linear foot of duct, based on quantities less than 1,000 pounds of prefabricated duct. Deduct 15% from the material cost for over 1,000 pounds. Add 50% to the material cost for lined duct. These costs include delivery, typical bracing, cleats, scrap, hangers, end closures.

Description	Craft@Hrs	Unit	Material $	Labor $	Equipment $	Total $

Galvanized steel 32" rectangular duct 22 gauge

Description	Craft@Hrs	Unit	Material $	Labor $	Equipment $	Total $
8"	S2@.374	LF	25.40	13.20	—	38.60
10"	S2@.393	LF	26.50	13.90	—	40.40
12"	S2@.412	LF	27.80	14.60	—	42.40
14" - 16"	S2@.442	LF	30.00	15.60	—	45.60
18" - 20"	S2@.478	LF	32.60	16.90	—	49.50
22" - 24"	S2@.517	LF	35.00	18.30	—	53.30
26" - 28"	S2@.553	LF	37.50	19.60	—	57.10
30" - 32"	S2@.589	LF	40.20	20.80	—	61.00
34" - 36"	S2@.628	LF	42.50	22.20	—	64.70
38" - 40"	S2@.664	LF	45.20	23.50	—	68.70
42" - 44"	S2@.702	LF	47.40	24.80	—	72.20
46" - 48"	S2@.739	LF	50.10	26.10	—	76.20
50" - 52"	S2@.777	LF	52.50	27.50	—	80.00
54"	S2@.807	LF	54.50	28.50	—	83.00

Description	Craft@Hrs	Unit	Material $	Labor $	Equipment $	Total $

Galvanized steel 34" rectangular duct 22 gauge

Description	Craft@Hrs	Unit	Material $	Labor $	Equipment $	Total $
8"	S2@.393	LF	26.50	13.90	—	40.40
10"	S2@.412	LF	27.80	14.60	—	42.40
12"	S2@.431	LF	29.00	15.20	—	44.20
14" - 16"	S2@.461	LF	31.10	16.30	—	47.40
18" - 20"	S2@.497	LF	34.20	17.60	—	51.80
22" - 24"	S2@.534	LF	36.00	18.90	—	54.90
26" - 28"	S2@.572	LF	38.90	20.20	—	59.10
30" - 32"	S2@.608	LF	41.30	21.50	—	62.80
34" - 36"	S2@.645	LF	44.00	22.80	—	66.80
38" - 40"	S2@.683	LF	46.40	24.20	—	70.60
42" - 44"	S2@.722	LF	49.10	25.50	—	74.60
46" - 48"	S2@.758	LF	51.60	26.80	—	78.40
50" - 52"	S2@.796	LF	53.80	28.10	—	81.90
54"	S2@.824	LF	56.00	29.10	—	85.10

Galvanized steel 36" rectangular duct 22 gauge

Description	Craft@Hrs	Unit	Material $	Labor $	Equipment $	Total $
8"	S2@.412	LF	27.80	14.60	—	42.40
10"	S2@.431	LF	29.00	15.20	—	44.20
12"	S2@.453	LF	30.00	16.00	—	46.80
14" - 16"	S2@.478	LF	32.60	16.90	—	49.50
18" - 20"	S2@.517	LF	35.00	18.30	—	53.30
22" - 24"	S2@.553	LF	37.50	19.60	—	57.10
26" - 28"	S2@.589	LF	40.20	20.80	—	61.00
30" - 32"	S2@.628	LF	42.50	22.20	—	64.70
34" - 36"	S2@.664	LF	45.20	23.50	—	68.70
38" - 40"	S2@.702	LF	47.40	24.80	—	72.20
42" - 44"	S2@.739	LF	50.10	26.10	—	76.20
46" - 48"	S2@.777	LF	52.50	27.50	—	80.00
50" - 52"	S2@.816	LF	54.80	28.90	—	83.70
54"	S2@.841	LF	57.40	29.70	—	87.10

Galvanized steel 38" rectangular duct 22 gauge

Description	Craft@Hrs	Unit	Material $	Labor $	Equipment $	Total $
8"	S2@.431	LF	29.00	15.20	—	44.20
10"	S2@.453	LF	30.80	16.00	—	46.80
12"	S2@.470	LF	31.70	16.60	—	48.30
14" - 16"	S2@.497	LF	34.20	17.60	—	51.80
18" - 20"	S2@.534	LF	36.00	18.90	—	54.90
22" - 24"	S2@.572	LF	38.90	20.20	—	59.10
26" - 28"	S2@.608	LF	41.30	21.50	—	62.80
30" - 32"	S2@.645	LF	44.00	22.80	—	66.80
34" - 36"	S2@.683	LF	46.40	24.20	—	70.60
38" - 40"	S2@.722	LF	49.10	25.50	—	74.60
42" - 44"	S2@.758	LF	51.60	26.80	—	78.40
46" - 48"	S2@.796	LF	53.80	28.10	—	81.90
50" - 52"	S2@.833	LF	56.60	29.50	—	86.10
54"	S2@.863	LF	58.40	30.50	—	88.90

Galvanized Steel Rectangular Ductwork

Description	Craft@Hrs	Unit	Material $	Labor $	Equipment $	Total $

Galvanized steel 40" rectangular duct 22 gauge

Description	Craft@Hrs	Unit	Material $	Labor $	Equipment $	Total $
8"	S2@.453	LF	30.80	16.00	—	46.80
10"	S2@.470	LF	31.70	16.60	—	48.30
12"	S2@.487	LF	33.30	17.20	—	50.50
14" - 16"	S2@.517	LF	35.00	18.30	—	53.30
18" - 20"	S2@.553	LF	37.50	19.60	—	57.10
22" - 24"	S2@.589	LF	40.20	20.80	—	61.00
26" - 28"	S2@.628	LF	42.50	22.20	—	64.70
30" - 32"	S2@.664	LF	45.20	23.50	—	68.70
34" - 36"	S2@.702	LF	47.40	24.80	—	72.20
38" - 40"	S2@.739	LF	50.10	26.10	—	76.20
42" - 44"	S2@.777	LF	52.50	27.50	—	80.00
46" - 48"	S2@.816	LF	54.80	28.90	—	83.70
50" - 52"	S2@.852	LF	58.00	30.10	—	88.10
54"	S2@.880	LF	59.40	31.10	—	90.50

Galvanized steel 42" rectangular duct 22 gauge

Description	Craft@Hrs	Unit	Material $	Labor $	Equipment $	Total $
8"	S2@.470	LF	31.70	16.60	—	48.30
10"	S2@.487	LF	33.30	17.20	—	50.50
12"	S2@.508	LF	34.60	18.00	—	52.60
14" - 16"	S2@.534	LF	36.00	18.90	—	54.90
18" - 20"	S2@.572	LF	38.90	20.20	—	59.10
22" - 24"	S2@.611	LF	41.30	21.60	—	62.90
26" - 28"	S2@.645	LF	44.00	22.80	—	66.80
30" - 32"	S2@.683	LF	46.40	24.20	—	70.60
34" - 36"	S2@.722	LF	49.10	25.50	—	74.60
38" - 40"	S2@.758	LF	51.60	26.80	—	78.40
42" - 44"	S2@.798	LF	53.80	28.20	—	82.00
46" - 48"	S2@.833	LF	56.60	29.50	—	86.10
50" - 52"	S2@.871	LF	59.10	30.80	—	89.90
54"	S2@.901	LF	60.80	31.90	—	92.70

Galvanized steel 44" rectangular duct 22 gauge

Description	Craft@Hrs	Unit	Material $	Labor $	Equipment $	Total $
8"	S2@.487	LF	33.20	17.20	—	50.40
10"	S2@.508	LF	34.30	18.00	—	52.30
12"	S2@.525	LF	35.40	18.60	—	54.00
14" - 16"	S2@.555	LF	37.10	19.60	—	56.70
18" - 20"	S2@.589	LF	40.00	20.80	—	60.80
22" - 24"	S2@.628	LF	42.10	22.20	—	64.30
26" - 28"	S2@.666	LF	44.90	23.50	—	68.40
30" - 32"	S2@.705	LF	47.30	24.90	—	72.20
34" - 36"	S2@.739	LF	49.90	26.10	—	76.00
38" - 40"	S2@.777	LF	52.30	27.50	—	79.80
42" - 44"	S2@.816	LF	54.70	28.90	—	83.60
46" - 48"	S2@.852	LF	57.60	30.10	—	87.70
50" - 52"	S2@.890	LF	59.80	31.50	—	91.30
54"	S2@.918	LF	61.80	32.50	—	94.30

Description	Craft@Hrs	Unit	Material $	Labor $	Equipment $	Total $

Galvanized steel 46" rectangular duct 22 gauge

Description	Craft@Hrs	Unit	Material $	Labor $	Equipment $	Total $
8"	S2@.508	LF	34.60	18.00	—	52.60
10"	S2@.525	LF	35.40	18.60	—	54.00
12"	S2@.542	LF	36.80	19.20	—	56.00
14" - 16"	S2@.572	LF	38.90	20.20	—	59.10
18" - 20"	S2@.611	LF	41.30	21.60	—	62.90
22" - 24"	S2@.649	LF	44.00	22.90	—	66.90
26" - 28"	S2@.683	LF	46.40	24.20	—	70.60
30" - 32"	S2@.722	LF	49.10	25.50	—	74.60
34" - 36"	S2@.758	LF	51.60	26.80	—	78.40
38" - 40"	S2@.796	LF	53.80	28.10	—	81.90
42" - 44"	S2@.833	LF	56.60	29.50	—	86.10
46" - 48"	S2@.871	LF	59.10	30.80	—	89.90
50" - 52"	S2@.909	LF	61.50	32.10	—	93.60
54"	S2@.935	LF	63.30	33.10	—	96.40

Galvanized steel 48" rectangular duct 22 gauge

Description	Craft@Hrs	Unit	Material $	Labor $	Equipment $	Total $
8"	S2@.525	LF	35.40	18.60	—	54.00
10"	S2@.542	LF	36.80	19.20	—	56.00
12"	S2@.564	LF	38.20	19.90	—	58.10
14" - 16"	S2@.589	LF	40.20	20.80	—	61.00
18" - 20"	S2@.628	LF	42.50	22.20	—	64.70
22" - 24"	S2@.664	LF	45.20	23.50	—	68.70
26" - 28"	S2@.702	LF	47.40	24.80	—	72.20
30" - 32"	S2@.739	LF	50.10	26.10	—	76.20
34" - 36"	S2@.777	LF	52.50	27.50	—	80.00
38" - 40"	S2@.816	LF	54.80	28.90	—	83.70
42" - 44"	S2@.852	LF	58.00	30.10	—	88.10
46" - 48"	S2@.890	LF	60.20	31.50	—	91.70
50" - 52"	S2@.927	LF	63.00	32.80	—	95.80
54"	S2@.956	LF	64.40	33.80	—	98.20

Galvanized steel 50" rectangular duct 22 gauge

Description	Craft@Hrs	Unit	Material $	Labor $	Equipment $	Total $
8"	S2@.542	LF	36.80	19.20	—	56.00
10"	S2@.564	LF	38.20	19.90	—	58.10
12"	S2@.581	LF	39.40	20.50	—	59.90
14" - 16"	S2@.611	LF	41.30	21.60	—	62.90
18" - 20"	S2@.645	LF	44.00	22.80	—	66.80
22" - 24"	S2@.683	LF	46.40	24.20	—	70.60
26" - 28"	S2@.722	LF	49.10	25.50	—	74.60
30" - 32"	S2@.758	LF	51.60	26.80	—	78.40
34" - 36"	S2@.798	LF	53.80	28.20	—	82.00
38" - 40"	S2@.833	LF	56.60	29.50	—	86.10
42" - 44"	S2@.871	LF	59.10	30.80	—	89.90
46" - 48"	S2@.909	LF	61.50	32.10	—	93.60
50" - 52"	S2@.946	LF	63.90	33.50	—	97.40
54"	S2@.974	LF	65.70	34.40	—	100.10

Galvanized Steel Rectangular Ductwork

Description	Craft@Hrs	Unit	Material $	Labor $	Equipment $	Total $

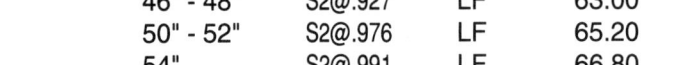

Galvanized steel 52" rectangular duct 22 gauge

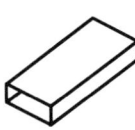

Description	Craft@Hrs	Unit	Material $	Labor $	Equipment $	Total $
8"	S2@.564	LF	38.20	19.90	—	58.10
10"	S2@.581	LF	39.40	20.50	—	59.90
12"	S2@.598	LF	40.70	21.10	—	61.80
14" - 16"	S2@.628	LF	42.50	22.20	—	64.70
18" - 20"	S2@.662	LF	45.20	23.40	—	68.60
22" - 24"	S2@.700	LF	47.40	24.80	—	72.20
26" - 28"	S2@.739	LF	50.10	26.10	—	76.20
30" - 32"	S2@.777	LF	52.50	27.50	—	80.00
34" - 36"	S2@.816	LF	54.80	28.90	—	83.70
38" - 40"	S2@.852	LF	58.00	30.10	—	88.10
42" - 44"	S2@.890	LF	60.20	31.50	—	91.70
46" - 48"	S2@.927	LF	63.00	32.80	—	95.80
50" - 52"	S2@.976	LF	65.20	34.50	—	99.70
54"	S2@.991	LF	66.80	35.00	—	101.80

Galvanized steel 54" rectangular duct 22 gauge

Description	Craft@Hrs	Unit	Material $	Labor $	Equipment $	Total $
8"	S2@.581	LF	39.40	20.50	—	59.90
10"	S2@.598	LF	40.70	21.10	—	61.80
12"	S2@.619	LF	41.80	21.90	—	63.70
14" - 16"	S2@.645	LF	44.00	22.80	—	66.80
18" - 20"	S2@.683	LF	46.40	24.20	—	70.60
22" - 24"	S2@.722	LF	49.10	25.50	—	74.60
26" - 28"	S2@.758	LF	51.60	26.80	—	78.40
30" - 32"	S2@.796	LF	53.80	28.10	—	81.90
34" - 36"	S2@.833	LF	56.60	29.50	—	86.10
38" - 40"	S2@.871	LF	59.10	30.80	—	89.90
42" - 44"	S2@.909	LF	61.50	32.10	—	93.60
46" - 48"	S2@.946	LF	63.90	33.50	—	97.40
50" - 52"	S2@.982	LF	66.10	34.70	—	100.80
54"	S2@1.01	LF	68.70	35.70	—	104.40

Costs for Galvanized 20 Gauge Rectangular Duct. Costs per linear foot of duct, based on quantities less than 1,000 pounds of prefabricated duct. Deduct 15% from the material cost for over 1,000 pounds. Add 50% to the material cost for lined duct. These costs include delivery, typical bracing, cleats, scrap, hangers, end closures, sealants and miscellaneous hardware.

Description	Craft@Hrs	Unit	Material $	Labor $	Equipment $	Total $

Galvanized steel 56" rectangular duct 20 gauge

Description	Craft@Hrs	Unit	Material $	Labor $	Equipment $	Total $
18" - 20"	S2@.828	LF	56.60	29.30	—	85.90
22" - 24"	S2@.871	LF	59.40	30.80	—	90.20
26" - 28"	S2@.916	LF	62.30	32.40	—	94.70
30" - 32"	S2@.959	LF	65.50	33.90	—	99.40
34" - 36"	S2@1.00	LF	68.70	35.40	—	104.10
38" - 40"	S2@1.05	LF	71.50	37.10	—	108.60
42" - 44"	S2@1.09	LF	74.70	38.50	—	113.20
46" - 48"	S2@1.14	LF	77.60	40.30	—	117.90
50" - 52"	S2@1.18	LF	80.70	41.70	—	122.40
54" - 56"	S2@1.23	LF	83.50	43.50	—	127.00
58" - 60"	S2@1.27	LF	86.70	44.90	—	131.60
62" - 64"	S2@1.31	LF	89.90	46.30	—	136.20
66" - 68"	S2@1.36	LF	92.60	48.10	—	140.70
70" - 72"	S2@1.40	LF	96.00	49.50	—	145.50

Galvanized steel 58" rectangular duct 20 gauge

Description	Craft@Hrs	Unit	Material $	Labor $	Equipment $	Total $
18" - 20"	S2@.850	LF	58.10	30.10	—	88.20
22" - 24"	S2@.895	LF	61.10	31.60	—	92.70
26" - 28"	S2@.937	LF	63.90	33.10	—	97.00
30" - 32"	S2@.980	LF	67.10	34.70	—	101.80
34" - 36"	S2@1.02	LF	70.00	36.10	—	106.10
38" - 40"	S2@1.07	LF	72.90	37.80	—	110.70
42" - 44"	S2@1.11	LF	76.10	39.20	—	115.30
46" - 48"	S2@1.16	LF	78.90	41.00	—	119.90
50" - 52"	S2@1.02	LF	82.40	36.10	—	118.50
54" - 56"	S2@1.25	LF	85.30	44.20	—	129.50
58" - 60"	S2@1.29	LF	87.70	45.60	—	133.30
62" - 64"	S2@1.33	LF	91.40	47.00	—	138.40
66" - 68"	S2@1.38	LF	94.40	48.80	—	143.20
70" - 72"	S2@1.42	LF	97.50	50.20	—	147.70

Galvanized steel 60" rectangular duct 20 gauge

Description	Craft@Hrs	Unit	Material $	Labor $	Equipment $	Total $
18" - 20"	S2@.871	LF	59.40	30.80	—	90.20
22" - 24"	S2@.916	LF	62.30	32.40	—	94.70
26" - 28"	S2@.959	LF	65.50	33.90	—	99.40
30" - 32"	S2@1.00	LF	68.70	35.40	—	104.10
34" - 36"	S2@1.05	LF	71.50	37.10	—	108.60
38" - 40"	S2@1.09	LF	74.70	38.50	—	113.20
42" - 44"	S2@1.14	LF	77.60	40.30	—	117.90
46" - 48"	S2@1.18	LF	80.70	41.70	—	122.40
50" - 52"	S2@1.23	LF	83.50	43.50	—	127.00
54" - 56"	S2@1.27	LF	86.70	44.90	—	131.60
58" - 60"	S2@1.31	LF	89.90	46.30	—	136.20
62" - 64"	S2@1.36	LF	92.60	48.10	—	140.70
66" - 68"	S2@1.40	LF	96.00	49.50	—	145.50
70" - 72"	S2@1.45	LF	99.00	51.30	—	150.30

Galvanized Steel Rectangular Ductwork

Description	Craft@Hrs	Unit	Material $	Labor $	Equipment $	Total $

Galvanized steel 62" rectangular duct 20 gauge

Description	Craft@Hrs	Unit	Material $	Labor $	Equipment $	Total $
18" - 20"	S2@.895	LF	61.10	31.60	—	92.70
22" - 24"	S2@.937	LF	63.90	33.10	—	97.00
26" - 28"	S2@.980	LF	67.10	34.70	—	101.80
30" - 32"	S2@1.02	LF	70.00	36.10	—	106.10
34" - 36"	S2@1.07	LF	72.90	37.80	—	110.70
38" - 40"	S2@1.11	LF	76.10	39.20	—	115.30
42" - 44"	S2@1.16	LF	78.90	41.00	—	119.90
46" - 48"	S2@1.20	LF	82.40	42.40	—	124.80
50" - 52"	S2@1.25	LF	85.30	44.20	—	129.50
54" - 56"	S2@1.29	LF	87.70	45.60	—	133.30
58" - 60"	S2@1.33	LF	91.40	47.00	—	138.40
62" - 64"	S2@1.38	LF	94.40	48.80	—	143.20
66" - 68"	S2@1.42	LF	97.50	50.20	—	147.70
70" - 72"	S2@1.47	LF	100.00	52.00	—	152.00

Galvanized steel 64" rectangular duct 20 gauge

Description	Craft@Hrs	Unit	Material $	Labor $	Equipment $	Total $
18" - 20"	S2@.916	LF	62.30	32.40	—	94.70
22" - 24"	S2@.959	LF	65.50	33.90	—	99.40
26" - 28"	S2@1.00	LF	68.70	35.40	—	104.10
30" - 32"	S2@1.05	LF	71.50	37.10	—	108.60
34" - 36"	S2@1.09	LF	74.70	38.50	—	113.20
38" - 40"	S2@1.14	LF	77.60	40.30	—	117.90
42" - 44"	S2@1.18	LF	80.70	41.70	—	122.40
46" - 48"	S2@1.23	LF	83.50	43.50	—	127.00
50" - 52"	S2@1.27	LF	86.70	44.90	—	131.60
54" - 56"	S2@1.31	LF	89.90	46.30	—	136.20
58" - 60"	S2@1.36	LF	92.60	48.10	—	140.70
62" - 64"	S2@1.40	LF	96.00	49.50	—	145.50
66" - 68"	S2@1.45	LF	99.00	51.30	—	150.30
70" - 72"	S2@1.49	LF	102.00	52.70	—	154.70

Galvanized steel 66" rectangular duct 20 gauge

Description	Craft@Hrs	Unit	Material $	Labor $	Equipment $	Total $
18" - 20"	S2@.937	LF	63.90	33.10	—	97.00
22" - 24"	S2@.980	LF	67.10	34.70	—	101.80
26" - 28"	S2@1.02	LF	70.00	36.10	—	106.10
30" - 32"	S2@1.07	LF	72.90	37.80	—	110.70
34" - 36"	S2@1.11	LF	76.10	39.20	—	115.30
38" - 40"	S2@1.16	LF	78.90	41.00	—	119.90
42" - 44"	S2@1.20	LF	82.40	42.40	—	124.80
46" - 48"	S2@1.25	LF	85.30	44.20	—	129.50
50" - 52"	S2@1.29	LF	87.70	45.60	—	133.30
54" - 56"	S2@1.33	LF	91.40	47.00	—	138.40
58" - 60"	S2@1.38	LF	94.40	48.80	—	143.20
62" - 64"	S2@1.42	LF	97.50	50.20	—	147.70
66" - 68"	S2@1.47	LF	100.00	52.00	—	152.00
70" - 72"	S2@1.51	LF	103.00	53.40	—	156.40

Description	Craft@Hrs	Unit	Material $	Labor $	Equipment $	Total $

Galvanized steel 68" rectangular duct 20 gauge

Description	Craft@Hrs	Unit	Material $	Labor $	Equipment $	Total $
18" - 20"	S2@.959	LF	65.50	33.90	—	99.40
22" - 24"	S2@1.00	LF	68.70	35.40	—	104.10
26" - 28"	S2@1.05	LF	71.50	37.10	—	108.60
30" - 32"	S2@1.09	LF	74.70	38.50	—	113.20
34" - 36"	S2@1.14	LF	77.60	40.30	—	117.90
38" - 40"	S2@1.18	LF	80.70	41.70	—	122.40
42" - 44"	S2@1.23	LF	83.50	43.50	—	127.00
46" - 48"	S2@1.27	LF	86.70	44.90	—	131.60
50" - 52"	S2@1.31	LF	89.90	46.30	—	136.20
54" - 56"	S2@1.36	LF	92.60	48.10	—	140.70
58" - 60"	S2@1.40	LF	96.00	49.50	—	145.50
62" - 64"	S2@1.45	LF	99.00	51.30	—	150.30
66" - 68"	S2@1.49	LF	102.00	52.70	—	154.70
70" - 72"	S2@1.53	LF	105.00	54.10	—	159.10

Galvanized steel 70" rectangular duct 20 gauge

Description	Craft@Hrs	Unit	Material $	Labor $	Equipment $	Total $
18" - 20"	S2@.980	LF	67.10	34.70	—	101.80
22" - 24"	S2@1.02	LF	70.00	36.10	—	106.10
26" - 28"	S2@1.07	LF	72.90	37.80	—	110.70
30" - 32"	S2@1.11	LF	76.10	39.20	—	115.30
34" - 36"	S2@1.16	LF	78.90	41.00	—	119.90
38" - 40"	S2@1.20	LF	82.40	42.40	—	124.80
42" - 44"	S2@1.25	LF	85.30	44.20	—	129.50
46" - 48"	S2@1.29	LF	87.70	45.60	—	133.30
50" - 52"	S2@1.33	LF	91.40	47.00	—	138.40
54" - 56"	S2@1.38	LF	94.40	48.80	—	143.20
58" - 60"	S2@1.42	LF	97.50	50.20	—	147.70
62" - 64"	S2@1.47	LF	100.00	52.00	—	152.00
66" - 68"	S2@1.51	LF	103.00	53.40	—	156.40
70" - 72"	S2@1.55	LF	106.00	54.80	—	160.80

Galvanized steel 72" rectangular duct 20 gauge

Description	Craft@Hrs	Unit	Material $	Labor $	Equipment $	Total $
18" - 20"	S2@1.00	LF	68.70	35.40	—	104.10
22" - 24"	S2@1.05	LF	71.50	37.10	—	108.60
26" - 28"	S2@1.09	LF	74.70	38.50	—	113.20
30" - 32"	S2@1.14	LF	77.60	40.30	—	117.90
34" - 36"	S2@1.18	LF	80.70	41.70	—	122.40
38" - 40"	S2@1.23	LF	83.50	43.50	—	127.00
42" - 44"	S2@1.27	LF	86.70	44.90	—	131.60
46" - 48"	S2@1.31	LF	89.90	46.30	—	136.20
50" - 52"	S2@1.36	LF	92.60	48.10	—	140.70
54" - 56"	S2@1.40	LF	96.00	49.50	—	145.50
58" - 60"	S2@1.45	LF	99.00	51.30	—	150.30
62" - 64"	S2@1.49	LF	102.00	52.70	—	154.70
66" - 68"	S2@1.53	LF	105.00	54.10	—	159.10
70" - 72"	S2@1.58	LF	107.00	55.90	—	162.90

Galvanized Steel Rectangular Elbows

Costs for Galvanized Rectangular 90 Degree Elbows. Costs per elbow, based on quantities less than 1,000 pounds. Deduct 15% from the material cost for over 1,000 pounds. Add 50% to the material cost for lined duct. These costs include delivery, typical bracing, cleats, scrap, hangers, end closures, sealants and miscellaneous hardware. For turning vanes, please see page 337.

Description	Craft@Hrs	Unit	Material $	Labor $	Equipment $	Total $

Galvanized steel 4" rectangular duct 90 degree elbow

Description	Craft@Hrs	Unit	Material $	Labor $	Equipment $	Total $
4"	S2@.042	Ea	5.74	1.49	—	7.23
6"	S2@.058	Ea	7.88	2.05	—	9.93
8"	S2@.078	Ea	10.60	2.76	—	13.36
10"	S2@.104	Ea	14.10	3.68	—	17.78
12"	S2@.132	Ea	18.00	4.67	—	22.67
14" - 16"	S2@.236	Ea	32.10	8.34	—	40.44
18" - 20"	S2@.339	Ea	46.40	12.00	—	58.40
22" - 24"	S2@.468	Ea	63.30	16.50	—	79.80
26" - 28"	S2@.615	Ea	83.10	21.70	—	104.80
30" - 32"	S2@.854	Ea	117.00	30.20	—	147.20
34" - 36"	S2@1.19	Ea	162.00	42.10	—	204.10
38" - 40"	S2@1.45	Ea	197.00	51.30	—	248.30
42" - 44"	S2@1.73	Ea	233.00	61.20	—	294.20
46" - 48"	S2@2.04	Ea	276.00	72.10	—	348.10
50"	S2@2.28	Ea	311.00	80.60	—	391.60

Galvanized steel 6" rectangular duct 90 degree elbow

Description	Craft@Hrs	Unit	Material $	Labor $	Equipment $	Total $
4"	S2@.058	Ea	7.88	2.05	—	9.93
6"	S2@.078	Ea	10.60	2.76	—	13.36
8"	S2@.104	Ea	14.10	3.68	—	17.78
10"	S2@.132	Ea	18.00	4.67	—	22.67
12"	S2@.148	Ea	20.10	5.23	—	25.33
14" - 16"	S2@.266	Ea	36.40	9.41	—	45.81
18" - 20"	S2@.370	Ea	50.10	13.10	—	63.20
22" - 24"	S2@.506	Ea	68.90	17.90	—	86.80
26" - 28"	S2@.655	Ea	88.90	23.20	—	112.10
30" - 32"	S2@.905	Ea	122.00	32.00	—	154.00
34" - 36"	S2@1.24	Ea	169.00	43.80	—	212.80
38" - 40"	S2@1.49	Ea	202.00	52.70	—	254.70
42" - 44"	S2@1.78	Ea	241.00	62.90	—	303.90
46" - 48"	S2@2.09	Ea	284.00	73.90	—	357.90
50"	S2@2.34	Ea	319.00	82.70	—	401.70

Description	Craft@Hrs	Unit	Material $	Labor $	Equipment $	Total $

Galvanized steel 8" rectangular duct 90 degree elbow

Description	Craft@Hrs	Unit	Material $	Labor $	Equipment $	Total $
4"	S2@.078	Ea	10.60	2.76	—	13.36
6"	S2@.104	Ea	14.10	3.68	—	17.78
8"	S2@.132	Ea	18.00	4.67	—	22.67
10"	S2@.148	Ea	20.10	5.23	—	25.33
12"	S2@.170	Ea	23.20	6.01	—	29.21
14" - 16"	S2@.300	Ea	40.70	10.60	—	51.30
18" - 20"	S2@.403	Ea	54.70	14.30	—	69.00
22" - 24"	S2@.549	Ea	74.60	19.40	—	94.00
26" - 28"	S2@.694	Ea	94.70	24.50	—	119.20
30" - 32"	S2@.950	Ea	131.00	33.60	—	164.60
34" - 36"	S2@1.28	Ea	174.00	45.30	—	219.30
38" - 40"	S2@1.54	Ea	211.00	54.50	—	265.50
42" - 44"	S2@1.83	Ea	248.00	64.70	—	312.70
46" - 48"	S2@2.13	Ea	290.00	75.30	—	365.30
50"	S2@2.39	Ea	330.00	84.50	—	414.50

Galvanized steel 10" rectangular duct 90 degree elbow

Description	Craft@Hrs	Unit	Material $	Labor $	Equipment $	Total $
4"	S2@.104	Ea	14.10	3.68	—	17.78
6"	S2@.132	Ea	18.00	4.67	—	22.67
8"	S2@.148	Ea	20.10	5.23	—	25.33
10"	S2@.170	Ea	23.20	6.01	—	29.21
12"	S2@.194	Ea	26.20	6.86	—	33.06
14" - 16"	S2@.336	Ea	45.30	11.90	—	57.20
18" - 20"	S2@.438	Ea	59.40	15.50	—	74.90
22" - 24"	S2@.587	Ea	79.90	20.80	—	100.70
26" - 28"	S2@.737	Ea	99.70	26.10	—	125.80
30" - 32"	S2@.997	Ea	136.00	35.30	—	171.30
34" - 36"	S2@1.33	Ea	181.00	47.00	—	228.00
38" - 40"	S2@1.59	Ea	216.00	56.20	—	272.20
42" - 44"	S2@1.87	Ea	254.00	66.10	—	320.10
46" - 48"	S2@2.18	Ea	297.00	77.10	—	374.10
50"	S2@2.43	Ea	333.00	85.90	—	418.90

Galvanized steel 12" rectangular duct 90 degree elbow

Description	Craft@Hrs	Unit	Material $	Labor $	Equipment $	Total $
4"	S2@.132	Ea	18.00	4.67	—	22.67
6"	S2@.148	Ea	20.10	5.23	—	25.33
8"	S2@.170	Ea	23.20	6.01	—	29.21
10"	S2@.194	Ea	26.20	6.86	—	33.06
12"	S2@.224	Ea	30.50	7.92	—	38.42
18" - 20"	S2@.478	Ea	64.40	16.90	—	81.30
22" - 24"	S2@.628	Ea	85.30	22.20	—	107.50
26" - 28"	S2@.775	Ea	106.00	27.40	—	133.40
30" - 32"	S2@1.04	Ea	141.00	36.80	—	177.80
34" - 36"	S2@1.38	Ea	187.00	48.80	—	235.80
38" - 40"	S2@1.64	Ea	221.00	58.00	—	279.00
42" - 44"	S2@1.92	Ea	261.00	67.90	—	328.90
46" - 48"	S2@2.23	Ea	305.00	78.90	—	383.90
50"	S2@2.48	Ea	336.00	87.70	—	423.70

Galvanized Steel Rectangular Elbows

Description	Craft@Hrs	Unit	Material $	Labor $	Equipment $	Total $

Galvanized steel 14" rectangular duct 90 degree elbow

Description	Craft@Hrs	Unit	Material $	Labor $	Equipment $	Total $
4"	S2@.212	Ea	28.70	7.50	—	36.20
6"	S2@.232	Ea	31.70	8.20	—	39.90
8"	S2@.260	Ea	35.30	9.19	—	44.49
10"	S2@.292	Ea	39.60	10.30	—	49.90
12"	S2@.328	Ea	44.50	11.60	—	56.10
14" - 16"	S2@.415	Ea	56.60	14.70	—	71.30
18" - 20"	S2@.527	Ea	71.50	18.60	—	90.10
22" - 24"	S2@.867	Ea	90.50	30.70	—	121.20
26" - 28"	S2@.816	Ea	110.00	28.90	—	138.90
30" - 32"	S2@1.09	Ea	149.00	38.50	—	187.50
34" - 36"	S2@1.43	Ea	195.00	50.60	—	245.60
38" - 40"	S2@1.69	Ea	229.00	59.80	—	288.80
42" - 44"	S2@1.97	Ea	269.00	69.70	—	338.70
46" - 48"	S2@2.28	Ea	310.00	80.60	—	390.60
50"	S2@2.53	Ea	346.00	89.50	—	435.50

Galvanized steel 16" rectangular duct 90 degree elbow

Description	Craft@Hrs	Unit	Material $	Labor $	Equipment $	Total $
4"	S2@.260	Ea	35.30	9.19	—	44.49
6"	S2@.310	Ea	42.10	11.00	—	53.10
8"	S2@.340	Ea	46.30	12.00	—	58.30
10"	S2@.380	Ea	51.90	13.40	—	65.30
12"	S2@.420	Ea	57.10	14.90	—	72.00
14" - 16"	S2@.480	Ea	65.10	17.00	—	82.10
18" - 20"	S2@.555	Ea	75.40	19.60	—	95.00
22" - 24"	S2@.707	Ea	96.30	25.00	—	121.30
26" - 28"	S2@.856	Ea	116.00	30.30	—	146.30
30" - 32"	S2@1.14	Ea	155.00	40.30	—	195.30
34" - 36"	S2@1.48	Ea	201.00	52.30	—	253.30
38" - 40"	S2@1.73	Ea	236.00	61.20	—	297.20
42" - 44"	S2@2.02	Ea	275.00	71.40	—	346.40
46" - 48"	S2@2.32	Ea	319.00	82.00	—	401.00
50"	S2@2.58	Ea	350.00	91.20	—	441.20

Galvanized steel 18" rectangular duct 90 degree elbow

Description	Craft@Hrs	Unit	Material $	Labor $	Equipment $	Total $
4"	S2@.312	Ea	42.50	11.00	—	53.50
6"	S2@.350	Ea	47.70	12.40	—	60.10
8"	S2@.390	Ea	53.20	13.80	—	67.00
10"	S2@.427	Ea	58.30	15.10	—	73.40
12"	S2@.470	Ea	63.30	16.60	—	79.90
14" - 16"	S2@.519	Ea	70.60	18.40	—	89.00
18" - 20"	S2@.604	Ea	82.20	21.40	—	103.60
22" - 24"	S2@.747	Ea	102.00	26.40	—	128.40
26" - 28"	S2@.895	Ea	121.00	31.60	—	152.60
30" - 32"	S2@1.18	Ea	161.00	41.70	—	202.70
34" - 36"	S2@1.53	Ea	209.00	54.10	—	263.10
38" - 40"	S2@1.78	Ea	241.00	62.90	—	303.90
42" - 44"	S2@2.07	Ea	282.00	73.20	—	355.20
46" - 48"	S2@2.37	Ea	326.00	83.80	—	409.80
50"	S2@2.63	Ea	357.00	93.00	—	450.00

Description	Craft@Hrs	Unit	Material $	Labor $	Equipment $	Total $

Galvanized steel 20" rectangular duct 90 degree elbow

Description	Craft@Hrs	Unit	Material $	Labor $	Equipment $	Total $
4"	S2@.370	Ea	50.10	13.10	—	63.20
6"	S2@.390	Ea	53.20	13.80	—	67.00
8"	S2@.415	Ea	56.60	14.70	—	71.30
10"	S2@.448	Ea	60.90	15.80	—	76.70
12"	S2@.487	Ea	65.80	17.20	—	83.00
14" - 16"	S2@.553	Ea	75.00	19.60	—	94.60
18" - 20"	S2@.658	Ea	89.50	23.30	—	112.80
22" - 24"	S2@.786	Ea	107.00	27.80	—	134.80
26" - 28"	S2@.931	Ea	126.00	32.90	—	158.90
30" - 32"	S2@1.23	Ea	166.00	43.50	—	209.50
34" - 36"	S2@1.58	Ea	215.00	55.90	—	270.90
38" - 40"	S2@1.83	Ea	250.00	64.70	—	314.70
42" - 44"	S2@2.11	Ea	287.00	74.60	—	361.60
46" - 48"	S2@2.42	Ea	332.00	85.60	—	417.60
50"	S2@2.67	Ea	367.00	94.40	—	461.40

Galvanized steel rectangular 22" duct 90 degree elbow

Description	Craft@Hrs	Unit	Material $	Labor $	Equipment $	Total $
22" - 24"	S2@.828	Ea	112.00	29.30	—	141.30
26" - 28"	S2@1.01	Ea	137.00	35.70	—	172.70
30" - 32"	S2@1.31	Ea	178.00	46.30	—	224.30
34" - 36"	S2@1.71	Ea	232.00	60.50	—	292.50
38" - 40"	S2@2.01	Ea	275.00	71.10	—	346.10
42" - 44"	S2@2.31	Ea	311.00	81.70	—	392.70
46" - 48"	S2@2.61	Ea	354.00	92.30	—	446.30
50" - 52"	S2@2.91	Ea	396.00	103.00	—	499.00
54" - 56"	S2@3.51	Ea	477.00	124.00	—	601.00
58" - 60"	S2@4.26	Ea	581.00	151.00	—	732.00
62" - 64"	S2@4.76	Ea	643.00	168.00	—	811.00
66" - 68"	S2@5.25	Ea	714.00	186.00	—	900.00
70" - 72"	S2@5.76	Ea	781.00	204.00	—	985.00

Galvanized steel 24" rectangular duct 90 degree elbow

Description	Craft@Hrs	Unit	Material $	Labor $	Equipment $	Total $
22" - 24"	S2@.922	Ea	125.00	32.60	—	157.60
26" - 28"	S2@1.10	Ea	150.00	38.90	—	188.90
30" - 32"	S2@1.41	Ea	192.00	49.90	—	241.90
34" - 36"	S2@1.81	Ea	246.00	64.00	—	310.00
38" - 40"	S2@2.10	Ea	287.00	74.30	—	361.30
42" - 44"	S2@2.41	Ea	330.00	85.20	—	415.20
46" - 48"	S2@2.71	Ea	368.00	95.80	—	463.80
50" - 52"	S2@3.01	Ea	407.00	106.00	—	513.00
54" - 56"	S2@3.62	Ea	492.00	128.00	—	620.00
58" - 60"	S2@4.38	Ea	594.00	155.00	—	749.00
62" - 64"	S2@4.87	Ea	658.00	172.00	—	830.00
66" - 68"	S2@5.38	Ea	728.00	190.00	—	918.00
70" - 72"	S2@5.87	Ea	799.00	208.00	—	1,007.00

Galvanized Steel Rectangular Elbows

Description	Craft@Hrs	Unit	Material $	Labor $	Equipment $	Total $

Galvanized steel 26" rectangular duct 90 degree elbow

Description	Craft@Hrs	Unit	Material $	Labor $	Equipment $	Total $
22" - 24"	S2@1.01	Ea	137.00	35.70	—	172.70
26" - 28"	S2@1.19	Ea	162.00	42.10	—	204.10
30" - 32"	S2@1.50	Ea	204.00	53.00	—	257.00
34" - 36"	S2@1.92	Ea	261.00	67.90	—	328.90
38" - 40"	S2@2.21	Ea	300.00	78.10	—	378.10
42" - 44"	S2@2.51	Ea	343.00	88.80	—	431.80
46" - 48"	S2@2.81	Ea	381.00	99.40	—	480.40
50" - 52"	S2@3.11	Ea	425.00	110.00	—	535.00
54" - 56"	S2@3.73	Ea	510.00	132.00	—	642.00
58" - 60"	S2@4.50	Ea	611.00	159.00	—	770.00
62" - 64"	S2@5.02	Ea	678.00	178.00	—	856.00
66" - 68"	S2@5.53	Ea	754.00	196.00	—	950.00
70" - 72"	S2@6.04	Ea	822.00	214.00	—	1,036.00

Galvanized steel 28" rectangular duct 90 degree elbow

Description	Craft@Hrs	Unit	Material $	Labor $	Equipment $	Total $
22" - 24"	S2@1.10	Ea	149.00	38.90	—	187.90
26" - 28"	S2@1.28	Ea	174.00	45.30	—	219.30
30" - 32"	S2@1.60	Ea	218.00	56.60	—	274.60
34" - 36"	S2@2.03	Ea	276.00	71.80	—	347.80
38" - 40"	S2@2.31	Ea	311.00	81.70	—	392.70
42" - 44"	S2@2.61	Ea	354.00	92.30	—	446.30
46" - 48"	S2@2.91	Ea	396.00	103.00	—	499.00
50" - 52"	S2@3.21	Ea	440.00	114.00	—	554.00
54" - 56"	S2@3.84	Ea	523.00	136.00	—	659.00
58" - 60"	S2@4.61	Ea	629.00	163.00	—	792.00
62" - 64"	S2@5.15	Ea	693.00	182.00	—	875.00
66" - 68"	S2@5.66	Ea	764.00	200.00	—	964.00
70" - 72"	S2@6.15	Ea	831.00	217.00	—	1,048.00

Galvanized steel 30" rectangular duct 90 degree elbow

Description	Craft@Hrs	Unit	Material $	Labor $	Equipment $	Total $
22" - 24"	S2@1.19	Ea	161.00	42.10	—	203.10
26" - 28"	S2@1.37	Ea	187.00	48.40	—	235.40
30" - 32"	S2@1.70	Ea	230.00	60.10	—	290.10
34" - 36"	S2@2.14	Ea	290.00	75.70	—	365.70
38" - 40"	S2@2.41	Ea	330.00	85.20	—	415.20
42" - 44"	S2@2.71	Ea	368.00	95.80	—	463.80
46" - 48"	S2@3.01	Ea	407.00	106.00	—	513.00
50" - 52"	S2@3.31	Ea	449.00	117.00	—	566.00
54" - 56"	S2@3.95	Ea	535.00	140.00	—	675.00
58" - 60"	S2@4.74	Ea	641.00	168.00	—	809.00
62" - 64"	S2@5.25	Ea	714.00	186.00	—	900.00
66" - 68"	S2@5.79	Ea	782.00	205.00	—	987.00
70" - 72"	S2@6.28	Ea	853.00	222.00	—	1,075.00

Description	Craft@Hrs	Unit	Material $	Labor $	Equipment $	Total $

Galvanized steel 32" rectangular duct 90 degree elbow

Description	Craft@Hrs	Unit	Material $	Labor $	Equipment $	Total $
22" - 24"	S2@1.53	Ea	209.00	54.10	—	263.10
26" - 28"	S2@1.73	Ea	233.00	61.20	—	294.20
30" - 32"	S2@1.93	Ea	262.00	68.20	—	330.20
34" - 36"	S2@2.25	Ea	308.00	79.60	—	387.60
38" - 40"	S2@2.51	Ea	343.00	88.80	—	431.80
42" - 44"	S2@2.81	Ea	381.00	99.40	—	480.40
46" - 48"	S2@3.11	Ea	421.00	110.00	—	531.00
50" - 52"	S2@3.41	Ea	464.00	121.00	—	585.00
54" - 56"	S2@4.06	Ea	548.00	144.00	—	692.00
58" - 60"	S2@4.85	Ea	657.00	171.00	—	828.00
62" - 64"	S2@5.38	Ea	728.00	190.00	—	918.00
66" - 68"	S2@5.89	Ea	800.00	208.00	—	1,008.00
70" - 72"	S2@6.40	Ea	870.00	226.00	—	1,096.00

Galvanized steel 34" rectangular duct 90 degree elbow

Description	Craft@Hrs	Unit	Material $	Labor $	Equipment $	Total $
22" - 24"	S2@1.67	Ea	227.00	59.10	—	286.10
26" - 28"	S2@1.91	Ea	261.00	67.50	—	328.50
30" - 32"	S2@2.12	Ea	287.00	75.00	—	362.00
34" 36"	S2@2.36	Ea	321.00	83.40	—	404.40
38" - 40"	S2@2.60	Ea	354.00	91.90	—	445.90
42" - 44"	S2@2.91	Ea	396.00	103.00	—	499.00
46" - 48"	S2@3.21	Ea	431.00	114.00	—	545.00
50" - 52"	S2@3.51	Ea	477.00	124.00	—	601.00
54" - 56"	S2@4.17	Ea	568.00	147.00	—	715.00
58" - 60"	S2@4.97	Ea	676.00	176.00	—	852.00
62" - 64"	S2@5.49	Ea	747.00	194.00	—	941.00
66" - 68"	S2@6.04	Ea	822.00	214.00	—	1,036.00
70" - 72"	S2@6.51	Ea	880.00	230.00	—	1,110.00

Galvanized steel 36" rectangular duct 90 degree elbow

Description	Craft@Hrs	Unit	Material $	Labor $	Equipment $	Total $
22" - 24"	S2@1.83	Ea	250.00	64.70	—	314.70
26" - 28"	S2@2.04	Ea	278.00	72.10	—	350.10
30" - 32"	S2@2.26	Ea	308.00	79.90	—	387.90
34" - 36"	S2@2.47	Ea	336.00	87.30	—	423.30
38" - 40"	S2@2.70	Ea	368.00	95.50	—	463.50
42" - 44"	S2@3.01	Ea	407.00	106.00	—	513.00
46" - 48"	S2@3.31	Ea	449.00	117.00	—	566.00
50" - 52"	S2@3.61	Ea	491.00	128.00	—	619.00
54" - 56"	S2@4.27	Ea	581.00	151.00	—	732.00
58" - 60"	S2@5.10	Ea	690.00	180.00	—	870.00
62" - 64"	S2@5.61	Ea	761.00	198.00	—	959.00
66" - 68"	S2@6.15	Ea	831.00	217.00	—	1,048.00
70" - 72"	S2@6.64	Ea	902.00	235.00	—	1,137.00

Galvanized Steel Rectangular Elbows

Description	Craft@Hrs	Unit	Material $	Labor $	Equipment $	Total $

Galvanized steel 38" rectangular duct 90 degree elbow

Description	Craft@Hrs	Unit	Material $	Labor $	Equipment $	Total $
22" - 24"	S2@1.98	Ea	269.00	70.00	—	339.00
26" - 28"	S2@2.18	Ea	297.00	77.10	—	374.10
30" - 32"	S2@2.38	Ea	326.00	84.20	—	410.20
34" - 36"	S2@2.58	Ea	353.00	91.20	—	444.20
38" - 40"	S2@2.81	Ea	381.00	99.40	—	480.40
42" - 44"	S2@3.11	Ea	421.00	110.00	—	531.00
46" - 48"	S2@3.41	Ea	464.00	121.00	—	585.00
50" - 52"	S2@3.71	Ea	502.00	131.00	—	633.00
54" - 56"	S2@4.38	Ea	594.00	155.00	—	749.00
58" - 60"	S2@5.21	Ea	710.00	184.00	—	894.00
62" - 64"	S2@5.74	Ea	776.00	203.00	—	979.00
66" - 68"	S2@6.28	Ea	853.00	222.00	—	1,075.00
70" - 72"	S2@6.75	Ea	919.00	239.00	—	1,158.00

Galvanized steel 40" rectangular duct 90 degree elbow

Description	Craft@Hrs	Unit	Material $	Labor $	Equipment $	Total $
40" - 42"	S2@3.06	Ea	417.00	108.00	—	525.00
44" - 46"	S2@3.36	Ea	453.00	119.00	—	572.00
48" - 50"	S2@3.66	Ea	498.00	129.00	—	627.00
52" - 54"	S2@3.98	Ea	540.00	141.00	—	681.00
56" - 58"	S2@5.17	Ea	700.00	183.00	—	883.00
60" - 62"	S2@5.59	Ea	760.00	198.00	—	958.00
64" - 66"	S2@6.11	Ea	828.00	216.00	—	1,044.00
68" - 70"	S2@6.66	Ea	905.00	235.00	—	1,140.00
72"	S2@7.05	Ea	957.00	249.00	—	1,206.00

Galvanized steel 42" rectangular duct 90 degree elbow

Description	Craft@Hrs	Unit	Material $	Labor $	Equipment $	Total $
40" - 42"	S2@3.21	Ea	440.00	114.00	—	554.00
44" - 46"	S2@3.52	Ea	479.00	124.00	—	603.00
48" - 50"	S2@3.83	Ea	523.00	135.00	—	658.00
52" - 54"	S2@4.16	Ea	566.00	147.00	—	713.00
56" - 58"	S2@5.34	Ea	724.00	189.00	—	913.00
60" - 62"	S2@5.81	Ea	785.00	205.00	—	990.00
64" - 66"	S2@6.32	Ea	858.00	223.00	—	1,081.00
68" - 70"	S2@6.85	Ea	930.00	242.00	—	1,172.00
72"	S2@7.22	Ea	987.00	255.00	—	1,242.00

Galvanized steel 44" rectangular duct 90 degree elbow

Description	Craft@Hrs	Unit	Material $	Labor $	Equipment $	Total $
40" - 42"	S2@3.36	Ea	453.00	119.00	—	572.00
44" - 46"	S2@3.69	Ea	501.00	130.00	—	631.00
48" - 50"	S2@4.01	Ea	544.00	142.00	—	686.00
52" - 54"	S2@4.32	Ea	587.00	153.00	—	740.00
56" - 58"	S2@5.51	Ea	750.00	195.00	—	945.00
60" - 62"	S2@6.00	Ea	812.00	212.00	—	1,024.00
64" - 66"	S2@6.51	Ea	880.00	230.00	—	1,110.00
68" - 70"	S2@7.05	Ea	962.00	249.00	—	1,211.00
72"	S2@7.39	Ea	1,000.00	261.00	—	1,261.00

Description	Craft@Hrs	Unit	Material $	Labor $	Equipment $	Total $

Galvanized steel 46" rectangular duct 90 degree elbow

Description	Craft@Hrs	Unit	Material $	Labor $	Equipment $	Total $
40" - 42"	S2@3.52	Ea	479.00	124.00	—	603.00
44" - 46"	S2@3.86	Ea	525.00	136.00	—	661.00
48" - 50"	S2@4.16	Ea	566.00	147.00	—	713.00
52" - 54"	S2@4.32	Ea	608.00	153.00	—	761.00
56" - 58"	S2@5.72	Ea	774.00	202.00	—	976.00
60" - 62"	S2@6.19	Ea	838.00	219.00	—	1,057.00
64" - 66"	S2@6.70	Ea	914.00	237.00	—	1,151.00
68" - 70"	S2@7.26	Ea	988.00	257.00	—	1,245.00
72"	S2@7.60	Ea	1,030.00	269.00	—	1,299.00

Galvanized steel 48" rectangular duct 90 degree elbow

Description	Craft@Hrs	Unit	Material $	Labor $	Equipment $	Total $
40" - 42"	S2@3.67	Ea	499.00	130.00	—	629.00
44" - 46"	S2@4.01	Ea	544.00	142.00	—	686.00
48" - 50"	S2@4.32	Ea	587.00	153.00	—	740.00
52" - 54"	S2@4.61	Ea	623.00	163.00	—	786.00
56" - 58"	S2@5.91	Ea	806.00	209.00	—	1,015.00
60" - 62"	S2@6.40	Ea	870.00	226.00	—	1,096.00
64" - 66"	S2@6.90	Ea	938.00	244.00	—	1,182.00
68" - 70"	S2@7.41	Ea	1,010.00	262.00	—	1,272.00
72"	S2@7.77	Ea	1,060.00	275.00	—	1,335.00

Galvanized steel 50" rectangular duct 90 degree elbow

Description	Craft@Hrs	Unit	Material $	Labor $	Equipment $	Total $
40" - 42"	S2@3.82	Ea	523.00	135.00	—	658.00
44" - 46"	S2@4.16	Ea	566.00	147.00	—	713.00
48" - 50"	S2@4.46	Ea	608.00	158.00	—	766.00
52" - 54"	S2@4.78	Ea	644.00	169.00	—	813.00
56" - 58"	S2@6.11	Ea	828.00	216.00	—	1,044.00
60" - 62"	S2@6.60	Ea	896.00	233.00	—	1,129.00
64" - 66"	S2@7.09	Ea	965.00	251.00	—	1,216.00
68" - 70"	S2@7.62	Ea	1,040.00	269.00	—	1,309.00
72"	S2@7.94	Ea	1,080.00	281.00	—	1,361.00

Galvanized steel 52" rectangular duct 90 degree elbow

Description	Craft@Hrs	Unit	Material $	Labor $	Equipment $	Total $
40" - 42"	S2@3.98	Ea	540.00	141.00	—	681.00
44" - 46"	S2@4.32	Ea	587.00	153.00	—	740.00
48" - 50"	S2@4.61	Ea	623.00	163.00	—	786.00
52" - 54"	S2@4.95	Ea	671.00	175.00	—	846.00
56" - 58"	S2@6.32	Ea	858.00	223.00	—	1,081.00
60" - 62"	S2@6.79	Ea	923.00	240.00	—	1,163.00
64" - 66"	S2@7.30	Ea	991.00	258.00	—	1,249.00
68" - 70"	S2@7.81	Ea	1,060.00	276.00	—	1,336.00
72"	S2@8.16	Ea	1,100.00	289.00	—	1,389.00

Galvanized Steel Rectangular Elbows

Description	Craft@Hrs	Unit	Material $	Labor $	Equipment $	Total $

Galvanized steel 54" rectangular duct 90 degree elbow

Description	Craft@Hrs	Unit	Material $	Labor $	Equipment $	Total $
40" - 42"	S2@4.16	Ea	566.00	147.00	—	713.00
44" - 46"	S2@4.46	Ea	608.00	158.00	—	766.00
48" - 50"	S2@4.78	Ea	644.00	169.00	—	813.00
52" - 54"	S2@5.15	Ea	693.00	182.00	—	875.00
56" - 58"	S2@6.51	Ea	880.00	230.00	—	1,110.00
60" - 62"	S2@7.00	Ea	951.00	248.00	—	1,199.00
64" - 66"	S2@7.49	Ea	1,020.00	265.00	—	1,285.00
68" - 70"	S2@8.01	Ea	1,090.00	283.00	—	1,373.00
72"	S2@8.41	Ea	1,130.00	297.00	—	1,427.00

Galvanized steel 56" rectangular duct 90 degree elbow

Description	Craft@Hrs	Unit	Material $	Labor $	Equipment $	Total $
40" - 42"	S2@5.10	Ea	690.00	180.00	—	870.00
44" - 46"	S2@5.49	Ea	747.00	194.00	—	941.00
48" - 50"	S2@5.85	Ea	796.00	207.00	—	1,003.00
52" - 54"	S2@6.26	Ea	849.00	221.00	—	1,070.00
56" - 58"	S2@6.68	Ea	913.00	236.00	—	1,149.00
60" - 62"	S2@7.15	Ea	975.00	253.00	—	1,228.00
64" - 66"	S2@7.66	Ea	1,040.00	271.00	—	1,311.00
68" - 70"	S2@8.18	Ea	1,100.00	289.00	—	1,389.00
72"	S2@8.58	Ea	1,170.00	303.00	—	1,473.00

Galvanized steel 58" rectangular duct 90 degree elbow

Description	Craft@Hrs	Unit	Material $	Labor $	Equipment $	Total $
58" - 60"	S2@7.15	Ea	987.00	253.00	—	1,240.00
62" - 64"	S2@7.62	Ea	1,040.00	269.00	—	1,309.00
66" - 68"	S2@8.13	Ea	1,110.00	287.00	—	1,397.00
70" - 72"	S2@8.69	Ea	1,190.00	307.00	—	1,497.00

Galvanized steel 60" rectangular duct 90 degree elbow

Description	Craft@Hrs	Unit	Material $	Labor $	Equipment $	Total $
58" - 60"	S2@7.37	Ea	1,010.00	261.00	—	1,271.00
62" - 64"	S2@7.88	Ea	1,080.00	279.00	—	1,359.00
66" - 68"	S2@8.41	Ea	1,160.00	297.00	—	1,457.00
70" - 72"	S2@8.95	Ea	1,220.00	316.00	—	1,536.00

Galvanized steel 62" rectangular duct 90 degree elbow

Description	Craft@Hrs	Unit	Material $	Labor $	Equipment $	Total $
58" - 60"	S2@7.62	Ea	1,040.00	269.00	—	1,309.00
62" - 64"	S2@8.13	Ea	1,100.00	287.00	—	1,387.00
66" - 68"	S2@8.69	Ea	1,180.00	307.00	—	1,487.00
70" - 72"	S2@9.18	Ea	1,250.00	325.00	—	1,575.00

Galvanized steel 64" rectangular duct 90 degree elbow

Description	Craft@Hrs	Unit	Material $	Labor $	Equipment $	Total $
58" - 60"	S2@7.88	Ea	1,070.00	279.00	—	1,349.00
62" - 64"	S2@8.41	Ea	1,140.00	297.00	—	1,437.00
66" - 68"	S2@8.95	Ea	1,210.00	316.00	—	1,526.00
70" - 72"	S2@9.44	Ea	1,280.00	334.00	—	1,614.00

Description	Craft@Hrs	Unit	Material $	Labor $	Equipment $	Total $

Galvanized steel 66" rectangular duct 90 degree elbow

Description	Craft@Hrs	Unit	Material $	Labor $	Equipment $	Total $
58" - 60"	S2@8.13	Ea	1,100.00	287.00	—	1,387.00
62" - 64"	S2@8.69	Ea	1,180.00	307.00	—	1,487.00
66" - 68"	S2@9.18	Ea	1,250.00	325.00	—	1,575.00
70" - 72"	S2@9.69	Ea	1,330.00	343.00	—	1,673.00

Galvanized steel 68" rectangular duct 90 degree elbow

Description	Craft@Hrs	Unit	Material $	Labor $	Equipment $	Total $
58" - 60"	S2@8.41	Ea	1,140.00	297.00	—	1,437.00
62" - 64"	S2@8.95	Ea	1,210.00	316.00	—	1,526.00
66" - 68"	S2@9.44	Ea	1,280.00	334.00	—	1,614.00
70" - 72"	S2@9.95	Ea	1,360.00	352.00	—	1,712.00

Galvanized steel 70" rectangular duct 90 degree elbow

Description	Craft@Hrs	Unit	Material $	Labor $	Equipment $	Total $
58" - 60"	S2@8.69	Ea	1,180.00	307.00	—	1,487.00
62" - 64"	S2@9.18	Ea	1,250.00	325.00	—	1,575.00
66" - 68"	S2@9.69	Ea	1,330.00	343.00	—	1,673.00
70" - 72"	S2@10.2	Ea	1,390.00	361.00	—	1,751.00

Galvanized steel 72" rectangular duct 90 degree elbow

Description	Craft@Hrs	Unit	Material $	Labor $	Equipment $	Total $
58" - 60"	S2@8.95	Ea	1,210.00	316.00	—	1,526.00
62" - 64"	S2@9.44	Ea	1,280.00	334.00	—	1,614.00
66" - 68"	S2@9.95	Ea	1,360.00	352.00	—	1,712.00
70" - 72"	S2@10.5	Ea	1,420.00	371.00	—	1,791.00

Galvanized Steel Rectangular Drops and Tees

Description	Craft@Hrs	Unit	Material $	Labor $	Equipment $	Total $

Costs for Galvanized Rectangular Drops and Tap-in Tees. Costs per drop or tap-in tee, based on quantities less than 1,000 pounds. Deduct 15% from the material cost for over 1,000 pounds. Add 50% to the material cost for the lined duct. These costs include delivery, typical bracing, cleats, scrap, hangers, end closures, sealants and miscellaneous hardware. For turning vanes, please see page 337.

Galvanized steel 6" rectangular duct drops and tap-in tees

Description	Craft@Hrs	Unit	Material $	Labor $	Equipment $	Total $
6"	S2@.072	Ea	10.70	2.55	—	13.25
8"	S2@.084	Ea	12.50	2.97	—	15.47
10"	S2@.097	Ea	14.20	3.43	—	17.63
12" - 16"	S2@.139	Ea	20.50	4.92	—	25.42
20" - 24"	S2@.216	Ea	31.90	7.64	—	39.54
28" - 32"	S2@.309	Ea	45.30	10.90	—	56.20
36" - 40"	S2@.412	Ea	60.90	14.60	—	75.50
44" - 48"	S2@.489	Ea	72.20	17.30	—	89.50
52" - 56"	S2@.591	Ea	87.10	20.90	—	108.00
60"	S2@.747	Ea	110.00	26.40	—	136.40

Galvanized steel 8" rectangular duct drops and tap-in tees

Description	Craft@Hrs	Unit	Material $	Labor $	Equipment $	Total $
6"	S2@.084	Ea	12.50	2.97	—	15.47
8"	S2@.097	Ea	14.20	3.43	—	17.63
10"	S2@.109	Ea	16.10	3.85	—	19.95
12" - 16"	S2@.152	Ea	22.70	5.37	—	28.07
20" - 24"	S2@.231	Ea	34.30	8.17	—	42.47
28" - 32"	S2@.326	Ea	48.50	11.50	—	60.00
36" - 40"	S2@.432	Ea	63.90	15.30	—	79.20
44" - 48"	S2@.506	Ea	74.70	17.90	—	92.60
52" - 56"	S2@.634	Ea	93.80	22.40	—	116.20
60"	S2@.760	Ea	112.00	26.90	—	138.90

Galvanized steel 10" rectangular duct drops and tap-in tees

Description	Craft@Hrs	Unit	Material $	Labor $	Equipment $	Total $
6"	S2@.097	Ea	14.20	3.43	—	17.63
8"	S2@.109	Ea	16.10	3.85	—	19.95
10"	S2@.121	Ea	18.00	4.28	—	22.28
12" - 16"	S2@.167	Ea	24.60	5.91	—	30.51
20" - 24"	S2@.246	Ea	36.70	8.70	—	45.40
28" - 32"	S2@.343	Ea	51.00	12.10	—	63.10
36" - 40"	S2@.450	Ea	66.10	15.90	—	82.00
44" - 48"	S2@.525	Ea	77.40	18.60	—	96.00
52" - 56"	S2@.653	Ea	96.50	23.10	—	119.60
60"	S2@.773	Ea	113.00	27.30	—	140.30

Description	Craft@Hrs	Unit	Material $	Labor $	Equipment $	Total $

Galvanized steel 12" rectangular duct drops and tap-in tees

Description	Craft@Hrs	Unit	Material $	Labor $	Equipment $	Total $
6"	S2@.109	Ea	16.10	3.85	—	19.95
8"	S2@.121	Ea	18.00	4.28	—	22.28
10"	S2@.133	Ea	19.70	4.70	—	24.40
12" - 16"	S2@.180	Ea	26.50	6.36	—	32.86
20" - 24"	S2@.262	Ea	38.90	9.26	—	48.16
28" - 32"	S2@.360	Ea	53.20	12.70	—	65.90
36" - 40"	S2@.470	Ea	69.00	16.60	—	85.60
44" - 48"	S2@.544	Ea	80.00	19.20	—	99.20
52" - 56"	S2@.675	Ea	99.60	23.90	—	123.50
60"	S2@.794	Ea	117.00	28.10	—	145.10

Galvanized steel 16" rectangular duct drops and tap-in tees

Description	Craft@Hrs	Unit	Material $	Labor $	Equipment $	Total $
6"	S2@.170	Ea	25.00	6.01	—	31.01
8"	S2@.184	Ea	27.50	6.51	—	34.01
10"	S2@.200	Ea	29.70	7.07	—	36.77
12" - 16"	S2@.231	Ea	34.30	8.17	—	42.47
20" - 24"	S2@.293	Ea	43.00	10.40	—	53.40
28" - 32"	S2@.396	Ea	58.30	14.00	—	72.30
36" - 40"	S2@.506	Ea	74.70	17.90	—	92.60
44" - 48"	S2@.581	Ea	85.80	20.50	—	106.30
52" - 56"	S2@.709	Ea	105.00	25.10	—	130.10
60"	S2@.841	Ea	124.00	29.70	—	153.70

Galvanized steel 20" rectangular duct drops and tap-in tees

Description	Craft@Hrs	Unit	Material $	Labor $	Equipment $	Total $
6"	S2@.200	Ea	29.70	7.07	—	36.77
8"	S2@.216	Ea	31.90	7.64	—	39.54
10"	S2@.231	Ea	34.30	8.17	—	42.47
12" - 16"	S2@.262	Ea	38.90	9.26	—	48.16
20" - 24"	S2@.324	Ea	47.90	11.50	—	59.40
28" - 32"	S2@.428	Ea	63.20	15.10	—	78.30
36" - 40"	S2@.544	Ea	80.00	19.20	—	99.20
44" - 48"	S2@.617	Ea	91.30	21.80	—	113.10
52" - 56"	S2@.758	Ea	111.00	26.80	—	137.80
60"	S2@.884	Ea	131.00	31.30	—	162.30

Galvanized Steel Rectangular Drops and Tees

Description	Craft@Hrs	Unit	Material $	Labor $	Equipment $	Total $

Galvanized steel 24" rectangular duct drops and tap-in tees

Description	Craft@Hrs	Unit	Material $	Labor $	Equipment $	Total $
6"	S2@.231	Ea	34.30	8.17	—	42.47
8"	S2@.247	Ea	36.70	8.73	—	45.43
10"	S2@.262	Ea	38.90	9.26	—	48.16
12" - 16"	S2@.293	Ea	43.00	10.40	—	53.40
20" - 24"	S2@.354	Ea	52.30	12.50	—	64.80
28" - 32"	S2@.462	Ea	67.60	16.30	—	83.90
36" - 40"	S2@.581	Ea	85.80	20.50	—	106.30
44" - 48"	S2@.655	Ea	96.80	23.20	—	120.00
52" - 56"	S2@.798	Ea	118.00	28.20	—	146.20
60"	S2@.927	Ea	137.00	32.80	—	169.80

Galvanized steel 28" rectangular duct drops and tap-in tees

Description	Craft@Hrs	Unit	Material $	Labor $	Equipment $	Total $
6"	S2@.262	Ea	38.90	9.26	—	48.16
8"	S2@.278	Ea	40.70	9.83	—	50.53
10"	S2@.292	Ea	43.00	10.30	—	53.30
12" - 16"	S2@.324	Ea	47.90	11.50	—	59.40
20" - 24"	S2@.385	Ea	56.80	13.60	—	70.40
28" - 32"	S2@.497	Ea	73.30	17.60	—	90.90
36" - 40"	S2@.617	Ea	91.30	21.80	—	113.10
44" - 48"	S2@.694	Ea	102.00	24.50	—	126.50
52" - 56"	S2@.837	Ea	124.00	29.60	—	153.60
60"	S2@.969	Ea	143.00	34.30	—	177.30

Galvanized steel 32" rectangular duct drops and tap-in tees

Description	Craft@Hrs	Unit	Material $	Labor $	Equipment $	Total $
6"	S2@.356	Ea	52.50	12.60	—	65.10
8"	S2@.374	Ea	55.90	13.20	—	69.10
10"	S2@.393	Ea	58.10	13.90	—	72.00
12" - 16"	S2@.432	Ea	63.90	15.30	—	79.20
20" - 24"	S2@.506	Ea	74.70	17.90	—	92.60
28" - 32"	S2@.581	Ea	85.80	20.50	—	106.30
36" - 40"	S2@.655	Ea	96.80	23.20	—	120.00
44" - 48"	S2@.730	Ea	107.00	25.80	—	132.80
52" - 56"	S2@.877	Ea	131.00	31.00	—	162.00
60"	S2@1.02	Ea	150.00	36.10	—	186.10

Description	Craft@Hrs	Unit	Material $	Labor $	Equipment $	Total $

Galvanized steel round duct snap-lock with slip joints

Description	Craft@Hrs	Unit	Material $	Labor $	Equipment $	Total $
3"	S2@.080	LF	2.43	2.83	—	5.26
4"	S2@.090	LF	2.50	3.18	—	5.68
5"	S2@.095	LF	2.68	3.36	—	6.04
6"	S2@.100	LF	3.14	3.54	—	6.68
7"	S2@.110	LF	4.82	3.89	—	8.71
8"	S2@.120	LF	4.97	4.24	—	9.21
9"	S2@.140	LF	5.21	4.95	—	10.16
10"	S2@.160	LF	5.43	5.66	—	11.09

Galvanized steel round duct 90 degree adjustable elbow

Description	Craft@Hrs	Unit	Material $	Labor $	Equipment $	Total $
3"	S2@.250	Ea	4.71	8.84	—	13.55
4"	S2@.300	Ea	6.48	10.60	—	17.08
5"	S2@.350	Ea	7.83	12.40	—	20.23
6"	S2@.400	Ea	10.20	14.10	—	24.30
7"	S2@.450	Ea	34.70	15.90	—	50.60
8"	S2@.500	Ea	38.90	17.70	—	56.60
9"	S2@.600	Ea	42.10	21.20	—	63.30
10"	S2@.700	Ea	46.10	24.80	—	70.90

Galvanized steel round duct 45 degree adjustable elbow

Description	Craft@Hrs	Unit	Material $	Labor $	Equipment $	Total $
3"	S2@.250	Ea	4.71	8.84	—	13.55
4"	S2@.300	Ea	6.48	10.60	—	17.08
5"	S2@.350	Ea	7.83	12.40	—	20.23
6"	S2@.400	Ea	10.20	14.10	—	24.30
7"	S2@.450	Ea	34.70	15.90	—	50.60
8"	S2@.500	Ea	38.90	17.70	—	56.60
9"	S2@.600	Ea	42.10	21.20	—	63.30
10"	S2@.700	Ea	46.10	24.80	—	70.90

Fiberglass Ductwork

Standard fiberglass ductwork, or duct board, should not be used in systems where air velocities exceed 2,500 feet per minute, or where static pressures are higher than 2 inches of water gauge.

Additionally, it should not be used for the following applications:

1) Equipment rooms

2) Underground or outdoors

3) Final connections to air handling equipment

4) Plenums or casings

5) Fume, heat or moisture exhaust systems

The costs for manufacturing and installing fiberglass ductwork are based on the net areas of the duct and fittings.

Here are approximate current material costs, per square foot:

Fiberglass Type	Up to 2,500 SF Material $/SF	Over 2,500 SF Material $/SF
475	4.04	3.68
800	4.43	4.02
1400	5.40	5.21

Additional costs: Approximately 25 cents per square foot should be added for the cost of tie rods, hangers, staples, tape and waste.

Type 475 fiberglass board is used almost exclusively for residential and small commercial air systems, while Types 800 and 1400 are used in larger commercial and industrial systems.

Fiberglass Ductwork Fabrication Labor*						
Size ranges (inches)						
	Up through 16		17 through 32		33 through 48	
Activity	SF/Hour Output	Labor $/SF	SF/Hour Output	Labor $/SF	SF/Hour Output	Labor $/SF
Duct only	60	.58	58	.60	56	.62
Fittings only	40	.87	38	.91	35	.99
Duct & 15% fittings	55	.63	53	.65	50	.69
Duct & 25% fittings	53	.65	51	.68	48	.72
Duct & 35% fittings	51	.68	49	.71	46	.75

*Labor rates are based on using a standard grooving machine and include all tie-rod reinforcing.

Labor correction factors:

1) Hand grooving in lieu of machine grooving: 1.58

2) Use of an auto-closer: .82

3) Tie-rods enclosed in conduit: 1.36

4) T-bar or channel reinforcing in lieu of tie-rods: 1.25

In the following installation table the measurements refer to the "semi-perimeter", the width plus depth of the ductwork. Example – 12" duct (8" wide plus 4" high) has 2.00 SF of area for each linear foot.

Description	Craft@Hrs	Unit	Material $	Labor $	Equipment $	Total $

Fiberglass ductwork installation

Description	Craft@Hrs	Unit	Material $	Labor $	Equipment $	Total $
12" (2.00 SF/LF)	S2@.070	LF	—	2.48	—	2.48
14" (2.35 SF/LF)	S2@.082	LF	—	2.90	—	2.90
16" (2.65 SF/LF)	S2@.093	LF	—	3.29	—	3.29
18" (3.00 SF/LF)	S2@.105	LF	—	3.71	—	3.71
20" (3.35 SF/LF)	S2@.117	LF	—	4.14	—	4.14
22" (3.65 SF/LF)	S2@.128	LF	—	4.53	—	4.53
24" (4.00 SF/LF)	S2@.140	LF	—	4.95	—	4.95
26" (4.35 SF/LF)	S2@.152	LF	—	5.37	—	5.37
28" (4.65 SF/LF)	S2@.163	LF	—	5.76	—	5.76
30" (5.00 SF/LF)	S2@.175	LF	—	6.19	—	6.19
32" (5.35 SF/LF)	S2@.187	LF	—	6.61	—	6.61
34" (5.65 SF/LF)	S2@.198	LF	—	7.00	—	7.00
36" (6.00 SF/LF)	S2@.210	LF	—	7.43	—	7.43
38" (6.35 SF/LF)	S2@.222	LF	—	7.85	—	7.85
40" (6.65 SF/LF)	S2@.232	LF	—	8.20	—	8.20
42" (7.00 SF/LF)	S2@.245	LF	—	8.66	—	8.66
44" (7.35 SF/LF)	S2@.257	LF	—	9.09	—	9.09
46" (7.65 SF/LF)	S2@.268	LF	—	9.48	—	9.48
48" (8.00 SF/LF)	S2@.280	LF	—	9.90	—	9.90
50" (8.35 SF/LF)	S2@.292	LF	—	10.30	—	10.30
52" (8.65 SF/LF)	S2@.303	LF	—	10.70	—	10.70
54" (9.00 SF/LF)	S2@.315	LF	—	11.10	—	11.10
56" (9.35 SF/LF)	S2@.327	LF	—	11.60	—	11.60
58" (9.65 SF/LF)	S2@.338	LF	—	12.00	—	12.00
60" (10.0 SF/LF)	S2@.350	LF	—	12.40	—	12.40
64" (10.6 SF/LF)	S2@.373	LF	—	13.20	—	13.20
68" (11.4 SF/LF)	S2@.397	LF	—	14.00	—	14.00
72" (12.0 SF/LF)	S2@.420	LF	—	14.90	—	14.90
76" (12.6 SF/LF)	S2@.443	LF	—	15.70	—	15.70
80" (13.4 SF/LF)	S2@.467	LF	—	16.50	—	16.50
84" (14.0 SF/LF)	S2@.490	LF	—	17.30	—	17.30
88" (14.6 SF/LF)	S2@.513	LF	—	18.10	—	18.10
92" (15.4 SF/LF)	S2@.537	LF	—	19.00	—	19.00
96" (16.0 SF/LF)	S2@.560	LF	—	19.80	—	19.80

For estimating purposes, when measuring the length of ductwork, include the fitting.
Then apply the labor per run foot to the total.

1-1/4" thick wire reinforced flexible fiberglass duct with vinyl cover

Description	Craft@Hrs	Unit	Material $	Labor $	Equipment $	Total $
4"	S2@.060	LF	6.89	2.12	—	9.01
5"	S2@.070	LF	7.14	2.48	—	9.62
6"	S2@.080	LF	7.80	2.83	—	10.63
8"	S2@.090	LF	9.92	3.18	—	13.10
9"	S2@.100	LF	10.60	3.54	—	14.14
10"	S2@.120	LF	11.10	4.24	—	15.34
12"	S2@.140	LF	14.30	4.95	—	19.25
14"	S2@.160	LF	17.10	5.66	—	22.76
16"	S2@.180	LF	19.80	6.36	—	26.16
18"	S2@.200	LF	22.70	7.07	—	29.77

Fiberglass Pipe Insulation

Description	Craft@Hrs	Unit	Material $	Labor $	Equipment $	Total $

Fiberglass Pipe Insulation with AP-T Plus (all purpose self sealing) Jacket. By nominal pipe diameter. R factor equals 2.56 at 300 degrees F. These costs do not include scaffolding, add for same, if required. Also see fittings, flanges and valves at the end of this section.

Jacketed fiberglass insulation for 1/2" pipe

Description	Craft@Hrs	Unit	Material $	Labor $	Equipment $	Total $
1/2" thick	P1@.038	LF	2.10	1.36	—	3.46
1" thick	P1@.038	LF	2.46	1.36	—	3.82
1½" thick	P1@.040	LF	5.03	1.43	—	6.46

Jacketed fiberglass insulation for 3/4" pipe

Description	Craft@Hrs	Unit	Material $	Labor $	Equipment $	Total $
1/2" thick	P1@.038	LF	2.27	1.36	—	3.63
1" thick	P1@.038	LF	2.80	1.36	—	4.16
1½" thick	P1@.040	LF	5.21	1.43	—	6.64

Jacketed fiberglass insulation for 1" pipe

Description	Craft@Hrs	Unit	Material $	Labor $	Equipment $	Total $
1/2" thick	P1@.040	LF	2.42	1.43	—	3.85
1" thick	P1@.040	LF	3.21	1.43	—	4.64
1½" thick	P1@.042	LF	5.50	1.50	—	7.00

Jacketed fiberglass insulation for 1-1/4" pipe

Description	Craft@Hrs	Unit	Material $	Labor $	Equipment $	Total $
1/2" thick	P1@.040	LF	2.69	1.43	—	4.12
1" thick	P1@.040	LF	3.24	1.43	—	4.67
1½" thick	P1@.042	LF	5.85	1.50	—	7.35

Jacketed fiberglass insulation for 1-1/2" pipe

Description	Craft@Hrs	Unit	Material $	Labor $	Equipment $	Total $
1/2" thick	P1@.042	LF	2.87	1.50	—	4.37
1" thick	P1@.042	LF	3.65	1.50	—	5.15
1½" thick	P1@.044	LF	6.31	1.57	—	7.88

Jacketed fiberglass insulation for 2" pipe

Description	Craft@Hrs	Unit	Material $	Labor $	Equipment $	Total $
1/2" thick	P1@.044	LF	3.02	1.57	—	4.59
1" thick	P1@.044	LF	4.19	1.57	—	5.76
1½" thick	P1@.046	LF	6.78	1.64	—	8.42

Jacketed fiberglass insulation for 2-1/2" pipe

Description	Craft@Hrs	Unit	Material $	Labor $	Equipment $	Total $
1" thick	P1@.047	LF	4.32	1.68	—	6.00
1½" thick	P1@.048	LF	6.13	1.72	—	7.85
2" thick	P1@.050	LF	9.77	1.79	—	11.56

Jacketed fiberglass insulation for 3" pipe

Description	Craft@Hrs	Unit	Material $	Labor $	Equipment $	Total $
1" thick	P1@.051	LF	4.75	1.82	—	6.57
1½" thick	P1@.054	LF	6.70	1.93	—	8.63
2" thick	P1@.056	LF	10.70	2.00	—	12.70

Description	Craft@Hrs	Unit	Material $	Labor $	Equipment $	Total $

Jacketed fiberglass insulation for 4" pipe

Description	Craft@Hrs	Unit	Material $	Labor $	Equipment $	Total $
1" thick	P1@.063	LF	5.60	2.25	—	7.85
1½" thick	P1@.066	LF	7.46	2.36	—	9.82
2" thick	P1@.069	LF	12.00	2.47	—	14.47

Jacketed fiberglass insulation for 6" pipe

Description	Craft@Hrs	Unit	Material $	Labor $	Equipment $	Total $
1" thick	P1@.077	LF	6.53	2.75	—	9.28
1½" thick	P1@.080	LF	8.98	2.86	—	11.84
2" thick	P1@.085	LF	15.10	3.04	—	18.14

Jacketed fiberglass insulation for 8" pipe

Description	Craft@Hrs	Unit	Material $	Labor $	Equipment $	Total $
1" thick	P1@.117	LF	8.98	4.18	—	13.16
1½" thick	P1@.123	LF	10.90	4.40	—	15.30
2" thick	P1@.129	LF	17.70	4.61	—	22.31

Jacketed fiberglass insulation for 10" pipe

Description	Craft@Hrs	Unit	Material $	Labor $	Equipment $	Total $
1" thick	P1@.140	LF	10.60	5.01	—	15.61
1½" thick	P1@.147	LF	13.10	5.26	—	18.66
2" thick	P1@.154	LF	21.40	5.51	—	26.91

Jacketed fiberglass insulation for 12" pipe

Description	Craft@Hrs	Unit	Material $	Labor $	Equipment $	Total $
1" thick	P1@.155	LF	12.30	5.54	—	17.84
1½" thick	P1@.163	LF	15.10	5.83	—	20.93
2" thick	P1@.171	LF	23.30	6.11	—	29.41
Add for .016" aluminum jacket						
Per SF of surface	P1@.035	SF	1.75	1.25	—	3.00

Pipe fittings and flanges
For each fitting or flange use the cost for 3 LF of pipe of the same pipe size

Valves
Body only: for each valve body use the cost for 5 LF of pipe of the same pipe size

Body and bonnet or yoke: for each valve use the cost for 10 LF of pipe of the same pipe size

Flanged valves: add the cost for flanges per above

Calcium Silicate Pipe Insulation with Aluminum Jacket

Calcium Silicate Pipe Insulation with .016" Aluminum Jacket. By nominal pipe diameter. R factor equals 2.20 at 300 degrees F. Manufactured by Pabco. These costs do not include scaffolding, add for same, if required. Also see fittings, flanges and valves below.

Description	Craft@Hrs	Unit	Material $	Labor $	Equipment $	Total $

Jacketed calcium silicate insulation for 6" pipe

Description	Craft@Hrs	Unit	Material $	Labor $	Equipment $	Total $
2" thick	P1@.183	LF	28.50	6.54	—	35.04
4" thick	P1@.218	LF	61.80	7.80	—	69.60
6" thick	P1@.288	LF	87.60	10.30	—	97.90

Jacketed calcium silicate insulation for 8" pipe

Description	Craft@Hrs	Unit	Material $	Labor $	Equipment $	Total $
2" thick	P1@.194	LF	34.20	6.94	—	41.14
4" thick	P1@.228	LF	72.60	8.15	—	80.75
6" thick	P1@.308	LF	110.00	11.00	—	121.00

Jacketed calcium silicate insulation for 10" pipe

Description	Craft@Hrs	Unit	Material $	Labor $	Equipment $	Total $
2" thick	P1@.202	LF	40.00	7.22	—	47.22
4" thick	P1@.239	LF	84.70	8.55	—	93.25
6" thick	P1@.337	LF	132.00	12.10	—	144.10

Jacketed calcium silicate insulation for 12" pipe

Description	Craft@Hrs	Unit	Material $	Labor $	Equipment $	Total $
2" thick	P1@.208	LF	46.40	7.44	—	53.84
4" thick	P1@.261	LF	94.10	9.33	—	103.43
6" thick	P1@.388	LF	149.00	13.90	—	162.90

Pipe fittings and flanges

For each fitting or flange use the cost for 3 LF of pipe of the same pipe size

Valves

Body only: for each valve body use the cost for 5 LF of pipe of the same pipe size

Body and bonnet or yoke: for each valve use the cost for 10 LF of pipe of the same pipe size

Flanged valves: add the cost for flanges per above

Closed Cell Elastomeric Pipe and Tubing Insulation

Description	Craft@Hrs	Unit	Material $	Labor $	Equipment $	Total $

Closed Cell Elastomeric Pipe and Tubing Insulation. Semi split, by nominal pipe or tube diameter with insulation wall thickness as shown, no cover. R factor equals 3.58 at 220 degrees F. Manufactured by Rubatex. These costs do not include scaffolding, add for same, if required. Also see fittings, flanges and valves below.

Description	Craft@Hrs	Unit	Material $	Labor $	Equipment $	Total $
1/4", 1/2" thick	P1@.039	LF	.84	1.39	—	2.23
3/8", 1/2" thick	P1@.039	LF	.86	1.39	—	2.25
1/2", 1/2" thick	P1@.039	LF	.96	1.39	—	2.35
3/4", 1/2" thick	P1@.039	LF	1.10	1.39	—	2.49
1", 1/2" thick	P1@.042	LF	1.23	1.50	—	2.73
1¼", 1/2" thick	P1@.042	LF	1.38	1.50	—	2.88
1½", 1/2" thick	P1@.042	LF	1.62	1.50	—	3.12
2", 1/2" thick	P1@.046	LF	2.04	1.64	—	3.68
2½", 3/8" thick	P1@.056	LF	2.59	2.00	—	4.59
2½", 1/2" thick	P1@.056	LF	3.48	2.00	—	5.48
2½", 3/4" thick	P1@.056	LF	4.49	2.00	—	6.49
3", 3/8" thick	P1@.062	LF	2.97	2.22	—	5.19
3", 1/2" thick	P1@.062	LF	3.91	2.22	—	6.13
3", 3/4" thick	P1@.062	LF	5.09	2.22	—	7.31
4", 3/8" thick	P1@.068	LF	3.48	2.43	—	5.91
4", 1/2" thick	P1@.068	LF	4.59	2.43	—	7.02
4", 3/4" thick	P1@.068	LF	5.97	2.43	—	8.40

Pipe fittings and flanges
For each fitting or flange use the cost for 3 LF of pipe of the same pipe size

Valves
Body only: for each valve body use the cost for 5 LF of pipe of the same pipe size

Body and bonnet or yoke: for each valve use the cost for 10 LF of pipe of the same pipe size

Flanged valves: add the cost for flanges per above

Description	Craft@Hrs	Unit	Material $	Labor $	Equipment $	Total $

Fiberglass flexible wrap duct insulation with foil facing vapor barrier to 250 degrees F

Description	Craft@Hrs	Unit	Material $	Labor $	Equipment $	Total $
1"	S2@.023	SF	.91	.81	—	1.72
1½"	S2@.023	SF	1.09	.81	—	1.90
2"	S2@.025	SF	1.47	.88	—	2.35
Deduct foil facing	—	%	-30.0	—	—	—

Fiberglass rigid board duct insulation plain, *without* foil facing vapor barrier to 450 degrees F

Description	Craft@Hrs	Unit	Material $	Labor $	Equipment $	Total $
1"	S2@.067	SF	1.12	2.37	—	3.49
1½"	S2@.067	SF	1.88	2.37	—	4.25
2"	S2@.075	SF	2.24	2.65	—	4.89
3"	S2@.085	SF	3.33	3.01	—	6.34

Fiberglass rigid board duct insulation with FSK (foil) facing vapor barrier to 250 degrees F

Description	Craft@Hrs	Unit	Material $	Labor $	Equipment $	Total $
1"	S2@.067	SF	1.47	2.37	—	3.84
1½"	S2@.067	SF	1.79	2.37	—	4.16
2"	S2@.075	SF	2.15	2.65	—	4.80
3"	S2@.085	SF	3.08	3.01	—	6.09

Fiberglass flexible blanket duct lining to 250 degrees F

Description	Craft@Hrs	Unit	Material $	Labor $	Equipment $	Total $
1/2"	S2@.032	SF	.68	1.13	—	1.81
1"	S2@.032	SF	.88	1.13	—	2.01
1½"	S2@.035	SF	1.25	1.24	—	2.49
2"	S2@.035	SF	1.65	1.24	—	2.89

Fiberglass rigid duct lining to 250 degrees F

Description	Craft@Hrs	Unit	Material $	Labor $	Equipment $	Total $
5/8"	S2@.032	SF	1.49	1.13	—	2.62
1"	S2@.032	SF	1.94	1.13	—	3.07
1½"	S2@.035	SF	2.55	1.24	—	3.79
2"	S2@.035	SF	3.25	1.24	—	4.49

Note: Material prices include all necessary tape, pins, mastic and 15% waste.

Description	Craft@Hrs	Unit	Material $	Labor $	Equipment $	Total $

Air balancing – air handling units

Description	Craft@Hrs	Unit	Material $	Labor $	Equipment $	Total $
Central station	S2@6.00	Ea	—	212.00	—	212.00
Multi-zone A/C	S2@4.00	Ea	—	141.00	—	141.00
Single-zone A/C	S2@3.00	Ea	—	106.00	—	106.00
Packaged A/C	S2@2.50	Ea	—	88.40	—	88.40
Built-up low pressure	S2@4.50	Ea	—	159.00	—	159.00
Built-up high pres.	S2@6.00	Ea	—	212.00	—	212.00
Built-up VAV	S2@6.00	Ea	—	212.00	—	212.00
Built-up dual duct	S2@6.00	Ea	—	212.00	—	212.00

Air balancing – terminal boxes

Description	Craft@Hrs	Unit	Material $	Labor $	Equipment $	Total $
Variable air volume	S2@0.40	Ea	—	14.10	—	14.10
Constant volume	S2@0.40	Ea	—	14.10	—	14.10
Dual duct box	S2@0.70	Ea	—	24.80	—	24.80
Terminal box and all downstream diffusers						
Cost per zone	S2@1.75	Ea	—	61.90	—	61.90

Air balancing – diffusers

Description	Craft@Hrs	Unit	Material $	Labor $	Equipment $	Total $
Ceiling diffusers to 500 CFM	S2@.250	Ea	—	8.84	—	8.84
Ceiling diffusers over 500 CFM	S2@.350	Ea	—	12.40	—	12.40
Ceiling diffusers perforated	S2@.370	Ea	—	13.10	—	13.10
Sidewall diffusers double deflection	S2@.250	Ea	—	8.84	—	8.84
Linear slot diffusers 2 slot	S2@.060	LF	—	2.12	—	2.12
Linear slot diffusers 3 slot	S2@.070	LF	—	2.48	—	2.48
Linear slot diffusers 4 slot	S2@.080	LF	—	2.83	—	2.83
Plug-in diffuser	S2@.250	Ea	—	8.84	—	8.84
Light troffers 1 slot	S2@.195	Ea	—	6.90	—	6.90
Light troffers 2 slot	S2@.220	Ea	—	7.78	—	7.78
Laminar flow diff.	S2@.750	Ea	—	26.50	—	26.50
Laminar flow diffusers with HEPA filter	S2@3.00	Ea	—	106.00	—	106.00

Balancing of HVAC Systems

Description	Craft@Hrs	Unit	Material $	Labor $	Equipment $	Total $

Air balancing – grilles

Description	Craft@Hrs	Unit	Material $	Labor $	Equipment $	Total $
Return air grilles to 800 CFM	S2@.250	Ea	—	8.84	—	8.84
Return air grilles over 800 CFM	S2@.350	Ea	—	12.40	—	12.40
Bar type return air grilles 2 slot	S2@.060	LF	—	2.12	—	2.12
Bar type return air grilles 4 slot	S2@.070	LF	—	2.48	—	2.48
Bar type return air grilles 6 slot	S2@.080	LF	—	2.83	—	2.83

Air balancing – centrifugal fans

Description	Craft@Hrs	Unit	Material $	Labor $	Equipment $	Total $
Supply and return air to 3,000 CFM	S2@1.00	Ea	—	35.40	—	35.40
Supply and return air to 5,000 CFM	S2@1.50	Ea	—	53.00	—	53.00
Supply and return air to 10,000 CFM	S2@2.00	Ea	—	70.70	—	70.70
Supply and return air to 20,000 CFM	S2@2.75	Ea	—	97.20	—	97.20
Roof exhaust to 6,000 CFM	S2@.725	Ea	—	25.60	—	25.60
Roof exhaust to 10,000 CFM	S2@.925	Ea	—	32.70	—	32.70
Roof exhaust to 20,000 CFM	S2@1.50	Ea	—	53.00	—	53.00
Range hood exhaust to 10,000 CFM	S2@1.75	Ea	—	61.90	—	61.90
Range hood exhaust to 20,000 CFM	S2@2.50	Ea	—	88.40	—	88.40

Air balancing – fan coil units

Description	Craft@Hrs	Unit	Material $	Labor $	Equipment $	Total $
Vertical floor model	S2@.500	Ea	—	17.70	—	17.70
Horiz. ceiling mtd	S2@.650	Ea	—	23.00	—	23.00

Air balancing – fume hoods

Description	Craft@Hrs	Unit	Material $	Labor $	Equipment $	Total $
Lab fume hood	S2@2.00	Ea	—	70.70	—	70.70
General purpose	S2@.450	Ea	—	15.90	—	15.90
Dust collectors	S2@1.50	Ea	—	53.00	—	53.00

Description	Craft@Hrs	Unit	Material $	Labor $	Equipment $	Total $

Wet balancing

Description	Craft@Hrs	Unit	Material $	Labor $	Equipment $	Total $
Boiler	S2@1.65	Ea	—	58.30	—	58.30
Chiller	S2@3.50	Ea	—	124.00	—	124.00
Cooling tower	S2@2.00	Ea	—	70.70	—	70.70
Circulating pump	S2@1.65	Ea	—	58.30	—	58.30
Heat exchanger	S2@1.75	Ea	—	61.90	—	61.90
AHU water coil	S2@2.50	Ea	—	88.40	—	88.40
Reheat coil	S2@.450	Ea	—	15.90	—	15.90
Wall fin radiation	S2@.350	Ea	—	12.40	—	12.40
Radiant heat panel	S2@.450	Ea	—	15.90	—	15.90
Fan coil unit, 2 pipe	S2@.350	Ea	—	12.40	—	12.40
Fan coil unit, 4 pipe	S2@.450	Ea	—	15.90	—	15.90
Cabinet heater	S2@.350	Ea	—	12.40	—	12.40
Unit ventilator	S2@.450	Ea	—	15.90	—	15.90
Unit heater	S2@.350	Ea	—	12.40	—	12.40

Temperature Controls

Description	Craft@Hrs	Unit	Material $	Labor $	Equipment $	Total $

Thermostat, electric. Install and connect. Make additional allowances for sub-subcontractor markups and control wiring as required.

Description	Craft@Hrs	Unit	Material $	Labor $	Equipment $	Total $
Heat/cool						
Low voltage	BE@.575	Ea	40.00	22.80	—	62.80
Heat Only						
Low voltage	BE@.500	Ea	33.30	19.80	—	53.10
Heat /cool						
Line voltage	BE@.650	Ea	121.00	25.70	—	146.70
Programmable – Residential						
Low voltage	BE@.850	LF	183.00	33.70	—	216.70
Programmable – commercial w/ subbase						
Low voltage	BE@1.10	LF	377.00	43.50	—	420.50

Thermostat, pneumatic. Install and connect. Make additional allowances for sub-subcontractor markups and pneumatic tubing as required.

Description	Craft@Hrs	Unit	Material $	Labor $	Equipment $	Total $
Direct acting						
55 to 85 Deg. F.	BE@.675	Ea	104.00	26.70	—	130.70
Reverse acting						
55 to 85 Deg. F.	BE@.675	Ea	86.20	26.70	—	112.90
Floating	BE@.675	Ea	183.00	26.70	—	209.70

Control valve, 2 way, electric. 6.9 Volt Input. Proportional Spring Return. Connect only. N.O. means: "normally open." Make additional allowances for sub-subcontractor markups, control wiring and valve installation by piping trades as required. (See Plumbing and Piping Specialties)

Description	Craft@Hrs	Unit	Material $	Labor $	Equipment $	Total $
1/2", 2 way, N.O.	BE@.400	Ea	466.00	15.80	—	481.80
3/4", 2 way, N.O.	BE@.400	Ea	495.00	15.80	—	510.80
1", 2 way, N.O.	BE@.400	Ea	534.00	15.80	—	549.80
1¼" 2 way, N.O.	BE@.400	Ea	583.00	15.80	—	598.80
1½" 2 way, N.O.	BE@.400	Ea	1,250.00	15.80	—	1,265.80
2", 2 way, N.O.	BE@.400	Ea	1,370.00	15.80	—	1,385.80

Control valve, 3 way, electric. 6.9 Volt Input, Proportional Spring Return. Connect only. Make additional allowances for sub-subcontractor markups, control wiring and valve installation by piping trades as required. (See Plumbing and Piping Specialties)

Description	Craft@Hrs	Unit	Material $	Labor $	Equipment $	Total $
1/2", 3 way	BE@.450	Ea	486.00	17.80	—	503.80
3/4", 3 way	BE@.450	Ea	529.00	17.80	—	546.80
1", 3 way	BE@.450	Ea	634.00	17.80	—	651.80

Description	Craft@Hrs	Unit	Material $	Labor $	Equipment $	Total $

Control valve, 2 way, pneumatic. Proportional Spring Return, 3 to 7 psi Input, Connect only. N.O. means "normally open." Make additional allowances for sub-subcontractor markups, pneumatic tubing and valve installation by piping trades as required. (See Plumbing and Piping Specialties)

Description	Craft@Hrs	Unit	Material $	Labor $	Equipment $	Total $
1/2", 2 way, N.O.	BE@.400	Ea	191.00	15.80	—	206.80
3/4", 2 way, N.O.	BE@.400	Ea	224.00	15.80	—	239.80
1", 2 way, N.O.	BE@.400	Ea	263.00	15.80	—	278.80
1¼" 2 way, N.O.	BE@.400	Ea	313.00	15.80	—	328.80
1½" 2 way, N.O.	BE@.400	Ea	557.00	15.80	—	572.80
2", 2 way, N.O.	BE@.400	Ea	665.00	15.80	—	680.80

Control valve, 3 way, pneumatic. Proportional Spring Return, 3 to 7 psi Input, Connect only. Make additional allowances for sub-subcontractor markups, pneumatic tubing and valve installation by piping trades as required. (See Plumbing and Piping Specialties)

Description	Craft@Hrs	Unit	Material $	Labor $	Equipment $	Total $
1/2", 3 way	BE@.450	Ea	214.00	17.80	—	231.80
3/4", 3 way	BE@.450	Ea	257.00	17.80	—	274.80
1", 3 way	BE@.450	Ea	359.00	17.80	—	376.80
1¼", 3 way	BE@.450	Ea	414.00	17.80	—	431.80
1½", 3 way	BE@.450	Ea	595.00	17.80	—	612.80
2", 3 way	BE@.450	Ea	695.00	17.80	—	712.80

Air compressor. Reciprocating piston type. Larger systems include HVAC actuator controls. Underwriters Laboratories and CSA rated. With ASME "U"-stamped pressure tank, pressure safety relief valve, silencer, lubrication system with lube cooling radiator and fan. Single phase 110 volt electric motor, drive belt and OSHA-rated safety belt guard. Add the cost of compressor hose and fittings and hose reel, if required. Set in place only. Add installation costs from the table below.

Description	Craft@Hrs	Unit	Material $	Labor $	Equipment $	Total $
1/2 hp	SL@3.00	Ea	360.00	107.00	—	467.00
3/4 hp	SL@3.00	Ea	393.00	107.00	—	500.00
1 hp	SL@3.50	Ea	456.00	124.00	—	580.00
1½ hp	SL@3.50	Ea	505.00	124.00	—	629.00
2 hp	SL@3.50	Ea	537.00	124.00	—	661.00

Installation of air compressors.

Description	Craft@Hrs	Unit	Material $	Labor $	Equipment $	Total $
Cut, frame support and gasket hole in sidewall	S2@1.00	Ea	35.00	35.40	—	70.40
Install piping for compressor line	P1@.100	LF	3.75	3.58	—	7.33
Install electrical wiring	BE@.150	LF	1.75	5.94	—	7.69
Install compressor controls	BE@.500	Ea	—	19.80	—	19.80
Commission and test	P1@2.00	Ea	—	71.50	—	71.50
Fine-tune system	P1@1.00	Ea	—	35.76	—	35.76

Ductile Iron, Class 153, Cement-Lined with Mechanical Joints

Description	Craft@Hrs	Unit	Material $	Labor $	Equipment $	Total $
Cement-lined Class 153 ductile iron pipe with mechanical joints						
4"	P1@.140	LF	10.80	5.01	—	15.81
6"	ER@.180	LF	15.60	6.86	1.70	24.16
8"	ER@.200	LF	21.40	7.62	1.80	30.82
10"	ER@.230	LF	31.80	8.77	2.10	42.67
12"	ER@.270	LF	43.30	10.30	2.50	56.10
14"	ER@.310	LF	65.70	11.80	2.80	80.30
16"	ER@.350	LF	84.80	13.30	3.20	101.30
Cement-lined Class 153 ductile iron 1/16 bend with mechanical joints						
4"	P1@.900	Ea	42.70	32.20	—	74.90
6"	ER@1.30	Ea	66.10	49.50	11.90	127.50
8"	ER@1.70	Ea	104.00	64.80	15.60	184.40
10"	ER@2.15	Ea	142.00	81.90	19.80	243.70
12"	ER@2.40	Ea	196.00	91.50	22.00	309.50
14"	ER@2.70	Ea	430.00	103.00	24.80	557.80
16"	ER@3.00	Ea	487.00	114.00	27.50	628.50
Cement-lined Class 153 ductile iron 1/8 bend with mechanical joints						
4"	P1@.900	Ea	44.20	32.20	—	76.40
6"	ER@1.30	Ea	65.70	49.50	11.90	127.10
8"	ER@1.70	Ea	105.00	64.80	15.60	185.40
10"	ER@2.15	Ea	144.00	81.90	19.80	245.70
12"	ER@2.40	Ea	199.00	91.50	22.00	312.50
14"	ER@2.70	Ea	427.00	103.00	24.80	554.80
16"	ER@3.00	Ea	501.00	114.00	27.50	642.50
Cement-lined Class 153 ductile iron 1/4 bend with mechanical joints						
4"	P1@.900	Ea	47.40	32.20	—	79.60
6"	ER@1.30	Ea	77.00	49.50	11.90	138.40
8"	ER@1.70	Ea	161.00	64.80	15.60	241.40
10"	ER@2.15	Ea	242.00	81.90	19.80	343.70
12"	ER@2.40	Ea	331.00	91.50	22.00	444.50
14"	ER@2.70	Ea	852.00	103.00	24.80	979.80
16"	ER@3.00	Ea	894.00	114.00	27.50	1,035.50
Cement-lined Class 153 ductile iron tee with mechanical joints						
4"	P1@1.65	Ea	71.80	59.00	—	130.80
6"	ER@2.35	Ea	105.00	89.60	21.60	216.20
8"	ER@3.15	Ea	161.00	120.00	29.00	310.00
10"	ER@4.00	Ea	242.00	152.00	36.70	430.70
12"	ER@4.45	Ea	331.00	170.00	40.90	541.90
14"	ER@4.90	Ea	852.00	187.00	44.90	1,083.90
16"	ER@5.30	Ea	894.00	202.00	48.60	1,144.60

Ductile Iron, Class 153, Cement-Lined with Mechanical Joints

Description	Craft@Hrs	Unit	Material $	Labor $	Equipment $	Total $

Cement-lined Class 153 ductile iron wye with mechanical joints

Description	Craft@Hrs	Unit	Material $	Labor $	Equipment $	Total $
4"	P1@1.65	Ea	161.00	59.00	—	220.00
6"	ER@2.35	Ea	200.00	89.60	21.60	311.20
8"	ER@3.15	Ea	302.00	120.00	29.00	451.00
10"	ER@4.00	Ea	563.00	152.00	36.70	751.70
12"	ER@4.45	Ea	847.00	170.00	40.90	1,057.90
14"	ER@4.90	Ea	1,730.00	187.00	45.70	1,962.70
16"	ER@5.30	Ea	2,160.00	202.00	48.60	2,410.60

Cement-lined Class 153 ductile iron reducer with mechanical joints

Description	Craft@Hrs	Unit	Material $	Labor $	Equipment $	Total $
6"	ER@1.45	Ea	52.60	55.30	13.30	121.20
8"	ER@2.00	Ea	83.20	76.20	18.30	177.70
10"	ER@2.55	Ea	110.00	97.20	23.50	230.70
12"	ER@3.05	Ea	142.00	116.00	28.10	286.10
14"	ER@3.60	Ea	318.00	137.00	33.00	488.00
16"	ER@4.20	Ea	386.00	160.00	38.50	584.50

Ductile Iron, Class 153, Double Cement-Lined with Mechanical Joints

Description	Craft@Hrs	Unit	Material $	Labor $	Equipment $	Total $

Double cement-lined Class 153 ductile iron pipe with mechanical joints

4"	P1@.140	LF	11.40	5.01	—	16.41
6"	ER@.180	LF	16.50	6.86	1.70	25.06
8"	ER@.200	LF	22.70	7.62	1.80	32.12
10"	ER@.230	LF	33.80	8.77	2.10	44.67
12"	ER@.270	LF	46.00	10.30	2.50	58.80
14"	ER@.310	LF	69.90	11.80	2.80	84.50
16"	ER@.350	LF	89.90	13.30	3.20	106.40

Double cement-lined Class 153 ductile iron 1/16 bend with mechanical joints

4"	P1@.900	Ea	50.40	32.20	—	82.60
6"	ER@1.30	Ea	78.10	49.50	11.90	139.50
8"	ER@1.70	Ea	124.00	64.80	15.60	204.40
10"	ER@2.15	Ea	167.00	81.90	19.80	268.70
12"	ER@2.40	Ea	229.00	91.50	22.00	342.50
14"	ER@2.70	Ea	511.00	103.00	24.80	638.80
16"	ER@3.00	Ea	579.00	114.00	27.50	720.50

Double cement-lined Class 153 ductile iron 1/8 bend with mechanical joints

4"	P1@.900	Ea	50.40	32.20	—	82.60
6"	ER@1.30	Ea	76.10	49.50	11.90	137.50
8"	ER@1.70	Ea	121.00	64.80	15.60	201.40
10"	ER@2.15	Ea	166.00	81.90	19.80	267.70
12"	ER@2.40	Ea	227.00	91.50	22.00	340.50
14"	ER@2.70	Ea	497.00	103.00	24.80	624.80
16"	ER@3.00	Ea	579.00	114.00	27.50	720.50

Double cement-lined Class 153 ductile iron 1/4 bend with mechanical joints

4"	P1@.900	Ea	54.60	32.20	—	86.80
6"	ER@1.30	Ea	88.40	49.50	11.90	149.80
8"	ER@1.70	Ea	137.00	64.80	15.60	217.40
10"	ER@2.15	Ea	214.00	81.90	19.80	315.70
12"	ER@2.40	Ea	278.00	91.50	22.00	391.50
14"	ER@2.70	Ea	661.00	103.00	24.80	788.80
16"	ER@3.00	Ea	751.00	114.00	27.50	892.50

Ductile Iron, 153, Double Cement-Lined with Mechanical Joints

Description	Craft@Hrs	Unit	Material $	Labor $	Equipment $	Total $

Double cement-lined Class 153 ductile iron tee with mechanical joints

Description	Craft@Hrs	Unit	Material $	Labor $	Equipment $	Total $
4"	P1@1.65	Ea	83.50	59.00	—	142.50
6"	ER@2.35	Ea	121.00	89.60	21.60	232.20
8"	ER@3.15	Ea	184.00	120.00	29.00	333.00
10"	ER@4.00	Ea	278.00	152.00	36.70	466.70
12"	ER@4.45	Ea	379.00	170.00	40.90	589.90
14"	ER@4.90	Ea	976.00	187.00	44.90	1,207.90
16"	ER@5.30	Ea	1,030.00	202.00	48.60	1,280.60

Double cement-lined Class 153 ductile iron wye with mechanical joints

Description	Craft@Hrs	Unit	Material $	Labor $	Equipment $	Total $
4"	P1@1.65	Ea	184.00	59.00	—	243.00
6"	ER@2.35	Ea	235.00	89.60	21.60	346.20
8"	ER@3.15	Ea	348.00	120.00	29.00	497.00
10"	ER@4.00	Ea	655.00	152.00	36.70	843.70
12"	ER@4.45	Ea	971.00	170.00	40.90	1,181.90
14"	ER@4.90	Ea	2,000.00	187.00	44.90	2,231.90
16"	ER@5.30	Ea	2,470.00	202.00	48.60	2,720.60

Double cement-lined Class 153 ductile iron reducer with mechanical joints

Description	Craft@Hrs	Unit	Material $	Labor $	Equipment $	Total $
6"	ER@1.45	Ea	60.60	55.30	13.30	129.20
8"	ER@2.00	Ea	86.30	76.20	18.30	180.80
10"	ER@2.55	Ea	112.00	97.20	23.50	232.70
12"	ER@3.05	Ea	164.00	116.00	28.10	308.10
14"	ER@3.60	Ea	353.00	137.00	27.50	517.50
16"	ER@4.20	Ea	443.00	160.00	38.50	641.50

Ductile Iron, Class 110, Cement-Lined with Mechanical Joints

Description	Craft@Hrs	Unit	Material $	Labor $	Equipment $	Total $

Cement-lined Class 110 ductile iron pipe with mechanical joints

Description	Craft@Hrs	Unit	Material $	Labor $	Equipment $	Total $
4"	P1@.180	LF	12.50	6.44	—	18.94
6"	ER@.220	LF	17.80	8.38	2.00	28.18
8"	ER@.240	LF	24.60	9.15	2.20	35.95
10"	ER@.270	LF	36.80	10.30	2.50	49.60
12"	ER@.320	LF	49.90	12.20	2.90	65.00
16"	ER@.400	LF	97.40	15.20	3.70	116.30

Cement-lined Class 110 ductile iron 1/16 bend with mechanical joints

Description	Craft@Hrs	Unit	Material $	Labor $	Equipment $	Total $
4"	P1@1.40	Ea	125.00	50.10	—	175.10
6"	ER@2.00	Ea	167.00	76.20	18.30	261.50
8"	ER@2.50	Ea	230.00	95.30	22.90	348.20
10"	ER@3.20	Ea	350.00	122.00	29.30	501.30
12"	ER@3.60	Ea	445.00	137.00	33.00	615.00
16"	ER@4.50	Ea	1,150.00	171.00	41.30	1,362.30

Cement-lined Class 110 ductile iron 1/8 bend with mechanical joints

Description	Craft@Hrs	Unit	Material $	Labor $	Equipment $	Total $
4"	P1@1.40	Ea	122.00	50.10	—	172.10
6"	ER@2.00	Ea	165.00	76.20	18.30	259.50
8"	ER@2.50	Ea	230.00	95.30	22.90	348.20
10"	ER@3.20	Ea	337.00	122.00	29.30	488.30
12"	ER@3.60	Ea	444.00	137.00	33.00	614.00
16"	ER@4.50	Ea	1,130.00	171.00	41.30	1,342.30

Cement-lined Class 110 ductile iron 1/4 bend with mechanical joints

Description	Craft@Hrs	Unit	Material $	Labor $	Equipment $	Total $
4"	P1@1.40	Ea	135.00	50.10	—	185.10
6"	ER@2.00	Ea	182.00	76.20	18.30	276.50
8"	ER@2.50	Ea	262.00	95.30	22.90	380.20
10"	ER@3.20	Ea	406.00	122.00	29.30	557.30
12"	ER@3.60	Ea	608.00	137.00	33.00	778.00
16"	ER@4.50	Ea	1,470.00	171.00	41.30	1,682.30

Cement-lined Class 110 ductile iron tee with mechanical joints

Description	Craft@Hrs	Unit	Material $	Labor $	Equipment $	Total $
4"	P1@2.50	Ea	205.00	89.40	—	294.40
6"	ER@3.50	Ea	273.00	133.00	32.10	438.10
8"	ER@4.70	Ea	389.00	179.00	43.10	611.10
10"	ER@6.00	Ea	671.00	229.00	55.00	955.00
12"	ER@6.70	Ea	852.00	255.00	60.50	1,167.50
16"	ER@8.00	Ea	2,240.00	305.00	73.30	2,618.30

Cement-lined Class 110 ductile iron reducer with mechanical joints

Description	Craft@Hrs	Unit	Material $	Labor $	Equipment $	Total $
6"	ER@2.20	Ea	135.00	83.80	20.20	239.00
8"	ER@3.00	Ea	188.00	114.00	27.50	329.50
10"	ER@3.80	Ea	308.00	145.00	34.80	487.80
12"	ER@4.60	Ea	363.00	175.00	42.20	580.20
16"	ER@6.30	Ea	994.00	240.00	57.80	1,291.80

Description	Craft@Hrs	Unit	Material $	Labor $	Equipment $	Total $

Class 150 cast iron pipe with mechanical joints

Description	Craft@Hrs	Unit	Material $	Labor $	Equipment $	Total $
1½"	P1@.080	LF	6.09	2.86	—	8.95
2"	P1@.100	LF	5.95	3.58	—	9.53
3"	P1@.120	LF	8.23	4.29	—	12.52
4"	P1@.140	LF	10.70	5.01	—	15.71
6"	ER@.180	LF	18.30	6.86	1.70	26.86
8"	ER@.200	LF	29.50	7.62	1.80	38.92

Class 150 cast iron 1/8 bend with mechanical joints

Description	Craft@Hrs	Unit	Material $	Labor $	Equipment $	Total $
1½"	P1@.750	Ea	6.71	26.80	—	33.51
2"	P1@.800	Ea	7.48	28.60	—	36.08
3"	P1@1.00	Ea	10.10	35.80	—	45.90
4"	P1@1.20	Ea	12.80	42.90	—	55.70
6"	ER@1.70	Ea	30.00	64.80	15.60	110.40
8"	ER@2.25	Ea	85.80	85.70	20.70	192.20

Class 150 cast iron 1/4 bend with mechanical joints

Description	Craft@Hrs	Unit	Material $	Labor $	Equipment $	Total $
2"	P1@.800	Ea	10.80	28.60	—	39.40
3"	P1@1.00	Ea	15.50	35.80	—	51.30
4"	P1@1.20	Ea	24.30	42.90	—	67.20
6"	ER@1.70	Ea	50.40	64.80	15.60	130.80
8"	ER@2.25	Ea	136.00	85.70	20.70	242.40

Class 150 cast iron sanitary tee with mechanical joints

Description	Craft@Hrs	Unit	Material $	Labor $	Equipment $	Total $
1½"	P1@.750	Ea	11.20	26.80	—	38.00
2"	P1@1.00	Ea	12.20	35.80	—	48.00
3"	P1@1.20	Ea	14.80	42.90	—	57.70
4"	P1@1.65	Ea	22.90	59.00	—	81.90
6"	ER@2.35	Ea	65.30	89.60	21.60	176.50
8"	ER@3.15	Ea	208.00	120.00	29.00	357.00

Class 150 cast iron wye with mechanical joints

Description	Craft@Hrs	Unit	Material $	Labor $	Equipment $	Total $
1½"	P1@.750	Ea	11.40	26.80	—	38.20
2"	P1@1.00	Ea	11.40	35.80	—	47.20
3"	P1@1.20	Ea	16.10	42.90	—	59.00
4"	P1@1.65	Ea	26.20	59.00	—	85.20
6"	ER@2.35	Ea	70.20	89.60	21.60	181.40
8"	ER@3.15	Ea	165.00	120.00	29.00	314.00

Class 150 cast iron reducer

Description	Craft@Hrs	Unit	Material $	Labor $	Equipment $	Total $
2" x 1½"	P1@.600	Ea	6.14	21.50	—	27.64
3" x 2"	P1@.800	Ea	6.08	28.60	—	34.68
4" x 2"	P1@1.00	Ea	9.48	35.80	—	45.28
6" x 4"	ER@1.45	Ea	25.30	55.30	13.30	93.90
8" x 6"	ER@2.00	Ea	43.30	76.20	18.30	137.80

Asbestos-Cement, Class 2400 or 3000 with Mechanical Joints

Description	Craft@Hrs	Unit	Material $	Labor $	Equipment $	Total $
Asbestos-cement pipe with mechanical joints						
4"	ER@.080	LF	10.20	3.05	.70	13.95
6"	ER@.100	LF	6.46	3.81	.90	11.17
8"	ER@.140	LF	10.20	5.34	1.30	16.84
10"	ER@.200	LF	15.90	7.62	1.80	25.32
12"	ER@.260	LF	21.10	9.91	2.40	33.41
14"	ER@.300	LF	28.30	11.40	2.80	42.50
16"	ER@.350	LF	35.00	13.30	3.20	51.50
Asbestos-cement 1/16 bend with mechanical joints						
4"	ER@1.00	Ea	144.00	38.10	9.20	191.30
6"	ER@1.25	Ea	191.00	47.60	11.50	250.10
8"	ER@1.70	Ea	264.00	64.80	15.60	344.40
10"	ER@2.20	Ea	402.00	83.80	20.20	506.00
12"	ER@2.70	Ea	460.00	103.00	24.80	587.80
14"	ER@3.40	Ea	923.00	130.00	31.20	1,084.20
16"	ER@3.90	Ea	1,330.00	149.00	35.80	1,514.80
Asbestos-cement 1/8 bend with mechanical joints						
4"	ER@1.00	Ea	139.00	38.10	9.20	186.30
6"	ER@1.25	Ea	190.00	47.60	11.50	249.10
8"	ER@1.70	Ea	264.00	64.80	15.60	344.40
10"	ER@2.20	Ea	386.00	83.80	20.20	490.00
12"	ER@2.70	Ea	514.00	103.00	24.80	641.80
14"	ER@3.40	Ea	913.00	130.00	31.20	1,074.20
16"	ER@3.90	Ea	1,320.00	149.00	35.80	1,504.80
Asbestos-cement 1/4 bend with mechanical joints						
4"	ER@1.00	Ea	156.00	38.10	9.20	203.30
6"	ER@1.25	Ea	211.00	47.60	11.50	270.10
8"	ER@1.70	Ea	302.00	64.80	15.60	382.40
10"	ER@2.20	Ea	459.00	83.80	20.20	563.00
12"	ER@2.70	Ea	608.00	103.00	24.80	735.80
14"	ER@3.40	Ea	1,130.00	130.00	31.20	1,291.20
16"	ER@3.90	Ea	1,670.00	149.00	35.80	1,854.80

Asbestos-Cement, Class 2400 or 3000 with Mechanical Joints

Description	Craft@Hrs	Unit	Material $	Labor $	Equipment $	Total $
Asbestos-cement tee or wye with mechanical joints						
4"	ER@1.40	Ea	235.00	53.40	12.80	301.20
6"	ER@1.75	Ea	311.00	66.70	16.00	393.70
8"	ER@2.40	Ea	444.00	91.50	22.00	557.50
10"	ER@3.10	Ea	765.00	118.00	28.40	911.40
12"	ER@3.80	Ea	974.00	145.00	34.80	1,153.80
14"	ER@4.75	Ea	1,760.00	181.00	43.60	1,984.60
16"	ER@5.45	Ea	2,540.00	208.00	50.10	2,798.10
Asbestos-cement reducer with mechanical joints						
6" x 4"	ER@1.20	Ea	165.00	45.70	11.00	221.70
8" x 6"	ER@1.50	Ea	242.00	57.20	13.80	313.00
10" x 8"	ER@1.95	Ea	353.00	74.30	17.90	445.20
12" x 10"	ER@2.45	Ea	491.00	93.40	22.60	607.00

Fiberglass Tanks

Description	Craft@Hrs	Unit	Material $	Labor $	Equipment $	Total $

Fiberglass septic tank. 2 compartments. Shallow burial (2 feet of earth cover). 32" manway entrance. One piece construction. Including excavation, backfill and equipment costs.

Description	Craft@Hrs	Unit	Material $	Labor $	Equipment $	Total $
750 gallon	P1@8.00	Ea	1,340.00	286.00	440.00	2,066.00
1,000 gallon	P1@9.00	Ea	1,790.00	322.00	495.00	2,607.00
1,250 gallon	P1@10.0	Ea	2,220.00	358.00	550.00	3,128.00
1,500 gallon	P1@11.5	Ea	2,680.00	411.00	633.00	3,724.00
1,750 gallon	P1@12.0	Ea	3,130.00	429.00	660.00	4,219.00
2,000 gallon	P1@12.5	Ea	3,570.00	447.00	688.00	4,705.00

Fiberglass septic tank. 2 compartments. Deep burial (7 feet of earth cover). 32" x 40" manway extension. One piece construction. Including excavation, backfill and equipment costs.

Description	Craft@Hrs	Unit	Material $	Labor $	Equipment $	Total $
750 gallon	P1@8.00	Ea	1,500.00	286.00	440.00	2,226.00
1,000 gallon	P1@9.00	Ea	1,920.00	322.00	495.00	2,737.00
1,250 gallon	P1@10.0	Ea	2,550.00	358.00	550.00	3,458.00
1,500 gallon	P1@11.5	Ea	3,050.00	411.00	633.00	4,094.00
1,750 gallon	P1@12.0	Ea	3,550.00	429.00	660.00	4,639.00
2,000 gallon	P1@12.5	Ea	4,070.00	447.00	688.00	5,205.00
Manway extension	P1@.350	LF	71.40	12.50	—	83.90

Fiberglass holding tank. Single compartment. Shallow burial (2 feet of earth cover). 32" manway entrance. One piece construction. Including excavation, backfill and equipment costs. Can be used for drink water storage or sewage holding.

Description	Craft@Hrs	Unit	Material $	Labor $	Equipment $	Total $
750 gallon	P1@8.00	Ea	1,200.00	286.00	440.00	1,926.00
1,000 gallon	P1@9.00	Ea	1,540.00	322.00	495.00	2,357.00
1,250 gallon	P1@10.0	Ea	1,970.00	358.00	550.00	2,878.00
1,500 gallon	P1@11.5	Ea	2,370.00	411.00	633.00	3,414.00
1,750 gallon	P1@12.0	Ea	2,750.00	429.00	660.00	3,839.00
2,000 gallon	P1@12.5	Ea	3,150.00	447.00	688.00	4,285.00

Fiberglass holding tank. Single compartment. Deep burial (7 feet of earth cover). 32" x 40" manway extension. One piece construction. Including excavation, backfill and equipment costs. Can be used for drink water storage or sewage holding.

Description	Craft@Hrs	Unit	Material $	Labor $	Equipment $	Total $
750 gallon	P1@8.00	Ea	1,290.00	286.00	440.00	2,016.00
1,000 gallon	P1@9.00	Ea	1,900.00	322.00	495.00	2,717.00
1,250 gallon	P1@10.0	Ea	2,550.00	358.00	550.00	3,458.00
1,500 gallon	P1@11.5	Ea	2,700.00	411.00	633.00	3,744.00
1,750 gallon	P1@12.0	Ea	3,220.00	429.00	660.00	4,309.00
2,000 gallon	P1@12.5	Ea	3,520.00	447.00	688.00	4,655.00
Manway extension	P1@.350	LF	71.40	12.50	—	83.90

Description	Craft@Hrs	Unit	Material $	Labor $	Equipment $	Total $

Plastic water tank. Suitable for potable water. Labor includes setting in place only. Freight, hoisting & rigging excluded.

Description	Craft@Hrs	Unit	Material $	Labor $	Equipment $	Total $
Vertical tank, D31" x H55"						
165 gallon	P1@2.00	Ea	561.00	71.50	—	632.50
Vertical tank, D48" x H102"						
500 gallon	P1@4.00	Ea	1,030.00	143.00	—	1,173.00
Vertical tank, D64" x H79"						
1,000 gallon	P1@4.00	Ea	1,480.00	143.00	—	1,623.00
Vertical tank, D87" x H65"						
1,500 gallon	P1@5.00	Ea	1,930.00	179.00	—	2,109.00
Vertical tank, D95" x H89"						
2,500 gallon	P1@5.00	Ea	2,020.00	179.00	—	2,199.00
Vertical tank, D95" x H107"						
3,000 gallon	P1@6.00	Ea	3,160.00	215.00	—	3,375.00
Vertical tank, D102" x H125"						
4,000 gallon	P1@6.00	Ea	3,370.00	215.00	—	3,585.00
Vertical tank, D102" x H150"						
5,000 gallon	P1@8.00	Ea	6,320.00	286.00	—	6,606.00
Vertical tank, D141" x H192"						
12,000 gallon	P1@9.00	Ea	9,790.00	322.00	—	10,112.00

Plastic holding tank. Single compartment. One piece polyethylene construction. Shallow burial (2 feet of earth cover). 28" manway entrance. Including excavation, backfill and equipment costs. Used for drink water storage.

Description	Craft@Hrs	Unit	Material $	Labor $	Equipment $	Total $
30 gallon	P1@3.00	Ea	188.00	107.00	165.00	460.00
40 gallon	P1@3.00	Ea	213.00	107.00	165.00	485.00
75 gallon	P1@3.00	Ea	238.00	107.00	165.00	510.00
300 gallon	P1@5.00	Ea	1,040.00	179.00	275.00	1,494.00
525 gallon	P1@6.00	Ea	1,510.00	215.00	330.00	2,055.00
1,150 gallon	P1@8.00	Ea	2,640.00	286.00	440.00	3,366.00
1,700 gallon	P1@11.0	Ea	3,400.00	393.00	605.00	4,398.00
1,950 gallon	P1@12.0	Ea	3,620.00	429.00	660.00	4,709.00

Plastic holding tank. Single compartment. One piece polyethylene construction. Above ground applications only. Used for drink water storage.

Description	Craft@Hrs	Unit	Material $	Labor $	Equipment $	Total $
180 gallon	P1@3.00	Ea	488.00	107.00	165.00	760.00
200 gallon	P1@3.75	Ea	725.00	134.00	207.00	1,066.00
300 gallon	P1@4.25	Ea	937.00	152.00	207.00	1,296.00
525 gallon	P1@5.00	Ea	1,380.00	179.00	275.00	1,834.00
1,000 gallon	P1@5.75	Ea	1,880.00	206.00	317.00	2,403.00
1,150 gallon	P1@6.25	Ea	2,300.00	224.00	345.00	2,869.00
1,700 gallon	P1@7.00	Ea	3,060.00	250.00	385.00	3,695.00

Plastic septic tank. 750 gallon. Single compartment. One piece polyethylene construction. Including excavation, backfill and equipment costs.

Description	Craft@Hrs	Unit	Material $	Labor $	Equipment $	Total $
Trickle tank	P1@8.00	Ea	2,130.00	286.00	440.00	2,856.00
Shallow buried	P1@8.00	Ea	2,130.00	286.00	440.00	2,856.00
Deep buried	P1@8.00	Ea	2,250.00	286.00	440.00	2,976.00
Manway extension						
	P1@.350	LF	85.10	12.50	—	97.60

Plastic Tanks

Description	Craft@Hrs	Unit	Material $	Labor $	Equipment $	Total $

Plastic sump pit. One piece polyethylene construction. Complete with radon gasket electrical and discharge grommets. Including excavation, backfill and equipment costs.

Description	Craft@Hrs	Unit	Material $	Labor $	Equipment $	Total $
30 gallon	P1@3.00	Ea	174.00	107.00	165.00	446.00
40 gallon	P1@3.00	Ea	213.00	107.00	165.00	485.00
75 gallon	P1@3.00	Ea	250.00	107.00	165.00	522.00
110 gallon	P1@3.25	Ea	399.00	116.00	180.00	695.00
150 gallon	P1@3.75	Ea	539.00	134.00	207.00	880.00
250 gallon	P1@4.75	Ea	851.00	170.00	262.00	1,283.00
380 gallon	P1@5.50	Ea	1,320.00	197.00	303.00	1,820.00

Plastic sewage lift tank. One piece polyethylene construction. Complete with radon gasket, inlet, vent, electrical and discharge grommets. Including excavation, backfill and equipment costs.

Description	Craft@Hrs	Unit	Material $	Labor $	Equipment $	Total $
30 gallon	P1@3.00	Ea	220.00	107.00	165.00	492.00
40 gallon	P1@3.00	Ea	244.00	107.00	165.00	516.00
75 gallon	P1@3.00	Ea	274.00	107.00	165.00	546.00
110 gallon	P1@3.25	Ea	462.00	116.00	180.00	758.00
150 gallon	P1@3.75	Ea	562.00	134.00	207.00	903.00
250 gallon	P1@4.75	Ea	924.00	170.00	262.00	1,356.00
380 gallon	P1@5.50	Ea	1,360.00	197.00	303.00	1,860.00

Plastic grey water treatment tank. 75 gallon. One piece polyethylene construction. Including excavation, backfill and equipment costs.

Description	Craft@Hrs	Unit	Material $	Labor $	Equipment $	Total $
Buried	P1@3.00	Ea	438.00	107.00	165.00	710.00
Non-buried	P1@1.50	Ea	399.00	53.60	83.00	535.60

Because of varying soil compositions and densities, trenching costs can sometimes be difficult to accurately estimate. The tables in this section are based on average soil conditions, which are loosely defined as dirt, soft clay or gravel mixed with rocks. The costs for trenching through limestone, sandstone or shale can increase by as much as 1,000 percent. For these conditions, it is best to consult with experts or to subcontract the work to specialists.

The primary consideration in excavations is to ensure the safety of the pipe layers. Current OSHA statutes require all excavations over 5 feet deep to be sloped, shored, sheeted or braced to prevent cave-ins. The use of shoring minimizes trenching costs by eliminating the need to slope the trench walls, but increases pipe-laying costs because of the difficulty in working between cross-bracing.

The estimator should carefully compare the costs for installing and removing shoring, and reduced pipe installation productivity, against the additional excavation costs for sloping the walls of the trench. For average soil conditions, not exceeding 6 to 8 feet in depth, it is usually more economical to slope the trench walls.

The current average cost for a backhoe with operator is $95.90 [average of all backhoes dailey rental costs from 2001 AED+ 41.43 operator/hr]per hour. Some trenching contractors additionally charge $25.00 move-on and $25.00 move-off costs, particularly for small projects.

Total trenching, backfill and compaction costs are calculated by multiplying the backhoe hours from the following charts by $95.90 per hour, and adding move-on and move-off charges, if applicable.

Backhoe Trenching					
Trench walls at a 90 degree angle, based on average soil, including trenching, backfill and compaction. Hours per 100 linear feet of trench					
Trench depth (feet)	Trench width (feet)				
	1	2	3	4	5	6
2	2.6	5.1	7.7	10.2	12.8	15.3
3	3.8	7.7	11.5	15.3	19.2	23.0
4	5.1	10.2	15.3	20.4	25.6	30.7
5	6.4	12.8	19.2	25.6	32.0	38.3
6	7.7	15.3	23.0	30.7	38.3	46.0
7	8.9	17.9	26.8	35.8	44.7	53.7
8	10.2	20.4	30.7	40.9	51.1	61.3
9	11.5	23.0	34.5	46.0	57.5	69.0
10	12.8	25.5	38.3	51.1	63.9	76.6

Notes:
1) Compaction is based on the use of a compactor wheel.
2) Additional labor costs must be added if watering-in, stone removal or excess spoil removal is required.

Trenching

Backhoe Trenching Trench walls at a 45 degree angle, based on average soil including trenching, backfill and compaction. Hours per 100 linear feet of trench						
Trench depth (feet)	Trench width (feet)					
	1	2	3	4	5	6
2	7.6	10.3	12.6	15.5	17.9	20.3
3	15.2	22.6	22.6	26.6	30.7	34.5
4	25.5	30.7	35.5	40.7	45.9	51.0
5	38.3	44.5	51.0	57.6	63.8	70.0
6	53.4	61.0	69.0	76.6	84.1	91.7
7	71.4	80.3	89.3	98.3	107.2	115.9
8	92.1	102.1	112.4	122.4	132.8	143.1
9	114.8	126.6	137.9	149.3	161.0	172.4
10	140.3	153.1	165.9	178.6	191.4	204.5

Notes:
1) Compaction is based on the use of a compactor wheel.
2) Additional labor costs must be added if watering-in, stone removal or excess spoil removal is required.

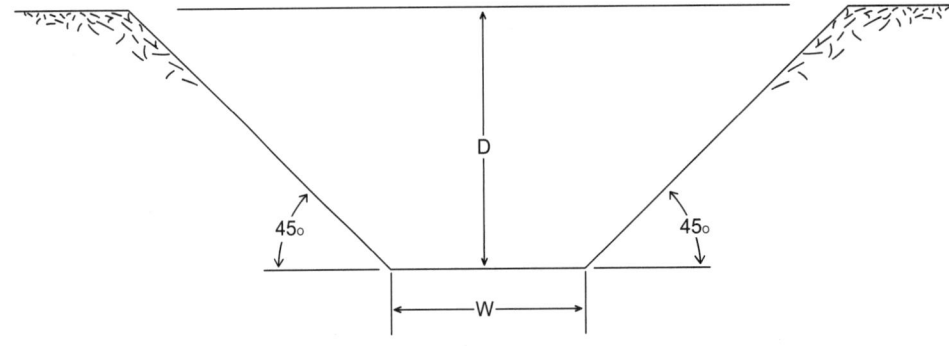

Equipment Rental			
Description	**Per day ($)**	**Per week ($)**	**Per month ($)**
Air compressor, gas or diesel, 100-125 CFM with 100' hose, air tool and 4 bits	174.00	500.00	1,380.00
Air compressor, gas or diesel, 165-185 CFM with 100' hose, air tool and 4 bits	206.00	601.00	1,680.00
Air compressor, gas or diesel, 100-125 CFM, no accessories	101.00	300.00	903.00
Air compressor, gas or diesel, 165-185 CFM, no accessories	133.00	401.00	1,200.00
Backhoe with bucket, 55 HP	410.00	1,150.00	3,400.00
Backhoe with bucket, 100 HP	485.00	1,480.00	4,350.00
Backhoe hydraulic compactor, 5-10 ton	150.00	450.00	1,300.00
Bender, hydraulic	100.00	375.00	1,000.00
Bevel machine with torch	20.00	61.00	177.00
Boom lift, 31' - 40'	235.00	660.00	1,850.00
Boom lift, 51' - 60'	225.00	670.00	1,950.00
Chain hoist, to 2 tons	40.00	140.00	370.00
Chain-hoist, 10 tons	80.00	160.00	480.00
Come-along, ½ to 1 ton	19.00	58.00	174.00
Come-along, 1½ to 2 tons	34.00	102.00	305.00
Come-along, 4-ton	55.00	163.00	489.00
Compactor, soil, gas or diesel, 125 lb	70.00	215.00	610.00
Crane, truck-mounted, 13,000 lb capacity	380.00	1,110.00	3,450.00
Crane, truck-mounted, 30,000 lb capacity	450.00	1,340.00	3,860.00
Forklift, 2,000 pound capacity	130.00	490.00	1,420.00
Forklift, 4,000 pound capacity	110.00	350.00	1,050.00
Forklift, 6,000 pound capacity	130.00	450.00	1,340.00
Front-end loader, 1 yard capacity	290.00	850.00	2,480.00
Front-end loader, 2 yard capacity	340.00	1,050.00	3,150.00
Front-end loader, 3¼ yard capacity	700.00	2,200.00	6,500.00
Pipe machine, up to 2" capacity	55.00	163.00	453.00
Pipe machine, 2½" to 4" capacity	61.00	211.00	646.00
Ramset (nail gun)	34.00	120.00	360.00
Scissors-lift, to 20', diesel, self-propelled	110.00	280.00	675.00
Scissors-lift, 21' – 30'	165.00	475.00	1,300.00
Skid Steer Loader 1,750 lb capacity	200.00	640.00	1,830.00
Trailer, office, 25'	55.00	180.00	420.00
Truck, 3/4 ton pickup	79.00	315.00	1,260.00
Truck, 1 ton pickup	129.00	410.00	1,290.00
Truck, 1 ton stake bed	166.00	497.00	1,460.00
Van, storage, 26'	—	—	315.00
Van, storage, 40'	—	—	383.00
Welding machine, gas, 200 amp, trailer-mounted with 50' bare lead	55.00	200.00	520.00
Welding machine, diesel, 400 amp, trailer-mounted with 50' bare lead	80.00	275.00	720.00

Close-out Items

Description	Craft@Hrs	Unit	Material $	Labor $	Equipment $	Total $

1½" brass valve tags with black filled numbers

Typical costs P1@.150	Ea	3.10	5.36	—	8.46

1½" blank brass valve tags

Typical costs P1@.150	Ea	1.54	5.36	—	6.90

1½" plastic valve tags with engraved numbers

Typical costs P1@.150	Ea	2.59	5.36	—	7.95

Engraved plastic equipment nameplates

Typical costs P1@.400	Ea	6.69	14.30	—	20.99

Pipe and duct markers

Typical costs P1@.250	Ea	3.82	8.94	—	12.76

As-built drawings

Typical costs P1@.500	Ea	—	17.90	—	17.90

Instructing owner's operating personnel

Typical costs P1@8.00	Ea	—	286.00	—	286.00

Description	Craft@Hrs	Unit	Material $	Labor $	Equipment $	Total $

Remove rectangular galvanized ductwork.

22 to 30 gauge galvanized sheet metal (dimension/gauge per SMACNA standards). Per linear foot of sheet metal duct work. The cost of removing ductwork includes the cost of removing the support hangers. Add the cost of disconnecting and removing heating coils, air volume zone boxes, air handing units, condensate drains, motorized dampers or other associated accessories, piping or wiring. These figures assume demolished materials are piled on site. Disposal costs or salvage values excluded. (SMACNA = Sheetmetal & Air Conditioning National Association.) Equipment may be needed for some sizes and locations. Refer to the equipment rental section at the end of this book.

Description	Craft@Hrs	Unit	Material $	Labor $	Equipment $	Total $
4" to 12" rectangular/square duct, remove	S2@.015	LF	—	.53	—	.53
14" to 24" rectangular/square duct, remove	S2@.020	LF	—	.71	—	.71
26" to 36" rectangular/square duct, remove	S2@.065	LF	—	2.30	—	2.30
38" to 60" rectangular/square duct, remove	S2@.090	LF	—	3.18	—	3.18

Remove round galvanized ductwork.

22 to 30 gauge galvanized sheet metal (dimension/gauge per SMACNA standards). Per linear foot of sheet metal duct work. The cost of removing ductwork includes the cost of removing the support hangers. Add the cost of disconnecting and removing heating coils, air volume zone boxes, air handing units, condensate drains, motorized dampers or other associated accessories, piping or wiring. These figures assume demolished materials are piled on site. Disposal costs or salvage values excluded. (SMACNA = Sheetmetal & Air Conditioning National Association.) Equipment may be needed for some sizes and locations. Refer to the equipment rental section at the end of this book.

Description	Craft@Hrs	Unit	Material $	Labor $	Equipment $	Total $
4" to 12" round duct, remove	S2@.012	LF	—	.42	—	.42
14" to 24" round duct, remove	S2@.016	LF	—	.57	—	.57
26" to 36" round duct, remove	S2@.052	LF	—	1.84	—	1.84
38" to 60" round duct, remove	S2@.072	LF	—	2.55	—	2.55

Description	Craft@Hrs	Unit	Material $	Labor $	Equipment $	Total $

Remove welded steel ductwork. 10 to 16 gauge mild steel. Per linear foot of welded mild steel duct work. The cost of removing ductwork includes the cost of removing the support hangers. Add the cost of disconnecting and removing exhaust fans, condensate drains, sprinkler heads or other associated accessories, piping or wiring. These figures assume demolished materials are piled on site. Disposal costs or salvage values excluded. Equipment may be needed for some sizes and locations. Refer to the equipment rental section at the end of this book.

Description	Craft@Hrs	Unit	Material $	Labor $	Equipment $	Total $
4" to 12", square or round duct, remove	S2@.045	LF	—	1.59	—	1.59
14" to 24", square or round duct, remove	S2@.060	LF	—	2.12	—	2.12
26" to 36", square or round duct, remove	S2@.095	LF	—	3.36	—	3.36
38" to 60", square or round duct, remove	S2@.150	LF	—	5.30	—	5.30

Remove diffuser or grille. These figures assume demolished materials are piled on site. Disposal costs or salvage values excluded. Equipment may be needed for some sizes and locations. Refer to the equipment rental section at the end of this book.

Description	Craft@Hrs	Unit	Material $	Labor $	Equipment $	Total $
6" to 12", square or round diffuser/grille, remove	S2@.150	Ea	—	5.30	—	5.30
14" to 24", square or round diffuser/grille, remove	S2@.180	Ea	—	6.36	—	6.36
26" to 36", square or round diffuser/grille, remove	S2@.400	Ea	—	14.10	—	14.10
38" to 60" square or round diffuser/grille, remove	S2@.800	Ea	—	28.30	—	28.30

Remove air mixing box. These figures include isolating and disconnecting hot water, chilled water, steam, refrigerant and/or pneumatic control tube or control wiring. These figures assume demolished materials are piled on site. Disposal costs or salvage values excluded. Equipment may be needed for some sizes and locations. Refer to the equipment rental section at the end of this book.

Description	Craft@Hrs	Unit	Material $	Labor $	Equipment $	Total $
Air Box, 50 to 150 CFM, remove	S2@.500	Ea	—	17.70	—	17.70
Air Box, 160 to 500 CFM, remove	S2@.650	Ea	—	23.00	—	23.00
Air Box, 550 to 750 CFM, remove	S2@.900	Ea	—	31.80	—	31.80

Description	Craft@Hrs	Unit	Material $	Labor $	Equipment $	Total $

Remove air handling unit. These figures include isolating and disconnecting hot water, chilled water, steam or gas and/or ductwork, electrical connections and unit disassembly when required for removal. Disposal costs or salvage values excluded. Equipment may be needed for some sizes and locations. Refer to the equipment rental section at the end of this book.

Description	Craft@Hrs	Unit	Material $	Labor $	Equipment $	Total $
AHU, 2,000- 5,000 CFM, remove	P1@8.00	Ea	—	286.00	—	286.00
AHU, 5,000-10,000 CFM, remove	P1@12.0	Ea	—	429.00	—	429.00
AHU,10,000-20,000 CFM, remove	P1@18.0	Ea	—	644.00	—	644.00
AHU, over 20,000 CFM, remove	P1@28.0	Ea	—	1,000.00	—	1,000.00

Remove roof top unit. These figures include isolating and disconnecting hot water, chilled water, steam or gas and/or ductwork, electrical connections and unit disassembly when required for removal. Disposal costs or salvage values excluded. Make additional hoisting allowances for roof heights above 40'. Equipment may be needed for some sizes and locations. Refer to the equipment rental section at the end of this book.

Description	Craft@Hrs	Unit	Material $	Labor $	Equipment $	Total $
RTU, 2,000- 5,000 CFM, remove	P1@6.00	Ea	—	215.00	—	215.00
RTU, 5,000-10,000 CFM, remove	P1@10.0	Ea	—	358.00	—	358.00
RTU. 10,000-20,000 CFM, remove	P1@16.0	Ea	—	572.00	—	572.00
RTU, over 20,000 CFM, remove	P1@26.0	Ea	—	930.00	—	930.00

Description	Craft@Hrs	Unit	Material $	Labor $	Equipment $	Total $

Remove boiler. These figures include isolating and disconnecting hot water, steam or gas and/or electrical connections and unit disassembly when required for removal. Disposal costs or salvage values excluded. Equipment may be needed for some sizes and locations. Refer to the equipment rental section at the end of this book.

Description	Craft@Hrs	Unit	Material $	Labor $	Equipment $	Total $
10-50 KW electric boiler, remove	P1@8.00	Ea	—	286.00	—	286.00
50-100 KW electric boiler, remove	P1@12.0	Ea	—	429.00	—	429.00
100-200 KW electric boiler, remove	P1@26.0	Ea	—	930.00	—	930.00
200-500 KW electric boiler, remove	P1@36.0	Ea	—	1,290.00	—	1,290.00
over 500 KW electric boiler, remove	P1@44.0	Ea	—	1,570.00	—	1,570.00
to 150 Mbh gas fired boiler, remove	P1@9.00	Ea	—	322.00	—	322.00
150-300 Mbh gas fired boiler, remove	P1@14.0	Ea	—	501.00	—	501.00
300-750 Mbh gas fired boiler, remove	P1@30.0	Ea	—	1,070.00	—	1,070.00
750-1500 Mbh gas fired boiler, remove	P1@46.0	Ea	—	1,640.00	—	1,640.00
1500-2000 Mbh gas fired boiler, remove	P1@65.0	Ea	—	2,320.00	—	2,320.00
over 2000 Mbh gas fired boiler, remove	P1@80.0	Ea	—	2,860.00	—	2,860.00

Remove furnace. These figures include isolating and disconnecting plenums, fuel and/or electrical connections. Disposal costs or salvage values excluded. Equipment may be needed for some sizes and locations. Refer to the equipment rental section at the end of this book.

Description	Craft@Hrs	Unit	Material $	Labor $	Equipment $	Total $
5-30 KW electric furnace, remove	P1@2.50	Ea	—	89.40	—	89.40
30-60 KW electric furnace, remove	P1@3.75	Ea	—	134.00	—	134.00
30-150 Mbh gas fired furnace, remove	P1@5.00	Ea	—	179.00	—	179.00
150-225 Mbh gas fired furnace, remove	P1@7.00	Ea	—	250.00	—	250.00

Description	Craft@Hrs	Unit	Material $	Labor $	Equipment $	Total $

Remove chiller. These figures include isolating and disconnecting chilled water and electrical connections, evacuating refrigerant to storage vessels and unit disassembly when required for removal. Disposal costs or salvage values excluded. Equipment may be needed for some sizes and locations. Refer to the equipment rental section at the end of this book.

Description	Craft@Hrs	Unit	Material $	Labor $	Equipment $	Total $
Chiller, 20-50 tons, remove	P1@34.0	Ea	—	1,220.00	—	1,220.00
Chiller, 50-100 tons, remove	P1@48.0	Ea	—	1,720.00	—	1,720.00
Chiller, 100-150 tons, remove	P1@56.0	Ea	—	2,000.00	—	2,000.00
Chiller, 150-200 tons, remove	P1@68.0	Ea	—	2,430.00	—	2,430.00
Chiller, 200-250 tons, remove	P1@80.0	Ea	—	2,860.00	—	2,860.00
Chiller, over 250 tons, remove	P1@100	Ea	—	3,580.00	—	3,580.00

Remove air cooled condenser. These figures include isolating and disconnecting refrigerant lines and electrical connections, evacuating refrigerant to storage vessels and unit disassembly when required for removal. Disposal costs or salvage values excluded. Make additional hoisting allowances for roof heights above 40'. Equipment may be needed for some sizes and locations. Refer to the equipment rental section at the end of this book.

Description	Craft@Hrs	Unit	Material $	Labor $	Equipment $	Total $
Air cooled condenser, 2-4 tons, remove	P1@4.00	Ea	—	143.00	—	143.00
Air cooled condenser, 5-10 tons, remove	P1@8.00	Ea	—	286.00	—	286.00
Air cooled condenser, 10-20 tons, remove	P1@20.0	Ea	—	715.00	—	715.00
Air cooled condenser, 20-30 tons, remove	P1@28.0	Ea	—	1,000.00	—	1,000.00
Air cooled condenser, 30-50 tons, remove	P1@36.0	Ea	—	1,290.00	—	1,290.00
Air cooled condenser, 50-150 tons, remove	P1@44.0	Ea	—	1,570.00	—	1,570.00
Air cooled condenser, 150-250 tons, remove	P1@52.0	Ea	—	1,860.00	—	1,860.00

Description	Craft@Hrs	Unit	Material $	Labor $	Equipment $	Total $

Remove cooling tower. These figures include isolating and disconnecting chilled water lines and electrical connections, and unit disassembly when required for removal. Disposal costs or salvage values excluded. Make additional hoisting allowances for roof heights above 40'. Equipment may be needed for some sizes and locations. Refer to the equipment rental section at the end of this book.

Description	Craft@Hrs	Unit	Material $	Labor $	Equipment $	Total $
Cooling tower, 5-10 tons, remove	P1@8.00	Ea	—	286.00	—	286.00
Cooling tower, 10-20 tons, remove	P1@16.0	Ea	—	572.00	—	572.00
Cooling tower, 20-30 tons, remove	P1@24.0	Ea	—	858.00	—	858.00
Cooling tower, 30-50 tons, remove	P1@32.0	Ea	—	1,140.00	—	1,140.00
Cooling tower, 50-150 tons, remove	P1@40.0	Ea	—	1,430.00	—	1,430.00
Cooling tower, 150-250 tons, remove	P1@48.0	Ea	—	1,720.00	—	1,720.00
Cooling tower, 250-500 tons, remove	P1@70.0	Ea	—	2,500.00	—	2,500.00

Remove fan coil. These figures include isolating and disconnecting chilled water and/or hot water lines and/or duct and electrical connections, and unit disassembly when required for removal. Disposal costs or salvage values excluded. Equipment may be needed for some sizes and locations. Refer to the equipment rental section at the end of this book.

Description	Craft@Hrs	Unit	Material $	Labor $	Equipment $	Total $
2 pipe fan coil, 300- 500 CFM, remove	P1@4.00	Ea	—	143.00	—	143.00
2 pipe fan coil, 550- 750 CFM, remove	P1@4.50	Ea	—	161.00	—	161.00
2 pipe fan coil, 800- 1200 CFM, remove	P1@6.00	Ea	—	215.00	—	215.00
4 pipe fan coil, 300- 500 CFM, remove	P1@4.50	Ea	—	161.00	—	161.00
4 pipe fan coil, 550- 750 CFM, remove	P1@5.00	Ea	—	179.00	—	179.00
4 pipe fan coil, 800- 1200 CFM, remove	P1@6.50	Ea	—	232.00	—	232.00

Description	Craft@Hrs	Unit	Material $	Labor $	Equipment $	Total $

Remove unit heater. These figures include isolating and disconnecting hot water, steam lines or gas and/or electrical connections. Disposal costs or salvage values excluded. Equipment may be needed for some sizes and locations. Refer to the equipment rental section at the end of this book.

Description	Craft@Hrs	Unit	Material $	Labor $	Equipment $	Total $
Gas fired unit heater, 50-150 Mbh, remove	P1@2.25	Ea	—	80.50	—	80.50
Gas fired unit heater, 200-350 Mbh, remove	P1@3.50	Ea	—	125.00	—	125.00
Steam unit heater, 75–200 Mbh, remove	P1@3.25	Ea	—	116.00	—	116.00
Steam unit heater, 250– 350 Mbh, remove	P1@4.75	Ea	—	170.00	—	170.00
Electric unit heater, 5-25 KW, remove	P1@2.00	Ea	—	71.50	—	71.50
Electric unit heater, 30-50 KW, remove	P1@2.75	Ea	—	98.30	—	98.30

Remove heat pump. These figures include isolating and disconnecting pipe, duct and electrical connections and evacuating refrigerant to storage vessels when required for removal. Disposal costs or salvage values excluded. Equipment may be needed for some sizes and locations. Refer to the equipment rental section at the end of this book.

Description	Craft@Hrs	Unit	Material $	Labor $	Equipment $	Total $
Heat pump, split system, 2-5 Ton, remove	P1@4.75	Ea	—	170.00	—	170.00
Heat pump, split system, 7.5–10 Ton, remove	P1@8.50	Ea	—	304.00	—	304.00
Heat pump, packaged system, 2-5 Ton, remove	P1@3.25	Ea	—	116.00	—	116.00
Heat pump, pkg. system, 7.5-10 Ton, remove	P1@5.50	Ea	—	197.00	—	197.00

HVAC & Plumbing Demolition

Description	Craft@Hrs	Unit	Material $	Labor $	Equipment $	Total $

Remove heat exchanger. These figures include isolating and disconnecting chilled water, hot water and/or steam line connections. Disposal costs or salvage values excluded. Equipment may be needed for some sizes and locations. Refer to the equipment rental section at the end of this book.

Description	Craft@Hrs	Unit	Material $	Labor $	Equipment $	Total $
Heat exchanger, 6" x 30", remove	P1@2.75	Ea	—	98.30	—	98.30
Heat exchanger, 10" x 48", remove	P1@4.00	Ea	—	143.00	—	143.00
Heat exchanger, 12" x 72", remove	P1@5.25	Ea	—	188.00	—	188.00
Heat exchanger, 16" x 90", remove	P1@6.00	Ea	—	215.00	—	215.00

Remove duct mounted coil. These figures include isolating and disconnecting steam, chilled water, hot water, condensate, refrigerant or electrical and duct connections. Disposal costs or salvage values excluded. Refrigerant coil removal costs include system refrigerant evacuation and storage. Equipment may be needed for some sizes and locations. Refer to the equipment rental section at the end of this book.

Description	Craft@Hrs	Unit	Material $	Labor $	Equipment $	Total $
Single row water coil, 12" to 18", remove	P1@2.00	Ea	—	71.50	—	71.50
Single row water coil, 20" to 30", remove	P1@3.50	Ea	—	125.00	—	125.00
Single water coil, 32" to 60", remove	P1@8.00	Ea	—	286.00	—	286.00
Single row steam coil, 12" to 18" remove	P1@2.25	Ea	—	80.50	—	80.50
Single row steam coil, 20" to 30" remove	P1@3.75	Ea	—	134.00	—	134.00
Single row steam coil, 32" to 60" remove	P1@9.00	Ea	—	322.00	—	322.00
Electric resistance, 12" to 18" remove	P1@1.25	Ea	—	44.70	—	44.70
Electric resistance, 20" to 30" remove	P1@1.75	Ea	—	62.60	—	62.60
Electric resistance, 32" to 60" remove	P1@4.00	Ea	—	143.00	—	143.00
Refrigerant coil, 2-3 ton remove	P1@4.00	Ea	—	143.00	—	143.00
Refrigerant coil, 4-5 ton remove	P1@6.50	Ea	—	232.00	—	232.00
Refrigerant coil, 6-10 ton remove	P1@12.0	Ea	—	429.00	—	429.00
Refrigerant coil, 12-15 ton remove	P1@16.0	Ea	—	572.00	—	572.00

Description	Craft@Hrs	Unit	Material $	Labor $	Equipment $	Total $

Remove steel piping. These figures include cutting out and removing threaded or welded steel piping. These figures assume demolished materials are piled on site. Equipment, fixture or apparatus disconnection costs and salvage values excluded. Equipment may be needed for some sizes and locations. Refer to the equipment rental section at the end of this book.

Description	Craft@Hrs	Unit	Material $	Labor $	Equipment $	Total $
Steel pipe, sch. 40, 1/2" – 1"						
remove	P1@.038	LF	—	1.36	—	1.36
Steel pipe, sch. 40, 1½" – 2"						
remove	P1@.068	LF	—	2.43	—	2.43
Steel pipe, sch. 40, 2½" – 4"						
remove	P1@.108	LF	—	3.86	—	3.86
Steel pipe, sch. 40, 6" – 8"						
remove	P1@.230	LF	—	8.22	—	8.22
Steel pipe, sch. 40, 10" – 12"						
remove	P1@.330	LF	—	11.80	—	11.80
Steel pipe, sch. 80, 1/2" – 1"						
remove	P1@.042	LF	—	1.50	—	1.50
Steel pipe, sch. 80, 1½" – 2"						
remove	P1@.075	LF	—	2.68	—	2.68
Steel pipe, sch. 80, 2½" – 4"						
remove	P1@.118	LF	—	4.22	—	4.22
Steel pipe, sch. 80, 6" – 8"						
remove	P1@.290	LF	—	10.40	—	10.40
Steel pipe, sch. 80, 10" – 12"						
remove	P1@.360	LF	—	12.90	—	12.90
Remove steel pipe, sch. 160,						
Add	—	%	—	—	—	30.00

Remove copper piping. These figures include cutting out and removing copper piping. These figures assume demolished materials are piled on site. Equipment, fixture or apparatus disconnection costs and salvage values excluded.

Description	Craft@Hrs	Unit	Material $	Labor $	Equipment $	Total $
Copper pipe, 1/2" – 1"						
remove	P1@.022	LF	—	.79	.07	.86
Copper pipe, 1½" – 2"						
remove	P1@.028	LF	—	1.00	.08	1.08
Copper pipe, 2½" – 4"						
remove	P1@.044	LF	—	1.57	.11	1.68

Remove plastic piping. These figures include cutting out and removing plastic piping. These figures assume demolished materials are piled on site. Equipment, fixture or apparatus disconnection costs and salvage values excluded. Equipment may be needed for some sizes and locations. Refer to the equipment rental section at the end of this book.

Description	Craft@Hrs	Unit	Material $	Labor $	Equipment $	Total $
plastic pipe, 1/2" – 1"						
remove	P1@.015	LF	—	.54	—	.54
plastic pipe, 1½" – 2"						
remove	P1@.030	LF	—	1.07	—	1.07
plastic pipe, 2½" – 4"						
remove	P1@.038	LF	—	1.36	—	1.36
plastic pipe, 6" – 8"						
remove	P1@.080	LF	—	.00	—	.00

Description	Craft@Hrs	Unit	Material $	Labor $	Equipment $	Total $

Remove pump. These figures include isolating and disconnecting pipe and electrical connections. Disposal costs or salvage values excluded. Equipment may be needed for some sizes and locations. Refer to the equipment rental section at the end of this book.

Description	Craft@Hrs	Unit	Material $	Labor $	Equipment $	Total $
Base mounted pump, 1/2 – 1 HP, remove	P1@1.00	Ea	—	35.80	—	35.80
Base mounted pump, 1½ – 3 HP, remove	P1@1.75	Ea	—	62.60	—	62.60
Base mounted pump, 5 – 10 HP, remove	P1@2.75	Ea	—	98.30	—	98.30
Base mounted pump, 15 – 25 HP, remove	P1@3.75	Ea	—	134.00	—	134.00
Base mounted pump, 30 – 50 HP, remove	P1@4.55	Ea	—	163.00	—	163.00
In-line mounted pump, 1/12 – 1/3 HP, remove	P1@.650	Ea	—	23.20	—	23.20
In-line mounted pump, 1/2 – 1 HP, remove	P1@.750	Ea	—	26.80	—	26.80
In-line mounted pump, 3 – 5 HP, remove	P1@1.85	Ea	—	66.20	—	66.20
In-line mounted pump, 7.5 – 10 HP, remove	P1@2.55	Ea	—	91.20	—	91.20
In-line mounted pump, 15 – 25 HP, remove	P1@3.90	Ea	—	139.00	—	139.00
In-line mounted pump, 30 – 50 HP, remove	P1@4.85	Ea	—	173.00	—	173.00

Remove valve. These figures include isolation and disconnection. Disposal costs or salvage values excluded. Equipment may be needed for some sizes and locations. Refer to the equipment rental section at the end of this book.

Description	Craft@Hrs	Unit	Material $	Labor $	Equipment $	Total $
Soldered valve, 1/2" – 1", remove	P1@.165	Ea	—	5.90	—	5.90
Soldered valve, 1½" – 2", remove	P1@.225	Ea	—	8.05	—	8.05
Soldered valve, 2-1/2" – 4", remove	P1@.650	Ea	—	23.20	—	23.20
Threaded valve, 1/2" – 1", remove	P1@.150	Ea	—	5.36	—	5.36
Threaded valve, 1½" – 2", remove	P1@.235	Ea	—	8.40	—	8.40
Threaded valve, 2½" – 4", remove	P1@.680	Ea	—	24.30	—	24.30
Flanged valve, 1/2" – 1", remove	P1@.200	Ea	—	7.15	—	7.15
Flanged valve, 1½" – 2", remove	P1@.280	Ea	—	10.00	—	10.00
Flanged valve, 2½" – 4", remove	P1@.625	Ea	—	22.40	—	22.40
Flanged valve, 6" – 8", remove	P1@1.70	Ea	—	60.80	—	60.80
Flanged valve, 10" – 12", remove	P1@2.65	Ea	—	94.80	—	94.80

Description	Craft@Hrs	Unit	Material $	Labor $	Equipment $	Total $

Remove hot water tank. These figures include isolation and disconnection. Disposal costs or salvage values excluded. Equipment may be needed for some sizes and locations. Refer to the equipment rental section at the end of this book.

Description	Craft@Hrs	Unit	Material $	Labor $	Equipment $	Total $
Gas fired hot water tank, 40 gallon, remove	P1@2.00	Ea	—	71.50	—	71.50
Gas fired hot water tank, 60 gallon, remove	P1@2.50	Ea	—	89.40	—	89.40
Gas fired hot water tank, 100 gallon, remove	P1@3.00	Ea	—	107.00	—	107.00
Electric hot water tank, 40 Gallon, remove	P1@1.60	Ea	—	57.20	—	57.20
Electric hot water tank, 60 Gallon, remove	P1@2.00	Ea	—	71.50	—	71.50
Electric hot water tank, 100 Gallon, remove	P1@2.60	Ea	—	93.00	—	93.00

Remove plumbing fixture. These figures include isolation and disconnection. Disposal costs or salvage values excluded. Equipment may be needed for some sizes and locations. Refer to the equipment rental section at the end of this book.

Description	Craft@Hrs	Unit	Material $	Labor $	Equipment $	Total $
Floor mount, tank type water closet, remove	P1@.850	ea	—	30.40	—	30.40
Wall mount, tank type water closet, remove	P1@1.35	Ea	—	48.30	—	48.30
Floor mount, flush valve water closet, remove	P1@.950	Ea	—	34.00	—	34.00
Wall mount, flush valve water closet, remove	P1@1.55	Ea	—	55.40	—	55.40
Wall mount, flush valve urinal, remove	P1@1.15	Ea	—	41.10	—	41.10
Wall mount, tank type urinal, remove	P1@1.65	Ea	—	59.00	—	59.00
Kitchen sink, remove	P1@.850	Ea	—	30.40	—	30.40
Bathroom sink, remove	P1@.850	Ea	—	30.40	—	30.40
Bathtub, remove	P1@2.25	Ea	—	80.50	—	80.50
Shower stall, remove	P1@2.00	Ea	—	71.50	—	71.50
Tub and shower combination, remove	P1@2.85	Ea	—	102.00	—	102.00
Laundry sink, remove	P1@.950	Ea	—	34.00	—	34.00
Mop sink, remove	P1@1.50	Ea	—	53.60	—	53.60
Slop sink, remove	P1@1.75	Ea	—	62.60	—	62.60
Drinking fountain, remove	P1@1.40	Ea	—	50.10	—	50.10

HVAC & Plumbing Demolition

Description	Craft@Hrs	Unit	Material $	Labor $	Equipment $	Total $

Disconnect plumbing fixture. These figures include isolation and disconnection only. Fixtures removed by others. Disposal costs or salvage values excluded. Equipment may be needed for some sizes and locations. Refer to the equipment rental section at the end of this book.

Description	Craft@Hrs	Unit	Material $	Labor $	Equipment $	Total $
Floor mount, tank type water closet, disconnect	P1@.450	Ea	—	16.10	—	16.10
Wall mount, tank type water closet, disconnect	P1@.650	Ea	—	23.20	—	23.20
Floor mount, flush valve water closet, disconnect	P1@.550	Ea	—	19.70	—	19.70
Wall mount, flush valve water closet, disconnect	P1@.650	Ea	—	23.20	—	23.20
Wall mount, flush valve urinal, disconnect	P1@.650	Ea	—	23.20	—	23.20
Wall mount, tank type urinal, disconnect	P1@.950	Ea	—	34.00	—	34.00
Kitchen sink, disconnect	P1@.450	Ea	—	16.10	—	16.10
Bathroom sink, disconnect	P1@.450	Ea	—	16.10	—	16.10
Bathtub, disconnect	P1@.650	Ea	—	23.20	—	23.20
Shower stall, disconnect	P1@.650	Ea	—	23.20	—	23.20
Tub and shower combination, disconnect	P1@.650	Ea	—	23.20	—	23.20
Laundry sink, disconnect	P1@.450	Ea	—	16.10	—	16.10
Mop sink, disconnect	P1@.450	Ea	—	16.10	—	16.10
Slop sink, disconnect	P1@.750	Ea	—	26.80	—	26.80
Drinking fountain, disconnect	P1@.650	Ea	—	23.20	—	23.20
Gas fired hot water tank, 40 gallon, disconnect	P1@.900	Ea	—	32.20	—	32.20
Gas fired hot water tank, 60 gallon, disconnect	P1@1.15	Ea	—	41.10	—	41.10
Gas fired hot water tank, 100 gallon, disconnect	P1@1.35	Ea	—	48.30	—	48.30
Electric hot water tank, 40 gallon, disconnect	P1@.600	Ea	—	21.50	—	21.50
Electric hot water tank, 60 gallon, disconnect	P1@.900	Ea	—	32.20	—	32.20
Electric hot water tank, 100 gallon, disconnect	P1@1.00	Ea	—	35.80	—	35.80

Description	Craft@Hrs	Unit	Material $	Labor $	Equipment $	Total $

Remove pipe insulation. These figures assume demolished materials are piled on site. Disposal costs or salvage values excluded.

Description	Craft@Hrs	Unit	Material $	Labor $	Equipment $	Total $
Fiberglass pipe insulation, 1/2" – 1", remove	P1@.020	LF	—	.72	—	.72
Fiberglass pipe insulation, 1½" – 2", remove	P1@.025	LF	—	.89	—	.89
Fiberglass pipe insulation, 2½" – 4", remove	P1@.035	LF	—	1.25	—	1.25
Fiberglass pipe insulation, 6" – 8", remove	P1@.060	LF	—	2.15	—	2.15
Fiberglass pipe insulation, 10" – 12", remove	P1@.090	LF	—	3.22	—	3.22
Pipe insul with canvas jacket - add		%	—	10.00	—	10.00
Pipe insul. with alum. jacket - add%			—	20.00	—	20.00
Calcium silicate pipe insul. - add%			—	15.00	—	15.00

Remove duct insulation. These figures assume demolished materials are piled on site. Disposal costs or salvage values excluded.

Description	Craft@Hrs	Unit	Material $	Labor $	Equipment $	Total $
Fiberglass blanket insulation, remove	S2@.020	SF	—	.71		.71
Fiberglass rigid board insulation, remove	S2@.025	SF	—	.88	—	.88
Insul with canvas jacket - add	%		—	8.00	—	8.00
Insul. with alum jacket - add	%		—	10.00	—	15.00

Budget Estimates for Plumbing and HVAC Work

General contractors and owners often ask mechanical subcontractors to prepare budget cost estima for use in planning projects. A budget estimate isn't a bid for the contract and you probably won't be p for preparing the estimate. That's why you won't want to spend too much time on these estima Accuracy within 8 to 10 percent of the final bid price will usually be acceptable. Normally you'll have c preliminary architectural drawings and possibly a guide specification to work from. The tables that fo will help you prepare budget estimates for nearly any plumbing or HVAC job.

The costs in these tables are averages and will vary considerably with design requirements and geographic area. Make it clear to the owner or architect requesting the figures that these are bud estimates and will be revised when final plans and specifications are available.

Budget Estimates for Plumbing: First, count the number of plumbing fixtures. Multiply the numbe fixtures of each type by the price per fixture shown in the table below. These prices include all the us labor and material for a complete average plumbing installation. Prices are based on branch pip lengths averaging 35 feet between the fixture and the main. Typical main piping is also included in prices.

Description	Craft@Hrs	Unit	Material $	Labor $	Equipment $	Total $
Budget plumbing estimate, cast iron DWV, copper supply pipe						
Bathtubs	P1@15.0	Ea	2,080.00	536.00	—	2,616.00
Showers	P1@9.00	Ea	1,130.00	322.00	—	1,452.00
Lavatories	P1@11.0	Ea	1,950.00	393.00	—	2,343.00
Kitchen sinks	P1@12.0	Ea	2,030.00	429.00	—	2,459.00
Service sinks	P1@10.0	Ea	1,750.00	358.00	—	2,108.00
Bar sinks	P1@9.00	Ea	1,480.00	322.00	—	1,802.00
Floor sinks	P1@7.00	Ea	913.00	250.00	—	1,163.00
Water closets						
Tank-type	P1@11.0	Ea	1,480.00	393.00	—	1,873.00
Flush valve	P1@12.0	Ea	1,660.00	429.00	—	2,089.00
Urinals	P1@8.00	Ea	2,130.00	286.00	—	2,416.00
Drinking fountains						
(refrigerated)	P1@12.0	Ea	2,460.00	429.00	—	2,889.00
Wash fountains	P1@25.0	Ea	5,300.00	894.00	—	6,194.00
Can washers	P1@18.0	Ea	2,710.00	644.00	—	3,354.00
Floor drains	P1@8.00	Ea	834.00	286.00	—	1,120.00
Area drains	P1@6.00	Ea	718.00	215.00	—	933.00
Roof drains	P1@18.0	Ea	1,890.00	644.00	—	2,534.00
Overflow drains	P1@18.0	Ea	1,890.00	644.00	—	2,534.00
Deck drains	P1@4.00	Ea	590.00	143.00	—	733.00
Cleanouts	P1@5.00	Ea	638.00	179.00	—	817.00
Trap primers	P1@2.00	Ea	324.00	71.50	—	395.50
Sump pumps	P1@6.00	Ea	3,150.00	215.00	—	3,365.00
Water heaters						
Gas, 40 gal.	P1@16.0	Ea	3,730.00	572.00	—	4,302.00
Gas, 80 gal.	P1@17.0	Ea	4,130.00	608.00	—	4,738.00
Gas, 120 gal.	P1@18.0	Ea	4,810.00	644.00	—	5,454.00

Description	Craft@Hrs	Unit	Material $	Labor $	Equipment $	Total $

Budget plumbing estimate, plastic DWV and plastic supply pipe

Description	Craft@Hrs	Unit	Material $	Labor $	Equipment $	Total $
Bathtubs	P1@12.0	Ea	1,680.00	429.00	—	2,109.00
Showers	P1@7.00	Ea	812.00	250.00	—	1,062.00
Lavatories	P1@9.00	Ea	1,560.00	322.00	—	1,882.00
Kitchen sinks	P1@10.0	Ea	1,740.00	358.00	—	2,098.00
Service sinks	P1@8.00	Ea	1,390.00	286.00	—	1,676.00
Bar sinks	P1@7.00	Ea	1,170.00	250.00	—	1,420.00
Floor sinks	P1@6.00	Ea	707.00	215.00	—	922.00
Water closets						
Tank-type	P1@8.00	Ea	1,090.00	286.00	—	1,376.00
Flush valve	P1@9.00	Ea	1,310.00	322.00	—	1,632.00
Urinals	P1@6.00	Ea	1,760.00	215.00	—	1,975.00
Drinking fountains						
(refrigerated)	P1@10.0	Ea	2,040.00	358.00	—	2,398.00
Wash fountains	P1@22.0	Ea	4,680.00	787.00	—	5,467.00
Can washers	P1@15.0	Ea	2,290.00	536.00	—	2,826.00
Floor drains	P1@7.00	Ea	584.00	250.00	—	834.00
Area drains	P1@5.00	Ea	476.00	179.00	—	655.00
Roof drains	P1@16.0	Ea	1,630.00	572.00	—	2,202.00
Overflow drains	P1@16.0	Ea	1,630.00	572.00	—	2,202.00
Deck drains	P1@3.00	Ea	407.00	107.00	—	514.00
Cleanouts	P1@4.00	Ea	496.00	143.00	—	639.00
Trap primers	P1@2.00	Ea	246.00	71.50	—	317.50
Sump pumps	P1@5.00	Ea	2,860.00	179.00	—	3,039.00
Water heaters						
Gas, 40 gal.	P1@15.0	Ea	3,460.00	536.00	—	3,996.00
Gas, 80 gal.	P1@16.0	Ea	3,790.00	572.00	—	4,362.00
Gas, 120 gal.	P1@17.0	Ea	4,400.00	608.00	—	5,008.00

Budget Estimating

Budget Estimates for HVAC Work: Budget estimates for heating, ventilating and air conditioning work are figured by the square foot of floor. Multiply the area of conditioned space by the cost per square foot in the table that follows. The result is the estimated cost including a complete heating, ventilating and air conditioning system and 20 percent overhead and profit. The table also shows average heating and cooling requirements for various types of buildings.

Description	Cooling (Btu/SF)	Heating (Btu/SF)	Total Selling Price ($) Per SF	Per Ton
Apartments and condominiums	25	25	12.40	4,570.00
Auditoriums	40	40	31.80	7,310.00
Banks	45	30	35.00	7,150.00
Barber shops	40	40	32.90	7,540.00
Beauty shops	60	30	38.50	5,900.00
Bowling alleys	40	40	25.40	5,820.00
Churches	35	35	27.50	7,190.00
Classrooms	45	40	50.00	10,200.00
Cocktail lounges	70	30	34.80	4,570.00
Computer rooms	145	20	70.90	4,480.00
Department stores	35	30	22.30	5,860.00
Laboratories	60	45	61.70	9,460.00
Libraries	45	35	36.80	7,520.00
Manufacturing plants	40	35	14.20	3,280.00
Medical buildings	35	35	24.40	6,410.00
Motels	30	30	17.00	5,210.00
Museums	35	40	27.20	7,130.00
Nursing homes	45	35	31.30	6,440.00
Office buildings, low-rise	35	35	24.90	6,510.00
Office buildings, high-rise	40	30	33.70	7,710.00
Residences	20	30	7.60	3,510.00
Retail shops	45	30	25.40	5,170.00
Supermarkets	30	30	13.20	4,040.00
Theaters	40	40	24.90	5,740.00

Note that these are budget estimates and can vary widely with conditions which may be unique to the job site. Many of these conditions are identified below. Avoid submitting even a budget estimate without either a site walk-through, a conference with the consulting engineer or architect or a careful review of the plans.

Be alert to any of the following conditions. Any of these can have a major impact on cost:

1. Do HVAC unit dimensions exceed the size of available openings?

2. If a crane is required, is there ample clearance for crane operations?

3. Will any doors, permanent ladders, scaffolding or load-bearing supports obstruct installation?

4. Do the plans show all vents, drains and cleanouts required by the code? Many items like these are not specified by the engineer but will be required by the inspector.

5. For retrofit work, consider carefully problems associated with demolition, including removal of the existing HVAC system, structural modifications and any encased asbestos that may be present.

6. In retrofit jobs, the type and size of HVAC equipment is limited by the window area and installed insulation.

7. How will available power, gas and water lines be extended to the new HVAC equipment?

The following pages contain business forms and letters most plumbing and HVAC contractors can use.

- Change Orders
  Change Estimate Takeoff
  Change Estimate Worksheet
  Change Estimate Summary
  Change Order Log
- Subcontract Forms
  Standard Form Subcontract
  Subcontract Change Order
- Purchase Order

- Construction Schedule
- Letter of Intent
- Submittal Data Cover Sheet
- Submittal Index
- Billing Breakdown Worksheet
- Monthly Progress Billing
- Warranty Letter

An explanation is provided on how the forms function in the contractor's office, how to fill them out, and a sample provided. Where appropriate, a blank copy of the form is provided for your use.

Permission is hereby given by the publisher for the duplication of any of the blank forms on a copy machine or at a printer, provided they are for your own use and not to be sold at a cost exceeding the actual printing and paper cost.

Space has been provided at the top of the forms so you can add your own letterhead or logo.

Change Estimates

Very few projects are completed without some changes to the original plans. Most changes in the scope of work are a result of either errors by the designer or requests by the owner.

Here's the usual sequence in handling changes.

1) The project architect usually initiates a change by sending a Request for Quotation (R.F.Q.) to the general contractor with a detailed description of the work to be modified.

2) The general contractor prepares a Change Estimate (C.E.) if the work is to be done by the general.

3) The subcontractor prepares a change estimate which shows the work to be added to or deleted from the contract and the cost of changes, including markup. The general contractor reviews the change estimate and forwards a summary to the architect for decision.

4) If the decision is to go ahead with the change, the architect issues a Change Order (C.O.) to the general contractor. This change order is authorization to proceed with the change at the price quoted.

Your costs for making changes will normally be higher than your costs for similar items on the original estimate. Here's why:

- Equipment and materials for changes are almost always purchased in smaller quantities. Volume discounts are seldom available.

- Special handling and delivery of small orders will further increase equipment and material costs.

- Scheduled work will be disrupted.

- Overhead costs will be higher because special purchase orders or subcontracts will be required.

- Vendor restocking charges for deleted items will range from 20 to 50 percent of the original cost.

- Large change orders may require additional manpower, which usually results in decreased efficiency.

- The cumulative effect of change orders will be to increase overhead costs if the project completion date is extended.

To compensate for these higher costs, most estimators use a higher markup on change estimates:

- Add an extra 10 percent markup to equipment and material costs to cover increased overhead and handling costs, prior to adding final Overhead and Profit markup.

- Increase competitive labor units (manhours required to perform a specific task) by 15 percent to compensate for inefficiencies such as disruptions in the work schedule, extra handling of equipment and material, and additional manpower.

- Mark up subcontractor prices by 10 percent.

- Figure overhead and profit markup at 15/10 or 10/10.

By following these suggestions, the mathematical gross profit will appear to be 40 to 45 percent, but the true gross will be closer to 25 percent because your costs for change order work are higher, as listed above.

Change Estimate Example

Let's assume that the general contractor on your job has sent a R.F.Q. to your attention. The extra work is a run of hot water piping that will connect to equipment being furnished and set by others. The first step is to list all material and labor required to do the work. Notice the items entered on the Change Estimate Take-off form on page 435.

When the take-off is complete, transfer the manhour total to the Change Estimate Worksheet on page 436. To complete this form:

1) Multiply manhours by the cost per hour.

2) Add the cost of supervision.

3) Total direct labor cost.

4) Add the cost of preparing the change order estimate (C.E.).

5) Add equipment and tool costs.

6) Add subcontract costs.

Then bring totals forward to the Change Estimate Summary on page 437:

1) Enter the material cost total from the Change Estimate Take-off form on page 439.

2) Enter direct and indirect labor cost totals from the Change Estimate Worksheet.

3) Enter equipment, tool and subcontract cost totals from the Change Estimate Worksheet.

Notice the additional bond charge on line M of the Change Estimate Summary. This is the extra premium charged by the bonding company to cover the increase in contract value. The next line, Service Reserve, is the allowance for service work (if any is required) during the warranty period. A sample Change Order Log, used to record the status of changes, is also included on page 438.

ACME
Mechanical Contractors

7600 Oak Avenue - Smallville, U.S.A. 12345-6789
Voice 1-234-5678 / FAX 1-234-5680

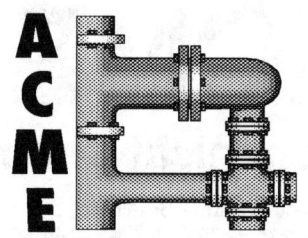

Change Estimate Take-off

Project CBC Bank Building   **C. E. Number** One

Take-off by DCO   **Job No** 7777

Date 2-2-11

	Equipment and Material		Material $		Labor Manhours	
Item	Description	Quantity	Unit	Total	Unit	Total
1	1" Type L hard copper pipe	40 LF	3.87	154.80	.038	1.52
2	1" 90 degree copper ell	6 Ea	4.21	25.26	.23	1.38
3	1" copper coupling	4 Ea	2.54	10.16	.24	.96
4	1" bronze ball valve	2 Ea	17.30	34.60	.36	.72
5	2" x 1" reducing copper tee	2 Ea	20.60	41.20	.40	.80
6	1" copper union	2 Ea	19.10	38.20	.26	.52
7	1" pipe hanger	8 Ea	4.33	34.64	.25	2.00
8	Miscellaneous material	1 Ea	LS	50.00	1.78	1.78
		Totals		388.86		9.68

435

ACME
Mechanical Contractors

7600 Oak Avenue - Smallville, U.S.A. 12345-6789
Voice 1-234-5678 / FAX 1-234-5680

Change Estimate Worksheet

Estimated by   DCO   **C. E. Number**   One

Date   2-3-11   **Job No**   7777

Direct Labor Expense (including benefits)

Plumber or pipefitter journeyman	9.68 manhours at $	41.85	per hour =	$ 405.11
Plumber or pipefitter supervisor	1.00 manhours at $	45.00	per hour =	$ 45.00
Sheet metal journeyman	manhours at $		per hour =	$
Sheet metal supervisor	manhours at $		per hour =	$
Laborer journeyman	manhours at $		per hour =	$
	manhours at $		per hour =	$

Total Direct Labor Cost $ 450.11

Indirect Costs

Travel		$
Subsistence		$
Labor	Cost Estimate Preparation (1 Hour)	$ 40.00

Total Indirect Cost $ 40.00

Equipment and Tool Costs

Ramset rental	$ 35.00
Propane	$ 4.50

Total Equipment and Tool Cost $ 39.50

Subcontract Costs

Quikrap Insulation Co.	$ 100.00
	$

Total Subcontract Cost $ 100.00

436

ACME
Mechanical Contractors

7600 Oak Avenue - Smallville, U.S.A. 12345-6789
Voice 1-234-5678 / FAX 1-234-5680

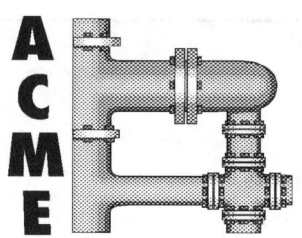

Change Estimate Summary

Quotation to   ABC General Contractors, Inc   **Date**   2-3-11

Address   380 First Street   **Change Order No.**   One

Address   **Job Number**   7777

City / State / Zip   Smallville, CA 63876

Attention: Mr. J.H. Smith

Job Name   CBC Bank Building

Reference: Your R.F.O. No. 7 Labor and Material to furnish and install additional hot water piping.

A. Materials and equipment: $ 500.00

B. Sales tax: 7% . $ 27.22

C. Direct labor: $ 450.11

D. Indirect costs: $ 40.00

E. Equipment and tools: $ 39.50

F.   **Subtotal $ 945.69**

G. Overhead at : 15% of line F: $ 141.85

H. Subcontracts: $ 100.00

I. Overhead at : 10% of line H: $ 10.00

J.   **Subtotal $ 1,197.54**

K. Profit at 10% of line J: $ 119.75

L.   **Subtotal $ 1,317.29**

M. Bond premium at 1% of line L: $ 13.17

N. Service reserve at 0.5% of line L: $ 6.59

O. **Total cost estimate, lines L thru N:**   Add   Deduct $ 1,337.05

P. Exclusions from this estimate: Per Base

Q. This quotation is valid for 30 days

R. We require 1 day(s) extension of the contract time.

S. We are proceeding with this work per your authorization.

T. Please forward your confirming change order.

Signed by: _____
Project Manager

Change Order Log

Job Name _____   **Job Number** _____

Change Number	Customer Number	Date	Amount	Date Approved	Amount Approved	Remarks

Use this space to insert your company logo and letterhead.

Change Estimate Take-off

Project _____ **C. E. Number** _____

Take-off by _____ **Job No** _____

Date _____

Item	Description	Quantity	Material $ Unit	Material $ Total	Labor Manhours Unit	Labor Manhours Total
		Equipment and Material		Material $		Labor Manhours
	Totals					

Use this space to insert your company logo and letterhead.

Change Estimate Worksheet

Estimated by _____   **C. E. Number** _____

Date _____   **Job No** _____

Direct Labor Expense (including benefits)

Plumber or pipefitter journeyman _____ manhours at $ _____ per hour = $ _____

Plumber or pipefitter supervisor _____ manhours at $ _____ per hour = $ _____

Sheet metal journeyman _____ manhours at $ _____ per hour = $ _____

Sheet metal supervisor _____ manhours at $ _____ per hour = $ _____

Laborer journeyman _____ manhours at $ _____ per hour = $ _____

_____ _____ manhours at $ _____ per hour = $ _____

Total Direct Labor Cost $ _____

Indirect Costs

Travel _____ $ _____

Subsistence _____ $ _____

Labor _____ $ _____

Total Indirect Cost $ _____

Equipment and Tool Costs

_____ $ _____

_____ $ _____

Total Equipment and Tool Cost $ _____

Subcontract Costs

_____ $ _____

_____ $ _____

Total Subcontract Cost $ _____

Use this space to insert your company logo and letterhead.

Change Estimate Summary

Quotation to _____	**Date** _____
Address _____	**Change Order No.** _____
Address _____	**Job Number** _____
City / State / Zip _____	
Attention: _____	
Job Name _____	
Reference: _____
_____

A.	Materials and equipment:	$ _____	
B.	Sales tax: _____ .	$ _____	
C.	Direct labor:	$ _____	
D.	Indirect costs:	$ _____	
E.	Equipment and tools:	$ _____	
F.			**Subtotal $ _____**
G.	Overhead at _____ of line F:	$ _____	
H.	Subcontracts:	$ _____	
I.	Overhead at _____ of line H:	$ _____	
J.			**Subtotal $ _____**
K.	Profit at _____ of line J:	$ _____	
L.			**Subtotal $ _____**
M.	Bond premium at 1% of line L:	$ _____	
N.	Service reserve at 0.5% of line L:	$ _____	
O.	**Total cost estimate, lines L thru N:**	Add Deduct	$ _____
P.	Exclusions from this estimate: _____
Q.	This quotation is valid for _____ days
R.	We require _____ day(s) extension of the contract time.
S.	We are proceeding with this work per your authorization.
T.	Please forward your confirming change order.

Signed by: _____

Project Manager

Subcontract Forms

Another term for the general contractor is *prime contractor*. Subcontractors working directly for the prime contractor are called *second-tier subcontractors*. Contractors working for these second-tier subs are usually known as *third-tier subcontractors*. On a large job, many second-tier subs may hire third-tier subs to perform specialized work such as insulation, fireproofing, test and balance, water treatment, instrumentation and trenching.

If a portion of your work will be subcontracted to others, I recommend drawing up a subcontract agreement for that work. Any subcontractors who furnish both material and labor for a job should have a written contract.

Most larger plumbing, heating and A/C subcontractors use a two-part *Standard Form Subcontract* published by the Associated General Contractors of America. This two-part form is used to write the base subcontract. The A.G.C. also publishes a *Subcontract Change Order* form which is used when adding or deleting items after the base subcontract is signed. These forms are sold at modest cost at most A.G.C. offices.

The first section of page one of the Standard Form Subcontract identifies the two contractors entering into the agreement. The party issuing the subcontract is the *contractor*. Recipient of the agreement is the *subcontractor*.

Under *Recitals,* the contractor must affirm the right to issue a subcontract by declaring that a legal contract exists with the *owner* (general contractor).

Section 1 - Entire Contract describes the contract documents: the project plans, specifications and addenda. This list of documents should always be the same as the documents identified in the second-tier subcontractor's contract with the general contractor.

Section 2 - Scope describes the specific scope of work to be performed.

Section 3 - Contract Price. Complete this section by inserting the agreed-upon price for doing the work described.

Section 4 - Payment Schedule on page two should show the same payment terms as your contract with the general contractor.

Section 7 - Special Provisions is used to list any special agreements or exclusions.

When the contract is typed, do not sign it. Instead, forward it to the third-tier subcontractor with instructions to sign it and return it intact. When the signed contract is received, sign it and send a copy to the third-tier subcontractor.

Standard Form Subcontract

Subcontract No. 7777-1

THIS AGREEMENT, made and entered into at Smallville, CA. this 3rd day of May, 2010, by and between   Acme Mechanical Contractors hereinafter called **CONTRACTOR,** with principal office at 7600 Oak Avenue, Smallville, California, and Quikrap Insulation Co., 320 Maple Blvd., Smallville, CA hereinafter called a **SUBCONTRACTOR**.

RECITALS

On or about the____2nd____day of____January____, 20_11_, CONTRACTOR entered into a prime contract with

ABC General Contractors, Inc._____ hereinafter called OWNER, whose address is
380 First Street, Smallville, California____ to perform the following construction work: The installation of all mechanical and plumbing systems for the proposed Mile-Hi office building to be located at 2600 Second, North, Smallville, California.

Said work is to be performed in accordance with the prime contract and the plans and specifications. Said plans and specifications have been prepared by or on behalf of Quik-Draw and Associates, Smallville, California____, **ARCHITECT**

SECTION 1 - ENTIRE CONTRACT

SUBCONTRACTOR certifies and agrees that he is fully familiar with all of the terms, conditions and obligations of the Contract Documents, as hereinafter defined, the location of the job site, and the conditions under which the work is to be performed, and that he enters into this Agreement based upon his investigation of all of such matters and is in no way relying upon any opinions or representations of CONTRACTOR. It is agreed that this Agreement represents the entire agreement. It is further agreed that the Contract Documents are incorporated in this Agreement by this reference, with the same force and effect as if the same were set forth at length herein, and that SUBCONTRACTOR and his subcontractors will be and are bound by any and all of said Contract Documents insofar as they relate in any part or in any way, directly or indirectly to the work covered by this Agreement. SUBCONTRACTOR agrees to be bound to CONTRACTOR in the same manner and to the same extent as CONTRACTOR is bound to OWNER under the Contract Documents, to the extent of the work provided for in this Agreement, and that where, in the Contract Documents, reference is made to CONTRACTOR and the work or specification therein pertains to SUBCONTRACTOR'S trade, craft, or type of work, then such work or specification shall be interpreted to apply to SUBCONTRACTOR instead of CONTRACTOR. The phrase "Contract Documents" is defined to mean and include:

Drawing Nos. A-1 thru A-17, C-1 thru C-4, S-1 thru S-7, E-1 thru E-6 and M-1 thru M-7, all dated
10-4-2009, Specifications dated 10-2-09 and Addendum No. 1 dated 10-18-10.

SECTION 2 - SCOPE

SUBCONTRACTOR agrees to furnish all labor, services, materials, installation, cartage, hoisting, supplies, insurance, equipment, scaffolding, tools and other facilities of every kind and description required for the prompt and efficient execution of the work described herein and to perform the work necessary or incidental to complete thermal insulation for all plumbing, HHW, CHW and condensate piping systems, including pumps P-1, P-2 and P-3 for the project in strict accordance with the Contract Documents and as more particularly, though not exclusively, specified in:

Division 15, Section 400 entitled "Thermal Insulation systems" and General Conditions, as applicable.

SECTION 3 - CONTRACT PRICE

CONTRACTOR agrees to pay SUBCONTRACTOR for the strict performance of his work, the sum of:

Forty-two thousand, three hundred dollars ($42,300.00), subject to additions and deductions for changes in the work as may be agreed upon, and to make payment in accordance with the Payment Schedule, Section 4.

SECTION 4 - PAYMENT SCHEDULE

CONTRACTOR agrees to pay SUBCONTRACTOR in monthly payments of ____90____% of labor and materials which have been placed in position and for which payment has been made by OWNER to CONTRACTOR. The remaining ____10____% shall be retained by CONTRACTOR until he receives final payment from OWNER, but not less than thirty-five days after the entire work required by the prime contract has been fully completed in conformity with the Contract Documents and has been delivered and accepted by OWNER, ARCHITECT, and CONTRACTOR. Subject to the provisions of the next sentence, the retained percentage shall be paid SUBCONTRACTOR promptly after CONTRACTOR receives his final payment from OWNER. SUBCONTRACTOR agrees to furnish, if and when required by CONTRACTOR, payroll affidavits, receipts, vouchers, release of claims for labor, material and subcontractors performing work or furnishing materials under this Agreement, all in form satisfactory to CONTRACTOR, and it is agreed that no payment hereunder shall be made, except at CONTRACTOR'S option, until and unless such payroll affidavits, receipts, vouchers or release; or any or all of them, have been furnished. And payment made hereunder prior to completion and acceptance of the work, as referred to above, shall not be construed as evidence of acceptance of any part of SUBCONTRACTOR'S work.

SECTION 5 - GENERAL SUBCONTRACT PROVISIONS

General Subcontract Provisions on back of Pages 1 and 2 are an integral part of this Agreement.

SECTION 6 - GENERAL PROVISIONS

1. *SUBCONTRACTOR* agrees to begin work as soon as instructed by the CONTRACTOR, and shall carry on said work promptly, efficiently and at a speed that will not cause delay in the progress of the CONTRACTOR'S work or work of other subcontractors. If, in the opinion of the CONTRACTOR, the SUBCONTRACTOR falls behind in the progress of the work, the CONTRACTOR may direct the SUBCONTRACTOR to take such steps as the CONTRACTOR deems necessary to improve the rate of progress, including, without limitation, requiring the SUBCONTRACTOR to increase the number of shifts, personnel, overtime operations, days of work, equipment, amount of plant, or other remedies and to submit to CONTRACTOR for CONTRACTOR'S approval an outline schedule demonstrating the manner in which the required rate of progress will be regained, without additional cost to the CONTRACTOR. CONTRACTOR may require SUBCONTRACTOR to prosecute, in preference to other parts of the work, such part or parts of the work as CONTRACTOR may specify.

The SUBCONTRACTOR shall complete the work as required by the progress schedule prepared by the CONTRACTOR, which may be amended from time to time. The progress schedule may be reviewed in the office of the CONTRACTOR and sequence of construction will be as directed by the CONTRACTOR.

The SUBCONTRACTOR agrees to have an acceptable representative (an officer of SUBCONTRACTOR if requested by the CONTRACTOR) present at all job meetings and to submit weekly progress reports in writing if requested by the CONTRACTOR. Any job progress schedules are hereby made a part of and incorporated herein by reference.

2. *Reserved Gate Usage*

SUBCONTRACTOR shall notify in writing, and assign its employees, material men and suppliers, to such gates or entrances as may be established for their use by CONTRACTOR and in accordance with such conditions and at such times as may be imposed by CONTRACTOR. Strict compliance with CONTRACTOR'S gate usage procedures shall be required by the SUBCONTRACTOR who shall be responsible for such gate usage by its employees, material men, suppliers, subcontractors, and their material men and suppliers.

3. *Staggered Days and Hours of Work and for Deliveries*

SUBCONTRACTOR shall schedule the work and the presence of its employees at the jobsite and any deliveries of supplies or materials by its material men and suppliers to the jobsite on such days, and at such times and during such hours, as may be directed by CONTRACTOR. SUBCONTRACTOR shall assume responsibility for such schedule compliance not only for its employees but for all its material men, suppliers and subcontractors, and their material men and suppliers.

SECTION 7 - SPECIAL PROVISIONS

     1. Pipe hanger inserts will be furnished and installed by Contractor.
     2. All trash pickup by Subcontractor; haul-away will be by General Contractor.

Contractors are required by law to be licensed and regulated by the Contractors' State License Board. Any questions concerning a contractor may be referred to the registrar of the board whose address is:

Contractors' State License Board -- P. O. Box 26000, Sacramento, California 95826

IN WITNESS WHEREOF: The parties hereto have executed this Agreement for themselves, their heirs, executors, successors, administrators, and assignees on the day and year first above written.

SUBCONTRACTOR	CONTRACTOR
Quikrap Insulation Company	Acme Mechanical Contractors
By _____	By _____
Name Title	Name Title
☐ Corporation ☐ Partnership ☐ Proprietorship	
(Seal)	
Contractor's State License No. _____	Contractor's State License No. _____

Revised 5-09   Published by AGC San Diego
Page 2 of 2

Subcontract Change Order

Number   One   g

Date 6-11-11	**Subcontract Number** 7777-1
To Quickrap Insulation Co.	**Our Job Number**
	Our Proposal Number
	Architect's C.O. Number
	Effective Date of Charge 6-17-11

Subject to all the provisions of this Change Order, you are hereby directed to make the following change(s)

Delete HHW and CHW piping insulation located above Room No. 322 per ABC General Contractor's Change Order No. 18, dated 6-5-11.

The following change(s) will alter the price provided in your subcontract by the __deduction__ *of $* __465.00__ g

Surety Consent		
This Change is Approved **Date** _____	Adjusted Subcontract Price Through C.O. Number 0	$ 42,300.00
Name: _____	Amount this C.O. Number:	$ (465.00)
By: _____		
Title: _____	Current Adjusted Subcontract Price:	$ 41,835.00

When this Change Order is signed by both parties (and by Subcontractor's surety is subcontract is bonded), it constitutes their agreement:

(A) That the subcontract price is adjusted as shown above and that no further adjustment in that price by reason of the change(s) provided herein shall be made; and (B) That all the terms and conditions of the subcontract, except as modified by this and any previous changes, shall remain in full force and effect and apply to the work as so changed.

Accepted and Agreed:	**Date:** _____
By _____	**By** _____
Authorized Signature – Subcontractor	Authorized Signature – Contractor
Title	**Title**

* Subcontractor: Sign pink copy and return immediately. If subcontract is bonded, obtain consent of surety endorsed thereon.
  No payments on account of this Change Order will be made until you have complied with the foregoing.

Published by San Diego Chapter A.G.C.

Purchase Order

A purchase order is a legal document used to order equipment and materials from a vendor, with the implied promise to pay for these goods within a specified time after receipt of the items. Purchase orders are sometimes used to write subcontracts whose value does not exceed one or two thousand dollars. Refer to the sample purchase order on page 447.

While most of the purchase order is self-explanatory, certain items need to be emphasized:

"Description" should include all pertinent data such as model numbers, arrangements, performance requirements, applicable standards, electrical characteristics and so on.

"Tag" is used to identify a particular piece of equipment to facilitate unloading it at the proper location at the jobsite. Tag numbers for all major equipment can usually be found on the contract drawings.

"48 hour delivery notice": This can be very important if the material or equipment being delivered is too heavy to be unloaded by hand. Time will be required to arrange for a crane or forklift to unload the truck.

"Sales tax" is the current state sales tax which must be included unless the goods are to be delivered and installed at a project located out-of-state.

"Terms": The usual payment terms are "Net 30" which means the entire amount of the purchase order must be paid within thirty days after receipt of all goods, undamaged. If items are received damaged, payment should be withheld until satisfactory repair or replacement is made.

ACME
Mechanical Contractors

7600 Oak Avenue - Smallville, U.S.A. 12345-6789
Voice 1-234-5678 / FAX 1-234-5680

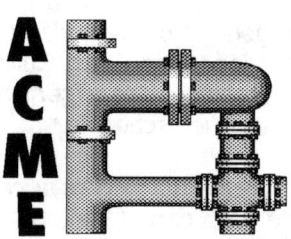

Purchase Order

P.O. Number: 7777-36   **Date:** 1-6-11

Job: CBC Bank Building

Job No: 7777

Seller:	Smallville Industrial Sales Co. 3657 Third Avenue Northwest Smallville, CA 63876	**Ship to:**	Acme Mechanical Contractors c/o CBC Bank Building 465 Commercial Street Smallville, CA 63876

Attn: Mr. L.H. Seller   **By:** RJS

Delivery date: 2-28-11   ☐ **24 Hr. Delivery Notice Required**   **Payment terms:** Net 30

Quantity	Description	Unit	Total
Two	Big-Blo Model No. 1JN 86-13 centrifugal	$443.00	$886.00
	fans with 480/3/60 one-half H.P., O.D.P.		
	motor. 3250 CFM @ 0.50 T.S.P.		
	Tag one EF-13 and one EF-23		

Number of copies of operation and maintenance
manuals required __2__ g

All shipments are to be F.O.B. jobsite or shop unless
otherwise noted.

Total before tax	$886.00
Sales tax	$62.02
Freight	Included
Total Order	$948.02

Buyer: RJS

Acceptance by seller _____   Date _____

Construction Schedules

The general contractor's construction schedule may range from a simple bar chart to a complicated computer-generated CPM (Critical Path Method) diagram. More complex jobs require more complete and detailed schedules. On larger jobs, lower-tier subcontractors may be required to prepare a bar chart showing their construction schedule.

If you have to prepare a construction schedule, make conservative estimates of the time required to complete each part of the job. A slower schedule makes it possible to use smaller crews which are usually more efficient.

You can't schedule your own work until the general contractor has supplied a schedule for the balance of the project. When you have that schedule, prepare a list of tasks your crews will perform. Show this list to the general contractor to be certain there are no conflicts with the master schedule.

When the list of tasks to be performed has been approved, begin recording the manhours required for each task. Take these manhours from your estimate. When the duration for each task has been determined, figure the crew sizes required.

For example, suppose the estimate shows that it will take 1,040 manhours to install the underground plumbing for a project. The work must be completed in two months. Assuming one worker averages 173 hours of work in one month, how many workers will be required to meet the schedule? Here's the solution:

$$\frac{1,040 \text{ manhours}}{173 \text{ MH} / \text{Mo. x 2 months}} = 3 \text{ workers}$$

The first of the two bar charts on the following pages shows the schedule a general contractor might expect to receive from a plumbing and HVAC subcontractor. The chart on page 449 is intended for use by the subcontractor. It includes information a mechanical subcontractor needs to determine proper crew sizes for each task. The *Cumulative Manhours* line shows the budgeted manhours for each month of the project. If labor costs are to be kept within budget, actual hours should not exceed estimated hours. Monitor labor costs each month by comparing the line *Actual MH Used* (actual manhours used) with the line *Cumulative MH* (estimated cumulative manhours).

Construction Schedule

Company: Acme Mechanical Contractors   **Project:** Mile-Hi Office Bldg   **Job Number:** ????   **Date:** 6/8/11

Activity	2011						2012					Crew Sizes
	Jul	Aug	Sep	Oct	Nov	Dec	Jan	Feb	Mar	Apr	May	
Submittals	▓											
Purchasing		▓	▓									
U/G Plumbing				▓								
A/G Plumbing					▓	▓	▓					
Fin. Plumbing								▓				
HVAC Piping						▓	▓	▓				
HVAC Ducting					▓	▓	▓	▓				
Equipment							▓	▓	▓			
Insulation						▓	▓	▓	▓			
Temp. Controls									▓			
Start-up										▓		

Construction Schedule

Company: Acme Mechanical Contractors   **Project:** Mile-Hi Office Bldg   **Job Number:** 7777   **Date:** 6/8/11

		2011					2012					
Activity	Jul	Aug	Sep	Oct	Nov	Dec	Jan	Feb	Mar	Apr	May	Crew Sizes
Submittals	▓											N/A
Purchasing		▓	▓									N/A
U/G Plumbing			▓	▓								1040 MH ÷ (173 x 2) = 3
A/G Plumbing				▓	▓	▓	▓	▓				2075 MH ÷ (173 x 4) = 3
Fin. Plumbing								▓				130 MH ÷ (173 x .5) = 1.5
HVAC Piping					▓	▓	▓	▓				2760 MH ÷ (173 x 4) = 4
HVAC Ducting					▓	▓	▓	▓	▓			2790 MH ÷ (173 x 4) = 4
Equipment								▓	▓			690 MH ÷ (173 x 2) = 2
Insulation					▓	▓	▓	▓	▓			Subcontract
Temp. Controls					▓	▓	▓	▓	▓			Subcontract
Start-up										▓		85 MH ÷ (173 x .5) = 1
Crew size/month			3	6	11	11	11	11.5	2	1		
Cumulative MH*			520	1159	3466	5373	7280	9143	9448	9573		MH Estimated = 9573
Actual MH used												
*Per estimate												
*Rounded												

450

On most large projects the contractor is required to submit drawings and technical data to the owner's representative for approval before work actually begins. Most of these submittals will be prepared by your suppliers and third-tier subs. Vendors and subs may be reluctant to prepare these documents without some assurance that they've been selected to do the work. That's the purpose of a letter of intent. It notifies a proposed vendor or subcontractor that you plan to contract with them when the owner approves the information submitted.

A letter of intent isn't a contract. It's an expressed intention to make a contract, which isn't a contract at all. But a vendor who gets your letter of intent and supplies all the information requested should get the contract. If you place the order with a different vendor, obviously, your letter of intent was worthless. That first vendor may be very reluctant to cooperate in the future.

A sample letter of intent follows this section.

ACME

Mechanical Contractors

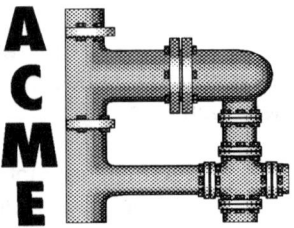

7600 Oak Avenue - Smallville, U.S.A. 12345-6789
Voice 1-234-5678 / FAX 1-234-5680

December 20, 2010

Quikrap Insulation Company
320 Maple Boulevard
Smallville, U.S.A. 12345-6789

Attn: Mr. R. S. Quikrap

Dear Mr. Quikrap:

Acme Mechanical Contractors has been issued a subcontract to furnish and install the mechanical and plumbing systems for the proposed Mile-Hi Office Building to be located at 2600 Second Street in Smallville.

It is my intention to award your firm a subcontract to provide the thermal insulation for these systems in accordance with your proposal dated 12-13-10 in the amount of $42,300.00.

Please prepare ten copies of your submittal data and forward them to me no later than January 15, 2011. Upon approval of your submittal, I will mail you the subcontract for your review and signature.

Thank you,

Bobby Thompson
Project Manager

CC: Project file

The construction specifications may require that you submit manufacturer's technical data on certain equipment you plan to install. You may have to submit six to ten copies of this technical data for approval before buying the equipment or contracting with third-tier subcontractors.

First, find out how many submittal copies will be required. Request that number of copies plus at least two additional sets for your use. A subcontractor who provides only services (no materials) may have to submit a detailed written explanation of the work to be done.

The second step is to label each submittal with the paragraph number of the specification where that equipment, material or service is described. Type or write this number in the upper right corner of each submittal sheet. From this set of numbered submittals, prepare a submittal index, a list of submittals identified by title and paragraph number. This index should be arranged in specification paragraph order. A blank submittal index form follows this section.

I prefer to use a three-ring binder to hold submittals. That makes it easier to add and delete sections as needed. Be sure to include a cover sheet, usually on company letterhead, which identifies the project, job number and scope of work being submitted. A sample submittal cover sheet follows this section.

Purchase orders and subcontracts should not be written until an approved copy of the submittal brochure has been received.

ACME
Mechanical Contractors

7600 Oak Avenue - Smallville, U.S.A. 12345-6789
Voice 1-234-5678 / FAX 1-234-5680

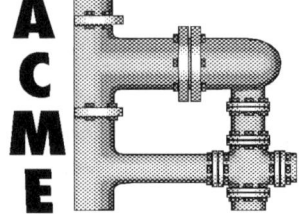

Division 15

Heating, Ventilating, Air Conditioning and Plumbing Systems

Submittal Data for

Mile-Hi Office Building
2600 Second Street
Smallville, U.S.A.

Submittal Index

Specification Section	Specification Paragraph	Item Description	Proposed Vendor or Sub	As Specified	See Submittal Data Page

Billing Breakdown Worksheet

The format to be used for monthly progress billings is seldom specified in the contract documents. In any case, your bill has to include enough detail to satisfy the owner and the lender.

Before preparing your bill, ask the general contractor how much detail is required. Some general contractors will want detailed cost breakdowns showing costs for each system on each building floor. Others will require much less detail.

Find out if billings can include the cost of equipment and materials delivered and stored either on site or off site but not yet installed. If billings can include materials stored off site, do these materials have to be stored in a bonded warehouse? Do vendor's invoices have to accompany progress billings?

When you understand what's required for monthly progress billings, prepare a billing breakdown worksheet. A sample worksheet follows this section.

Column one, Activity Lists each work item in the order work will be performed. The first item listed will usually be mobilization. This includes the cost of the job site trailer, electrical and telephone hookups, office furniture and supplies and initial labor costs. Billings for mobilization costs are sometimes denied by the lender. Many contractors minimize these costs to increase the chance of approval. It's better to receive partial payment for these costs than none at all.

Column two, Cost ($) Lists the actual contractor's costs for each activity, including labor, material, equipment and sales taxes. Markup for overhead and profit are not included.

Column three, Factor This column shows the markup assigned to each activity. Activities scheduled to be completed first are usually assigned the highest markups. This is called front loading and is common in the construction industry. Most bills won't be paid for 30 to 60 days. When the bill is paid, the amount received probably won't include the percentage allowed for retention. Generally, retention is 10 percent and isn't released until the project is complete and accepted by the owner.

Front loading helps contractors carry the financial burden of work in progress. Few subcontractors have enough cash to pay all their bills when due and still wait months to collect for work that's been completed. Employees have to be paid in full and on time. Assigning a higher markup to work completed first accelerates the payment schedule and helps spread payments more evenly over the entire project. Front loading is routine with subcontractors and unpopular with owners and lenders. In practice, subcontractors have no choice but to place a higher price or higher markup on work completed first.

Note that figures in the Factor column are less than 1.0 for work done near the end of the project. That's because front loading doesn't change the contract price. It changes only when that money is received.

Column four, Sell ($) Values for this column are the product of the cost and factor columns. Use the figures in this column when submitting monthly progress invoices.

ACME
Mechanical Contractors

7600 Oak Avenue - Smallville, U.S.A. 12345-6789
Voice 1-234-5678 / FAX 1-234-5680

Billing Breakdown Worksheet
(CBC Bank Building - Job No. 7777)

Activity	Cost ($)	Factor	Sell ($)
Mobilization	2,000	1.50	3,000
Underground chilled and heated water piping	120,000	1.35	162,000
Underground plumbing & piping	87,000	1.35	117,450
Above ground chilled and heating water piping	115,000	1.20	138,000
Above ground plumbing & piping	87,000	1.20	104,400
HVAC ducting	96,000	1.20	115,200
Equipment	312,000	1.10	343,200
Duct and pipe insulation	119,000	1.00	119,000
Temperature controls	53,000	.90	47,700
Plumbing fixtures	43,000	.65	27,950
Water treatment	1,800	.55	990
Test and balance	13,000	.55	7,150
Validation	500	.50	250
	1,049,300		1,186,290

$$\text{Estimated gross profit} = \frac{\$1,186,290 - \$1,049,300}{\$1,049,300} = 13\%$$

457

Monthly Progress Billings

When the billing worksheet has been completed, prepare the billing breakdown as shown on the next page.

The general contractor will usually have a cutoff date each month for progress billings. If the invoice has not been received by that date, it won't be considered until the following month. The cutoff date is usually either the 20th or 25th of the month. The general contractor may permit you to include in each monthly billing work that's not yet complete but should be by the last day of the month.

Column one, Activity List each work item on the billing breakdown worksheet.

Column two, Amount ($) List costs from the Sell ($) column of the billing breakdown worksheet.

Column three, % Complete Show the percentage of completion for each activity listed. Obviously, this figure has to be based on estimates. Be prepared to support your estimate of completion if asked by the general contractor or the lender. If you anticipate a dispute on the figures in this column, call the lender or general contractor to explain the percentage you claim before submitting the bill.

Column four, $ Complete Multiply the percentage in the % Complete column by the figure in the Amount ($) column. Enter the result in the $ Complete column.

Total the $ Complete column. Then subtract the retention percentage and progress billings received to date. The new total is the progress billing for the current month.

ACME
Mechanical Contractors

7600 Oak Avenue - Smallville, U.S.A. 12345-6789
Voice 1-234-5678 / FAX 1-234-5680

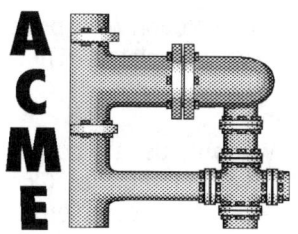

Monthly Progress Billing

(CBC Bank Building — Job No. 7777)

Activity	Amount ($)	% Complete	$ Complete
Mobilization	3,000		
Underground chilled and heated water piping	162,000		
Underground plumbing & piping	117,450		
Above ground chilled and heating water piping	138,000		
Above ground plumbing piping	104,400		
HVAC ducting	115,200		
Equipment	343,200		
Duct and pipe insulation	119,000		
Temperature controls	47,700		
Plumbing fixtures	27,950		
Water treatment	990		
Test and balance	7,150		
Validation	250		
Change order numbers _____ through _____			
Contract price to date	$1,186,290		

Total completed to date	$
Less _____ % retention	$
Subtotal	$
Less previous billings	$
Amount this billing	$

459

Warranty Letters

Upon completion of a project, all subcontractors may be required to provide a warranty letter for the work they've completed. These letters are addressed to the general contractor and will probably be forwarded to the owner.

The warranty usually covers defects in labor or material discovered during the first year after the project is turned over to the owner. Occasionally, the owner will request occupancy of a portion of the project before the entire project is completed. This is called a period of beneficial occupancy. If there will be a period of beneficial occupancy, a separate warranty letter may be required covering only the areas of the building occupied by the owner and only the portion of the plumbing and HVAC systems used in those areas.

Before preparing a warranty letter, get warranty letters from third-tier subs such as the environmental control, thermal insulation, test and balance subs. Any exclusions or limitations in these warranty letters should also appear as exclusions and limitations in your warranty letter. Otherwise you may be accepting responsibility for repairs on work you never performed.

A sample warranty letter follows.

ACME
Mechanical Contractors

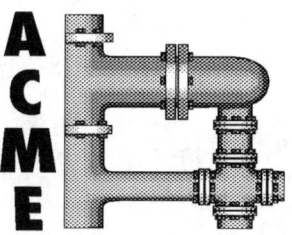

7600 Oak Avenue - Smallville, U.S.A. 12345-6789
Voice 1-234-5678 / FAX 1-234-5680

Date: _____

To: _____
_____
_____
_____

Project:_____

Gentlemen:

In complianoo with the project specifications, we submit our guarantee of equipment, materials and workmanship furnished by Acme Mechanical Contractors for a period of one year. The warranty period is as follows:

From _____ to _____

Equipment and materials furnished by others, but installed by Acme Mechanical Contractors, are not covered by this warranty, except for the installation work performed by Acme Mechanical Contractors.

Ordinary wear is not covered by this warranty. The owner's abuse, neglect or failure to perform recommended maintenance procedures will void this warranty.

Should any problems occur during the specified warranty period, due to faulty equipment, materials or workmanship, Acme Mechanical Contractors will correct the problems, without charge, to the satisfaction of the owner.

Sincerely,

_____
Project Manager

Index

S

Practical References for Builders

Basic Plumbing with Illustrations, Revised

This completely revised edition brings this comprehensive manual fully up-to-date with all the latest plumbing codes. It is the journeyman's and apprentice's guide to installing plumbing, piping, and fixtures in residential and light commercial buildings: how to select the right materials, lay out the job and do professional-quality plumbing work, use essential tools and materials, make repairs, maintain plumbing systems, install fixtures, and add to existing systems. Includes extensive study questions at the end of each chapter, and a section with all the correct answers.
384 pages, 8½ x 11, $33.00

Construction Contract Writer

Relying on a "one-size-fits-all" boilerplate construction contract to fit your jobs can be dangerous — almost as dangerous as a handshake agreement. *Construction Contract Writer* lets you draft a contract in minutes that precisely fits your needs and the particular job, and meets both state and federal requirements. You just answer a series of questions — like an interview — to construct a legal contract for each project you take on. Anticipate where disputes could arise and settle them in the contract before they happen. Include the warranty protection you intend, the payment schedule, and create subcontracts from the prime contract by just clicking a box. Includes a feedback button to an attorney on the Craftsman staff to help should you get stumped — *No extra charge.* **$99.95.** Download the *Construction Contract Writer* at http://www.constructioncontractwriter.com

CD Estimator

If your computer has *Windows*™ and a CD-ROM drive, CD Estimator puts at your fingertips over 150,000 construction costs for new construction, remodeling, renovation & insurance repair, home improvement, framing & finish carpentry, electrical, concrete & masonry, painting, earthwork and heavy equipment and plumbing & HVAC. Quarterly cost updates are available at no charge on the Internet. You'll also have the *National Estimator* program — a stand-alone estimating program for *Windows*™ that *Remodeling* magazine called a "computer wiz," and *Job Cost Wizard*, a program that lets you export your estimates to QuickBooks Pro for actual job costing. A 60-minute interactive video teaches you how to use this CD-ROM to estimate construction costs. And to top it off, to help you create professional-looking estimates, the disk includes over 40 construction estimating and bidding forms in a format that's perfect for nearly any *Windows*™ word processing or spreadsheet program. **CD Estimator is $108.50**

Plumber's Handbook Revised

This new edition shows what will and won't pass inspection in drainage, vent, and waste piping, septic tanks, water supply, graywater recycling systems, pools and spas, fire protection, and gas piping systems. All tables, standards, and specifications are completely up-to-date with recent plumbing code changes. Covers common layouts for residential work, how to size piping, select and hang fixtures, practical recommendations, and trade tips. It's the approved reference for the plumbing contractor's exam in many states. Includes an extensive set of multiple-choice questions after each chapter, with answers and explanations in the back of the book, along with a complete sample plumber's exam. **352 pages, 8½ x 11, $41.50**

California Accessibility Reference Manual (CARM) 2009, 3ʳᵈ Ed.

Significant changes have been made recently to California standards for accessible building and facility design. As a result, this new edition has been completely updated, with hundreds of new illustrations, tables and features. Here you'll find construction details, tables, and complete checklists for virtually every type of access feature required. Includes an extremely detailed index to help you quickly and easily find specific requirements. Includes a CD-ROM with the entire book in Adobe Acrobat plus PDF versions of selected Accessibility chapters from California's Title 24, and the ADA Standard. **444 pages, 8½ x 11, $79.95**

Markup & Profit: A Contractor's Guide

In order to succeed in a construction business, you have to be able to price your jobs to cover all labor, material and overhead expenses, *and* make a decent profit. The problem is knowing what markup to use. You don't want to lose jobs because you charged too much, and you don't want to work for free because you charged too little. If you know how to calculate markup, you can apply it to your job costs to find the right sales price for your work. This book gives you tried and tested formulas, with step-by-step instructions and easy-to-follow examples, so you can easily figure the markup that's right for *your* business. Includes a CD-ROM with forms and checklists for your use. **320 pages, 8½ x 11, $32.50**

National Construction Estimator

Current building costs for residential, commercial, and industrial construction. Estimated prices for every common building material. Provides manhours, recommended crew, and gives the labor cost for installation. Includes a CD-ROM with an electronic version of the book with *National Estimator*, a stand-alone *Windows*™ estimating program, plus an interactive multimedia video that shows how to use the disk to compile construction cost estimates. **672 pages, 8½ x 11, $62.50. Revised annually**

Contractor's Guide to the Plumbing Code

This new publication by ICBO explains the 2003 *International Plumbing Code* in plain English. It's filled with illustrations and figures to help contractors meet code on all plumbing work. Covers history of the code, conventional DWV systems, cost factors that may affect plumbing design and much more. **112 pages, 8½ x 11, $26.25**

Contractor's Guide to *QuickBooks Pro* 2010

This user-friendly manual walks you through *QuickBooks Pro*'s detailed setup procedure and explains step-by-step how to create a first-rate accounting system. You'll learn in days, rather than weeks, how to use *QuickBooks Pro* to get your contracting business organized, with simple, fast accounting procedures. On the CD included with the book you'll find a *QuickBooks Pro* file for a construction company. Open it, enter your own company's data, and add info on your suppliers and subs. You also get a complete estimating program, including a database, and a job costing program that lets you export your estimates to *QuickBooks Pro*. It even includes many useful construction forms to use in your business.
344 pages, 8½ x 11, $57.00

Also available: **Contractor's Guide to *QuickBooks Pro* 2009, $56.50**
Contractor's Guide to *QuickBooks Pro* 2008, $54.75
Contractor's Guide to *QuickBooks Pro* 2007, $53.00
Contractor's Guide to *QuickBooks Pro* 2005, $49.75
Contractor's Guide to *QuickBooks Pro* 2004, $48.50
Contractor's Guide to *QuickBooks Pro* 2003, $47.75
Contractor's Guide to *QuickBooks Pro* 2002, $46.50
Contractor's Guide to *QuickBooks Pro* 2001, $45.25

Plumbing & HVAC Manhour Estimates

Hundreds of tested and proven manhours for installing just about any plumbing and HVAC component you're likely to use in residential, commercial, and industrial work. You'll find manhours for installing piping systems, specialties, fixtures and accessories, ducting systems, and HVAC equipment. If you estimate the price of plumbing, you shouldn't be without the reliable, proven manhours in this unique book.
224 pages, 5½ x 8½, $28.25

2009 *International Building Code*

Updated means of egress and interior finish requirements, comprehensive roof provisions, seismic engineering provisions, innovative construction technology, revamped structural provisions, reorganized occupancy classifications and the latest industry standards in material design.
680 pages, 8½ x 11, $109.25

Also available: **2006 *International Building Code*, $105.00**
(for prior years, see Checklist)

National Renovation & Insurance Repair Estimator

Current prices in dollars and cents for hard-to-find items needed on most insurance, repair, remodeling, and renovation jobs. All price items include labor, material, and equipment breakouts, plus special charts that tell you exactly how these costs are calculated. Includes a CD-ROM with an electronic version of the book with *National Estimator*, a stand-alone Windows™ estimating program, plus an interactive multimedia video that shows how to use the disk to compile construction cost estimates. **576 pages, 8½ x 11, $64.50. Revised annually**

Construction Forms for Contractors

This practical guide contains 78 practical forms, letters and checklists, guaranteed to help you streamline your office, organize your jobsites, gather and organize records and documents, keep a handle on your subs, reduce estimating errors, administer change orders and lien issues, monitor crew productivity, track your equipment use, and more. Includes accounting forms, change order forms, forms for customers, estimating forms, field work forms, HR forms, lien forms, office forms, bids and proposals, subcontracts, and more. All are also on the CD-ROM included, in Excel spreadsheets, as formatted Rich Text that you can fill out on your computer, and as PDFs. **360 pages, 8½ x 11, $48.50**

Estimating Home Building Costs, Revised

Estimate every phase of residential construction from site costs to the profit margin you include in your bid. Shows how to keep track of man hours and make accurate labor cost estimates for site clearing and excavation, footings, foundations, framing and sheathing finishes, electrical, plumbing, and more. Provides and explains sample cost estimate worksheets with complete instructions for each job phase. This practical guide to estimating home construction costs has been updated with digital Excel estimating forms and worksheets that ensure accurate and complete estimates for your residential projects. Enter your project information on the worksheets and Excel automatically totals each material and labor cost from every stage of construction to a final cost estimate worksheet. Load the enclosed CD-ROM into your computer and create your own estimate as you follow along with the step-by-step techniques in this book. **336 pages, 8½ x 11, $38.00**

Troubleshooting Guide to Residential Construction

How to solve practically every construction problem — before it happens to you! With this book you'll learn from the mistakes other builders made as they faced 63 typical residential construction problems. Filled with clear photos and drawings that explain how to enhance your reputation as well as your bottom line by avoiding problems that plague most builders. Shows how to avoid, or fix, problems ranging from defective slabs, walls and ceilings, through roofing, plumbing & HVAC, to paint. **304 pages, 8½ x 11, $32.50**

2009 *International Residential Code*

Replacing the *CABO One- and Two-Family Dwelling Code*, this book has the latest technological advances in building design and construction. Among the changes are provisions for steel framing and energy savings. Also contains mechanical, fuel gas and plumbing provisions that coordinate with the *International Mechanical Code* and *International Plumbing Code*. **874 pages, 8½ x 11, $88.00**

Also available: **2003 *International Residential Code*, $72.50**
2000 *International Residential Code*, $59.00

Illustrated Guide to the *International Plumbing & Fuel Gas Codes*

A comprehensive guide to the *International Plumbing* and *Fuel Gas Codes* that explains the intricacies of the code in easy-to-understand language. Packed with plumbing isometrics and helpful illustrations, it makes clear the code requirements for the installation methods and materials for plumbing and fuel gas systems. Includes code tables for pipe sizing and fixture units, and code requirements for just about all areas of plumbing, from water supply and vents to sanitary drainage systems. Covers the principles and terminology of the codes, how the various systems work and are regulated, and code-compliance issues you'll likely encounter on the job. Each chapter has a set of self-test questions for anyone studying for the plumber's exam, and tells you where to look in the code for the details. Written by a former plumbing inspector, this guide has the help you need to install systems in compliance with the *IPC* and the *IFGC*. **312 pages, 8½ x 11, $37.00**

Builder's Guide to Drainage & Retaining Walls

Creating an adequate drainage system is the critical first step in most construction projects. Retaining walls are integral to the design plan when used to achieve level grade. Here you'll find details for determining slope drainage, doing the grading, complying with street drainage requirements, and dealing with lots with septic systems. Many types of retaining walls are covered, including modular unit walls, bin and crib walls, and more. Also provides tables for estimating concrete quantities for both poured-in-place and masonry and concrete walls. Includes detailed recommendations for retaining walls in most of the situations you're likely to encounter. Includes a CD-ROM that brings you the entire book in an Adobe PDF file, for jobsite reference and quick word search. (This file has no print capability.) A second PDF file includes 100 sample details you can print and carry along on jobs. **294 pages, 8½ x 11, $59.95**

Contractor's Guide to the Building Code

Explains in plain, simple English just what the 2006 *International Building Code* and *International Residential Code* require. Building codes are elaborate laws, designed for enforcement; they're not written to be helpful how-to instructions for builders. Here you'll find down-to-earth, easy-to-understand descriptions, helpful illustrations, and code tables that you can use to design and build residential and light commercial buildings that pass inspection the first time. Written by a former building inspector, it tells what works with the inspector to allow cost-saving methods, and warns what common building shortcuts are likely to get cited. Filled with the tables and illustrations from the *IBC* and *IRC* you're most likely to need, fully explained, with examples to guide you. Includes a CD-ROM with the entire book in PDF format, with an easy search feature. **406 pages, 8½ x 11, $66.75**

Craftsman's Construction Installation Encyclopedia

Step-by-step installation instructions for just about any residential construction, remodeling or repair task, arranged alphabetically, from *Acoustic tile* to *Wood flooring*. Includes hundreds of illustrations that show how to build, install, or remodel each part of the job, as well as manhour tables for each work item so you can estimate and bid with confidence. Also includes a CD-ROM with all the material in the book, handy look-up features, and the ability to capture and print out for your crew the instructions and diagrams for any job. **792 pages, 8½ x 11, $65.00**

Residential Construction Performance Guidelines, 4th Ed.

Created and reviewed by more than 300 builders and remodelers, this guide gives cut-and-dried construction standards that should apply to new construction and remodeling. It defines corrective action necessary to bring all construction up to standards. Standards are listed for sitework, foundations, interior concrete slabs, basement and crawl spaces for block walls and poured walls, wood-floor framing, beams, columns and posts, plywood and joists, walls, wall insulation, windows, doors, exterior finishes and trim, 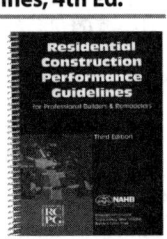 roofs, roof sheathing, roof installation and leaks, plumbing, sanitary and sewer systems, electrical, interior climate control, HVAC systems, cabinets and countertops, floor finishes and more. Published by: NAHB Remodelers Council. **120 pages, 6½ x 8½, $44.95**

Pipe & Excavation Contracting

Shows how to read plans and compute quantities for both trench and surface excavation, figure crew and equipment productivity rates, estimate unit costs, bid the work, and get the bonds you need. Explains what equipment will deliver maximum productivity for a job, how to lay all types of water and sewer pipe, and how to switch your business to excavation work when you don't have pipe contracts. Covers asphalt and rock removal, working on steep slopes or in high groundwater, and how to avoid the pitfalls that can wipe out your profits on any job. **400 pages, 5½ x 8½, $29.00**